Universitext

Universitext

Aksoy/Khamsi: Nonstandard Methods in Fixed Point Theory
Aupetit: A Primer on Spectral Theory
Booss/Bleecker: Topology and Analysis
Carleson/Gamelin: Complex Dynamics
Cecil: Lie Sphere Geometry: With Applications to Submanifolds
Chae: Lebesgue Integration (2nd ed.)
Charlap: Bieberbach Groups and Flat Manifolds
Chern: Complex Manifolds Without Potential Theory
Cohn: A Classical Invitation to Algebraic Numbers and Class Fields
Curtis: Abstract Linear Algebra
Curtis: Matrix Groups
DiBenedetto: Degenerate Parabolic Equations
Dimca: Singularities and Topology of Hypersurfaces
Edwards: A Formal Background to Mathematics I a/b
Edwards: A Formal Background to Mathematics II a/b
Foulds: Graph Theory Applications
Gardiner: A First Course in Group Theory
Gårding/Tambour: Algebra for Computer Science
Goldblatt: Orthogonality and Spacetime Geometry
Hahn: Quadratic Algebras, Clifford Algebras, and Arithmetic Witt Groups
Holmgren: A First Course in Discrete Dynamical Systems
Howe/Tan: Non-Abelian Harmonic Analysis: Applications of $SL(2, \mathbf{R})$
Howes: Modern Analysis and Topology
Humi/Miller: Second Course in Ordinary Differential Equations
Hurwitz/Kritikos: Lectures on Number Theory
Jennings: Modern Geometry with Applications
Jones/Morris/Pearson: Abstract Algebra and Famous Impossibilities
Kelly/Matthews: The Non-Euclidean Hyperbolic Plane
Kostrikin: Introduction to Algebra
Luecking/Rubel: Complex Analysis: A Functional Analysis Approach
MacLane/Moerdijk: Sheaves in Geometry and Logic
Marcus: Number Fields
McCarthy: Introduction to Arithmetical Functions
Meyer: Essential Mathematics for Applied Fields
Mines/Richman/Ruitenburg: A Course in Constructive Algebra
Moise: Introductory Problems Course in Analysis and Topology
Morris: Introduction to Game Theory
Porter/Woods: Extensions and Absolutes of Hausdorff Spaces
Ramsay/Richtmyer: Introduction to Hyperbolic Geometry
Reisel: Elementary Theory of Metric Spaces
Rickart: Natural Function Algebras
Rotman: Galois Theory
Sagan: Space-Filling Curves
Samelson: Notes on Lie Algebras
Schiff: Normal Families

(continued after index)

Norman R. Howes

Modern Analysis and Topology

Springer-Verlag
New York Berlin Heidelberg London Paris
Tokyo Hong Kong Barcelona Budapest

Norman R. Howes
Institute for Defense Analyses
1801 N. Beauregard Street
Alexandria, VA 22311-1772
USA

Mathematics Subject Classifications (1991): 26-02, 54-02, 54D20, 54D60, 54E15, 28A05, 28C10, 46Exx

Library of Congress Cataloging-in-Publication Data
Howes, Norman R.
Modern analysis and topology / Norman R. Howes.
p. cm. – (Universitext)
Includes bibliographical references (p. -) and index.
ISBN 0-387-97986-7 (softcover : acid-free)
1. Mathematical analysis. 2. Topology I. Title.
QA300.H69 1995
515′.13–dc20 95-3995

Printed on acid-free paper.

Production managed by Bill Imbornoni; manufacturing supervised by Joe Quatela.
Camera-ready copy prepared by the author.
Printed and bound by R.R. Donnelley & Sons, Harrisonburg, VA.
Printed in the United States of America.

9 8 7 6 5 4 3 2 1

ISBN 0-387-97986-7 Springer-Verlag New York Berlin Heidelberg

In Memory of Hisahiro Tamano

Preface

The purpose of this book is to provide an integrated development of modern analysis and topology through the integrating vehicle of uniform spaces. It is intended that the material be accessible to a reader of modest background. An advanced calculus course and an introductory topology course should be adequate. But it is also intended that this book be able to take the reader from that state to the frontiers of modern analysis and topology in-so-far as they can be done within the framework of uniform spaces.

Modern analysis is usually developed in the setting of metric spaces although a great deal of harmonic analysis is done on topological groups and much of functional analysis is done on various topological algebraic structures. All of these spaces are special cases of uniform spaces. Modern topology often involves spaces that are more general than uniform spaces, but the uniform spaces provide a setting general enough to investigate many of the most important ideas in modern topology, including the theories of Stone-Čech compactification, Hewitt Real-compactification and Tamano-Morita Para-compactification, together with the theory of rings of continuous functions, while at the same time retaining a structure rich enough to support modern analysis.

There is probably more material in this book than can comfortably be covered in a one year course. It is intended that a subset of the book could be used for an upper-level undergraduate course, whereas much of the full text would be suitable for a one year graduate class. The more advanced chapters are suitable for seminar work and contain many central unsolved problems suitable for a research program. The book emphasizes theory as opposed to application and does not stray far from the setting of uniform spaces, although some of the classical development of convergence theory and measure theory is done in a more abstract setting.

The uniform structure is what is needed to give topological spaces enough structure to support modern analysis. The lack of coverage of more general topological spaces that are currently popular in set theoretic and geometric topology and the minimal coverage of topological algebraic structures (like topological groups, vector spaces, etc.) that are more specialized will no doubt make this book inappropriate for some types of analysis or topology courses. But for a middle-of-the-road approach that covers the central theories of modern analysis and topology, the uniform space setting has a lot to offer. It also allows us to cut a path through this literal forest of subjects that starts out at first principles, follows a common thread and arrives at the frontier, albeit in an area that not many researchers venture into. Part of the reason for this is that analysis on uniform spaces is not as easy to develop as on metric spaces, and part of the reason is that uniform spaces are widely perceived to be harder than they really are. We will attempt to change that perception in this book.

The book is roughly divided into two parts. The first seven chapters are essentially topology and the last five are mostly analysis with some more topology added where needed to support the development of analysis. It is not necessary to cover the chapters in order. A strictly analysis course could cover Chapters 1, 2 and 4 and then skip to Chapter 8 and work as far as time or choice permitted. A strictly topology course could cover just the first seven chapters. Chapters 3, 6, 7, 10, and 12 could be skipped for an undergraduate introduction to modern analysis and topology, while Chapters 6, 7, 9, 10 and 12 contain advanced material suitable for research seminars.

Over half the material in this text has never appeared in book form. There is much from the recent literature of the 1980s and 1990s including answers to some unsolved uniform space questions from the 1960s, an extension of the concept of Haar measure to the Borel sets of an isogeneous uniform space (which generalizes and corrects work done during the 1940s through the 1970s), a necessary and sufficient condition for a locally compact space to have a topological group structure, and a development of uniform measures and uniform differentiation. There is also much from the theory of uniform spaces from the 1960s and 1970s including the concepts of uniform paracompactness and paracompactifications. It is the intent that there be much in this book to interest the expert as well as the novice.

An attempt has been made to document the history of all the central ideas and references so historical notes are embedded in the text. These can lead the interested reader to the foundational sources where these ideas emerged.

The author is indebted to Prof. Arthur Stone whose encouragement through the years has helped make this work possible, to Prof. John Mack whose discussions have shed light on several problems presented herein and to Prof. Hisahiro Tamano who taught the author most of what he knows and transferred to him the uniform space viewpoint.

Alexandria, Virginia 1995 Norman R. Howes

CONTENTS

PREFACE vii
INTRODUCTION: TOPOLOGICAL BACKGROUND xvii

CHAPTER 1: METRIC SPACES 1

1.1 Metric and Pseudo-Metric Spaces 1

Distance Functions, Spheres, Topology of Pseudo-Metric Spaces, The Ring C*(X), Real Hilbert Space, The Distance from a Point to a Set, Partitions of Unity

1.2 Stone's Theorem 6

Refinements, Star Refinements and Δ-Refinements, Full Normality, Paracompactness, Shrinkable Coverings, Stone's Theorem

1.3 The Metrization Problem 13

Functions That Can Distinguish Points from Sets, σ-Local Finiteness, Urysohn's Metrization Theorem, The Nagata-Smirnov Metrization Theorem, Local Starrings, Arhangel'skiĭ's Metrization Theorem

1.4 Topology of Metric Spaces 20

Complete Normality and Perfect Normality, First and Second Countable Spaces, Separable Spaces, The Diameter of a Set, The Lebesgue Number, Precompact Spaces, Countably Compact and Sequentially Compact Spaces

1.5 Uniform Continuity and Uniform Convergence 25

Uniform Continuity, Uniform Homeomorphisms and Isomorphisms, Isometric Functions, Uniform Convergence

1.6 Completeness 28

Convergence and Clustering of Sequences, Cauchy Sequences and Cofinally Cauchy, Sequences, Complete and Cofinally Complete Spaces, The Lebesgue Property, Borel Compactness, Regularly Bounded Metric Spaces

1.7 Completions 38

The Completion of a Metric Space, Uniformly Continuous Extensions

CHAPTER 2: UNIFORMITIES 43

2.1 Covering Uniformities 43

Uniform Spaces, Normal Sequences of Coverings, Bases and Subbases for Uniformities, Normal Coverings, Uniform Topology

2.2 Uniform Continuity 48

Uniform Continuity, Uniform Homeomorphisms, Pseudo-Metrics Determined by Normal Sequences

2.3 Uniformizability and Complete Regularity 52

Uniformizable Spaces, The Equivalence of Uniformizability and Complete Regularity, Regularly Open Sets and Coverings, Open and Closed Bases of Uniformities, Regularly Open Bases of Uniformities, Universal or Fine Uniformities

2.4 Normal Coverings 56

The Unique Uniformity of a Compact Hausdorff Space, Tukey's Characterization of Normal Spaces, Star-Finite Coverings, Precise Refinements, Some Results of K. Morita, Some Corrections of Tukey's Theorems by Morita

CHAPTER 3: TRANSFINITE SEQUENCES 62

3.1 Background 62

3.2 Transfinite Sequences in Uniform Spaces 63

Cauchy and Cofinally Cauchy Transfinite Sequences, A Characterization of Paracompactness in Terms of Transfinite Sequences, Shirota's e Uniformity, Some Characterizations of the Lindelöf Property in Terms of Transfinite Sequences, The β Uniformity, A Characterization of Compactness in Terms of Transfinite Sequences

3.3 Transfinite Sequences and Topologies 75

Characterizations of Open and Closed Sets in Terms of Transfinite Sequences, A Characterization of the Hausdorff Property in Terms of Transfinite Sequences, Cluster Classes and the Characterization of Topologies, A Characterization of Continuity in Terms of Transfinite Sequences

CHAPTER 4: COMPLETENESS, COFINAL COMPLETENESS AND UNIFORM PARACOMPACTNESS 83

4.1 Introduction 83

4.2 Nets 84

Convergence and Clustering of Nets, Characterizations of Open and Closed Sets in Terms of Nets, A Characterization of the Hausdorff Property in Terms of Nets, Subnets, A Characterization of

Compactness in Terms of Nets, A Characterization of Continuity in Terms of Nets, Convergence Classes and the Characterization of Topologies, Universal Nets, Characterizations of Paracompactness, the Lindelöf Property and Compactness in Terms of Nets

4.3 Completeness, Cofinal Completeness and Uniform Paracompactness 92

Cauchy and Cofinally Cauchy Nets, Completeness and Cofinal Completeness, The Lebesgue Property, Precompactness, Uniform Paracompactness

4.4 The Completion of a Uniform Space 97

Fundamental Nets, Completeness in Terms of Fundamental Nets, The Construction of the Completion with Fundamental Nets, The Uniqueness of the Completion

4.5 The Cofinal Completion or Uniform Paracompactification 103

The Topological Completion, Preparacompactness, Countable Boundedness and the Lindelöf Property, A Necessary and Sufficient Condition for a Uniform Space to Have a Paracompact Completion, A Necessary and Sufficient Condition for a Uniform Space to Have a Lindelöf Completion, The Existence of the Cofinal Completion, A Characterization of Preparacompactness

Chapter 5: FUNDAMENTAL CONSTRUCTIONS 110

5.1 Introduction 110

5.2 Limit Uniformities 111

Infimum and Supremum Topologies, Infimum and Supremum Uniformities, Projective and Inductive Limit Topologies, Projective and Inductive Limit Uniformities

5.3 Subspaces, Sums, Products and Quotients 114

Uniform Product Spaces, Uniform Subspaces, Quotient Uniform Spaces, The Uniform Sum

5.4 Hyperspaces 119

The Hyperspace of a Uniform Space, Supercompleteness, Burdick's Characterization of Supercompleteness, Other Characterizations of Supercompleteness, Supercompleteness and Cofinal Completeness, Paracompactness and Supercompleteness

5.5 Inverse Limits and Spectra 126

Inverse Limit Sequences, Inverse Limit Systems, Inverse Limit Systems of Uniform Spaces, Morita's Weak Completion, The Spectrum of Weakly Complete Uniform Spaces, Morita's and Pasynkov's Characterizations of Closed Subsets of Products of Metric Spaces

5.6 The Locally Fine Coreflection 133

Uniformly Locally Uniform Coverings, Locally Fine Uniform Spaces, The Derivative of a Uniformity, Partially Cauchy Nets, Injective Uniform Spaces, Subfine Uniform Spaces, The Subfine Coreflection

5.7 Categories and Functors 146

Concrete Categories, Objects, Morphisms, Covariant Functors, Isomorphisms, Monomorphisms, Duality, Subcategories, Reflection, Coreflection

CHAPTER 6: PARACOMPACTIFICATIONS 156

6.1 Introduction 156

Some Problems of K. Morita and H. Tamano, Topological Completion, Paracompactifications, Compactifications, Samuel Compactifications, The Stone-Čech Compactification, Uniform Paracompactifications, Tamano's Paracompactification Problem

6.2 Compactifications 159

Extensions of Open Sets, Extensions of Coverings, The Extent of a Covering, Stable Coverings, Star-Finite Partitions of Unity

6.3 Tamano's Completeness Theorem 171

The Radical of a Uniform Space, Tamano's Completeness Theorem, Necessary and Sufficient Conditions for Topological Completeness

6.4 Points at Infinity and Tamano's Theorem 178

Points and Sets at Infinity, Some Characterizations of Paracompactness by Tamano, Tamano's Theorem

6.5 Paracompactifications 182

Completions of Uniform Spaces as Subsets of βX, A Solution of Tamano's Paracompactification Problem, The Tamano-Morita Paracompactification, Characterizations of Paracompactness, the Lindelöf Property and Compactness in Terms of Supercompleteness, Another Necessary and Sufficient Condition for a Uniform Space to Have a Paracompact Completion, Another Necessary and Sufficient Condition for a Uniform Space to Have a Lindelöf Completion, The Definition and Existence of the Supercompletion

6.6 The Spectrum of βX 192

The Spectrum of βX, The Spectrum of uX, Morita's Weak Completion

6.7 The Tamano-Morita Paracompactification 197

M-spaces, Perfect and Quasi-perfect Mappings, The Topological Completion of an M-space, The Tamano-Morita Paracompactification of an M-space

CHAPTER 7: REALCOMPACTIFICATIONS 202

7.1 Introduction 202

Another Characterization of βX, Q-spaces, CZ-maximal Families

7.2 Realcompact Spaces 203

Realcompact Spaces, The Hewitt Realcompactification, Characterizations of Realcompactness, Properties of Realcompact Spaces, Pseudo-metric Uniformities, The c and c^* Uniformities

7.3 Realcompactifications 210

Realcompactifications, The Equivalence of υX and eX, The Uniqueness of the Hewitt Realcompactification, Characterizations of υX, Properties of υX, Hereditary Realcompactness

7.4 Realcompact Spaces and Lindelöf Spaces 217

Tamano's Characterization of Realcompact Spaces, A Necessary and Sufficient Condition for the Realcompactification to be Lindelöf, Tamano's Characterization of Lindelöf Spaces

7.5 Shirota's Theorem 221

Measurable Cardinals, {0,1} Measures, The Relationship of Non-Zero {0,1} Measures and CZ-maximal Families, A Necessary and Sufficient Condition for Discrete Spaces to be Realcompact, Closed Classes of Cardinals, Shirota's Theorem

CHAPTER 8: MEASURE AND INTEGRATION 229

8.1 Introduction 229

Riemann Integration, Lebesgue Integration, Measures, Invariant Integrals

8.2 Measure Rings and Algebras 230

Rings, Algebras, σ-Rings, σ-Algebras, Borel Sets, Baire Sets, Measures, Measure Rings, Measurable Sets, Measure Algebras, Measure Spaces, Complete Measures, The Completion of a Measure, Borel Measures, Lebesgue Measure, Baire Measures, The Lebesgue Ring, Lebesgue Measurable Sets, Finite Measures, Infinite Measures

8.3 Properties of Measures 235

Monotone Collections, Continuous from Below, Continuous from Above

8.4 Outer Measures 238

Hereditary Collections, Outer Measures, Extensions of Measures, μ*-Measurability

8.5 Measurable Functions 243

Measurable Spaces, Measurable Sets, Measurable Functions, Borel Functions, Limits Superior, Limits Inferior, Point-wise Limits of Functions, Simple Functions, Simple Measurable Functions

8.6 The Lebesgue Integral 249

Development of the Lebesgue Integral

8.7 Negligible Sets 256

Negligible Sets, Almost Everywhere, Complete Measures, Completion of a Measure

8.8 Linear Functionals and Integrals 257

Linear Functionals, Positive Linear Functionals, Lower Semi-continuous, Upper Semi-continuous, Outer Regularity, Inner Regularity, Regular Measures, Almost Regular Measures, The Riesz Representation Theorem

CHAPTER 9: HAAR MEASURE IN UNIFORM SPACES 264

9.1 Introduction 264

Isogeneous Uniform Spaces, Isomorphisms, Homogeneous Spaces, Translations, Rotations, Reflections, Haar Integral, Haar Measure

9.2 Haar Integrals and Measures 267

Development of the Haar integral on Locally Compact Isogeneous Uniform Spaces

9.3 Topological Groups and Uniqueness of Haar Measures 271

Topological Groups, Abelian Topological Groups, Open at 0, Right Uniformity, Left Uniformity, Right Coset, Left Coset, Quotient of a Topological Group, A Necessary and Sufficient Condition for a Locally Compact Space to Have a Topological Group Structure

CHAPTER 10: UNIFORM MEASURES 284

10.1 Introduction 284

Uniform Measures, The Congruence Axiom, Loomis Contents

10.2 Prerings and Loomis Contents 285

Prerings, Hereditary Open Prerings, Loomis Contents, Uniformly Separated, Left Continuity, Invariant Loomis Contents, Zero-boundary Sets

10.3 The Haar Functions 292

The Haar Covering Function, The Haar Function, Extension of Loomis Contents to Finitely Additive Measures

10.4 Invariance and Uniqueness of Loomis Contents and Haar Measures 299

Invariance with Respect to a Uniform Covering, Invariance on Compact Spheres, Development of Loomis Contents on Suitably Restricted Uniform Spaces.

10.5 Local Compactness and Uniform Measures 304

Almost Uniform Measures, Uniform Measures, Jordan Contents, Monotone Sequences of Sets, Monotone Classes, Development of Uniform Measures on Suitably Restricted Uniform Spaces

CHAPTER 11: SPACES OF FUNCTIONS 317

11.1 L^p-spaces 317

Conjugate Exponents, L^p-norm, The Essential Supremum, Essentially Bounded, Minkowski's Inequality, Hölder's Inequality, The Supremum Norm, The Completion of $C_K(X)$ with Respect to the L^p-norm

11.2 The Space $L^2(\mu)$ and Hilbert Spaces 326

Square Integrable Functions, Inner Product, Schwarz Inequality, Hilbert Space, Orthogonality, Orthogonal Projections, Linear Combinations, Linear Independence, Span, Basis of a Vector Space,

Orthonormal Sets, Orthonormal Bases, Bessel's Inequality, Riesz-Fischer Theorem, Hilbert Space Isomorphism

11.3 The Space $L^p(\mu)$ and Banach Spaces 340

Normed Linear Space, Banach Space, Linear Operators, Kernel of a Linear Operator, Bounded Linear Operators, Dual Spaces, Hahn-Banach Theorem, Second Dual Space, Baire's Category Theorem, Nowhere Dense Sets, Open Mapping Theorem, Closed Graph Theorem, Uniform Boundedness Principle, Banach-Steinhaus Theorem

11.4 Uniform Function Spaces 355

Uniformity of Pointwise Convergence, Uniformity of Uniform Convergence, Joint Continuity, Uniformity of Uniform Convergence on Compacta, Topology of Compact Convergence, Compact-Open Topology, Joint Continuity on Compacta, Ascoli Theorem, Equicontinuity

CHAPTER 12: UNIFORM DIFFERENTIATION 370

12.1 Complex Measures 370

Complex Measure, Total Variation, Absolute Continuity, Concentration of a Measure on a Subset, Orthogonality of Measures

12.2 The Radon-Nikodym Derivative 373

Radon-Nikodym Derivative and its Applications

12.3 Decompositions of Measures and Complex Integration 380

Polar Decomposition, Lebesgue Decomposition, Complex Integration

12.4 The Riesz Representation Theorem 386

Regular and Almost Regular Complex Measures, The Riesz Representation Theorem

12.5 Uniform Derivatives of Measures 389

Differentiation of a Measure at a Point, Differentiable Measures, L^1-differentiable Measures, Uniformly Differentiable Measures, Fubini's Theorem

INDEX 394

Introduction

TOPOLOGICAL BACKGROUND

This book is intended for readers with a basic understanding of general topological spaces and advanced calculus on Euclidean spaces who perhaps know little or nothing about the uniform structure these spaces may possess. Not all topological spaces have a uniform structure but a broad class of them, known as the Tychonoff spaces, do. A uniform structure in a space can be used to measure the "nearness" of one point to another in much the same way a metric is used to measure the distance of one point to another in a metric space. Consequently, many theorems about metric spaces have their counterparts in the theory of uniform spaces and much of the analysis performed on metric spaces can be generalized to uniform spaces.

The concept of a uniformity on a space is far less restrictive than the concept of a metric as there are many uniform spaces that are not metric. On the other hand, all metric spaces are uniform spaces. One of the purposes of this book is to indicate to the reader how much of the mathematical analysis of metric spaces can be extended to uniform spaces.

In this introduction, the topological background the reader is assumed to possess will be reviewed. All the needed definitions are here but the propositions and theorems are stated without proof. Most of the proofs are elementary and can be deduced with very little effort. A few are hard. Those proofs can be found in the elegant little book *Introduction to General Topology* by S. T. Hu published in 1966. The reader is also assumed to have some knowledge of Set Theory including cardinal arithmetic and the most widely known equivalents of the Axiom of Choice. A development such as found in the first chapter of K. Kunen's *Set Theory*, published in 1983, or the appendix of J. L. Kelly's *General Topology*, published in 1955, is sufficient.

Topological Spaces

There are several ways to approach the subject of topological spaces. In our approach we will use our axioms to characterize the behavior of the so-called "open" subsets of a given set X. Toward this end we let X be a set and define a **topology** in X to be a collection τ of subsets of X that satisfy the following axioms:

(1) The empty set $\varnothing$ belongs to τ,
(2) The set X belongs to τ,
(3) The union of a family of members of τ belongs to τ and
(4) The intersection of finitely many members of τ belongs to τ.

When working with topological spaces it is customary to call the members of X **points** and the members of τ **open sets.** A set X is said to be **topologized** by the topology τ and the pair (X,τ) is called a **topological space** or simply a **space** if the topology is understood. If we wish to be as concise as possible, we can define a topology in X without using axioms 1 and 2. Both of them follow from axioms 3 and 4 since the empty union of subsets of X is $\varnothing$ and the empty intersection is X. But it is customary to include axioms 1 and 2 in the definition of a topology.

Since topologies are sets of subsets of a given set X, we can compare topologies in the same set X by means of the inclusion relation. Let σ and τ be topologies in X. If $\sigma \subset \tau$ we say σ is **coarser** than τ and that τ is **finer** than σ.

A **basis** for a topology τ is a subcollection β of τ such that each open set U in τ is a union of members of β. In other words, for each U in τ and point $p \in U$, there is a V in β with $p \in V \subset U$. The members of β are called **basic open sets.** A **sub-basis** for a topology τ in X is a subcollection σ of τ such that the finite intersections of members of σ form a basis for τ. In other words, for each $U \in \tau$ and $p \in U$ there are finitely many members of σ, say $V_1 \cdots V_n$ such that

$$p \in V_1 \cap \cdots \cap V_n \subset U.$$

The members of σ will be called **sub-basic** open sets. Clearly, a topology is completely determined by any given basis or sub-basis. A space is said to satisfy the **second axiom of countability** or simply be **second countable,** denoted 2°, if it has a countable basis. From the definition of sub-basis, it is easily seen that a space is second countable if and only if it has a countable sub-basis.

Let X be a space and $p \in X$. $N \subset X$ is said to be a **neighborhood** of p if there is an open set U such that $p \in U \subset N$. It is an easy exercise to show:

PROPOSITION 0.1 A set U in a space X is open if and only if it contains a neighborhood of each of its points.

For any point p in a space X, the collection N of all neighborhoods of p is called the **neighborhood system** for p. The neighborhood system of p behaves in accordance with the following proposition:

PROPOSITION 0.2 Finite intersections of members of N belong to N and each set in X containing a member of N belongs to N.

By a **neighborhood basis** or **local basis** for a point p we mean a collection B of neighborhoods of p such that every neighborhood of p contains a member of B. The members of B will be referred to as **basic neighborhoods** of p. A space is said to satisfy the **first axiom of countability** or simply be **first countable**, denoted 1°, if it has a countable local basis at each of its points.

PROPOSITION 0.3 If a space is second countable then it is also first countable.

The converse of Proposition 0.3 is clearly false since every uncountable discrete space fails to satisfy the second axiom of countability. A point is said to be an **interior** point of a set A if there is a neighborhood of p contained in A. p is an **exterior** point of A if there is a neighborhood of p that contains no point of A. p is said to be a **boundary** point of A if every neighborhood of p contains a point of A and a point not in A. The set Int(A) of all interior points of A is called the **interior** of A while the set Ext(A) of all exterior points of A is called the **exterior** of A. The **boundary** of A, denoted ∂A, is the set of all boundary points of A.

PROPOSITION 0.4 The interior of A is the largest open set contained in A.

COROLLARY 0.1 A set is open if and only if it contains none of its boundary points.

A set in a space X is said to be **closed** if it contains all of its boundary points. Of course, in most spaces there is an abundance of sets that contain some, but not all of their boundary points, and consequently are neither open or closed.

PROPOSITION 0.5 A set is closed if and only if its complement is open.

As one might expect, since closed sets are merely complements of open sets, rules for the behavior of closed sets under union and intersection (similar to the defining axioms for a topological space) ought to exist. The next proposition enumerates them.

PROPOSITION 0.6 The closed sets of a space satisfy the following four conditions:

(1) The empty set is closed,
(2) The space itself (i.e., the underlying set) is closed,
(3) The intersection of a family of closed sets is closed and
(4) Any finite union of closed sets is closed.

Notice that for a space X, $\varnothing$ and X are both open and closed at the same time. Again, as might be expected, the four conditions of Proposition 0.6 could have been taken as the axioms for closed sets and then the axioms for open sets could have been proved as a proposition. Consequently, one has a choice of viewpoints and approaches when dealing with topological spaces. Yet another important approach to topological spaces is by means of the so-called "closure operators." If $A \subset X$, the **closure** of A, denoted Cl(A), is defined to be the smallest closed set containing A. Condition 3 of Proposition 0.6 guarantees the existence of a smallest closed set containing A and we see that Cl(A) is the intersection of all closed sets containing A.

PROPOSITION 0.7 For $A \subset X$ *we have* $Cl(A) = A \cup \partial A = Int(A) \cup \partial A$.

PROPOSITION 0.8 For any $A \subset X$ *we have:*
(1) A is open if and only if $A = Int(A)$;
(2) A is closed if and only if $A = Cl(A)$.

By a **closure operator** in X we mean a function f that assigns to each subset A of X a subset $f(A)$ of X such that the following four conditions are satisfied:

(1) $f(\varnothing) = \varnothing$,
(2) $A \subset f(A)$ for each $A \subset X$,
(3) $f(f(A)) = f(A)$ for each $A \subset X$ and
(4) $f(A \cup B) = f(A) \cup f(B)$ for each $A, B \subset X$.

PROPOSITION 0.9 Let f be a closure operator on the space X and let τ *be the collection of subsets A of X such that* $f(X - A) = X - A$. *Then:*
(1) τ *is a topology in X;*
(2) For each $A \subset X, f(A)$ *is the closure of A in X with respect to* τ.

Yet another approach to topology is via the "limit point." This approach is rooted in classical analysis and will be expounded in Chapter 1 on metric spaces. For the time being, we only introduce the concept of a limit point.

PROPOSITION 0.10 A point p is in Cl(A) if and only if each neighborhood of p meets A.

The concept of a limit point is motivated by the above proposition. If $p \in Cl(A)$ but p does not belong to A, each neighborhood of p must meet A, so $p \in \partial A$. Moreover, since p does not belong to A, each neighborhood of p must contain a point of A distinct from p. p is said to be a **limit point** of A if each neighborhood of p contains a point of A distinct from p.

PROPOSITION 0.11 A set is closed if and only if it contains all of its limit points.

As the name "limit point" implies, limit points also have to do with a limiting process similar to the process of taking limits of sequences of real numbers. These ideas will be pursued in detail in Chapters 1 and 3. A set is said to be **dense** in X if Cl(A) = X.

PROPOSITION 0.12 A is dense in X if and only if X has a basis such that every nonempty basic open set meets A.

A space is said to be **separable** if it contains a countable subset that is dense in X.

PROPOSITION 0.13 Every second countable space is separable.

Mappings

Let $f:X \to Y$ be a function from a space X into a space Y. f is said to be **continuous at the point** $p \in X$ if for each open set U in Y containing $f(p)$ there is an open set V in X containing p such that $f(V) \in U$. f is said to be **continuous** if it is continuous at every point of X. Continuous functions will also be called **mappings or maps.**

PROPOSITION 0.14 If $f:X \to Y$ is a function from a space X into a space Y then the following statements are equivalent:

(1) $f:X \to Y$ is a mapping,
(2) if U is a neighborhood of $f(p)$ then $f^{-1}(U)$ is a neighborhood of p,
(3) if U is a basic open set in Y then $f^{-1}(U)$ is open in X,
(4) if U is a sub-basic open set in Y then $f^{-1}(U)$ is open in X,
(5) if F is closed in Y then $f^{-1}(F)$ is closed in X,
(6) $f(Cl(A)) \subset Cl(f(A))$ for each $A \subset X$ and
(7) $f^{-1}(Cl(A))$ contains $Cl(f^{-1}(A))$ for each $A \subset Y$.

A mapping $f:X \to Y$ is said to be **open** if the image $f(U)$ of each open set U in X is open in Y. Similarly, f is **closed** if $f(F)$ is closed in Y for each closed F in X. f is sometimes called **bijective** if it is one-to-one and **surjective** if it is onto. Consequently, bijective mappings have well-defined inverse functions defined on the image $f(X) \subset Y$.

PROPOSITION 0.15 For any bijective mapping $f:X \to Y$, the following statments are equivalent:

(1) $f^{-1}:f(X) \subset Y \to X$ is continuous,
(2) f is open and
(3) f is closed.FR

If the mapping $f:X \to Y$ has a continuous inverse $f^{-1}:Y \to X$, then f is said to be a **homeomorphism**. This is equivalent to f being both bijective and surjective and satisfying one of the equivalent properties of Proposition 0.14. If $h:X \to Y$ is a homeomorphism then the spaces X and Y are said to be **homeomorphic** or **topologically equivalent**. A property that is preserved under homeomorphisms (i.e., if P is a property such that whenever X has property P then Y = h(X) also has property P) is said to be a **topological property**.

PROPOSITION 0.16 If $f:X \to Y$ and $g:Y \to Z$ are continuous, then the composition g © $f:X \to Z$ is also continuous.

PROPOSITION 0.17 If $f:X \to Y$ and $g:Y \to Z$ are homeomorphisms then so is the composition g © $f:X \to Z$.

Let $f:X \to Y$ be a function from a set X into the set Y. If τ is a topology in X we can define a collection σ of subsets of Y as follows: $U \subset Y$ is in σ if $f^{-1}(U) \in \tau$. It is easily verified that σ is a topology in Y. In fact, σ is the finest topology in Y such that f is continuous. σ is called the topology **induced by** f **and** τ or simply the **induced topology** if f and τ are understood.

Conversely, if ζ is a topology in Y, then we can define a collection ξ in X by $U \in \xi$ if $U = f^{-1}(V)$ for some $V \in \zeta$. Then ξ is the coarsest topology in X that makes f continuous. Similarly, ξ is said to be the topology induced by f and ζ and is denoted $f^{-1}(\zeta)$.

PROPOSITION 0.18 For any mapping $f:X \to Y$ the following two statements are equivalent:

(1) $A \subset Y$ is open in Y if and only if $f^{-1}(A)$ is open in X;
(2) $A \subset Y$ is closed in Y if and only if $f^{-1}(A)$ is closed in X.

A continuous surjection $f:X \to Y$ is said to be an **identification** if it satisfies the equivalent conditions (1) and (2) of Proposition 0.18.

PROPOSITION 0.19 If $f:X \to Y$ is either an open or closed surjection then f is an identification.

PROPOSITION 0.20 If $f:X \to Y$ is an identification and $g:Y \to Z$ is a function from Y into Z, then a necessary and sufficient condition for the continuity of g is that of the composition g © f.

If f:X $\rightarrow$ Y is a surjective function from the space X onto the set Y then as already discussed, f induces a topology in Y. When f is a surjection, this induced topology is usually called the **identification topology** in Y with respect to f because when Y is topologized in this manner, f becomes an identification.

On the other hand, a continuous bijection f:X $\rightarrow$ Y is said to be an **imbedding** if it satisfies the equivalent conditions (1) and (2) of Proposition 0.18. In this case, X is said to be an imbedding of X into Y and f(X) is said to be **imbedded** in Y.

Let A be a subset of a topological space (X,τ). The inclusion mapping i:A $\rightarrow$ X induces a topology $i^{-1}(\tau)$ in A as defined above. This topology in A is called the **relative** topology in A with respect to τ. When A is topologized in this manner, A will be referred to as a **subspace** of X. It follows from the definition of $i^{-1}(\tau)$ that a set U $\subset$ A is open in A if and only if there is an open set V in X with U = i^{-1}(V) = A $\cap$ V. Similarly, F $\subset$ A is closed in A if and only if there is a closed set K in X with F = A $\cap$ K.

PROPOSITION 0.21 A subspace S of a space X is open in X if and only if the inclusion mapping i:S $\rightarrow$ X is an open mapping. Consequently, every open subset of an open subspace is open in X.

PROPOSITION 0.22 A subspace S of a space X is closed if and only if the inclusion mapping i:S $\rightarrow$ X is closed. Hence, every closed subset of a closed subspace S is closed in X.

For any mapping f:X $\rightarrow$ Y the composition mapping g = f © i:S $\rightarrow$ Y is called the **restriction** of f onto the subspace S of X denoted by $g = f|S$ or $g = f_S$. For every point x $\in$ S we have $g(x) = f[i(x)] = f(x)$. Consequently, g is merely the function f "cut down" to the subspace S. On the other hand, if g:S $\rightarrow$ Y, an **extension** of g over X is a mapping f:X $\rightarrow$ Y such that $f|S = g$.

Sums, Products and Quotients of Spaces

Let $F = \{X_\alpha | \alpha \in A\}$ be a family of topological spaces such that if $\alpha \neq \beta$ then $X_\alpha \cap X_\beta = \emptyset$. Then the family F is said to be **disjoint**. Let S be the union of the sets X_α in F. Define a collection τ of subsets of S as follows: U $\subset$ S belongs to τ if U $\cap$ X_α is open in X_α for each $\alpha \in$ A. Clearly τ satisfies the defining axioms 1 through 4 of a topology. The space (S,τ) is called the **topological sum** of the family F and is usually denoted by $S = \Sigma X_\alpha$.

Most topological sums of interest are formed from disjoint families of spaces and these are often called **disjoint topological sums.** In other instances, families of spaces that are not disjoint are considered to be formally disjoint for the purpose of forming the disjoint topological sum.

PROPOSITION 0.23 For each $\alpha \in A$, the inclusion function $i_\alpha : X_\alpha \to S$ is continuous. If F is disjoint, each i_α is both open and closed.

COROLLARY 0.2 If F is disjoint, each i_α is an imbedding of X_α as an open and closed subspace of S.

Let P denote the Cartesian product set $P = \Pi\{X_\alpha \mid \alpha \in A\}$. Then P consists of all functions $f{:}A \to S$ such that $f(\alpha) \in X_\alpha$ for each $\alpha \in A$. Note that members of P are actually choice functions, so the existence of the product of an arbitrary collection of sets is dependent on the Axiom of Choice. The product topology in P is defined by defining which subsets of P are the sub-basic open sets. For any $\alpha \in A$ and open set U_α of X_α, the set

$$U^*_\alpha = \{f \in P \mid f(\alpha) \in U_\alpha\}$$

will be called a sub-basic open set in P. Let σ be the collection of all sub-basic open sets in P and let τ be the smallest topology in P containing σ. Then τ is called the **product topology** in P. The space (P,τ) is known as the **topological product** of F, often denoted by ΠX_α. Clearly σ is a sub-base for τ. When all the members of F are the same space X then the topological product is denoted by X^A.

In order that the topological product of a family F be non-trivial we always assume that for some $\alpha \in A$ that $X_\alpha \neq \varnothing$. For each $\alpha \in A$ let $p_\alpha{:}P \to X_\alpha$ denote the function defined by $p_\alpha(f) = f(\alpha)$ for each $f \in P$. Then p_α is a surjection for each $\alpha \in A$. p_α will be called the **canonical projection** or simply the **projection** of P onto its α-th coordinate space X_α. Notice that $U^*_\alpha = p_\alpha^{-1}(U_\alpha)$.

PROPOSITION 0.24 The projection $p_\alpha : P \to X_\alpha$ is an open mapping from P into X_α for each $\alpha \in A$.

PROPOSITION 0.25 A function $f{:}Y \to P$ from a space Y into the topological product P is continuous if and only if the composition p_α © f is continuous for each $\alpha \in A$.

Now let $\Phi = \{f_\alpha : X_\alpha \to Y_\alpha \mid \alpha \in A\}$ be a family of mappings and let $\mathbf{f} = \Pi f_\alpha{:}P \to Y$ be the Cartesian **product function** (i.e., $\mathbf{f}(a) = (\ldots, f_\alpha(a_\alpha), \ldots)$ where P is the topological product of the spaces X_α and Y is the topological product of the spaces Y_α.

PROPOSITION 0.26 The Cartesian product of a family of continuous functions is continuous with respect to the product topologies.

By a **decomposition** of a space X we mean a disjoint collection D of subsets of X that covers X. Then each member of X belongs to one and only one member of the decomposition. The function $p\colon X \to D$ defined by $p(x) = d$ where d is the unique member of D containing x is called the **canonical projection** of X onto D. If we give D the identification topology with respect to p then D is called a **decomposition space** of X.

If an equivalence relation R in the space X is given, then the equivalence classes of R are a decomposition of X. Let Q denote the disjoint collection of equivalence classes of R. Then the decomposition space Q obtained by topologizing Q with the identification topology with respect to the canonical projection of X onto Q is called the **quotient space** of X with respect to R and is denoted by X/R.

Conversely, if we are given some identification $j\colon X \to Y$ we can define an equivalence relation R in X by defining xRy in X if $j(x) = j(y)$ for any pair of points $x,y \in X$. It is left as an exercise to show that R is indeed an equivalence relation in X. If we let Q be the collection of equivalence classes with respect to R then the quotient space $Q = X/R$ is the decomposition space formed by taking the identification topology with respect to the canonical projection $p\colon X \to R$. Here, $p\colon X \to R$ is the same as the projection $p\colon X \to X/R = Q$ since the members of X/R are the decomposition (partition) induced by R.

Note that the members of Q are precisely the sets $j^{-1}(y)$ for the points $y \in Y$. We can define a one-to-one function $k\colon Y \to Q$ by $k(y) = j^{-1}(y)$ for each $y \in Y$. Since both j and p are identifications, the triangle

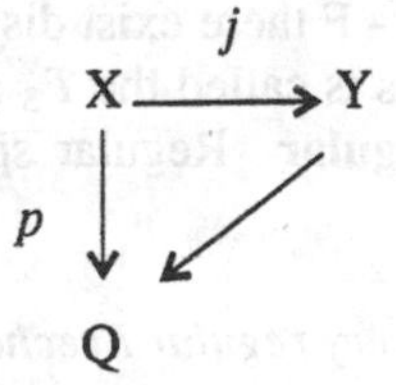

is commutative so by Proposition 0.20, k is a homeomorphism. Consequently, every identification mapping is essentially a canonical projection of a space onto its quotient space.

Separation Axioms

One of the first separation properties to be studied extensively, and perhaps the most natural separation property to arise, is the so-called Hausdorff property. Two points x and y in a space X are said to be **separated** if there exist disjoint open sets U and V in X such that $x \in U$ and $y \in V$. In this case the sets U and V are said to **separate** the points x and y. X is said to be **Hausdorff** if each pair of distinct points can be separated. Hausdorff spaces are often said to satisfy the T_2 separation axiom and are often called T_2 spaces or simply said to be T_2.

PROPOSITION 0.27 Every subspace of a Hausdorff space is Hausdorff.

PROPOSITION 0.28 The topological sum of a disjoint collection of Hausdorff spaces is Hausdorff.

PROPOSITION 0.29 The topological product of a collection of Hausdorff spaces is Hausdorff.

It is easily shown that in a Hausdorff space every point is a closed set. Spaces that satisfy this property are said to be **Frechet** spaces or to satisfy the T_1 separation axiom. The T_1 separation axiom is usually stated in the following equivalent form: if x and y are distinct points of X, then there exist two open subsets U and V of X such that $x \in$ U and $y \in$ V and such that x does not belong to V and y does not belong to U. Since points are closed sets in Hausdorff spaces, a natural strengthening of the T_2 separation axiom is the following: given two disjoint closed sets E and F in X, there exist disjoint open sets U and V in X such that E $\subset$ U and F $\subset$ V. This separation axiom is called the T_4 separation axiom, and spaces satisfying it are said to be **normal**. Normal spaces are said to be able to separate disjoint closed sets.

In the special case where one of the closed sets is a single point we obtain a separation axiom that lies between the T_2 and the T_4 axioms as follows: given a closed set F and a point $p \in$ X - F there exist disjoint open sets U and V in X such that $p \in$ U and F $\subset$ V. This is called the T_3 separation axiom and spaces that satisfy it are said to be **regular**. Regular spaces are said to be able to separate points from closed sets.

PROPOSITION 0.30 Every regular Frechet space is Hausdorff.

THEOREM 0.1 (Urysohn's Lemma, 1925) A space X is normal if and only if whenever E and F are disjoint closed sets in X there exists a continuous function $f:X \to [0,1]$ such that $f(E) = 0$ and $f(F) = 1$.

THEOREM 0.2 (Tietze's Extension Theorem) A space X is normal if and only if whenever E is a closed subset of X, every continuous function $f:E \to$ [0,1] can be extended to a continuous function $f^*:X \to$ [0,1] (i.e., a function f^* such that $f^*(x) = f(x)$ for each $x \in E$).

On the surface, normality appears to be a very natural topological property and Theorems 0.1 and 0.2 provide powerful tools with which to work in normal spaces. However, even more powerful tools will not help solve some of the fundamental problems involving normal spaces. It appears that normality is a property that is so "set theoretic" in nature that additional axioms for set theory are needed to settle these questions. In recent years, some fundamental questions regarding whether normal spaces satisfy certain topological properties or not have been shown to be independent of current set theoretic axioms.

J. W. Tukey in his 1940 Annals of Mathematics Studies (Princeton) monograph titled *Convergence and uniformity in general topology* concluded that normality might not prove to be as important a property as full normality (paracompactness). Essentially, Tukey's prediction has proven true, as we will discuss in later chapters.

By far the most useful separation axiom has proved to be an axiom that lies between regularity and normality called **complete regularity** although sometimes it is referred to by the rather clumsy name of $T_{3\text{-}1/2}$. The so-called $T_{3\text{-}1/2}$ separation axiom is stated as follows: given a point p and a closed set E in X such that $x \in X - E$ then there exists a continuous function $f: X \to [0,1]$ such that $f(p) = 1$ and $f(E) = 0$. Completely regular T_1 spaces are called **Tychonoff** spaces. It is this class of spaces we will be most interested in.

Covering Properties

A **covering** of a space X is a collection of subsets $\mathcal{U}$ of X such that $X = \cup\{U \mid U \in \mathcal{U}\}$. An **open covering** is a covering consisting of open subsets and a **closed covering** is one consisting of closed sets. X is said to be **compact** if every open covering has a finite subcovering. X is said to be **locally compact** if each point of X is contained in a compact neighborhood. $A \subset X$ is said to be covered by a collection $\mathcal{V}$ of subsets of X if $A \subset \cup\{V \mid V \in \mathcal{V}\}$. A is said to be a compact subset of X if each collection $\mathcal{V}$ of open subsets of X possesses a finite subcollection that also covers A. An equivalent definition of compactness of a subset A is that A is compact in the subspace topology when it is considered as a subspace of X.

PROPOSITION 0.31 A closed subset of a compact space is compact.

PROPOSITION 0.32 A compact Hausdorff space is normal.

Two useful generalizations of compactness are countable compactness and the Lindelöf property. A space is said to be **countably compact** if each countable open covering has a finite subcovering. It is said to be **Lindelöf** if each open covering has a countable subcovering.

PROPOSITION 0.33 A space is compact if and only if it is both Lindelöf and countably compact.

PROPOSITION 0.34 Every second countable space is Lindelöf.

A family F of sets is said to have the *finite intersection property* if the intersection of the members of each finite subfamily of F is nonempty.

PROPOSITION 0.35 A space is compact if and only if every family of closed sets with the finite intersection property has a non-void intersection.

Part I: Topology

Chapter 1

METRIC SPACES

The study of metric spaces preceded the study of topological spaces. The emerging awareness of the significance of the so-called *open* sets in metric spaces led to the concept of a topological space. Although many properties of topological spaces that have been studied extensively are motivated by our understanding of metric spaces, there are many topological spaces that are quite different from metric spaces. When we relax the conditions on a space so that we no longer have a metric we may get some surprising (and unpleasant) properties.

Metric spaces have a wonderful *uniformity* or *homogeneity* about them. Each point in a metric space has a local base with the same structure as every other point in the space. Also there is a uniform measure of nearness throughout metric spaces. In subsequent chapters, we will study a class of topological spaces (namely the uniformizable spaces) that retain much of the uniform or homogeneous nature of metric spaces while at the same time being considerably more general than metric spaces.

1.1 Metric and Pseudo-Metric Spaces

The concept of a pseudo-metric space is a minor abstraction of a metric space and since there are as many important examples of pseudo-metric spaces in modern analysis and topology as there are metric spaces, we may as well include them in our study since they share all the uniform properties of metric spaces. In fact, pseudo-metric spaces behave exactly like metric spaces except for the fact that they need not be Hausdorff.

By a **pseudo-metric** on a set X we mean a real valued function $d: X \times X \to \mathbf{R}$ such that for any points $x, y, z \in X$ we have:

(1) $d(x, z) \leq d(x, y) + d(y, z)$,
(2) $d(x, x) = 0$, and
(3) $d(x, y) = d(y, x)$.

Pseudo-metrics are often referred to as **distance functions** and property (1) is called the **triangle inequality** for pseudo-metrics.

LEMMA 1.1 If d is a distance function on X then $d(x, y) \geq 0$ for each pair $x, y \in X$.

Proof: By property (1), $d(x, x) + d(x, y) \leq d(x, y)$ which implies $d(x, x) \leq 2d(x,y)$. Since $d(x, x) = 0$ we have $0 \leq 2d(x, y)$ so that $d(x, y) \geq 0$. ∎

By a **metric** on a set X we mean a pseudo-metric d on X that satisfies an additional property:

(4) for each pair $p, q \in X$, $d(p, q) = 0$ implies $p = q$.

There are many interesting examples of metric spaces that arise naturally, the most familiar of which is probably the **usual metric** defined on the real numbers **R** by $d(x,y) = |x - y|$. But it is also instructive to consider certain less interesting pseudo-metrics such as the **trivial** pseudo-metric d defined on an arbitrary set X by $d(x,y) = 0$ for every pair of points $x,y \in X$ or the **discrete** metric defined by $d(x,y) = 1$ if $x \neq y$ and $d(x,y) = 0$ if $x = y$.

If $d: X \times X \to \mathbf{R}$ is a distance function on the set X then for any p in X and positive number $\varepsilon > 0$ the set $S(p, \varepsilon)$ called the **sphere about p of radius** ε or the **ε-sphere about p** is defined as:

$$S(p, \varepsilon) = \{x \in X \mid d(p, x) < \varepsilon\}.$$

Using these spheres as a basis for a topology is a natural idea. Define $U \subset X$ to be **open** if for each p in X there exists an $\varepsilon > 0$ such that $S(p, \varepsilon) \subset U$ and let τ be the collection of all open sets in X. Then

PROPOSITION 1.1 The collection τ of open sets in a metric space X is a topology for X.

Proof: Clearly ∅ and X are open by the above definition of τ. To show that unions of open sets are open let $\{U_\alpha \mid \alpha \in A\} \subset \tau$ and put $U = \cup U_\alpha$. Let $p \in U$. Then $p \in U_\beta \subset U$ for some $\beta \in A$. Now $U_\beta \in \tau$ implies the existence of an $\varepsilon > 0$ such that $S(p, \varepsilon) \subset U_\beta \subset U$. But then $U \in \tau$.

To show finite intersections of open sets are open, notice that it suffices to show that if $U, V \in \tau$ then $U \cap V \in \tau$ since if it can be shown for two elements, U and V, the result can be extended via induction to finitely many elements. To show $U \cap V \in \tau$ let $p \in U \cap V$. Since $U, V \in \tau$ there exists an $\varepsilon_1 > 0$ and an $\varepsilon_2 > 0$ with $S(p, \varepsilon_1) \subset U$ and $S(p, \varepsilon_2) \subset V$. Let $\varepsilon = min\{\varepsilon_1, \varepsilon_2\}$. Then $S(p, \varepsilon) \subset U \cap V$ so $U \cap V \in \tau$. ∎

By a **pseudo-metric space** we mean a set X and a topology τ on X that is defined by some pseudo-metric $d: X^2 \to \mathbf{R}$. Such a pseudo-metric space is usually denoted by (X, d) instead of (X, τ). (X, d) will be called a **metric space** if d is a metric.

PROPOSITION 1.2 A pseudo-metric space is a metric space if and only if it is T_1.

Proof: Suppose the pseudo-metric space (X, d) is T_1. Let p and q be distinct points of X. Then $\{q\} \subset X$ is closed which implies $X - \{q\}$ is open. Since $p \in X - \{q\}$ there is an $\varepsilon > 0$ with $S(p, \varepsilon) \subset X - \{q\}$. Since q is not contained in $S(p,\varepsilon)$, $d(p, q) \geq \varepsilon > 0$. Therefore $p \neq q$ implies $d(p, q) > 0$ so d is a metric.

Conversely, assume (X, d) is a metric space. Let $p \in X$. For each q in X distinct from p we have $d(p, q) > 0$. Consequently there exists an $\varepsilon_q > 0$ such that $0 < \varepsilon_q < d(p, q)$ which implies p is not contained in $S(q, \varepsilon_q)$. Clearly $X - \{p\} = \cup\{S(q, \varepsilon_q) \mid q \neq p\}$ and hence $X - \{p\}$ is open which implies $\{p\}$ is closed. Therefore X is T_1. ∎

There are many important examples of metric spaces (other than Euclidean n-space with which the reader is no doubt familiar). Examples of two of these spaces will be considered here. The first is the so called **ring of real-valued bounded continuous functions** on a topological space X which is denoted by C*(X). If f:X → **R** and g:X → **R** are bounded and continuous then the function $f+g$ defined by $[f+g](x) = f(x) + g(x)$ is bounded and continuous and hence $f+g$ belongs to C*(X). Moreover, for any pair of real numbers a and b, it is known that the function $af+bg$ defined by $[af+bg](x) = af(x) + bg(x)$ and the function fg defined by $[fg](x) = f(x)g(x)$ are bounded and continuous so that C*(X) does indeed form a ring with respect to these operations of "functional addition" and "functional multiplication."

If f and g belong to C*(X), the function $f - g$ defined by $[f - g](x) = f(x) - g(x)$ is bounded so we can define the function d:C*(X)×C*(X) → **R** by

$$d(f, g) = sup\{|f(x) - g(x)| \mid x \in X\}$$

for each pair $f,g \in$ C*(X). Then for any three functions f,g,h in C*(X) and any point x in X we have:

$$d(f, g) + d(g,h) \geq |f(x) - g(x)| + |g(x) - h(x)| \geq |f(x) - h(x)|.$$

Since this holds for all x in X we have

$$d(f, g) + d(g,h) \geq sup\{|f(x) - h(x)| \mid x \in X\} = d(f, h)$$

so that d satisfies the triangle inequality for a distance function. Clearly d also satisfies properties (2), (3) and (4) of the definition of a metric. Therefore (C*(X), d) is a metric space.

The next example is one we will meet again in later chapters. It is known as the **real Hilbert space** H and it has many important applications in modern analysis and topology. Let H denote the set of all sequences $x = \{x_n\}$ of real numbers such that the series Σx_n^2 converges. The number x_i in $\{x_n\}$ is called the ith coordinate of x. Let x and y be any two members of H. Then for any pair of real numbers a and b and positive integers m and n, the Schwarz inequality yields:

$$\Sigma_{i=m}^{n}(ax_i+by_i)^2 \leq a^2\Sigma_{i=m}^{n}x_i^2 + 2|ab|(\sqrt{\Sigma_{i=m}^{n}x_i^2})(\sqrt{\Sigma_{i=m}^{n}y_i^2}) + b^2\Sigma_{i=m}^{n}y_i^2$$

which can be used to show that the sequences $ax + by = \{ax_n + by_n\}$ belongs to H. If we put $a = 1$ and $b = -1$ then $\Sigma(x_n - y_n)^2$ is a convergent series so we can define a function d on H × H by $d(x,y) = \sqrt{\Sigma(x_n - y_n)^2}$. For any points $x, y, z \in$ H we have

$$d(x,y) + d(y,z) \geq \sqrt{\Sigma_{i=1}^{n}(x_i - y_i)^2} + \sqrt{\Sigma_{i=1}^{n}(y_i - z_i)^2} \geq \sqrt{\Sigma_{i=1}^{n}(x_i - z_i)^2}$$

because of the triangle inequality that holds for the Euclidean metric in $\mathbf{R}^n$. Since this inequality holds for all positive integers n we have $d(x,y) + d(y,z) \geq d(x,z)$ and hence the triangle inequality holds for d. Finally, we note that properties (2), (3) and (4) in the definition of a metric follow directly from the definition of d so d can be seen to be a metric on H.

Next we turn our attention to the fact that distance functions are continuous in the topologies they generate and certain frequently used consequences of this fact. As seen from Proposition 1.1, a distance function $d:X^2 \to \mathbf{R}$ generates a topology τ on X. If we denote the product topology on X × X by π the statement that d **is continuous in the metric topology** will mean that d is continuous with respect to π.

PROPOSITION 1.3 A distance function $d:X \times X \to R$ is continuous with respect to the topology π.

Proof: Since the collection of all sets of the form $H(z) = \{x \in \mathbf{R} | x < z\}$ and $K(z) = \{x \in \mathbf{R} | z < x\}$ is known to be a sub-basis for the topology for $\mathbf{R}$ it will suffice to show that $d^{-1}(H(z))$ and $d^{-1}(K(z))$ are open in X × X for any z in $\mathbf{R}$. To show $d^{-1}(H(z))$ is open first note that if $z \leq 0$ then $d^{-1}(H(z)) = \emptyset$ which is open in X so assume $z > 0$. Pick (p, q) in $d^{-1}(H(z))$. Then $d(p, q) = \delta < z$ for some $\delta \geq 0$. Put $\varepsilon = (z - \delta)/2$ and let $U = S(p, \varepsilon) \times S(q, \varepsilon)$ which is open in X × X. For each (x,y) in U we have $d(x, y) \leq d(x, p) + d(p, q) + d(q, y) < \delta + 2\varepsilon = \delta + (z - \delta) = z$. But then $d(x, y) < z$ which implies $(x,y) \in d^{-1}(H(z))$. Consequently $U \subset d^{-1}(H(z))$. We conclude that $d^{-1}(H(z))$ is open in X × X. The proof that $d^{-1}(K(z))$ is open in X × X is similar. ∎

When $d{:}X^2 \to \mathbf{R}$ is a distance function and (p, q) is a point in $X \times X$, $d(p,q)$ is called the **distance** between p and q. If H and K are two non-void subsets of X the **distance** between H and K is defined to be

$$d(H, K) = \inf\{d(x, y) \mid x \in \mathrm{H} \text{ and } y \in \mathrm{K}\}.$$

If $\mathrm{H} = \{p\}$ then $d(H, K)$ is simply denoted by $d(p, K)$ and is called the **distance from the point p to the set K.**

PROPOSITION 1.4 If $d{:}X \times X \to R$ is a distance function and E is a non-void subset of X, then the function $d_E{:}X \to R$ defined by $d_E(x) = d(x, E)$ for each $x \in X$ is continuous.

Proof: Analogously to the proof of Proposition 1.3, it suffices to show that the sets $d_E^{-1}(\mathrm{H}(z))$ and $d_E^{-1}(\mathrm{K}(z))$ are open for each $z \in \mathbf{R}$. To show $d_E^{-1}(\mathrm{H}(z))$ is open note that if $z \le 0$ then $d_E^{-1}(\mathrm{H}(z)) = \varnothing$ which is open. Consequently we may assume $z > 0$. Let $p \in d_E^{-1}(\mathrm{H}(z))$. Then $d_E(p) = \delta < z$. Put $\varepsilon = (z - \delta)/2$. Since $d_E(p) = d(p, E) = \delta$ there exists a point $q \in E$ with $d(p,q) < \delta + \varepsilon$. Let $\mathrm{U} = S(p,\varepsilon)$. Then $x \in \mathrm{U}$ implies that

$$d_E(x) = d(x, E) \le d(x, q) \le d(x, p) + d(p, q) < \varepsilon + \delta + \varepsilon = z.$$

Consequently $d_E(x) < z$ which implies $x \in d_E^{-1}(\mathrm{H}(z))$ so that $\mathrm{U} \subset d_E^{-1}(\mathrm{H}(z))$. Therefore $d_E^{-1}(\mathrm{H}(z))$ is open. The proof that $d_E^{-1}(\mathrm{K}(z))$ is open is similar. ■

PROPOSITION 1.5 Let $d{:}X \times X \to R$. Then for any $E \subset X$ we have $Cl(E) = \{x \in X \mid d(x, E) = 0\}$ with respect to the topology τ generated by d.

Proof: By Proposition 1.4 it is clear that the set $\mathrm{F} = \{x \in \mathrm{X} \mid d(x, \mathrm{E}) = 0\}$ is closed. Since $d(x, \mathrm{E}) = 0$ for each $x \in \mathrm{E}$ it follows that $\mathrm{E} \subset \mathrm{F}$ and hence $Cl(\mathrm{E}) \subset \mathrm{F}$. Conversely, suppose $x \in \mathrm{F}$. Let N be a neighborhood of x with respect to τ. Since d generates τ, there exists an $\varepsilon > 0$ such that $S(x, \varepsilon) \subset \mathrm{N}$. Now x in F implies $d(x, \mathrm{E}) = 0 < \varepsilon$ so there exists a y in E with $d(x,y) < \varepsilon$. But then y belongs to N. Consequently $\mathrm{N} \cap \mathrm{E} \neq \varnothing$ for each neighborhood N of x. Therefore $x \in Cl(\mathrm{E})$ which implies $\mathrm{F} \subset Cl(\mathrm{E})$. ■

Let (X,τ) be a space and let $P = \{f_\alpha \mid \alpha \in \mathrm{A}\}$ be a family of mappings from X to the unit interval [0,1]. If for each $x \in \mathrm{X}$ there is a neighborhood N of x such that for all but a finite number of the f_α's, $f_\alpha(\mathrm{X} - \mathrm{N}) = 0$ and such that $\Sigma\{f_\alpha(x) \mid \alpha \in \mathrm{A}\} = 1$ then P is called a **partition of unity**. If $\mathcal{U}$ is a covering of X, P is said to be **subordinate** to $\mathcal{U}$ (denoted $P < \mathcal{U}$) if for each $\alpha \in \mathrm{A}$ the **support** of f_α defined by $\{y \in \mathrm{X} \mid f_\alpha(y) > 0\}$ is contained in some member of $\mathcal{U}$. If for each $p \in \mathrm{X}$ there exists a neighborhood of p that meets only finitely many members of $\mathcal{U}$ then $\mathcal{U}$ is said to be **locally finite**.

PROPOSITION 1.6 For every locally finite open covering $\mathcal{U}$ of a pseudo-metric space, there exists a partition of unity $P = \{f_U \mid U \in \mathcal{U}\}$ on X such that the support of f_U is precisely the open set $U \in \mathcal{U}$.

Proof: Let $\mathcal{U}$ be a locally finite open covering of the pseudo-metric space (X,*d*). For each $U \in \mathcal{U}$ define f_U by

$$f_U(x) = \frac{d(x,X-U)}{\Sigma\{d(x,X-V) \mid V \in \mathcal{U}\}}.$$

Since the denominator of the above expression is positive for each $x \in X$, f_U is well defined for each $U \in \mathcal{U}$. Clearly $\Sigma\{f_U(x) \mid U \in \mathcal{U}\} = 1$ for each $x \in X$ and $f_U(x) > 0$ if and only if $x \in U$. Therefore $P = \{f_U \mid U \in \mathcal{U}\}$ is the desired partition of unity subordinate to $\mathcal{U}$. ∎

EXERCISES

1. Show that $d(Cl(A),Cl(B)) = d(A, B)$ for any pair of subsets A and B of the pseudo-metric space (X, *d*).

2. Show that if A and B are compact subsets of the pseudo-metric space (X, *d*) that there exists points $a,b \in X$ such that $a \in A$, $b \in B$ and $d(A, B) = d(a, b)$.

3. Let (X, *d*) be a pseudo-metric space and $\{x_i \mid i = 1 \ldots n\}$ be a finite subset of X. Show that $d(x_1, x_n) \leq \Sigma_{i=1}^{n-1} d(x_i, x_{i+1})$.

4. Two pseudo-metrics d and d' on a set X are said to be **equivalent** if both define the same topology for X. For each $x \in X$ let $S(x, \varepsilon)$ and $S'(x,\varepsilon)$ be the spheres about x of radius ε with respect to d and d' respectively. Show that d and d' are equivalent if and only if for each $\varepsilon > 0$ there exists positive numbers δ and δ' such that $S(x, \delta) \subset S'(x, \varepsilon)$ and $S'(x, \delta') \subset S(x, \varepsilon)$.

5. Let (X, *d*) be a pseudo-metric space and $x \in X$. Show that if $\varepsilon > 0$ the **closed sphere** of radius ε about x defined by $\{y \in X \mid d(x,y) \leq \varepsilon\}$ is a closed set in the topology generated by d.

6. A space is said to be **locally compact** if each point is contained in a compact neighborhood. Show that real Hilbert space H is not locally compact.

1.2 Stone's Theorem

The study of topology was fairly well developed by the mid forties. A topologist of that day probably would not have guessed the development that was to take place in the next two decades. The beginning of this extremely productive era was marked by a theorem due to A. H. Stone (1948) which provided the insight and techniques for much of this major advance.

The concept of full normality was studied by J. W. Tukey, who showed in 1940 that every metric space is fully normal. In 1944, J. Dieudonne introduced the important concept of paracompactness. Stone's achievement consisted of showing that full normality and paracompactness are the same thing and consequently every metric space is paracompact. Stone's Theorem elevated the importance of paracompactness. Originally investigated by Dieudonne as a generalization of compactness, para- compactness was now seen to be a simultaneous generalization of both metric spaces and compact spaces.

One of the key open questions at the time was the "metrization problem" i.e., the problem of characterizing which topological spaces have topologies that are generated by a metric (which spaces are metrizable). Stone's technique produced the important concept of "decomposing" an open covering into a countable family of locally finite sets, which led to the simultaneous solution of the metrization problem by J. Nagata and Y. M. Smirnov in 1951 as acknowledged by Nagata in his book *Modern Dimension Theory* published in 1964. It was also the basis for the proof of Shirota's celebrated theorem for uniform spaces in 1951.

Furthermore, according to Nagata, "[Stone's Theorem] made an epoch not only for modern general topology but for modern dimension theory. On the foundation of the developed covering theory for metric spaces, M. Katetov in 1952 and K. Morita in 1954 independently succeeded in extending the principle results of the classical dimension theory to general metric spaces...".

In this section, the original version of Stone's Theorem will be given. It is customary in topology books to obtain the proof of this theorem using arguments that rely on results that were discovered later, thereby not only disguising the historical development, but more importantly, camouflaging some of the strength of full normality, which is the side of paracompactness that plays an important role in uniform spaces. Since the emphasis of this book is on concepts rather than conciseness, the original approach will be the preferred one in this particular case.

A covering $\mathcal{U}$ of a set X is called a **refinement** of a covering $\mathcal{V}$ if for each $U \in \mathcal{U}$ there is a $V \in \mathcal{V}$ such that $U \subset V$. In this case we write $\mathcal{U} < \mathcal{V}$ and observe that $<$ is a partial ordering (Note: in this case the ordering "proceeds to the left" in the sense that $\mathcal{U}$ is a successor of $\mathcal{V}$ if $\mathcal{U} < \mathcal{V}$). If $H \subset X$, the **star of H with respect to $\mathcal{U}$** is the union of all elements of $\mathcal{U}$ that meet H and is denoted by $Star(H,\mathcal{U})$. Formally,

$$Star(H,\mathcal{U}) = \cup\{U \in \mathcal{U} \mid U \cap H \neq \varnothing\}.$$

In the event H is a single point say $H = \{x\}$ then $Star(H,\mathcal{U})$ is simply written as

Star $(x, \mathcal{U})$. The covering $\mathcal{U}^* = \{Star(U,\mathcal{U}) \mid U \in \mathcal{U}\}$ is called the **star** of $\mathcal{U}$. If $\mathcal{U}^*$ refines $\mathcal{V}$ then $\mathcal{U}$ is called a **star refinement** of $\mathcal{V}$ in which case we write $\mathcal{U} <^* \mathcal{V}$ (equivalently $\mathcal{U}^* < \mathcal{V}$) and observe that $<^*$ is also a partial ordering. If the covering $\{Star(x,\mathcal{U}) | x \in X\}$ refines $\mathcal{V}$ then we say $\mathcal{U}$ is a **Δ-refinement** of $\mathcal{V}$ and write $\mathcal{U}\, \Delta\, \mathcal{V}$. A space X is said to be **fully normal** if each open covering has an open Δ-refinement. When the concepts of star refinement and Δ-refinement first arose is obscure. Tukey used them in 1940 with the remark that the concepts had been known since Urysohn's earliest work on the metrization problem and how much further back they could be traced he did not know.

PROPOSITION 1.7 *If $\mathcal{U}$ is a Δ-refinement of $\mathcal{V}$ and $\mathcal{V}$ is a Δ-refinement of $\mathcal{W}$ then $\mathcal{U}$ is a star refinement of $\mathcal{W}$.*

Proof: Let $U_1 \in \mathcal{U}$. Pick $u_1 \in U_1$. For each $U \in \mathcal{U}$ that meets U_1 choose $u \in U_1 \cap U$. Since $\mathcal{U}\, \Delta\, \mathcal{V}$ there is a $V \in \mathcal{V}$ containing $Star(u,\mathcal{U})$ and hence containing $U_1 \cup U$. Since this holds for each $U \in \mathcal{U}$ we conclude that $Star(U_1,\mathcal{U}) \subset Star(u_1,\mathcal{V})$. But since $\mathcal{V}\, \Delta\, \mathcal{W}$ there is a $W \in \mathcal{W}$ containing $Star(u_1,\mathcal{V})$ and hence $Star(U_1,\mathcal{U})$. But then for each $U_1 \in \mathcal{U}$ there is a $W \in \mathcal{W}$ with $Star(U_1,\mathcal{U}) \subset W$ so $\mathcal{U} <^* \mathcal{W}$. ■

COROLLARY 1.1 *A space is fully normal if and only if each open covering has an open star refinement.*

Stone's Theorem relies on two lemmas that will be established first. A Hausdorff space is said to be **paracompact** if each open covering has a **locally finite** open refinement (i.e., a refinement such that each point in the space is contained in some neighborhood that meets only finitely many members of the refinement). The notion of paracompactness was introduced by J. Dieudonné in 1944 in a paper titled *Une généralization des espaces compacts* published in Jour. Math. Pures & Appl., Volume 23, pp. 67-76. In this paper he showed that paracompact spaces have several properties, one of which is:

LEMMA 1.2 *(J. Dieudonné, 1944) Every paracompact space is normal.*

Proof: Let X be paracompact. We first show X is regular. For this let p be a point contained in an open set U of X. For each q in X - U let V(q) and U(q) be disjoint open sets containing p and q respectively, and let $\mathcal{U}$ be the open covering consisting of U and all the U(q)'s for each $q \in$ X - U. Then $\mathcal{U}$ has a locally finite open refinement $\mathcal{W}$. Let N be an open neighborhood of p that meets only finitely many members of $\mathcal{W}$ and let $W_1 \ldots W_n$ denote those members of $\mathcal{W}$ which meet N but are not contained in U. Then there are n points $q_1 \ldots q_n$ in X - U such that $W_i \subset U(q_i)$ for each $i = 1 \ldots n$. Put V =

$N \cap V(q_1) \cap \ldots \cap V(q_n)$. It remains to show that $Cl(V) \subset U$. For this let W be the union of all members of $\mathcal{W}$ which are not contained in U. Then

$$V \cap W \subset N \cap V(q_1) \cap \ldots \cap V(q_n) \cap [U(q_1) \cup \ldots \cup U(q_n)] = \emptyset$$

since $V(q_1) \cap U(q_i) = \emptyset$ for each $i = 1 \ldots n$. This implies V is contained in the closed set $X - W \subset U$. But then $p \in V \subset Cl(V) \subset U$ which shows that X is regular.

To show X is normal let H and K be disjoint closed sets in X. For each $x \in H$ let $U(x)$ be an open set containing x such that $Cl(U(x))$ is contained in X - K and let $\mathcal{U}$ be the open covering consisting of X - H together with the sets $U(x)$ for each $x \in H$. Then $\mathcal{U}$ has a locally finite open refinement $\mathcal{W}$. Next let U be the union of all members of $\mathcal{W}$ that are not contained in X - H. Then U is an open set containing H.

Let $y \in K$. Since $\mathcal{W}$ is locally finite there is an open neighborhood $N(y)$ of y which meets only finitely many members of $\mathcal{W}$. Let $W_1 \ldots W_n$ be the members of $\mathcal{W}$ that meet $N(y)$ which are not contained in X - H. Then there are points $x_1 \ldots x_n$ in H such that $W_i \subset U(x_i)$ for each $i = 1 \ldots n$. Put

$$V(y) = N(y) \cap [X - Cl(U(x_1))] \cap \ldots \cap [X - Cl(U(x_n))].$$

Then $V(y)$ is an open neighborhood of y such that $U \cap V(q) = \emptyset$. Let V be the union of the open sets $V(y)$ for each y in K. Then V is an open set containing K such that $U \cap V = \emptyset$. Therefore X is normal. ∎

If $\mathcal{U} = \{U_\alpha \mid \alpha < \gamma\}$ is an open covering of a space X for some cardinal γ, and $\mathcal{V} = \{V_\alpha \mid \alpha < \gamma\}$ is an open covering such that $Cl(V_\alpha) \subset U_\alpha$ for each $\alpha < \gamma$ then $\mathcal{V}$ is said to be a **shrink** of $\mathcal{U}$ and $\mathcal{U}$ is said to be **shrinkable.** $\mathcal{U}$ is said to be **point finite** if each point of X is contained in only finitely many members of $\mathcal{U}$.

LEMMA 1.3 A space is normal if and only if each point finite open covering is shrinkable.

Proof: Let X be a normal space and $\mathcal{U} = \{U_\alpha \mid \alpha < \gamma\}$ be a point finite open covering of X. For each $\alpha < \gamma$, a covering $\mathcal{U}_\alpha$ will be defined inductively such that $\mathcal{U}_\alpha = \{V_\beta \mid \beta < \gamma\}$ where $Cl(V_\beta) \subset U_\beta$ for $\beta \leq \alpha$ and $V_\beta = U_\beta$ for each $\beta > \alpha$. Put $W_1 = U_1 - \cup\{U_\alpha \mid \alpha > 1\}$. If $W_1 = \emptyset$ let $V_1 = W_1$. Otherwise W_1 is closed since $\mathcal{U}$ covers X. In this case choose an open V_1 such that $W_1 \subset V_1 \subset Cl(V_1) \subset U_1$. Then let $V_\alpha = U_\alpha$ for each $\alpha > 1$. Clearly $\mathcal{U}_1 = \{V_\alpha \mid \alpha < \gamma\}$ is an open covering of X with the desired property.

Next suppose $\mathcal{U}_\beta$ has been defined for each $\beta < \alpha$. Let $W_\alpha = U_\alpha - [\cup_{\beta<\alpha} V_\beta] \cup [\cup_{\beta>\alpha} U_\beta]$. If $W_\alpha = \emptyset$ put $V_\alpha = W_\alpha$. Otherwise W_α can be shown to be closed in X. It will suffice to show that $U_\alpha \cup \{V_\beta \mid \beta < \alpha\} \cup \{U_\beta \mid \beta > \alpha\}$

covers X. For this let $x \in X$. Since $\mathcal{U}$ is point finite, there is a greatest index $\delta < \gamma$ such that $x \in U_\delta$. If $\alpha \leq \delta$ then $x \in U_\alpha \cup \{U_\beta \mid \beta > \alpha\}$. If $\delta < \alpha$ then $\mathcal{U}_\delta$ has been defined and hence there is a member of $\mathcal{U}_\delta = \{V_\beta \mid \beta \leq \delta\} \cup \{U_\beta \mid \beta > \delta\}$ containing x. Since x cannot belong to $\cup\{U_\beta \mid \beta > \delta\}$, $x \in V_\beta$ for some $\beta < \delta < \alpha$ which implies $x \in \cup\{V_\beta \mid \beta < \alpha\}$. Thus W_α is a closed subset of U_α. Let V_α be an open subset of X such that $W_\alpha \subset V_\alpha \subset Cl(V_\alpha) \subset U_\alpha$. Now put $\mathcal{U}_\alpha = \{V_\beta \mid \beta \leq \alpha\} \cup \{U_\beta \mid \beta > \alpha\}$. Clearly $\mathcal{U}_\alpha$ is an open covering of X with the desired property. This completes the induction argument.

Now put $\mathcal{V} = \{V_\alpha \mid \alpha < \gamma\}$ and let $x \in X$. Since $\mathcal{U}$ is point finite there is a greatest index δ such that $x \in U_\delta$. Since $\mathcal{U}_\delta$ covers X there is a member of $\{V_\beta \mid \beta \leq \delta\} \cup \{U_\beta \mid \beta > \delta\}$ containing x. Since x cannot belong to $\cup\{U_\beta \mid \beta > \delta\}$ it follows that $x \in V_\alpha$ for some $\alpha \leq \delta$. Consequently $\mathcal{V}$ covers X and is clearly a shrink of $\mathcal{U}$. Therefore, every point finite open covering of a normal space is shrinkable.

Conversely, suppose each point finite open covering of X is shrinkable and let H and K be disjoint closed sets in X. Put U = X - H and V = X - K. Then {U,V} is a point finite open covering of X. Let {A, B} be a shrink of {U,V}. Then $Cl(A) \subset U$ and $Cl(B) \subset V$. Let $U_1 = X - Cl(A)$ and $V_1 = X - Cl(B)$. Then U_1 and V_1 are open sets in X with $H \subset U_1$ and $K \subset V_1$. To show U_1 and V_1 are disjoint, suppose $x \in U_1 \cap V_1$. Then $x \in X - Cl(B)$ which implies x does not belong to either A or B which is impossible since {A, B} covers X. Therefore X is normal. ∎

If $\mathcal{U}$ is an open covering of the fully normal space X then there exists an open covering $\mathcal{U}_1$ of X such that $\mathcal{U}_1$ Δ-refines $\mathcal{U}$. By induction we can obtain a sequence of open coverings $\{\mathcal{U}_n\}$ such that $\mathcal{U}_{n+1}$ Δ-refines $\mathcal{U}_n$ for each positive integer n. Then for each $H \subset X$ we have the following:

$$\text{(1.1)} \qquad X - Star(X - H, \mathcal{U}_n) = \{x \in X \mid Star(x, \mathcal{U}_n) \subset H\},$$

$$\text{(1.2)} \qquad Star(X - Star(X - H, \mathcal{U}_n), \mathcal{U}_n) \subset H \text{ and}$$

$$\text{(1.3)} \qquad Star(Star(H, \mathcal{U}_{n+1}), \mathcal{U}_{n+1}) \subset Star(H, \mathcal{U}_n).$$

THEOREM 1.1 (A. H. Stone, 1948) A T_1 space is paracompact if and only if it is fully normal.

Proof: Assume X is fully normal and let $\mathcal{U} = \{U_\alpha \mid \alpha < \gamma\}$ be an open covering of X for some cardinal γ. Then there exists a sequence $\{\mathcal{U}_n\}$ of open coverings such that $\mathcal{U}_{n+1} \, \Delta \, \mathcal{U}_n$ for each positive integer n and $\mathcal{U}_1 \, \Delta \, \mathcal{U}$.

For each $\alpha < \gamma$ define the sequence $\{V_{n_\alpha}\}$ inductively as follows:

$$\text{(1.4)} \qquad V_{1_\alpha} = X - Star(X - U_\alpha, \mathcal{U}_1) \text{ and}$$

(1.5) $$V_{n\alpha} = Star(V_{n-1\alpha}, \mathcal{U}_n) \quad \text{for } n \geq 2.$$

Clearly $V_{n\alpha} \subset V_{n+1\alpha}$ for each n and $V_{n\alpha}$ is open if $n \geq 2$. An easy induction argument using (1.2) and (1.3) shows that $V_{n\alpha} \subset U_\alpha$ for each n. Put $V_\alpha = \cup\{V_{n\alpha}\} \subset U_\alpha$ and let $\mathcal{V} = \{V_\alpha | \alpha < \gamma\}$. To show $\mathcal{V}$ covers X let $x \in$ X. Then $Star(x, \mathcal{U}_1) \subset U_\alpha$ for some $\alpha < \gamma$ since $\mathcal{U}_1 \Delta \mathcal{U}$. By (1.1)

$$x \in X - Star(X - U_\alpha, \mathcal{U}_1) = V_{1\alpha} \subset V_\alpha$$

so that $\mathcal{V}$ covers X. If $x \in V_\alpha$ there exists some $n \geq 2$ such that $x \in V_{n-1\alpha}$. Then $Star(x, \mathcal{U}_n) \subset Star(V_{n-1\alpha}, \mathcal{U}_n) = V_{n\alpha}$. Consequently

(1.6) If $x \in V_\alpha$ there is an $n > 0$ with $Star(x, \mathcal{U}_n) \subset V_\alpha$.

Next we define for each $n > 0$ a transfinite sequence $\{H_{n\alpha}\}$ of closed sets by

(1.7) $$H_{n_1} = X - Star(X - V_1, \mathcal{U}_1) \text{ and}$$

(1.8) $$H_{n\alpha} = X - Star(X - [V_\alpha - \cup\{H_{n\beta} | \beta < \alpha\}], \mathcal{U}_n) \subset V_\alpha.$$

To show that if $\alpha \neq \delta$ then no U_n in $\mathcal{U}_n$ can meet both $H_{n\alpha}$ and $H_{n\delta}$, suppose $\delta < \alpha$ and $U_n \cap H_{n\alpha} \neq \emptyset$ for some U_n in $\mathcal{U}_n$. Pick x in $U_n \cap H_{n\alpha}$. By (1.1)

$$x \in H_{n\alpha} \text{ implies } x \in \{x \in X | Star(x, \mathcal{U}_n) \subset V_\alpha - \cup\{H_{n\beta} | \beta < \alpha\}\}.$$

Consequently x does not belong to $H_{n\delta}$. Clearly $x \in H_{n\delta}$ implies x does not belong to $H_{n\alpha}$. To show H covers X let $x \in$ X. Then there is a first $\alpha < \gamma$ with x in V_α. By (1.6) there is an $n > 0$ with $Star(x, \mathcal{U}_n) \subset V_\alpha$. Suppose x does not belong to $H_{n\alpha}$. From (1.1) and (1.8), $Star(x, \mathcal{U}_n)$ contains a point y such that y does not belong to $V_\alpha - \cup\{H_{n\beta} | \beta < \alpha\}$. But then $y \in \cup\{H_{n\beta} | \beta < \alpha\}$ which implies $y \in H_{n\beta}$ for some $\beta < \alpha$. Therefore $x \in Star(H_{n\beta}, \mathcal{U}_n) \subset V_\beta$ which contradicts α being the first member of A with x in V_α. Thus H covers X.

For each $\alpha < \gamma$ and positive integer n put

(1.9) $$E_{n\alpha} = Star(H_{n\alpha}, \mathcal{U}_{n+3}) \text{ and } G_{n\alpha} = Star(H_{n\alpha}, \mathcal{U}_{n+2}).$$

Clearly $H_{n\alpha} \subset E_{n\alpha} \subset Cl(E_{n\alpha}) \subset G_{n\alpha}$ and since $\alpha \neq \delta$ implies no U_n in $\mathcal{U}_n$ can meet both $H_{n\alpha}$ and $H_{n\delta}$ we have:

(1.10) $\alpha \neq \delta$ implies no U_{n+2} in $\mathcal{U}_{n+2}$ meets both $G_{n\alpha}$ and $G_{n\delta}$.

For each n put $F_n = \cup\{Cl(E_{n\alpha}) | \alpha < \gamma\}$. To show F_n is closed let $x \in Cl(F_n)$.

Then every neighborhood N of x meets some $Cl(E_{n_\alpha})$ and hence meets E_{n_α}. Suppose $N \subset Star(x, \mathcal{U}_{n+2})$. By (1.10) N can meet at most one G_{n_α} and hence at most one E_{n_α}. Therefore each neighborhood of x must meet E_{n_α} which implies $x \in Cl(E_{n_\alpha}) \subset F_n$. But then F_n is closed.

For each $\alpha < \gamma$ and positive integer n, define $W_{n_\alpha} = G_{n_\alpha} - (F_1 \cup F_2 \cup \ldots \cup F_{n-1})$ where W_{1_α} is considered to be G_{1_α}. Then each W_{n_α} is open. It will be shown that $\mathcal{W} = \{W_{n_\alpha}\}$ is the desired locally finite open refinement of $\mathcal{U}$. For this let $x \in X$. Then $x \in H_{n_\alpha} \subset Cl(E_{n_\alpha})$ for some $\alpha < \gamma$ and some positive integer n. Let m be the smallest positive integer for which there exists a $Cl(E_{m_\beta})$ containing x. Then $x \in G_{m_\beta}$ and x is not contained in $F_1 \cup F_2 \cup \ldots \cup F_{m-1}$ so that $x \in W_{m_\beta}$. Hence $\mathcal{W}$ covers X. Furthermore,

$$W_{n_\alpha} \subset G_{n_\alpha} \subset Star(H_{n_\alpha}, \mathcal{U}_n) \subset V_\alpha \subset U_\alpha$$

so $\mathcal{W}$ is a refinement of $\mathcal{U}$. It only remains to show that $\mathcal{W}$ is locally finite.

Let $x \in X$. Again, we have $x \in H_{n_\alpha}$ for some $\alpha < \gamma$ and positive integer n. Therefore $Star(x, \mathcal{U}_{n+3}) \subset E_{n_\alpha} \subset F_n$. Consequently $Star(x, \mathcal{U}_{n+3}) \cap W_{k_\beta} = \varnothing$ for each $k > n$. Furthermore, for a given $k \leq n$, $Star(x, \mathcal{U}_{n+3}) \subset U_{n+2} \subset U_{k+2}$ for some U_{n+2} in $\mathcal{U}_{n+2}$ and U_{k+2} in $\mathcal{U}_{k+2}$. Then by (1.9) and (1.10) $Star(x, \mathcal{U}_{n+3}) \cap G_{k_\beta} \neq \varnothing$ for at most one β in A. But then $Star(x, \mathcal{U}_{n+3}) \cap W_{k_\beta} \neq \varnothing$ for at most one β in A. Hence $Star(x, \mathcal{U}_{n+3})$ is a neighborhood of x that meets at most n of the sets of $\mathcal{W}$ so $\mathcal{W}$ is locally finite. This shows X is paracompact.

Conversely, assume X is paracompact and $\mathcal{U} = \{U_\alpha \mid \alpha \in A\}$ is a locally finite open covering of X. It will suffice to show that $\mathcal{U}$ has a Δ-refinement. By Lemma 1.2, X is normal and by Lemma 1.3, the covering $\mathcal{U}$ is shrinkable (since a locally finite covering is clearly point finite). Let $\mathcal{V} = \{V_\alpha \mid \alpha \in A\}$ be a shrink of $\mathcal{U}$. Then $Cl(V_\alpha) \subset U_\alpha$ for each $\alpha \in A$ and $\mathcal{V}$ is an open covering of X.

By hypothesis each $x \in X$ has an open neighborhood N(x) meeting only finitely many of the U_α's. Put $A(x) = \{\alpha < \gamma \mid U_\alpha \cap N(x) \neq \varnothing\}$. Let $B(x) = \{\alpha \in A(x) \mid x \in U_\alpha\}$ and $C(x) = \{\alpha \in A(x) \mid x \text{ is not contained in } Cl(V_\alpha)\}$. Clearly $B(x) \cup C(x) = A(x)$. Now put

$$W(x) = N(x) \cap [\cap\{U_\alpha \mid \alpha \in B(x)\}] \cap [\cap\{X - Cl(V_\alpha) \mid \alpha \in C(x)\}]$$

for each $x \in X$ and let $\mathcal{W} = \{W(x) \mid x \in X\}$. Since W($x$) is an open set containing x for each $x \in X$, $\mathcal{W}$ is an open covering of X. To show that $\mathcal{W}$ is a Δ-refinement of $\mathcal{U}$ let $y \in X$. Then there exists a V_β containing y. If $y \in W(x)$ for some $x \in X$ then $W(x) \cap Cl(V_\beta) \neq \varnothing$ so $\beta \in A(x)$ but is not contained in C(x). Thus $\beta \in B(x)$ which implies $W(x) \subset U_\beta$. Therefore $Star(y, \mathcal{W}) \subset U_\beta$

which shows that $\mathcal{W}$ is a Δ-refinement of $\mathcal{U}$. ∎

The concept of full normality was studied by J. W. Tukey in 1940 in his Annals of Mathematics Studies (Princeton) monograph titled *Convergence and uniformity in general topology*. In that work he introduced the notion of a normal open covering as follows: Let $\{\mathcal{U}_n\}$ be a sequence of open coverings such that $\mathcal{U}_{n+1}$ star refines $\mathcal{U}_n$ for each positive integer n. Then $\{\mathcal{U}_n\}$ is said to be a **normal sequence** of open coverings. If $\mathcal{U}$ is an open covering such that there exists a normal sequence $\{\mathcal{V}_n\}$ of open coverings with $\mathcal{V}_1$ refining $\mathcal{U}$ then $\mathcal{U}$ is said to be a **normal covering**.

THEOREM 1.2 (J. W. Tukey, 1940) Every pseudo-metric space is fully normal.

Proof: Let (X, d) be a pseudo-metric space and let $\mathcal{U} = \{U_\alpha \mid \alpha < \gamma\}$ be an open covering of X. For each $x \in X$ there exists an $\varepsilon > 0$ such that $S(x, \varepsilon) \subset U_\alpha$ for some $\alpha < \gamma$. Choose $\varepsilon(x)$ and $\alpha(x)$ such that $0 < \varepsilon(x) < 1$ and $S(x, 4\varepsilon(x)) \subset U_{\alpha(x)}$. Then $\mathcal{V} = \{S(x, \varepsilon(x)) \mid x \in X\}$ is an open covering of X. It will be shown that $\mathcal{V}$ Δ-refines $\mathcal{U}$. For this let $y \in X$ and consider $H = \{x \mid y \in S(x, \varepsilon(x))\}$. Now $H \neq \varnothing$ since $y \in H$. Therefore choose $z \in H$ such that $\varepsilon(z) > 2/3 sup\{\varepsilon(x) \mid x \in H\}$. Then if $x \in H$ we have:

$$S(x, \varepsilon(x)) \subset S(y, 2\varepsilon(x)) \subset S(y, 3\varepsilon(z)) \subset S(z, 4\varepsilon(z)).$$

Therefore $Star(y, \mathcal{V}) \subset S(z, 4\varepsilon(z)) \subset U_{\alpha(z)}$. Consequently $\mathcal{V}\ \Delta\ \mathcal{U}$. Let $\mathcal{U}_1 = \mathcal{U}$. By successive applications of the above argument it is possible to construct a sequence $\{\mathcal{U}_n\}$ of open coverings of X such that $\mathcal{U}_{n+1}\ \Delta\ \mathcal{U}_n$ for each n. But then $\{\mathcal{U}_{2n}\}$ is a normal sequence and hence $\mathcal{U}$ is normal. Therefore every open covering in a pseudo-metric space is normal which means every pseudo-metric space is fully normal. ∎

COROLLARY 1.1 (A. H. Stone) Every pseudo-metric space is paracompact.

EXERCISE

1. Let $\{U_\alpha\}$ be a covering of a space X. Another covering $\{V_\alpha\}$ with the same index set is said to be a **precise** refinement of $\{U_\alpha\}$ if $V_\alpha \subset U_\alpha$ for each α. Show that if $\{U_\alpha\}$ has a point finite (locally finite) refinement $\{W_\beta\}$, then there exists a precise point finite (locally finite) refinement $\{V_\alpha\}$ of $\{U_\alpha\}$.

1.3 The Metrization Problem

The metrization problem is the problem of determining which topological spaces have topologies that are generated by metrics (pseudo-metrics). It dates back to the earliest studies of topology. One of the early researchers of the problem was P. Urysohn whose metrization theorem will be proved in this section. Urysohn's Metrization Theorem states that regular second countable spaces are metrizable. Urysohn's Metrization Theorem is not a solution to the metrization problem because it is not both necessary and sufficient.

Although there are a number of known solutions to the metrization problem, many topologers consider the problem to be "essentially solved" by the Nagata-Smirnov Metrization Theorem. Certainly the Nagata-Smirnov solution is the most successful and widely known solution. None-the-less, new solutions appear from time to time. There is, however, something unsatisfying about the Nagata-Smirnov solution. Since a basis for a topology "defines" the topology, placing a metrization requirements on the basis is closely akin to defining the space to be metric. One would like a characterization of metrizable spaces in terms of something like a covering or separation property that does not explicitly involve a basis. In Chapter 2 it will be shown that uniform spaces can be characterized by a separation property.

The purpose of this section is not to give a comprehensive treatment of the metrization problem but to give a representative sample that will provide motivation for our study of uniform spaces in later chapters. Following the historical precedent we will first establish the Urysohn Metrization Theorem. Two lemmas are needed for this.

LEMMA 1.4 A space is metrizable (pseudo-metrizable) if and only if there exists an imbedding of the space into a metric (pseudo-metric) space.

Proof: Assume X is metrizable and that d is a metric that generates the topology of X. Let Y denote this metric space. Then the identity mapping i:X → Y is a homeomorphism and hence an imbedding of X into the metric space Y.

Conversely, let f:X → Y be an imbedding of the space X into the metric space Y. Let d denote the metric for Y. Define the function d':X × X → **R** by $d'(x,y) = d[f(x),f(y)]$ for each pair of points x and y in X. That d' is a metric that generates the topology of X is left as an exercise (Exercise 1). ∎

COROLLARY 1.2 Every subspace of a metric (pseudo-metric) space is metrizable (pseudo-metrizable). In particular, the closed unit interval I = [0,1] is metrizable.

LEMMA 1.5 The topological product of a countable family of metric (pseudo-metric) spaces is metrizable (pseudo-metrizable).

Proof: It will be shown that the topological product of a finite number of metric spaces $X = X_1 \times X_2 \times \ldots \times X_n$ is metrizable. For each $i = 1 \ldots n$ let d_i be the metric for X_i. Define $d{:}X \times X \rightarrow \mathbf{R}$ by

(1.11) $$d(x,y) = \sqrt{\Sigma_{i=1}^{n}[d_i(x_i,y_i)]^2}$$

for each pair of points $x = (x_1 \ldots x_n)$ and $y = (y_1 \ldots y_n)$ in X. That d is a metric that generates the topology of X is left as an exercise (Exercise 2(a)). This shows X is metrizable.

Next it will be shown that the topological product X of a sequence $\{X_n\}$ of metric spaces is metrizable. For each positive integer n let d_n denote the metric for X_n. First we want to show that for each n it is possible to construct a new metric f_n from d_n such that $f_n(x_n,y_n) \leq 1$ for each pair of points x_n and y_n in X_n. For this let

(1.12) $$f_n(x_n,y_n) = min\{1, d_n(x_n,y_n)\}$$

for each pair of points x_n and y_n in X_n. That f_n is a metric for X_n that generates the topology of X_n is left as an exercise (Exercise 2(b)). We can now define a metric for X by

(1.13) $$d(x,y) = \Sigma_{n=1}^{\infty} 2^{-n} f_n(x_n,y_n)$$

for any pair of points $x = (x_1, x_2, \ldots)$ and $y = (y_1, y_2, \ldots)$ of X. That d is a metric that defines the topology of X is left as an exercise (Exercise 2(c)). ∎

The subspace I(ω) of real Hilbert space H (defined in Section 1.1) consists of all points $\{x_n\}$ such that $0 \leq x_n \leq 1/n$ is known as the **Hilbert Cube.** It can be shown (Exercise 3) that I(ω) is homeomorphic to the countable product

$$I^N = \Pi\{I_n \mid n = 1, 2, 3 \ldots\} \text{ and } I_n = [0,1] \text{ for each } n$$

of unit intervals under the homeomorphism $f{:}\mathrm{I}(\omega) \rightarrow I_N$ defined by $f(\{x_n\}) = \{x_1, x_2, x_3 \ldots\}$.

COROLLARY 1.3 The Hilbert Cube is metrizable.

Let F be a family $\{f_\alpha{:}X \rightarrow Y_\alpha \mid \alpha < \gamma\}$ of mappings from a space X into spaces Y_α for each $\alpha < \gamma$. Let Y denote the topological product of the family $\{Y_\alpha\}$. Let $f{:}X \rightarrow Y$ denote the product mapping defined by

$$[f(x)]_\alpha = f_\alpha(x) \text{ for each } x \in X \text{ and } \alpha < \gamma.$$

Then f is a continuous mapping from X into Y. The family F is able to **distinguish points** of X if for any two distinct points x and y of X there exists an $\alpha < \gamma$ with $f_\alpha(x) \neq f_\alpha(y)$. F is able to **distinguish points from closed sets** if for any closed set K in X and point p in X - K there is an $\alpha < \gamma$ such that $f_\alpha(p)$ is not contained in $Cl(f_\alpha(K))$.

LEMMA 1.6 If the family F can distinguish points of X and can distinguish points from closed sets, then the product mapping $f:X \to Y$ is an imbedding.

Proof: Assume F can distinguish points and can distinguish points from closed sets. If x and y are distinct points of X then there is an $\alpha < \gamma$ with $f_\alpha(x) \neq f_\alpha(y)$ which implies $f(x) \neq f(y)$. Therefore f is a one-to-one mapping. Let U be an open subset of X. It will be shown that $f(U)$ is open in Y. For this let $p \in U$ and let $q = f(p)$. Since X - U is closed and p does not belong to X - U, there is an $\alpha < \gamma$ such that $f_\alpha(p)$ does not belong to $Cl(f_\alpha(X - U))$. Let

$$V = \{y \in Y \mid y_\alpha \text{ is not contained in } Cl(f_\alpha(X - U))\}.$$

Then V is a basic open set in Y which implies $V \cap f(X)$ is open in $f(X)$. Now $q = f(p) \in V \cap f(X) \subset f(U)$. It follows that $f(U)$ is open in $f(X)$. Therefore f is a one-to-one continuous open mapping and hence an imbedding. ∎

THEOREM 1.3 (Urysohn's Imbedding Theorem) A regular T_1 space with a countable basis can be imbedded as a subspace of the Hilbert Cube.

Proof: In view of Lemma 1.6 and the fact that I(ω) is homeomorphic to I^N, it will suffice to show the existence of a countable family F of continuous functions from X into the unit interval I = [0,1] which can distinguish points from closed sets. For this let B be a countable basis for X and let C be the subset of $B \times B$ which consists of pairs of open sets (U,V) such that $Cl(V) \subset U$. Then C is countable. By Urysohn's Lemma (Theorem 0.1) we can then obtain a countable family $F = \{f_{(U,V)}:X \to I \mid (U,V) \in C\}$ of continuous functions which map $Cl(V)$ into zero and X - U into one.

To show that F can distinguish points from closed sets let $p \in X - K$ where K is closed in X. Since X is regular and B is a basis, it is possible to find U and V in B such that $p \in V \subset Cl(V) \subset U \subset X - K$. Then $f_{(U,V)}(p) = 0$ and $f_{(U,V)}(K) = 1$. Consequently F can distinguish points from closed sets. ∎

THEOREM 1.4 (Urysohn's Metrization Theorem) Every regular T_1 space with a countable basis is metrizable.

Proof: By Theorem 1.3, X can be imbedded as a subspace of the Hilbert Cube. By Corollary 1.3 the Hilbert Cube is metrizable. Hence by Lemma 1.4, X is metrizable. ∎

The Nagata-Smirnov Metrization Theorem was discovered independently by J. Nagata and Y. M. Smirnov. The first to appear was Nagata's paper in the

Journal of the Polytechnical Institute at Osaka City University in 1950, Volume 1, pp. 93-100. Smirnov's paper appeared in 1951 in Doklady Akad. Nauk S.S.S.R.N.S. Volume 77, pp. 197-200.

In the proof of Stone's Theorem it was shown that if an open covering $\mathcal{U}$ is normal then an associated normal sequence $\{\mathcal{U}_n\}$ could be used to decompose $\mathcal{U}$ into a countable sequence of families $\{G_{n_\alpha} \mid \alpha < \gamma\}$ for $n = 1, 2, 3 \ldots$ where for each positive integer m, the family $\{G_{m_\alpha}\}$ is locally finite. A family F of subsets of a space X is said to be **σ-locally finite** if it is the union of countably many locally finite subfamilies. Using this terminology, the proof of Stone's Theorem shows that full normality implies each open covering has a σ-locally finite open refinement.

With this in mind, it is natural to wonder what a space would be like if it had a σ-locally finite basis. Clearly metric spaces have a σ-locally finite basis for if (X, d) is a metric space and $B_n = \{S(x, 2^{-n}) \mid x \in X\}$ is the covering consisting of the spheres of radius 2^{-n} for some positive integer n, then since X is paracompact (Corollary 1.1) there is a locally finite open refinement $\mathcal{U}_n$ of B_n. Put $\mathcal{U} = \cup\{\mathcal{U}_n\}$. Then $\mathcal{U}$ is a σ-locally finite collection of open sets and since $B = \cup\{B_n\}$ is a basis, so is $\mathcal{U}$. The Nagata-Smirnov Theorem shows that among regular T_1 spaces, the spaces with the σ-locally finite bases are precisely the metrizable ones.

THEOREM 1.5 (Nagata-Smirnov Metrization Theorem) A T_1 space is metrizable if and only if it is regular and has a σ-locally finite basis.

Proof: Let X be a regular space with a σ-locally finite basis $B = \{B_n\}$ where each B_n is locally finite. To show X is normal let H and K be disjoint closed subsets of X. Since X is regular, if $p \in$ H there is an open set W(p) in B such that $p \in W(p) \subset Cl(W(p)) \subset X - H$. Define the open coverings $\mathcal{U}$ and $\mathcal{V}$ of H and K respectively by $\mathcal{U} = \{W(p) \mid p \in H\}$ and $\mathcal{V} = \{W(q) \mid q \in K\}$. For each positive integer n put $\mathcal{U}_n = \mathcal{U} \cap B_n$ and $\mathcal{V}_n = \mathcal{V} \cap B_n$. Since $\mathcal{U}_n$ and $\mathcal{V}_n$ are locally finite collections we have:

$$Cl(U_n) = \cup\{Cl(U) \mid U \in \mathcal{U}_n\} \subset X - K \text{ and}$$

$$Cl(V_n) = \cup\{Cl(V) \mid V \in \mathcal{V}_n\} \subset X - H.$$

For each positive integer n, define the open sets

$$W_n = U_n - \cup\{Cl(V_i) \mid i \le n\} \text{ and } G_n = V_n - \cup\{Cl(U_i) \mid i \le n\}.$$

Since $W_m \cap V_n$ is empty whenever $m \ge n$ it follows that $W_m \cap G_n = \varnothing$ whenever $m \ge n$. Similarly, since $U_m \cap G_n = \varnothing$ whenever $m \le n$ we have $W_m \cap G_n = \varnothing$ whenever $m \le n$. Hence $W_m \cap G_n = \varnothing$ for all positive integers m and n. Now put

$W = \cup\{W_n\}$ and $G = \cup\{G_n\}$. Clearly $H \subset W$, $K \subset G$ and $W \cap G = \varnothing$. Thus X is normal.

Next it will be shown that there exists a metric for X that generates the topology τ of X. For this let N denote the positive integers and $A = N \times N$. Let $\alpha = (m,n)$ be a member of A. For each U in B_m let $W = \cup\{V \in B_n \mid Cl(V) \subset U\}$. Since B_n is locally finite $Cl(W) \subset U$. By Urysohn's Lemma (Theorem 0.1) there exists a continuous $f_U: X \to [0,1]$ such that $f_U(X - U) = 0$ and $f_U(Cl(W)) = 1$. Next, define the pseudo-metric d_α for X by

$$d_\alpha(x,y) = \Sigma\{\,|f_U(x) - f_U(y)|\,|\,U \in B_m\}$$

for any two points x and y of X. Since B_m is locally finite the summation is finite, so d_α is well defined. For each $\alpha < \gamma$, (X, d_α) is a pseudo-metric space which generates some topology τ_α for X. The identity function $i_\alpha:(X,\tau) \to (X,\tau_\alpha)$ can be shown to be continuous by showing that $S(x_0,\varepsilon) \in \tau$ for any x_0 in X. For this put $c_0 = f_U(x_0)$, $g_U(y) = |c_0 - f_U(y)|$ and $F(y) = \Sigma\{g_U(y) \mid U \in B_m\}$. Since f_U is continuous for each U in B_m, so is g_U. Since B_m is locally finite $F: X \to \mathbf{R}$ is also continuous. Now $S(x_0, \varepsilon) = \{y \in Y \mid F(y) < \varepsilon\}$. Consequently $S(x_0,\varepsilon) \in \tau$ which shows that i_α is continuous. Next let

$$F = \{i_\alpha:(X,\tau) \to (X,d_\alpha) \mid \alpha \in A\}.$$

Clearly F can distinguish points of X. To show that F can distinguish points from closed sets let K be a closed subset of X and p a point in X - K. Since X is regular, there are basic open sets U and V such that $p \in V \subset Cl(V) \subset U \subset X - K$. Let m and n be positive integers such that $U \in B_m$ and $V \in B_n$. Then $\alpha = (m,n) \in A$ and $d_\alpha(p, K) \geq 1$. By Proposition 1.5, $i_\alpha(p)$ is not in the closure of $i_\alpha(K)$. Thus F can distinguish points from closed sets.

By Lemma 1.6, the Cartesian product $\pi: X \to \Pi(X, d_\alpha)$ of the family F is an imbedding of X into the topological product of the (X, d_α)'s. Since A is countable, it follows from Lemma 1.5 that this product is pseudo-metrizable and by Corollary 1.2 this implies X is pseudo-metrizable. Since X is T_1 it is metrizable.

It has already been shown that if X is metrizable then it has a σ-locally finite basis. That X is regular follows from the fact that every metric space is paracompact (Corollary 1.1) and hence normal (Lemma 1.2). ∎

COROLLARY 1.4 A space is pseudo-metrizable if and only if it is regular and has a σ-locally finite basis.

The remainder of this section will be devoted to another solution to the metrization problem that does not require regularity in the hypothesis and does not state the metrization criteria in terms of a basis. Let $\mathcal{U}$ be an open covering of the space X. A sequence $\{\mathcal{U}_n\}$ of open coverings of X is said to be **locally starring** for $\mathcal{U}$ if for each $x \in$ X there is a neighborhood N of x and a positive integer n such that $Star(N, \mathcal{U}_n) \subset$ U for some U $\in \mathcal{U}$.

THEOREM 1.6 (A. V. Arhangel'skiĭ) A T_1 space X is metrizable if and only if there exists a sequence $\{V_n\}$ of open coverings that is locally starring for all open coverings of X.

Proof: First suppose (X, d) is a metric space. Let n be a positive integer and put $B_n = \{S(x, 2^{-n}) \mid x \in X\}$. Then $\cup\{B_n\}$ is a basis for the topology generated by d. Let $\mathcal{U}$ be an open covering of X and pick U $\in \mathcal{U}$ such that $x \in$ U. Then there is a positive integer m such that $S(x, 2^{-m}) \subset$ U. Now

$$Star(S(x, 2^{-(m+2)}), B_{n+2}) \subset S(x, 2^{-n}) \subset U$$

so that $\{B_n\}$ is locally starring for all open coverings of X.

Conversely, assume X is T_1 and there exists a sequence $\{V_n\}$ of open coverings of X that is locally starring for all open coverings of X. It will first be shown that X is paracompact. Let $\{\mathcal{U}_n\}$ be the sequence of open coverings defined by $\mathcal{U}_1 = V_1$ and $\mathcal{U}_n = \{V_1 \cap \ldots \cap V_n \mid V_i \in V_i \text{ for } i = 1 \ldots n\}$. Clearly $\{\mathcal{U}_n\}$ is also locally starring for all open coverings of X and $\mathcal{U}_{n+1}$ refines $\mathcal{U}_n$ for each positive integer n. Let $\mathcal{U}$ be an open covering of X and put

$$\mathcal{W} = \{W \subset X \mid W \text{ is open and for some positive integer } n, W \subset U_n \text{ for some } U_n \in \mathcal{U}_n \text{ and } Star(W, \mathcal{U}_n) \subset U \text{ for some } U \in \mathcal{U}\}.$$

For each W $\in \mathcal{W}$ let $n(W)$ be the least positive integer such that W $\subset U_{n(W)}$ for some $U_{n(W)}$ in $\mathcal{U}_{n(W)}$ and $Star(W, \mathcal{U}_{n(W)}) \subset$ U for some U $\in \mathcal{U}$. Since $\{\mathcal{U}_n\}$ is locally starring for $\mathcal{U}$, $\mathcal{W}$ is an open covering of X. It will be shown that $\mathcal{W}$ Δ-refines $\mathcal{U}$. Let $x \in$ X and put $n(y) = min\{n(W) \mid y \in W \in \mathcal{W}\}$ and let V $\in \mathcal{W}$ such that $n(V) = n(x)$. For each W $\in \mathcal{W}$ containing x it is clear that $n(W) \geq n(x)$. Thus

$$Star(x, \mathcal{W}) \subset \cup\{Star(x, \mathcal{U}_n) \mid n \geq n(y)\}.$$

But since $\mathcal{U}_{n+1}$ refines $\mathcal{U}_n$ for each positive integer n, we have

$$Star(x, \mathcal{W}) \subset Star(x, \mathcal{U}_{n(x)}) \subset Star(V, \mathcal{U}_{n(V)}) \subset U$$

for some U $\in \mathcal{U}$. Therefore $\mathcal{W}$ Δ-refines $\mathcal{U}$ and hence X is fully normal. By Stone's Theorem, X is paracompact.

For each n let B_n be a locally finite open refinement of $\mathcal{U}_n$. It will be shown that $B = \cup\{B_n\}$ is a basis for the topology of X. For this let A be an open subset of X and let $x \in$ A. Let W be an open set such that $x \in W \subset Cl(W) \subset$ A and put $B = X - Cl(W)$. Since $\{\mathcal{U}\}$ is locally starring for the covering $\{A, B\}$ and x is not contained in B, there is a neighborhood N of x and a positive integer n such that $Star(N, \mathcal{U}_n) \subset A$. But since B_n refines $\mathcal{U}_n$ we have:

$$x \in Star(x, B_n) \subset Star(x, \mathcal{U}_n) \subset Star(N, \mathcal{U}_n) \subset A.$$

Therefore B is a basis for the topology of X. Since each B_n is a locally finite collection, B is a σ-locally finite basis. Consequently, by the Nagata-Smirnov Theorem X is metrizable. ∎

EXERCISES

1. Complete the proof of Lemma 1.4 by showing d' is a metric.

2. Complete the proof of Lemma 1.5 by showing:

   (a) d defined by (1.11) is a metric that generates the topology of $X = X_1 \times \ldots \times X_n$,
   (b) the function f_n defined by (1.12) is a metric that generates the topology of X_n and
   (c) the function d defined by (1.13) is a metric that generates the product topology of X.

3. Show that the Hilbert Cube I(ω) is homeomorphic to a countable product of unit intervals.

1.4 Topology of Metric Spaces

This section is devoted to the behavior of metric (pseudo-metric) topologies. From Section 1.2 we know metric spaces are paracompact. We also know paracompact spaces are normal. Metric spaces also satisfy a variety of other separation properties such as being completely normal, perfectly normal and collectionwise normal. The verification that this is the case is left as an exercise (Exercise 2(a) through 2(c)). The definitions of these properties are given below.

A space is said to be **completely normal** if every subspace is normal. An alternate definition is that for each pair of subsets A and B such that $A \cap Cl(B) = \varnothing = Cl(A) \cap B$ there exist open sets U and V containing A and B respectively with $U \cap V = \varnothing$. That these definitions are equivalent is left as an exercise (Exercise 1). A set H is called a G_δ (pronounced "G-delta") if it is the intersection of a countable collection of open sets. A space is said to be **perfectly normal** if it is normal and every closed subset is a G_δ. A collection $F = \{F_\alpha \mid \alpha \in A\}$ of subsets is said to be **discrete** if each point of the space is contained in a neighborhood that meets at most one of the members of F. A space is said to be **collectionwise normal** if for each discrete collection $F = \{F_\alpha | \alpha \in A\}$ of closed sets with $\cup F$ being closed, there exists a collection $G = \{G_\alpha | \alpha \in A\}$ of disjoint open sets with $F_\alpha \subset G_\alpha$ for each $\alpha \in A$.

Clearly all pseudo-metrizable spaces are first countable since for each point x, the sequence $\{S(x, 2^{-n})\}$ is a local basis for x. Recall from Section 1.1 that discrete spaces are metrizable and consequently there exist metric spaces that are not second countable or separable. One of the distinct features of metric spaces is that the concepts of second countability and separability are equivalent in metric spaces.

THEOREM 1.7 A metric (pseudo-metric) space is second countable if and only if it is separable.

Proof: Assume (X, d) is a second countable metric space and let $B = \{B_n\}$ be a countable basis for X. For each positive integer n pick $x_n \in B_n$. Clearly the sequence $\{x_n\}$ is countable and dense in X. Therefore X is separable.

Conversely, assume (X, d) is separable and let K be a countable dense subset. Put $B = \{S(x, 2^{-n}) | x \in K$ and n is a positive integer$\}$. Then B is countable. It will be shown that B is a basis for the topology of X. For this let U be open in X and let $p \in U$. Then there exists an $\varepsilon > 0$ such that $S(p, \varepsilon) \subset U$. Let m be a positive integer such that $0 < 2(2^{-m}) < \varepsilon$. Since K is dense in X there is a $q \in K$ such that $q \in S(p, 2^{-m})$. Then $p \in S(q, 2^{-m})$. To show $S(q, 2^{-m}) \subset U$ let $x \in S(q, 2^{-m})$. Then $d(p, x) \leq d(p, q) + d(q, x) < 2^{-m} + 2^{-m} = 2(2^{-m}) < \varepsilon$. Therefore $x \in S(p, \varepsilon) \subset U$. Consequently $S(q, 2^{-m}) \subset S(p, \varepsilon) \subset U$ which shows that B is a basis for the topology of X. ∎

By Exercise 6 of Section 1.1 we see that a metric space may fail to be locally compact. This is one way in which an arbitrary metric space differs from Euclidean n-space. However, many of the properties of E^n are preserved in arbitrary metric spaces. For a non-empty subset E of a metric space (X, d) we define the **diameter** $\delta(E)$ by

$$\delta(E) = sup\{d(x,y) | x,y \in E\}.$$

If the diameter $\delta(E)$ of E is finite, E is said to be **bounded.** A subset K of E^n is compact if and only if it is closed and bounded. In an arbitrary metric space we have:

THEOREM 1.8 Every compact subset of a metric space is closed and bounded.

Proof: Let K be a compact subset of the metric space (X, d). Suppose $p \in X - K$. Then for each $q \in K$ there exists open sets U(q) and V(q) containing p and q

respectively such that $U(q) \cap V(q) = \emptyset$. Then $\mathcal{V} = \{V(q) \mid q \in K\}$ is a covering of K and therefore has a finite subcovering $\{V(q_i) \mid i = 1 \ldots n\}$. Then $U = \cap_{i=1}^n U(q_i)$ is a neighborhood of p such that $U \cap K = \emptyset$. Therefore p cannot be a limit point of K which implies K is closed.

To show K is bounded consider the family $\mathcal{U} = \{S(x,1) \mid x \in K\}$. Then $\mathcal{U}$ is an open covering of K and as such has a finite subcovering say $\{S(x_1, 1) \ldots S(x_m, 1)\}$. Let $k = max\{d(x_i, x_j) \mid i,j \leq m\}$. If $x,y \in K$ then there exist i and j such that $x \in S(x_i,1)$ and $y \in S(x_j,1)$. Then by the triangle inequality we have

$$d(x,y) \leq d(x, x_1) + d(x_i, x_j) + d(x_j,y) \leq 1 + k + 1 = k + 2.$$

Since x and y were chosen arbitrarily, we conclude that $\delta(K) \leq k + 2$. Therefore K is bounded. ▪

THEOREM 1.9 (Lebesgue Covering Theorem) For each open covering $\mathcal{U}$ of a compact metric (pseudo-metric) space (X, d) there exists a positive number λ such that every subset E of X with $\delta(E) < \lambda$ is contained in some member of $\mathcal{U}$.

Proof: Since X is compact there is a finite subcovering $\{U_1 \ldots U_n\}$ of $\mathcal{U}$. For each $i = 1 \ldots n$ define a function $f_i{:}X \to \mathbf{R}$ by $f_i(x) = d(x, X - U_i)$. By Proposition 1.4, each f_i is continuous and hence the function $f{:}X \to \mathbf{R}$ defined by $f(x) = \Sigma_{i=1}^n f_i(x)$ is continuous. Moreover, $f(x) > 0$ for each x. Since X is compact, so is $f(X)$. Therefore there is a positive number p such that $f(x) \geq p$ for each $x \in X$. Let $\lambda = p/n$ and let $y \in X$. Then $\Sigma_{i=1}^n f_i(y) = f(y) \geq p = n\lambda$. Therefore there is an i such that $f_i(y) \geq \lambda$. But then $d(y, X - U_i) \geq \lambda$ which implies $S(y, \lambda) \subset U_i$. Let $E \subset X$ such that $\delta(E) < \lambda$. Pick $e \in E$ and let x be an arbitrary point of E. $\delta(E) < \lambda$ implies $d(e, x) < \lambda$. Thus $x \in S(e, \lambda)$ which implies $E \subset S(e, \lambda) \subset U_i$ for some $i = 1 \ldots n$. ▪

The positive number λ in Theorem 1.9 is said to be a **Lebesgue number** for the covering $\mathcal{U}$. A subset K of a metric space is said to be **precompact** if for each $\varepsilon > 0$ there is a finite $F \subset K$ such that $K \subset \cup\{S(x, \varepsilon) \mid x \in F\}$. The proofs of the following propositions are left as exercises (Exercises 6 and 7).

PROPOSITION 1.8 Every compact subset of a metric (pseudo-metric) space is precompact and every precompact subset is bounded.

PROPOSITION 1.9 A precompact metric (pseudo-metric) space is separable.

COROLLARY 1.5 A compact metric (pseudo-metric) space is separable.

There are several generalizations of compactness besides paracompactness and local compactness that have been studied extensively such as countable compactness, sequential compactness, precompactness and pseudocompactness. Some of these (e.g., countable compactness and sequential compactness) are equivalent to compactness in metric (pseudo-metric) spaces. A space X is said to be **countably compact** if every countable open covering has a finite subcovering. It is said to be **sequentially compact** if every sequence has a convergent subsequence. Sequential compactness is not a generalization of compactness as the name seems to imply, for there are compact spaces that fail to be sequentially compact.

THEOREM 1.10 For an arbitrary metric space X the following statements are equivalent:

(1) X is compact,
(2) X is countably compact and
(3) X is sequentially compact.

Proof: Clearly (a) → (b). To show (b) → (c) let (X, *d*) be a countably compact metric space and let $\{x_n\}$ be a sequence in X. Suppose $\{x_n\}$ has no limit point. Then $\{x_n\}$ is closed so X-$\{x_n\}$ is open. For each $x_i \in \{x_n\}$, x_i is not a limit point of $\{x_n\}$ so there is an open set U(x_i) containing x_i but no other point of $\{x_n\}$. Then $\mathcal{U}$ = (X-$\{x_n\}$)∪{U(x_i)} is a countable open covering of X with no finite subcovering, which is a contradiction, so $\{x_n\}$ has a limit point after all.

Let p be a limit point of $\{x_n\}$. We will construct a subsequence of $\{x_n\}$ that converges to p. Since X is metric, the family $\{S(p, 2^{-n})\}$ is a local basis for p. Let x_{m_1} be the first element of $\{x_n\}$ contained in $S(p, 2^{-1})$. For each positive integer $k > 1$ let x_{m_k} be the first element of $\{x_n\}$ greater than $x_{m_{k-1}}$ that belongs to $S(p, 2^{-k})$. From this definition of $\{x_{m_n}\}$ it is easily seen that $\{x_{m_n}\}$ converges to p. Consequently X is sequentially compact.

To show (c) → (a) let (X, *d*) be a sequentially compact metric space. We first show X is countably compact. Suppose it is not. Then there is a countable open covering $\mathcal{U} = \{U_n\}$ with no finite subcovering. For each positive integer n pick $x_n \in$ X such that x_n is not contained in $\cup\{U_m \mid m \le n\}$. Let $p \in$ X. Since $\mathcal{U}$ covers X there is a positive integer n such that $p \in U_n$. By the definition of the sequence $\{x_n\}$ it is clear that U_n can contain at most finitely many members of $\{x_n\}$. Hence no subsequence of $\{x_n\}$ can converge to p. But this means X is not sequentially compact which is a contradiction. Therefore X is countably compact.

Next we will show X is precompact. Suppose it is not. Then there is an $\varepsilon > 0$ such that no finite F ⊂ X satisfies X = $\cup\{S(x, \varepsilon) \mid x \in F\}$. Therefore it is possible to construct a sequence $\{x_n\}$ in X of distinct points such that $\{x_n\} \cap S(x_i, \varepsilon) = \{x_i\}$ for each positive integer i. Define the open set G and the closed set H by

$$H = X - \cup\{S(x_n, \varepsilon)\} \text{ and } G = \cup\{S(x, \varepsilon/2) \mid x \in H\}.$$

Then $G = G\cup\{S(x_n, \varepsilon)\}$ is a countable open covering with no finite subcovering which is a contradiction. Therefore X must be precompact. By Proposition 1.9, X is separable and hence by Theorem 1.7, second countable. But then X is Lindelöf and a countably compact Lindelöf space is compact so (c) $\rightarrow$ (a). ■

Theorem 1.10 hints at the role sequences play in metric and pseudo-metric spaces. From it we see that a metric space is compact if and only if each sequence has a convergent subsequence. What is more striking is the fact that all topological properties in metric spaces can be determined by the behavior of the sequences in the space. In fact, the topology itself can be characterized in terms of which sequences converge to which points.

PROPOSITION 1.10 A set U in a pseudo-metric space (X, d) is open if and only if no sequence in X - U converges to a point of U.

Proof: Assume U is open. If $\{x_n\}$ is a sequence in X - U then $\{x_n\}$ cannot be eventually in U and consequently cannot converge to any point in U since U is open. Conversely, assume no sequence in X - U can converge to a point of U and suppose U is not open. Then there is a p in U such that each neighborhood of p meets X - U . For each positive integer n let $x_n \in S(p, 2^{-n})\cap(X\text{-}U)$. Then $\{x_n\} \subset X - U$. But clearly $\{x_n\}$ is eventually in $S(p, 2^{-n})$ for each n so $\{x_n\}$ converges to p which is a contradiction. Therefore U must be open after all. ■

PROPOSITION 1.11 A point p in a pseudo-metric space (X, d) belongs to the closure of a set $F \subset X$ if and only if there is a sequence in F that converges to p.

The proof of this proposition and the next are left as exercises (Exercises 9 and 10).

COROLLARY 1.6 A set F in a pseudo-metric space (X, d) is closed if and only if for each p in F there is a sequence in F that converges to p.

PROPOSITION 1.12 A function $f:X \rightarrow Y$ from a pseudo-metric space (X, d) to a pseudo-metric space (Y, ρ) is continuous if and only if for each sequence $\{x_n\} \subset X$ that converges to some $p \in X$, the sequence $\{f(x_n)\}$ in Y converges to $f(p)$.

EXERCISES

1. Show that a space X is completely normal if and only if for any pair of subsets A and B such that $A\cap Cl(B)$ and $B\cap Cl(A)$ are both empty, there exist disjoint open sets U and V with $A \subset U$ and $B \subset V$.

2. Show that pseudo-metric spaces are:
   (a) completely normal,
   (b) perfectly normal, and
   (c) collectionwise normal.

3. Show that the product of uncountably many unit intervals is not perfectly normal and hence not metrizable.

4. A subset of a space is said to be an F_σ (pronounced "F-sigma") if it is the union of a countable collection of closed sets. Show that a normal space is perfectly normal if and only if every open set is an F_σ. Hence open sets in metric spaces are F_σ's.

5. If an open set U in a metric space (X, *d*) contains a compact subset K, show that $d(K, X - U) > 0$ if both K and X - U are non-empty.

6. Prove Proposition 1.8.

7. Prove Proposition 1.9.

8. Show that precompactness is a hereditary property in metric spaces; i.e., show that every subspace of a precompact metric space is precompact.

9. Prove Proposition 1.11.

10. Prove Proposition 1.12.

11. Show that a sequence $\{x_n\}$ in a metric space (X, *d*) converges to a point $p \in$ X if and only if $\lim_{n\to\infty} d(p, x_n) = 0$.

12. Show that the real Hilbert space is not locally compact. Hence a metrizable space may fail to be locally compact.

1.5 Uniform Continuity and Uniform Convergence

Recall from elementary analysis that a real-valued function $f:\mathbf{R} \to \mathbf{R}$ from the reals **R** is continuous at a point p if for each $\varepsilon > 0$ there exists a $\delta > 0$ such that for each $q \in \mathbf{R}$ with $|p - q| < \delta$ we have $|f(p) - f(q)| < \varepsilon$. Then f is defined to be continuous if it is continuous at each point of **R**. It should be noted that this definition of continuity is equivalent to the definition given in the preface for the metric space **R**. In fact, for any pair of metric spaces (X, *d*) and (Y, ρ) it is easy to show that:

PROPOSITION 1.13 A function $f:X \to Y$ is continuous at a point $p \in X$ if and only if for each $\varepsilon > 0$ there is a $\delta > 0$ such that whenever $q \in X$ with $d(p,q) < \delta$ then $\rho(f(p), f(q)) < \varepsilon$.

The proof of this proposition is left as an exercise (Exercise 1). In the above definition of continuity, the δ that is assumed to exist depends on both ε and the point p. If for each $\varepsilon > 0$ it is possible to find a single $\delta > 0$ such that for any pair of points $p, q \in \mathbf{R}$ with $|p - q| < \delta$ then $|f(p) - f(q)| < \varepsilon$ then f is said to be **uniformly continuous**. This definition is extended to arbitrary metric spaces as follows: a function $f:X \to Y$ from a metric space (X, d) to a metric space (Y, ρ) is said to be uniformly continuous if for each $\varepsilon > 0$ there exists a $\delta > 0$ such that whenever $p, q \in X$ with $d(p, q) < \delta$ then $\rho(f(p), f(q)) < \varepsilon$. Clearly, a uniformly continuous function is continuous.

An example of a continuous function that is not uniformly continuous is the function $f:(0,1) \to \mathbf{R}$ defined by $f(x) = 1/x$ for each $x \in (0,1)$. However, if we define f on $[0,1]$ instead, we find that f is now uniformly continuous. It is impossible to find a continuous real valued function on a closed interval that is not uniformly continuous. In fact for arbitrary metric spaces we have:

PROPOSITION 1.14 Every continuous function from a compact metric space into a metric space is uniformly continuous.

Proof: Let $f:X \to Y$ be a continuous function from a compact metric space (X,d) into a metric space (Y,ρ) and let $\varepsilon > 0$. Then $\mathcal{U} = \{S(y, \varepsilon/2)) \mid y \in Y\}$ is an open covering of Y so $f^{-1}(\mathcal{U}) = \{f^{-1}(S(y, \varepsilon/2)) \mid y \in Y\}$ is an open covering of X. Since X is compact, by Theorem 1.9, is a Lebesgue number $\delta > 0$ for this covering. Let p and q be two points in X with $d(p, q) < \delta$. Then the set $\{p,q\}$ having diameter $< \delta$ must lie entirely in one of the sets $f^{-1}(S(y, \varepsilon/2))$. Thus $f(p)$ and $f(q)$ both belong to $S(y, \varepsilon/2)$ which implies $\rho(f(p), f(q)) < \varepsilon$. This establishes the uniform continuity of f. ▪

If $f:X \to Y$ is a uniformly continuous one-to-one function from X onto Y then the inverse image function f^{-1} is well defined. If f^{-1} is continuous then, of course, X and Y are homeomorphic topologically. If in addition f^{-1} is uniformly continuous then f is said to be a **uniform homeomorphism**.

Properties preserved by homeomorphisms are said to be topological properties whereas properties preserved by uniform homeomorphisms (and not by homeomorphisms) are called **uniform properties**. There is a special type of uniform homeomorphism that preserves distance; that is, for each pair of points $p, q \in X$ $d(p, q) = \rho(f(p), f(q))$. Such a uniform homeomorphism is called an **isomorphism**. Properties preserved by isomorphisms are called **metric properties**. Clearly the concepts of uniform continuity, uniform homeomorphism and isomorphism also apply to pseudo-metric spaces.

There exist uniformly continuous functions that preserve distance but are not isomorphisms. Functions that preserve distance are called **isometric functions.** An isometric function may fail to be an isomorphism for the following reasons: the inverse function does not exist (in case f is not one-to-one) or the inverse function is not uniformly continuous. An important example of an isometric mapping is the following. Let (X, d) be a pseudo-metric space and define an equivalence relation ~ in X by x~y if $d(x, y) = 0$. Let Y be the quotient space X/~ and p:X → Y the canonical projection that maps each point of X onto the equivalence class that contains it. Define the metric ρ in Y by

$$\rho(a,b) = d(p^{-1}(a), p^{-1}(b))$$

for each pair $a, b \in$ Y. It is easily shown that ρ is indeed a metric and ρ defines the quotient topology on Y. (Y, ρ) is called the metric space **associated** with (X, d). If X is a metric space it is clear that X = Y and p is the identity mapping. Since $\rho(p(x), p(y)) = d(x, y)$ for each pair $x, y \in$ X we see that the canonical projection is an isometric mapping. Topologically, isometric mappings are very well behaved as can be seen by the following:

THEOREM 1.11 If f:X → Y is an isometric function from the metric space X to the metric space Y then f is an imbedding of X into Y.

Proof: Let d denote the metric in X and ρ the metric in Y. We first show that f is an open mapping of X onto f(X). For this let U be an open set in X and put V = f(U) ⊂ f(X). Pick $p \in$ U and let $q = f(p)$. Choose $\varepsilon > 0$ such that $S(p, \varepsilon) \subset$ U. Let $y \in S(q, \varepsilon) \cap f$(X). Then there exists an $x \in$ X with $f(x) = y$. Since f is isometric we have $d(p, x) = \rho(q, y) < \varepsilon$. Consequently $x \in S(p, \varepsilon) \subset$ U so $y \in$ V. Therefore $S(q, \varepsilon) \cap f$(X) ⊂ V so V is open in f(X). This establishes that f is an open mapping. It remains to show that f is a one-to-one mapping which is left as an exercise (Exercise 2). ∎

In calculus, another "uniform" concept of significance is that of uniform convergence. A sequence $\{f_n\}$ of real valued functions from the reals is said to **converge uniformly** to a function f if for every $\varepsilon > 0$ there is a positive integer m such that for each $x \in$ X and each $k > m$, $|f_k(x) - f(x)| < \varepsilon$. Furthermore, the following classical theorem can be proved.

THEOREM 1.12 A function f:R → R is continuous if there is a sequence $\{f_n\}$ of continuous functions that converges uniformly to f.

Proof: Let $x_0 \in$ **R** and $\varepsilon > 0$. Put $\varepsilon' = \varepsilon/4$. Since $\{f_n\}$ converges uniformly to f there is a positive integer k such that if $m \geq k$ then $|f_m(x) - f(x)| < \varepsilon'$ for each $x \in$ **R**. In particular $|f_k(x_0) - f(x_0)| < \varepsilon'$. Since f_k is continuous there is a $\delta > 0$ such that if $|x_0 - y| < \delta$ then $|f_k(x_0) - f_k(y)| < \varepsilon'$. Also note that $|f_k(y) - f(y)| < \varepsilon'$. Now

$$|f(x_0) - f(y)| \le |f(x_0) - f_k(x_0)| + |f_k(x_0) - f_k(y)| + |f_k(y) - f(y)| < 3\varepsilon' < \varepsilon.$$

Consequently $f(S(x_0, \delta)) \subset S(f(x_0), \varepsilon)$ which means f is continuous at x_0. Since x_0 was chosen arbitrarily, f is continuous. ∎

The concept of uniform convergence is easily extended to the case of a function $f:X \to Y$ where X is a topological space and Y is a metric (pseudo-metric) space as follows: A sequence $\{f_n\}$ of functions from X into Y is said to converge uniformly to the function $f:X \to Y$ if for each $\varepsilon > 0$ there is a positive integer m such that for each $x \in X$ and each $k > m$, $d(f_k(x), f(x)) < \varepsilon$ where d is the metric on Y. Furthermore, the method of proof of Theorem 1.12 can be modified to prove:

THEOREM 1.13 A function $f:X \to Y$ from a topological space X to a metric (pseudo-metric) space Y is continuous if there is a sequence $\{f_n\}$ of continuous function that converges uniformly to f.

EXERCISES

1. Prove Proposition 1.13

2. Finish the proof of Theorem 1.11; i.e., that f is one-to-one.

3. Prove that an isometric function $f:X \to X$ of a compact metric space into itself is an onto function.

4. Prove the following:

   (a) compositions of isometric functions are isometric,
   (b) compositions of isomorphisms are isomorphisms and
   (c) the inverse of an isomorphism is an isomorphism.

1.6 Completeness

The concept of completeness occupies a central role in the theory of metric spaces and as we shall see, it is equally significant in the theory of uniform spaces. We devote this section to its study. We will also spend some time on a stronger form of completeness that has not previously been investigated in metric spaces. This stronger form of completeness, referred to as **cofinal completeness**, arose in the study of paracompactness in uniform spaces. Because of its importance in the following chapters it seems appropriate to introduce it in the setting of metric spaces where it takes on a simpler form.

To facilitate our study of completeness, we first introduce some terminology that will help streamline the discussion. Let (X, d) be a metric space and A a subset of X. If $\{x_n\}$ is a sequence of points in X we say $\{x_n\}$ is **eventually** in A if there exists a positive integer m such that $x_n \in$ A whenever $n \geq m$. Also, we say $\{x_n\}$ is **frequently** in A if for each positive integer m there is a positive integer $n \geq m$ such that $x_n \in$ A. Clearly if $\{x_n\}$ is eventually in A it must be frequently in A. It is also easily seen that $\{x_n\}$ is frequently in A if and only if some subsequence of $\{x_n\}$ is eventually in A.

In Chapters 3 and 4, more general objects than sequences will be introduced for the purpose of studying completeness in uniform spaces. They will be seen to have certain interesting subsets called **cofinal** subsets. Although we will not need this concept for our study of metric spaces, it will be seen in these later chapters that a subsequence of a sequence will satisfy the definition of a cofinal subset in the special case of sequences. For this reason we say subsequences are cofinal in sequences. This terminology will motivate our terminology of cofinal completeness a little later in this section.

We can restate the classic definitions of convergence and clustering that one encounters in elementary analysis in terms of *eventually* and *frequently* defined above. We say a sequence $\{x_n\}$ **converges** to a point p if it is eventually in each neighborhood of p and that $\{x_n\}$ **clusters** to p or that p is a **cluster point** of $\{x_n\}$ if $\{x_n\}$ is frequently in each neighborhood of p. By now the reader has no doubt observed that in a metric space, one neighborhood base for a point p consists of the sequence $\{S(p, 2^{-n})\}$ of spheres about p of radius 2^{-n} for each positive integer n. Consequently, $\{x_n\}$ converges to p if and only if for each positive integer n there exists a positive integer m such that $d(p, x_k) < 2^{-n}$ for each positive integer $k \geq m$.

In elementary analysis, the sequence $\{x_n\}$ is said to be **Cauchy** if for each $\varepsilon > 0$ there is a positive integer k such that $d(x_m, x_n) < \varepsilon$ for each $m \geq k$ and $n \geq k$. The following proposition is easily verifiable and left as an exercise.

PROPOSITION 1.15 A sequence $\{x_n\}$ is Cauchy if and only if for each $\varepsilon > 0$, $\{x_n\}$ is eventually in some sphere of radius ε.

The concept of a *cofinally Cauchy* sequence can be obtained by generalizing the property in Proposition 1.15. A sequence $\{x_n\}$ is said to be **cofinally Cauchy** if for each $\varepsilon > 0$, $\{x_n\}$ is frequently in some sphere of radius ε. But this is equivalent to defining $\{x_n\}$ to be cofinally Cauchy if for each $\varepsilon > 0$, $\{x_n\}$ has a subsequence that is eventually in some sphere of radius ε. As mentioned previously, subsequences are cofinal subsets of sequences. Hence the name cofinally Cauchy.

PROPOSITION 1.16 A convergent sequence in a metric (pseudometric) space is Cauchy.

Proof: Assume the sequence $\{x_n\}$ converges to the point p in the metric space (X, d). Then $\{x_n\}$ is eventually in each neighborhood of p. Let $\varepsilon > 0$. Then $\{x_n\}$ is eventually in $S(p, \varepsilon)$. By Proposition 1.15 this implies $\{x_n\}$ is Cauchy. ∎

The converse of Proposition 1.16 is not true in general. For example, the sequence $\{x_n\}$ defined by $x_n = 1/n$ in the open interval (0,1) of real numbers with the usual metric is Cauchy but does not converge to a point of (0,1). This example motivates the concept of completeness. A metric (pseudo-metric) space is said to be **complete** if each Cauchy sequence converges.

PROPOSITION 1.17 *A sequence that clusters in a metric (pseudo-metric) space is cofinally Cauchy.*

The proof of Proposition 1.17 is similar to the proof of Proposition 1.16. The example above used to show the existence of a Cauchy sequence that does not converge is also an example of a cofinally Cauchy sequence that does not cluster. This is due to the fact that by definition all Cauchy sequences are cofinally Cauchy and also the following:

PROPOSITION 1.18 *If a Cauchy sequence clusters to a point p then it also converges to p.*

Proof: Assume the Cauchy sequence $\{x_n\}$ clusters to p in the metric space (X,d). Let $\varepsilon > 0$. Then $\{x_n\}$ is frequently in $S(p, \varepsilon/3)$. Since $\{x_n\}$ is Cauchy there is a q in X with $\{x_n\}$ eventually in $S(q, \varepsilon/3)$. Let k be a positive integer such that $x_m \in S(q, \varepsilon/3)$ for each $m \geq k$. Then there exists a positive integer j such that $k < j$ and $x_j \in S(p, \varepsilon/3)$. But then $x_j \in S(p, \varepsilon/3) \cap S(q, \varepsilon/3)$. Let m be a positive integer such that $m \geq k$. Then

$$d(p, x_m) \leq d(p, x_j) + d(x_j, q) + d(q, x_m) < \varepsilon/3 + \varepsilon/3 + \varepsilon/3 < \varepsilon.$$

Therefore $x_m \in S(p, \varepsilon)$ which implies $\{x_n\}$ is eventually in each neighborhood of p so $\{x_n\}$ converges to p. ∎

A metric (pseudo-metric) space will be called **cofinally complete** if each cofinally Cauchy sequence clusters. Since, as noted above, all Cauchy sequences are cofinally Cauchy we have the following:

COROLLARY 1.7 *A cofinally complete metric (pseudo-metric) space is complete.*

PROPOSITION 1.19 *A sequence in a metric space clusters to a point p if and only if it has a subsequence that converges to p.*

Proof: Let $\{x_n\}$ be a sequence in the metric space (X, d) that clusters to $p \in$ X.

The sequence $B = \{S(p, 2^{-n})\}$ is a neighborhood base for p so $\{x_n\}$ is frequently in $S(p, 2^{-m})$ for each positive integer m. For each positive integer n let x_{k_n} denote the first element of x_n such that $k_n > n$ and $x_{k_n} \in S(p, 2^{-n})$. Then $\{x_{k_n}\}$ is a subsequence of $\{x_n\}$ and $\{x_{k_n}\}$ clearly converges to p.

Conversely, assume $\{x_n\}$ has a subsequence $\{x_{k_n}\}$ that converges to some p in X. Then $\{x_{k_n}\}$ is eventually in each neighborhood of p. As previously noted, this means $\{x_n\}$ is frequently in each neighborhood of p and hence $\{x_n\}$ clusters to p. ■

COROLLARY 1.8 A metric (pseudo-metric) space is sequentially compact if and only if each sequence clusters.

There is another property that can be sandwiched in between the properties of compactness and cofinal completeness. Using the Lebesgue Covering Theorem for motivation, we define a metric space to have the **Lebesgue property** if for each open covering $\mathcal{U}$, there is an $\varepsilon > 0$ such that $\mathcal{U}$ can be refined by the covering consisting of spheres of radius ε. By the Lebesgue Covering Theorem it is clear that compact metric spaces have the Lebesgue property.

PROPOSITION 1.20 A metric space with the Lebesgue property is cofinally complete.

Proof: Assume the metric space (X, d) has the Lebesgue property and suppose X is not cofinally complete. Then there exists a cofinally Cauchy sequence $\{x_n\}$ that does not cluster. For each $x \in$ X pick an open set U(x) containing x such that $\{x_n\}$ is eventually in X - U(x) and put $\mathcal{U} = \{U(x) \mid x \in X\}$. Since (X, d) has the Lebesgue property, there exists an $\varepsilon > 0$ such that $\{S(x, \varepsilon) \mid x \in X\}$ refines $\mathcal{U}$. But then $\{x_n\}$ is eventually in X - $S(x, \varepsilon)$ for each $x \in$ X which implies $\{x_n\}$ is not cofinally Cauchy which is a contradiction. We conclude (X, d) is cofinally complete. ■

These results yield the following implication diagram for metric (pseudo-metric) spaces:

compact

↓

Lebesgue property

↓

cofinally complete

↓

complete

PROPOSITION 1.21 A sequence in a metric space has a limit point p if and only if it clusters to p.

Proof: Assume the sequence $\{x_n\}$ in the metric space (X, d) has a limit point p. Then for each positive integer m, $S(p, 2^{-m})$ contains a point of $\{x_n\}$. In fact, $S(p,2^{-m})$ contains infinitely many points of $\{x_n\}$. If $S(p, 2^{-m})$ only contained finitely many points of $\{x_n\}$ say $\{x_{k_1} \ldots x_{k_j}\}$ then we could put $\varepsilon = min\{d(p, x_{k_i}) \mid i = 1 \ldots j\}$ and pick a positive integer b such that $2^{-b} < \varepsilon$. Then $S(p, 2^{-b})$ would contain no points of $\{x_n\}$ which would be a contradiction. For each positive integer m let k_m denote the first positive integer greater than m such that $x_{k_m} \in S(p, 2^{-m})$. Then $\{x_{k_n}\}$ is a subsequence of $\{x_n\}$ that converges to p. By Proposition 1.19 $\{x_n\}$ clusters to p.

The converse is obvious since if a sequence clusters to a point, it is frequently in each neighborhood of the point. ▪

The Lebesgue property in a metric space is very close to the property of compactness. In what follows it will be shown that if a metric space has the Lebesgue property then it is the union of a compact set and a discrete set.

THEOREM 1.14 (Kasahara, 1956) If (X, d) is a metric space with the Lebesgue property such that S(x, ε) contains at least two distinct points for each ε > 0 and x in X then every sequence in X clusters.

Proof: Assume $\{x_n\}$ is a sequence in X that does not cluster. By Proposition 1.21, $\{x_n\}$ has no limit point. Therefore V = X - $\{x_n\}$ is an open set. Moreover, for each positive integer m there is an $\varepsilon_m > 0$ such that $S(x_m, \varepsilon_m)$ contains no x_n for each $n \neq m$. Furthermore, it is possible to pick $\delta_m > 0$ such that:

(1) $S(x_m, \delta_m)$ is a proper subset of $S(x_m, 2^{-m})$ and
(2) $\delta_m < \varepsilon_m$.

Put $\mathcal{U} = V \cup \{S(x_n, \delta_n)\}$. Then $\mathcal{U}$ is an open covering of X. Since (X, d) has the Lebesgue property there is an $\varepsilon > 0$ such that $\{S(x, \varepsilon) \mid x \in X\}$ refines $\mathcal{U}$. Let k be a positive integer such that $2^{-k} < \varepsilon$. Then $\{S(x, 2^{-k}) \mid x \in X\}$ refines $\mathcal{U}$. Clearly $S(x_k, 2^{-k})$ is not contained in V. Suppose n is a positive integer such that $n \neq k$. Since x_k does not belong to $S(x_n,\delta_n)$ and $2^k < \varepsilon \leq \delta_n$ we see that $S(x_k, 2^{-k})$ is not a subset of $S(x_n,\delta_n)$ for each $n \neq k$.

Finally, $S(x_k, 2^{-k})$ is not a subset of $S(x_k,\delta_k)$ since by (1) above, $S(x_k,\delta_k)$ is a proper subset of $S(x_k, 2^{-k})$. Consequently $S(x_k, 2^{-k})$ is not a subset of U for each U $\in \mathcal{U}$ which implies $\{S(x, 2^{-k}) \mid x \in X\}$ does not refine $\mathcal{U}$ which is a contradiction. We conclude that $\{x_n\}$ must cluster. ▪

COROLLARY 1.9 (Kasahara) If a metric space has the Lebesgue property then it is the union of a compact set and a discrete set.

Proof: Let (X, d) be a metric space having the Lebesgue property and let Y be the subspace of X consisting of all points p such that $S(p, \varepsilon)$ contains at least two distinct points for each $\varepsilon > 0$. Then (Y, d) is a metric space satisfying the hypothesis of Theorem 1.14 and so every sequence in Y clusters. By Corollary 1.8 (Y, d) is sequentially compact so by Theorem 1.10, Y is compact. Now X - Y must consist of discrete points by the definition of Y. Therefore X is the union of a compact set and a discrete set. ∎

We now consider some examples of complete metric spaces. Consider first the real numbers **R** with the usual metric. It should come as no surprise that **R** is complete. In fact **R** is cofinally complete. Since **R** is not a union of a compact set and a discrete set, **R** cannot have the Lebesgue property by Corollary 1.8.

To show **R** is cofinally complete, let $\{x_n\}$ be a cofinally Cauchy sequence in **R**. Then there is a subsequence $\{x_{k_n}\}$ of $\{x_n\}$ that is eventually in $S(p,1)$ for some p in **R**. Since $Cl(S(p,1))$ is compact in **R**, by Corollary 1.8 and Theorem 1.10, $\{x_{k_n}\}$ clusters. By Proposition 1.18, $\{x_{k_n}\}$ has a limit point which implies $\{x_n\}$ has a limit point so $\{x_n\}$ clusters.

Next consider the real Hilbert space H introduced in Section 1.1. Let $h = \{h_n\}$ be a Cauchy sequence in H. For each positive integer j let p_j:H $\to$ **R** denote the canonical projection of H onto its jth coordinate subspace defined by $p_j(x) = x_j$ where $x = \{x_n\} \in$ H. For each positive integer j let $y_j = \{p_j(h_n)\}$. To show y_j is a Cauchy sequence of real numbers let $\varepsilon > 0$. Since h is Cauchy in H there is a positive integer N such that $m,n > N$ implies $d(h_m,h_n) < \varepsilon$. Since

$$d(h_m, h_n) = \sqrt{\Sigma_{i=1}^{\infty}(h_{m_i} - h_{n_i})^2}$$

we conclude that $|p_j(h_m) - p_j(h_n)| = |h_{m_j} - h_{n_j}| < \varepsilon$ so that y_j is Cauchy. We have just seen that **R** is complete so y_j converges to some $r_j \in$ **R**. Let $r = \{r_j\}$. We will show $r \in$ H and $\{h_n\}$ converges to r. Since $\{h_n\}$ is Cauchy in H, if $\varepsilon > 0$ there exists a positive integer N such that $m,n > N$ implies

$$\sqrt{\Sigma_{j=1}^{k}|p_j(h_m) - p_j(h_n)|^2} \leq d(h_m, h_n) < \varepsilon$$

for each positive integer k. If we hold n and k fixed and let n increase we have $\{p_j(h_m)\}$ converges to r_j. Therefore

$$\{\sqrt{\Sigma_{j=1}^{k}|p_j(h_m) - p_j(h_n)|^2}\} \text{ converges to } \{\sqrt{\Sigma_{j=1}^{k}|r_j - p_j(h_n)|^2}\}.$$

Consequently $\sqrt{\Sigma_{j=1}^{k}|r_j - p_j(h_n)|^2} \leq \varepsilon$ for each positive integer k. But then $\sqrt{\Sigma_{j=1}^{\infty}|r_j - p_j(h_n)|^2} \leq \varepsilon$ for each $n > N$. If $r \in$ H we see that this implies

$$d(r, h_n) = \sqrt{\Sigma_{j=1}^{\infty}|r_j - p_j(h_n)|^2} \leq \varepsilon$$

for each $n > N$. This shows h converges to r if $r \in$ H. Therefore it only remains to show that $r \in$ H. For this note that

$$r_j^2 = (r_j - p_j(h_n) + p_j(h_n))^2 = (r_j - p_j(h_n))^2 + 2[r_j - p_j(h_n)][p_j(h_n)] + [p_j(h_n)]^2 \leq$$
$$2((r_j - p_j(h_n))^2 + [p_j(h_n)]^2)$$

because the inequality $2xy \leq x^2 + y^2$ holds for each pair of real numbers x and y. Therefore

$$\Sigma_{j=1}^k r_j^2 \leq 2(\sqrt{\Sigma_{j=1}^k (r_j - p_j(h_n))^2} + \Sigma_{j=1}^k [p_j(h_n)]^2) \leq 2(\varepsilon + \Sigma_{j=1}^k [p_j(h_n)]^2)$$

for each integer $k > 0$ and $n > N$. Since $h_n \in$ H the series on the far right converges. Therefore the series consisting of the terms $\{r_j^2\}$ converges to $r \in$ H.

Consequently, the real Hilbert space is complete. However, unlike our previous example **R**, H is not cofinally complete. To demonstrate this we will exhibit a cofinally Cauchy sequence that does not cluster. For each positive integer n let $h_n = \{x_j^n\}$ where $x_1^n = n$ and $x_j^n = 0$ for each $j > 1$. Then for each n, $h_n \in$ H. Next, for each positive integer n let $K_n = \{y_i^n\}$ where $y_i^n = \{z_j^{n_i}\}$ and where

$$z_j^{n_i} = x_j^n + 2^{-n/2} \text{ if } i = j \text{ or } z_j^{n_i} = x_j^n \text{ if } i \neq j.$$

Then $y_i^n \in$ H for each positive integer i. Put Y $= \cup K_n$. Then Y is countable. Well order Y by $<$ such that each initial interval of Y has cardinality less than Y. Then Y can be represented as some sequence $\{a_j\} \subset$ H. For each positive integer i

$$d(h_n, y_i^n) = \Sigma_{m=1}^{\infty} |h_{n_m} - y_{i_m}^n|^2 = \Sigma_{m=1}^{\infty} |x_m^n - z_m^{n_i}|^2 = |x_i^n - z_i^{n_i}|^2 =$$
$$|x_i^n - x_i^n - 2^{-n/2}|^2 = 2^{-n} < 2^{-(n-1)}.$$

Therefore, $y_i^n \in S(h_n, 2^{-(n-1)})$ for each positive integer n. Thus countably many members of $\{a_j\}$ are contained in $S(h_n, 2^{-(n-1)})$ for each positive integer n. Hence $\{a_j\}$ is cofinally Cauchy.

It is clear that Y $\subset \cup\{S(h_n, 2^{-(n-1)}) \mid n$ is a positive integer$\}$. Then if p is a limit point of Y, $p \in S(h_n, 2^{-(n-1)})$ for some positive integer n. Now for each pair of positive integers i,j with $i \neq j$

$$d(y_i^n, y_j^n) = \Sigma_{m=1}^{\infty} |y_{i_m}^n - y_{j_m}^n|^2 = \Sigma_{m=1}^{\infty} |z_m^{n_i} - z_m^{n_j}|^2 =$$
$$|z_i^{n_i} - z_i^{n_j}|^2 + |z_j^{n_j} - z_j^{n_i}|^2 + |x_i^n + 2^{-n/2} - x_j^n|^2 + |x_i^n - x_j^n - 2^{-n/2}|^2 \geq$$

$$|2^{-n/2}|^2 + |2^{-n/2}|^2 = 2(2^{-n}) = 2^{-(n-1)}.$$

Then $S(p, 2^{-(n+1)})$ can contain at most one y_j^n which implies p is not a limit point after all. Therefore $\{a_j\}$ does not cluster so H is not cofinally complete.

At this point it may be interesting to consider under what conditions cofinal completeness and completeness are identical in metric spaces. Theorem 1.8 states that compact subsets of metric spaces are closed and bounded. Some metric spaces have the property that all closed and bounded subsets are compact. We will call such metric spaces **Borel compact**. Borel compact metric spaces are cofinally complete. In fact, if $\{x_n\}$ is a cofinally Cauchy sequence in a Borel compact metric space (X, d) then there exists a subsequence $\{x_{m_n}\}$ of $\{x_n\}$ that is contained in $S(p, 2^{-1})$ for some $p \in$ X. Since $Cl(S(p, 2^{-1}))$ is compact in X, $\{x_{m_n}\}$ has a convergent subsequence. But then $\{x_n\}$ has a convergent subsequence so $\{x_n\}$ clusters.

The Borel compactness property can be "factored" into the properties of cofinal completeness and **regularly bounded**; i.e., into a pair of properties that together imply Borel compactness and are both implied by Borel compactness. The property of being regularly bounded is defined by the rule that every bounded closed subset is totally bounded. Regularly bounded metric spaces are also of interest because in them, the concepts of completeness and cofinal completeness are identical. The proofs of these assertions are left as exercises.

PROPOSITION 1.22 A metric space is precompact if and only if each sequence has a Cauchy subsequence.

Proof: Assume each sequence in the metric space (X, d) has a Cauchy subsequence and suppose X is not precompact. Then there exists an $\varepsilon > 0$ such that $\{S(x, \varepsilon) \mid x \in X\}$ has no finite subcovering. Pick $p_1, p_2 \in$ X such that p_2 does not belong to $S(p_1, \varepsilon)$. Next assume $\{p_i \mid i = 1 \dots n\}$ has been defined such that for each pair i,j with $i < j \leq n$, we have p_j does not belong to $S(p_i, \varepsilon)$. Since $\cup\{S(p_i, \varepsilon) \mid i = 1 \dots n\} \neq X$ we can choose some $p_{n+1} \in$ X with p_{n+1} not belonging to $S(p_i, \varepsilon)$ for each $i = 1 \dots n$. Therefore it is possible to construct a sequence $\{p_n\} \subset$ X inductively such that for each pair i,j of positive integers with $i < j$ we have p_j does not belong to $S(p_i, \varepsilon)$. Since $\{p_n\}$ has a Cauchy subsequence, say $\{p_{m_n}\}$ there is a positive integer N such that for each $n \geq N$, $p_{m_n} \in S(p_0, \varepsilon/2)$ for some $p_0 \in$ X. But then

$$d(p_{m_N}, p_{m_{N+1}}) \leq d(p_{m_N}, p_0) + d(p_0, p_{m_{N+1}}) < \varepsilon/2 + \varepsilon/2 = \varepsilon.$$

Hence $p_{m_{N+1}} \in S(p_{m_N}, \varepsilon)$ which is a contradiction.

Conversely, assume (X, d) is precompact and let $\{x_n\}$ be a sequence in X. Since the covering $\{S(x, 2^{-1}) \mid x \in X\}$ has a finite subcovering there is a subsequence $\{x_{1_n}\}$ of $\{x_n\}$ such that $\{x_{1_n}\} \subset S(p_1, 2^{-1})$ for some $p_1 \in$ X.

Assume that for each positive integer $i \leq m$ for some positive integer m a subsequence $\{x_{i_n}\}$ has been defined such that for each pair of positive integers i,j with $i < j \leq m$, $\{x_{j_n}\}$ is a subsequence of $\{x_{i_n}\}$ and $\{x_{j_n}\} \subset S(p_j, 2^{-j})$ for some $p_j \in$ X. Since $\{S(x, 2^{-(m+1)}) | x \in X\}$ has a finite subcovering there is a subsequence $\{x_{m+1_n}\}$ of $\{x_{m_n}\}$ that is contained in $S(p_{m+1}, 2^{-(m+1)})$ for some $p_{m+1} \in$ X. By induction it is possible to construct a subsequence $\{x_{m_n}\}$ of $\{x_n\}$ for each positive integer m such that if i and j are positive integers with $i < j$ then $\{x_{j_n}\}$ is a subsequence of $\{x_{i_n}\}$ and $\{x_{m_n}\} \subset S(p_m, 2^{-m})$ for some $p_m \in$ X. Now construct a new subsequence $\{y_n\}$ of $\{x_n\}$ by the rule $y_n = \{x_{n_n}\}$ for each positive integer n. To show $\{y_n\}$ is Cauchy let $\varepsilon > 0$ and choose a positive integer m such that $2^{-m} < \varepsilon$. Then for each pair of positive integers $i,j > m$ we have $d(y_i, y_j) =$

$$d(x_{i_i}, x_{j_j}) \leq d(x_{i_i}, p_{m+1}) + d(p_{m+1}, x_{j_j}) < 2^{-(m+1)} + 2^{-(m+1)} + 2^{-m} < \varepsilon.$$

Thus $\{y_n\}$ is a Cauchy subsequence of $\{x_n\}$. ∎

PROPOSITION 1.23 A closed subspace of a complete metric space is complete.

Proof: Let F be a closed subspace of the complete metric space (X, d) and let $\{x_n\}$ be a Cauchy sequence in F. Then $\{x_n\}$ converges to some $p \in$ X and since F is closed $p \in$ F. Since $\{x_n\}$ was chosen arbitrarily, we conclude F is complete. ∎

PROPOSITION 1.24 A complete subspace of a metric space is closed.

Proof: Let A be a complete subspace of a metric space M. Let $p \in Cl$(A). By Proposition 1.11 there is a sequence $\{x_n\} \subset$ A that converges to p. Then by Proposition 1.16, $\{x_n\}$ is Cauchy. Since A is complete, $\{x_n\}$ converges to some $a \in$ A. Since M is Hausdorff, the only way $\{x_n\}$ can converge to both a and p is if $a = p$. Hence $p \in$ A which implies A is closed. ∎

PROPOSITION 1.25 A metric (pseudo-metric) space is compact if and only if it is complete and precompact.

Proof: Assume (X, d) is a compact metric space. Clearly X is precompact. To show X is complete let $\{x_n\}$ be a Cauchy sequence in X. By Theorem 1.10 $\{x_n\}$ has a convergent subsequence $\{x_{m_n}\}$. Let p denote a point in X to which $\{x_{m_n}\}$ converges. By Proposition 1.19 $\{x_n\}$ clusters to p. Then by Proposition 1.18 we see that $\{x_n\}$ must also converge to p. Since $\{x_n\}$ was chosen arbitrarily we conclude X is complete.

Conversely, assume (X, d) is complete and precompact. Let $\{x_n\}$ be a sequence in X. Since X is precompact, by Proposition 1.22, $\{x_n\}$ has a Cauchy subsequence $\{x_{m_n}\}$. Since X is complete $\{x_{m_n}\}$ converges to some $p \in$ X. Hence every sequence in X has a convergent subsequence so X is sequentially compact. By Theorem 1.10 X is compact. ▪

PROPOSITION 1.26 Let $(X_1, d_1) \ldots (X_n, d_n)$ be a finite collection of complete metric (pseudo-metric) spaces. Define the metric (pseudo-metric)

$$d(a, b) = \sqrt{\Sigma_{i=1}^{n} [d_i(a_i, b_i)]^2}$$

for any two points $a = (a_1 \ldots a_n)$ and $b = (b_1 \ldots b_n)$ in $X = X_1 \times X_2 \times \ldots \times X_n$. Then the product space (X, d) is also complete.

Proof: That d is a metric that generates the topology of X is the result of Exercise 2(a) in Section 1.3. To show (X, d) is complete, let $\{x_m\}$ be a Cauchy sequence in X. Since $d(a, b) \geq d_i(a_i, b_i)$ for each pair of points $a = (a_1 \ldots a_n)$ and $b = (b_1 \ldots b_n)$ in X and all integers $i = 1 \ldots n$, it follows that the sequence $\{y_m^i\}$ defined by $y_m^i = p_i(x_m)$, where p_i:X $\rightarrow X_i$ is the canonical projection of X onto its ith coordinate subspace, is a Cauchy sequence in X_i for each $i = 1 \ldots n$. Since each X_i is complete, $\{y_m^i\}$ converges to some $z_i \in X_i$. Then $z = (z_1 \ldots z_n)$ is a point of X.

To show $\{x_m\}$ converges to z let $\varepsilon > 0$. Since $\{y_m^i\}$ converges to z_i there is a positive integer k_i such that $d_i(z_i, y_j^i) < \sqrt{\varepsilon/n}$ for each integer $j > k_i$. Put $k = max\{k_1 \ldots k_n\}$. Then

$$d(z, x_j) = \sqrt{\Sigma_{i=1}^{n} [d_i(z_i, y_j^i)]^2} < \sqrt{\Sigma_{i=1}^{n} (\sqrt{\varepsilon/n})^2} = \varepsilon$$

for every $j > k$. Therefore $\{x_m\}$ converges to z. We conclude X is complete. ▪

EXERCISES

1. Show that Borel compact metric spaces are regularly bounded.

2. Show that a metric space is regularly bounded if and only if each bounded sequence has a Cauchy subsequence.

3. Show that a metric space is Borel compact if and only if it is both regularly bounded and complete.

4. Show that a metric space is Borel compact if and only if each bounded sequence clusters.

5. Show that the metric space C*(X) of all bounded real valued continuous functions on the space X (introduced in Section 1.1) is complete.

6. Show that in a regularly bounded metric space that completeness and cofinal completeness are equivalent.

1.7 Completions

Certain metric spaces, although not complete, are "almost" complete. Consider the metric space M constructed from the Euclidean plane E^2 by removing the point at the origin. Then M = E^2 - (0, 0). Let d be the Euclidean metric on M considered as a subspace of E^2. Let $\{x_n\}$ be the sequence of points in M such that $x_n = (2^{-n}, 0)$. Clearly $\{x_n\}$ is Cauchy with respect to d, but $\{x_n\}$ does not converge in M. Of course $\{x_n\}$ converges to (0, 0) in E^2 but since we have removed (0, 0) from E^2 to get M, $\{x_n\}$ cannot converge to a point of M.

This is a special case of a completion of a metric space. Here M is a metric space that is not complete, E^2 is a metric space that is complete, M is a subspace of E^2 and Cl(M) = E^2. We say that E^2 is a *completion* of M. More generally, we define the completion of a metric space as follows: A metric space is said to be **isometrically imbedded** in a metric space Y if there exists an isomorphism h:X → Y. If X is isomorphically imbedded in a complete metric space Y and h(X) is dense in Y, then Y is called a **completion** of X.

Notice that E^2 satisfies this definition of a completion. In this case the isomorphism h:M → E^2 is the identity mapping. We can construct infinitely many other subsets of E^2 that are not complete but for which E^2 is the completion. In fact, if we let A denote the set of all points in E^2 that have at least one irrational coordinate, then X = E^2 - A is a metric subspace of E^2 that is not complete. X is the set R × R where R represents the rationals and is therefore countable. However, we still have Cl(X) = E^2. Again, the identity mapping is the required isomorphism h:X → E^2 so that E^2 is a completion of X.

Notice that in this case the cardinality of the completion is greater than the cardinality of the original metric space X. The following theorem is of central importance in the theory of metric spaces. It not only guarantees the existence of a completion, but its constructive proof shows that this completion is essentially constructed by appending all the "missing" limit points of Cauchy sequences that do not converge in the original metric space.

THEOREM 1.15 Every metric space has a completion.

Proof: Let (X, d) be a metric space. Two Cauchy sequences in X will be said to be equivalent, denoted $\{x_n\} \sim \{y_n\}$ if $lim_{n\to\infty} d(x_n, y_n) = 0$. Clearly ~ is an equivalence relation. Let X* be the set of all equivalence classes with respect to ~. We will denote the members of X* by $\{x_n\}^*$ where $\{x_n\}$ is any member of $\{x_n\}^*$. A metric for X* can be defined by

$$d^*(\{x_n\}^*, \{y_n\}^*) = lim_{n\to\infty} d(x_n, y_n)$$

where $\{x_n\}$ and $\{y_n\}$ are any representatives of $\{x_n\}^*$ and $\{y_n\}^*$. To show this definition is independent of the choice of the representatives $\{x_n\}$ and $\{y_n\}$, let $\{w_n\}$ be another representative of $\{x_n\}^*$ and $\{z_n\}$ another representative of $\{y_n\}^*$. Then

$$lim_{n\to\infty}d(x_n,y_n) \le lim_{n\to\infty}[d(x_n,w_n) + d(w_n,z_n) + d(z_n,y_n))] = lim_{n\to\infty}d(w_n,z_n)$$

$$lim_{n\to\infty}d(w_n,z_n) \le lim_{n\to\infty}[d(w_n,x_n) + d(x_n,y_n) + d(y_n,z_n))] = lim_{n\to\infty}d(x_n,y_n).$$

Hence $lim_{n\to\infty}d(x_n,y_n) = lim_{n\to\infty}d(w_n,z_n)$. The verification that d^* is indeed a metric on X* is left as an exercise (Exercise 1).

Next, define the mapping h that carries $p \in$ X onto the class $\{x_n\}^* \in$ X* where $\{x_n\}$ converges to p. $\{x_n\}^* \ne \varnothing$ since the sequence $\{y_n\}$ defined by $y_n = p$ for each n belongs to $\{x_n\}^*$. Clearly h:X $\to$ X* is a one-to-one mapping onto h(X) $\subset$ X*. That h is an isomorphism is left as an exercise.

To show X* is complete, let $\{\{x_n^m\}^*_m\}$ be a Cauchy sequence in X*. Then we have a sequence of sequences

$$\begin{matrix} x_1^1, & x_2^1, & x_3^1, \dots \\ x_1^2, & x_2^2, & x_3^2, \dots \\ x_1^3, & x_2^3, & x_3^3, \dots \\ & \vdots & \end{matrix}$$

For the kth sequence there is, by definition, an integer n_k such that $d(x_{n_k}^k, x_i^k) < 1/k$ for each $i > n_k$. For each positive integer k we can define the constant sequence $\{y_n^k\}$ by $y_n^k = x_{n_k}^k$ for each n. From the definition of d^* we have

$$d^*(\{x_n^k\}^*,\{y_n^k\}^*) < 1/k.$$

Hence $lim_{n\to\infty}d^*(\{x_n^k\}^*,\{y_n^k\}^*) = 0$ so $\{y_n^k\} \sim \{x_n^k\}$ in X*. Let $\varepsilon > 0$. Since $\{\{y_n^k\}^*_k\} = \{\{x_n^k\}^*_k\}$ we see that $\{\{y_n^k\}^*_k\}$ is Cauchy in X* so there exists a positive integer j such that

$$d^*(\{\{y_n^k\}^*_k\},\{\{y_n^l\}^*_l\}) < \varepsilon \text{ whenever } k,l > j.$$

But then $d(x_{n_k}^k, x_{n_l}^l) < \varepsilon$ whenever $k,l > j$. Consequently, the sequence $\{x_{n_k}^k\}$ is Cauchy in X. We want to show that $\{\{x_n^m\}^*_m\} \subset$ X* converges to $\{x_{n_k}^k\}^*$. Now

$$lim_{m\to\infty}d^*(\{x_{n_k}^k\}^*,\{x_n^m\}^*_m) = lim_{k\to\infty}d^*(\{x_{n_k}^k\}^*,\{x_n^k\}^*_k\}) =$$

$$lim_{k\to\infty}[lim_{n\to\infty}d(x_{n_k}^k, x_n^k)] \le lim_{k\to\infty}[1/k] = 0.$$

By Exercise 11 of Section 1.4 $\{\{x_n^m\}^*_m\}$ converges to $\{x_{n_k}^k\}^*$ so X* is complete. To show h(X) is dense in X* let $\{x_n\}^* \in$ X* and let $\varepsilon > 0$. Since $\{x_n\}$ is Cauchy in X there is a positive integer k such that $d(x_m, x_n) < \varepsilon/2$ whenever $m,n > k$. Select a particular $m > k$ and let $p = x_m$. Let $\{y_n\}$ be the constant sequence defined by $y_n = p$ for each n. Then $\{y_n\}^* = h(p)$. Now $d(y_n, x_n) = d(x_m, x_n) <$ $\varepsilon/2$ whenever $n > k$. Thus

$$d^*(\{y_n\}^*, \{x_n\}^*) = lim_{n\to\infty} d(y_n, x_n) \leq \varepsilon/2 < \varepsilon.$$

Therefore $h(p) = \{y_n\}^* \in S(\{x_n\}^*, \varepsilon)$. Since $\{x_n\}$ * and ε were chosen arbitrarily we conclude h(X) is dense in X* so X* is a completion of X. ∎

Theorem 1.15 as well as the rest of the theorems that follow in this chapter were the motivating results in the early study of uniform spaces. Each of these theorems will have its counterpart in the theory of uniform spaces.

THEOREM 1.16 If f is a uniformly continuous function on a subset A of a metric space X into a complete metric space Y then f has a unique uniformly continuous extension f^* to Cl(A).

Proof: Let d represent the metric of X and ρ the metric of Y. For each $x \in A$ let $\{x_n\}$ be the constant sequence ($x_n = x$ for each n) that converges to x and for each $x \in Cl(A) - A$ pick a sequence $\{x_n\}$ in A that converges to x. This is of course possible by Proposition 1.11, and by Proposition 1.16 $\{x_n\}$ is Cauchy for each x.

Now let $\varepsilon > 0$. By the uniform continuity of f there is a $\delta > 0$ such that whenever $a,b \in A$ with $d(a,b) < \delta$ then $\rho(f(a), f(b)) < \varepsilon$. Since each $\{x_n\}$ is Cauchy there exists a $k(x)$ such that if $m,n > k(x)$ then $d(x_m, x_n) < \delta$ which implies $\rho(f(x_m), f(x_n)) < \varepsilon$. But then the sequence $\{f(x_n)\} \subset$ Y is Cauchy in Y for each $x \in Cl(A)$. Since Y is complete $\{f(x_n)\}$ converges to some point x^*. Define $f^*: Cl(A) \to$ Y by $f^*(x) = x^*$ for each $x \in Cl(A)$. Clearly $f^*(a) = f(a)$ for each $a \in A$.

To show the definition of f^* is independent of the choice of sequences $\{x_n\}$ let $\{y_n\}$ be another sequence converging to x. Suppose $x^* \neq y^*$. Then there exists an $\varepsilon > 0$ with $S(x^*, \varepsilon) \cap S(y^*, \varepsilon) = \emptyset$. Again, by the uniform continuity of f there is a $\delta > 0$ such that $d(a,b) < \delta$ implies $\rho(f(a), f(b)) < \varepsilon$. Since both $\{x_n\}$ and $\{y_n\}$ converge to x there is a positive integer k such that $n > k$ implies $d(x_n, y_n) < \delta$ which in turn implies $\rho(f(x_n), f(y_n)) < \varepsilon$. But this is impossible since $\{f(x_n)\}$ is eventually in $S(x^*, \varepsilon)$ and $\{f(y_n)\}$ is eventually in $S(y^*, \varepsilon)$. Hence $x^* = y^*$ so f^* is well defined.

To show f^* is uniformly continuous on $Cl(A)$ let $\varepsilon > 0$ and pick $\delta > 0$ such that whenever $a,b \in A$ with $d(a,b) < \delta$ then $\rho(f(a), f(b)) < \varepsilon/3$. Let $x,y \in Cl(A)$ with $d(x,y) < \delta/2$. Since $\{x_n\}$ converges to x, $\{y_n\}$ converges to y, $\{f(x_n)\}$ converges to x^* and $\{f(y_n)\}$ converges to y^* it is possible to choose a positive integer k such that whenever $n > k$ we have $d(x_n, x) < \delta/4$, $d(y_n, y) < \delta/4$, $\rho(x^*, f(x_n)) < \varepsilon/3$ and $\rho(f(y_n), y^*) < \varepsilon/3$. Then $d(x_n, y_n) < \delta$ and hence $\rho(f(x_n), f(y_n)) < \varepsilon/3$. Consequently

$$\rho(f^*(x), f^*(y)) \leq \rho(x^*, f(x_n)) + \rho(f(x_n), f(y_n)) + \rho(f(y_n), y^*) < \varepsilon.$$

Thus f^* is uniformly continuous on $Cl(A)$ and therefore a uniformly continuous extension of f from A to $Cl(A)$. To show f^* is unique, suppose f' is another uniformly continuous extension of f from A to $Cl(A)$ and let $x \in Cl(A) - A$. Then $\{x_n\} \subset A$ converges to x. Since both f' and f^* are continuous it follows from Proposition 1.12 that $\{f'(x_n)\}$ and $\{f^*(x_n)\}$ converge to $f'(x)$ and $f^*(x)$ respectively. But since $\{x_n\} \subset A$ we have $\{f'(x_n)\} = \{f(x_n)\} = \{f^*(x_n)\}$. Since a sequence in a metric space cannot converge to two distinct points it follows that $f'(x) = f^*(x)$ so that f^* is unique. ∎

The theorem (Theorem 1.17) will show that completions of metric spaces are essentially unique. To prove it we will make use of the following pair of lemmas.

LEMMA 1.7 Let X and Y be metric spaces. Let A be a subset of X and let $f: A \to Y$ be a continuous function with a continuous extension $f^*: Cl(A) \to Y$. If f is isometric then so is f^*.

Proof: Let x and y be two points of $Cl(A)$. Pick sequences $\{x_n\}$ and $\{y_n\}$ in A that converge to x and y respectively. Let d_X denote the metric in X and d_Y the metric in Y. By Proposition 1.12, $\{f^*(x_n)\}$ converges to $f^*(x)$ and $\{f^*(y_n)\}$ converges to $f^*(y)$. Consequently, the sequence $\{(f^*(x_n), f^*(y_n))\}$ in $Y \times Y$ converges to the point $(f^*(x), f^*(y))$ in $Y \times Y$. Since by Proposition 1.3, d_Y is continuous in the product topology we have:

$$d_Y(f^*(x), f^*(y)) = lim_{n\to\infty} d_Y(f^*(x_n), f^*(y_n)) = lim_{n\to\infty} d_Y(f(x_n), f(y_n))$$

since $x_n, y_n \in A$ for each n. Since f is isometric $d_X(x_n, y_n) = d_Y(f(x_n), f(y_n))$ and since d_X is also continuous we have:

$$lim_{n\to\infty} d_Y(f(x_n), f(y_n)) = lim_{n\to\infty} d_X(x, y) = d_X(x, y).$$

Hence $d_Y(f^*(x), f^*(y)) = d_X(x, y)$ so f^* is isometric. ∎

LEMMA 1.8 An isometric image of a complete metric space is complete.

Proof: Let $f: X \to Y$ be an isometric function from a complete metric space X onto a metric space Y. Let $\{y_n\}$ be a Cauchy sequence in Y. Since f is onto, there exists a sequence $\{x_n\}$ in X with $y_n = f(x_n)$ for each n. Let d_X denote the metric in X and d_Y the metric in Y. Since f is isometric we have:

$$d_Y(y_m, y_n) = d_Y(f(x_m), f(x_n)) = d_X(x_m, x_n)$$

for each pair of positive integers m,n. Since $\{y_n\}$ is Cauchy, this implies $\{x_n\}$ is Cauchy in X. Therefore $\{x_n\}$ converges to some $p \in$ X. Put $q = f(p)$. Then

$$d_Y(y_n, q) = d_Y(f(x_n), f(p)) = d_X(x_n, p)$$

for each positive integer n. Since $\{x_n\}$ converges to p, this implies $\{y_n\}$ converges to q. Hence Y is complete. ▪

THEOREM 1.17 Let h:X → Y denote a completion of a metric space X and let j:X → W be another completion of X. Then there exists an isomorphism i:W → Y such that $i \circledcirc j = h$.

Proof: Define a function f:j(X) → Y by $f(w) = h(j^{-1}(w))$ for each $w \in j$(X). By Theorem 1.16, f has a unique uniformly continuous extension i:W → Y. Since h and j are isomorphisms, f is an isomorphism of j(W) onto h(X) [Exercise 4, Section 1.5]. By Lemma 1.7, i is isometric. By Lemma 1.8, i(W) is complete. By Proposition 1.21, i(W) is closed in Y.

Since $f(j(X)) \subset i(W)$ and $f(j(X)) = h(j^{-1}(j(X)) = h(X)$ we have that $h(X) \subset i(W)$. Since h(X) is dense in Y and i(W) is closed in Y we have i(W) = Y. Hence i:W → Y is a uniformly continuous isometric mapping of W onto Y. By Theorem 1.11, i is one-to-one and hence an isomorphism. Since i is an extension of f and $f \circledcirc j = h$ we have $i \circledcirc j = h$. ▪

EXERCISES

1. Complete the details of the proof of Theorem 1.15.

2. A metric space X is said to be **absolutely closed** if every isometric image of X into a metric space Y is closed in Y. Show that a metric space is complete if and only if it is absolutely closed.

Chapter 2

UNIFORMITIES

2.1 Covering Uniformities

The concept of a metric space leads naturally to the concept of a uniform space, especially if one approaches the topic from the point of view of covering uniformities. The first development of uniform spaces by A. Weil titled *Sur les espaces à structure uniforme et sur la topologie générale* published in Actualities Sci. Ind. 551, Paris, 1937 took a different approach. It involved a family of pseudo-metrics that generate the topology of the space as opposed to a family of coverings. Weil's original approach was rather unwieldy and was soon replaced by two others.

In 1940 J. W. Tukey, in his Annals of Mathematics Studies (Princeton) monograph titled *Convergence and uniformity in general topology*, presented an elegant equivalent development of uniform spaces based on covering uniformities. The covering approach was popularized by K. Morita and T. Shirota in the 1950s and by J. R. Isbell in an American Mathematical Society publication of 1964 titled *Uniform Spaces*. Also, in 1940, N. Bourbaki advanced a development of uniform spaces based on *entourages* which are subsets of the product space of the given uniform space. The entourage approach is still favored by many. An English translation of Bourbaki's *Uniform Spaces* was published by Addison-Wesley in 1966.

In what follows uniform spaces will be developed from the covering point of view as being a more natural generalization of a metric space. Covering uniformities will also be much easier for us to use for the material we will be developing. In a metric space M, the metric function is the measure of distance, but in most proofs, one is usually concerned with the spheres $S(x, \varepsilon)$ of radius $\varepsilon > 0$ about some point x. Furthermore, the uniform coverings of the form $\mathcal{U} = \{S(x, 2^{-n}) \mid x \in M\}$, for each positive integer n, provide a uniform measure of nearness throughout the entire space.

A broad class of topological spaces admit a *uniform structure* that provides a "uniform measure of nearness" throughout the entire space. Even in the absence of a metric, this structure allows one to apply many metric space techniques to non-metric topological spaces. A family μ of coverings of a set X is called a **Hausdorff uniformity** if it satisfies the following conditions:

(1) If $\mathcal{U}, \mathcal{V} \in \mu$ there is a $\mathcal{W} \in \mu$ with $\mathcal{W} < \mathcal{U}$ and $\mathcal{W} < \mathcal{V}$,
(2) if $\mathcal{U} \in \mu$ and $\mathcal{U} < \mathcal{V}$ then $\mathcal{V} \in \mu$,
(3) every element of μ has a star refinement in μ and
(4) if $x \neq y \in X$ there is a $\mathcal{U} \in \mu$ with $Star(x, \mathcal{U}) \cap Star(y, \mathcal{U}) = \emptyset$.

The set X together with the Hausdorff uniformity μ is called a **Hausdorff uniform space.** Then members of μ are called **uniform coverings.** We will often use the alternative notation $\mathcal{U}^* < \mathcal{V}$ instead of $\mathcal{U} <^* \mathcal{V}$ where $\mathcal{U}$ and $\mathcal{V}$ are coverings and $\mathcal{U}^* = \{Star(\mathrm{U}, \mathcal{U}) \mid \mathrm{U} \in \mathcal{U}\}$. This notation has the advantage of conveying the notion of $\mathcal{U}^{**} < \mathcal{V}$ far more economically than the standard notation. Here, $\mathcal{U}^{**} = (\mathcal{U}^*)^*$.

A family of coverings of X that satisfy only the first three of these properties is often called a **preuniformity** or a **non-separating uniformity.** In what follows, we will adopt a less formal terminology and refer to both Hausdorff and non-Hausdorff uniformities simply as uniformities and leave it to the reader to infer from the context whether condition (4) applies or not. If it is not clear from the context, we will be careful to use the expressions Hausdorff and non-Hausdorff to make the context clear.

A uniformity μ that contains a uniformity ν is said to be **finer** than ν and ν is said to be **coarser** that μ. Clearly, the uniformities for a given set X form a partially ordered set with respect to the partial ordering of set inclusion.

A family of coverings that satisfy condition (3) above is called a **normal family.** If $\{\mathcal{U}_n\}$ is a sequence of coverings such that $\mathcal{U}^*_{n+1} < \mathcal{U}_n$ for each n, then $\{\mathcal{U}_n\}$ is called a **normal sequence.** A **basis** for a uniformity is a family of coverings that satisfies conditions (1) and (3) of the definition of a uniformity. A **sub-basis** for a uniformity is a family of coverings whose finite intersections form a basis. By the intersection of two coverings $\mathcal{U}$ and $\mathcal{V}$ we mean the covering $\mathcal{U} \cap \mathcal{V} = \{\mathrm{U} \cap \mathrm{V} \mid \mathrm{U} \in \mathcal{U} \text{ and } \mathrm{V} \in \mathcal{V}\}$.

THEOREM 2.1 Every normal family of coverings is a sub-basis for a uniformity.

Proof: Let μ be a normal family of coverings and let μ' be the set of all finite intersections of members of μ. Then let ν be the family of all coverings that can be refined by a member of μ'. Clearly ν satisfies condition (2) of the definition of a uniformity.

To show that ν satisfies condition (1), let $\mathcal{V}$, $\mathcal{V}' \in \nu$. Then there exists $\mathcal{U}_1 \ldots \mathcal{U}_n \in \mu$ such that $\cap_{i=1}^{n} \mathcal{U}_i < \mathcal{V}$ and $\mathcal{U}_{n+1} \ldots \mathcal{U}_{n+m} \in \mu$ such that $\cap_{i=n+1}^{n+m} \mathcal{U}_i < \mathcal{V}'$. Put $\mathcal{W} = \cap_{i=1}^{n+m} \mathcal{U}_i$. Then $\mathcal{W} \in \nu$, $\mathcal{W} < \mathcal{V}$ and $\mathcal{W} < \mathcal{V}'$. To show that ν satisfies condition (3), let $\mathcal{V}$ be an element of ν and observe that $\mathcal{V}$ must be refined by some $\mathcal{V}_1 \cap \ldots \cap \mathcal{V}_n$ where $\mathcal{V}_i \in \mu$ for each i. Since μ is a normal family, μ contains a $\mathcal{U}_i$ that is a star refinement of $\mathcal{V}_i$ for each i. It is easily shown that

$$\mathcal{U} = \mathcal{U}_1 \cap \ldots \cap \mathcal{U}_n <^* \mathcal{V}_1 \cap \ldots \cap \mathcal{V}_n < \mathcal{V}.$$

Consequently $\mathcal{U} \in \nu$ is the desired star refinement of $\mathcal{V}$. Therefore ν is the uniformity we are looking for. Also, we note that μ^* is a basis for ν. ∎

The union of normal families of coverings is also a normal family of coverings. Therefore, every family μ of coverings contains a largest normal subfamily ν. The members of ν are said to be **normal in** μ. A covering $\mathcal{U}$ is said to be **normal with respect to** μ if there exists a normal sequence $\{\mathcal{U}_n\}$ of coverings such that $\mathcal{U} = \mathcal{U}_1$ and for each positive integer n, $\mathcal{U}_n$ has a refinement in μ.

THEOREM 2.2 If μ is a family of coverings satisfying (1) and (2) of the definition of a uniformity, then there is a finest uniformity contained in μ consisting of all coverings normal with respect to μ.

Proof: Let ν be the set of all coverings that are normal with respect to μ. Clearly ν satisfies conditions (2) and (3) of the definition of a uniformity. To show ν satisfies condition (1) let $\mathcal{U},\mathcal{V} \in \nu$. Then there exist normal sequences $\{\mathcal{U}_n\}$ and $\{\mathcal{V}_n\}$ such that $\mathcal{U} = \mathcal{U}_1$, $\mathcal{V} = \mathcal{V}_1$ and for each positive integer n, $\mathcal{U}_n$ and $\mathcal{V}_n$ each have refinements in μ. For each positive integer n put $\mathcal{W}_n = \mathcal{U}_n \cap \mathcal{V}_n$ and put $\mathcal{W} = \mathcal{W}_1$. Clearly $\mathcal{W} < \mathcal{U}$ and $\mathcal{W} < \mathcal{V}$. For each positive integer n, $\mathcal{U}^*_{n+1} < \mathcal{U}_n$ and $\mathcal{V}^*_{n+1} < \mathcal{V}_n$ implies $\mathcal{W}_{n+1} <^* \mathcal{W}_n$ so $\{\mathcal{W}_n\}$ is a normal sequence. Also, for each n, $\mathcal{U}_n$ and $\mathcal{V}_n$ have refinements $\mathcal{U}'_n$ and $\mathcal{V}'_n$ in μ. Since μ satisfies condition (1) there is a $\mathcal{W}'_n \in \mu$ such that $\mathcal{W}'_n < \mathcal{U}'_n$ and $\mathcal{W}'_n < \mathcal{V}'_n$. But then

$$\mathcal{W}'_n < \mathcal{U}'_n \cap \mathcal{V}'_n < \mathcal{U}_n \cap \mathcal{V}_n = \mathcal{W}_n.$$

Therefore each $\mathcal{W}_n$ has a refinement in μ so $\mathcal{W} \in \nu$. Consequently ν is a uniformity.

To show that ν is the finest uniformity contained in μ, suppose ν' is another uniformity contained in μ. By condition (3) of the definition of a uniformity, if $\mathcal{U}' \in \nu'$ it is possible to construct inductively, a sequence $\{\mathcal{U}'_n\}$ of members of ν' such that $\mathcal{U}' = \mathcal{U}'_1$ and $\mathcal{U}'_{n+1} <^* \mathcal{U}'_n$ for each n. But then $\mathcal{U}'$ is normal with respect to μ so $\mathcal{U}' \in \nu$. Therefore ν is the finest uniformity contained in μ. ∎

A uniformity μ on a set X gives rise to a topology τ called the **uniform topology** or the **topology associated with the uniformity** μ. The topology τ is said to admit the uniformity μ and μ is said to **generate** τ. As we will see, distinct uniformities can give rise to the same topology. If $\mathcal{U} \in \mu$ and p is a point in X then a point $x \in$ X is said to be **$\mathcal{U}$-close** to p or x is said to be **within $\mathcal{U}$ of** p, denoted by $|p - x| < \mathcal{U}$, if there exists a U $\in \mathcal{U}$ such that both p and x belong to U. The **sphere (or ball) about** p **of radius** $\mathcal{U}$ denoted $S(p, \mathcal{U})$ is the set of all points of X that are $\mathcal{U}$ close to p. Clearly $S(p, \mathcal{U}) = Star(p, \mathcal{U})$. The uniform topology is defined as follows: a subset N of X is a neighborhood of $p \in X$ if there is a $\mathcal{U} \in \mu$ such that $S(p, \mathcal{U}) \subset N$. N is open if it is a neighborhood of each of its points. The proof that this really generates a topology is left as an exercise (Exercise 2) as is the proof of the following:

PROPOSITION 2.1 If (X, μ) is a uniform space and X is T_1 then μ is a Hausdorff uniformity.

EXERCISES

1. Show that in each metric space X, the collection of coverings of the form $\{S(x, 2^{-n}) \mid x \in X\}$ are a uniformity that generate the metric topology.

2. Prove Proposition 2.1.

ENTOURAGE UNIFORMITIES

3. Let X be a set. An **entourage** on X is a subset U of $X \times X$ containing the diagonal $\Delta = \{(x, x) \mid x \in X\}$. The inverse U^{-1} is the set of all pairs (x,y) such that $(y, x) \in U$. If $U = U^{-1}$ then U is said to be **symmetric.** If U and V are two entourages, the **composition** $U \circledcirc V$ is the set of all pairs (x,z) such that for some $y \in X$, $(x,y) \in V$ and $(y,z) \in U$. For each subset A of X, define the set $U[A]$ to be $\{y \in X \mid (x,y) \in U \text{ for some } x \in A\}$. If A consists of a single point p then we write $U[A] = U[p]$. If U,V,W are entourages show the following:

   (a) $U \circledcirc (V \circledcirc W) = (U \circledcirc V) \circledcirc W$,
   (b) $(U \circledcirc V)^{-1} = V^{-1} \circledcirc U^{-1}$,
   (c) if $A \subset X$ then $(U \circledcirc V)[A] = U[V[A]]$ and
   (d) if V is symmetric then $V \circledcirc U \circledcirc V = \cup\{V[x] \times V[y] \mid (x,y) \in U\}$.

4. An **entourage uniformity** for a set X is a family $\mathcal{U}$ of entourages on X such that:

   (a) if $U \in \mathcal{U}$ then $U^{-1} \in \mathcal{U}$,
   (b) if $U \in \mathcal{U}$ then $V \circledcirc V \subset U$ for some $V \in \mathcal{U}$,
   (c) if $U, V \in \mathcal{U}$ then $U \cap V \in \mathcal{U}$ and
   (d) if $U \in \mathcal{U}$ and $U \subset V$ then $V \in \mathcal{U}$.

Using this definition of a uniformity, $(X, \mathcal{U})$ is said to be a uniform space. $\mathcal{U}$ generates a topology as follows: $N \subset X$ is a neighborhood of the point p if there is a $U \in \mathcal{U}$ such that $U[p] \subset N$. Show the following: If $\mathcal{U}$ is an entourage uniformity, for each $U \in \mathcal{U}$ let $\mathcal{V}_U = \{U[p] \mid p \in X\}$ and put $\mu = \{\mathcal{V}_U \mid U \in \mathcal{U}\}$. Then μ is a covering uniformity that generates the same topology as $\mathcal{U}$.

5. A subfamily B of an entourage uniformity $\mathcal{U}$ is said to be a **base** for $\mathcal{U}$ if each member of $\mathcal{U}$ contains a member of B. Show that a non-void family B of subsets of $X \times X$ is a base for some entourage uniformity for X if and only if it satisfies the following four properties:

(a) each member of B contains the diagonal Δ,
(b) if $U \in B$ then U^{-1} contains a member of B,
(c) if $U \in B$ then $V \circledcirc V \subset U$ for some $V \in B$ and
(d) the intersection of two members of B contains a member of B.

6. A subfamily S of an entourage uniformity $\mathcal{U}$ is said to be a **sub-base** for $\mathcal{U}$ if the family of finite intersections of members of S is a base for $\mathcal{U}$. Show that a family S of subsets of $X \times X$ is a subbase for some entourage uniformity for X if it satisfies the following conditions:

(a) each member of S contains the diagonal Δ,
(b) for each $U \in S$, U^{-1} contains a member of S and
(c) for each $U \in S$ there is a $V \in S$ with $V \circledcirc V \subset U$.

In particular, this shows that the union of any collection of entourage uniformities for X is the subbase for a uniformity on X.

7. Let $\mathcal{U}$ be an entourage uniformity for a space X. Show that the interior of a subset A of X relative to the uniform topology (i.e., the topology generated by $\mathcal{U}$) is the set of all points x such that $U[x] \subset \mathrm{A}$ for some $U \in \mathcal{U}$.

8. If B is a base (subbase) for the entourage uniformity $\mathcal{U}$ show that for each x the family of sets of the form $U[x]$ for some $U \in B$ is a base (subbase) for the neighborhood system of x.

9. If $\mathcal{U}$ is an entourage uniformity for the space X show that the closure of a subset A of X relative to the uniform topology is $\cap\{U[\mathrm{A}] \mid U \in \mathcal{U}\}$.

10. Show that the family of closed symmetric members of an entourage uniformity is a base for the uniformity.

11. Show that the uniform topology generated by an entourage uniformity $\mathcal{U}$ is Hausdorff if and only if the intersection of all members of $\mathcal{U}$ is the diagonal Δ.

12. Let (X, μ) be a covering uniform space. For each $\mathcal{U} \in \mu$, put $V[\mathcal{U}] = \cup\{U \times U \mid U \in \mathcal{U}\}$ and let $\mathcal{V} = \{V[\mathcal{U}] \mid \mathcal{U} \in \mu\}$. Show that $\mathcal{V}$ is an entourage uniformity for X that generates the same topology as μ.

2.2 Uniform Continuity

A function $f:X \to Y$ from a uniform space (X, μ) into a uniform space (Y,ν) is said to be **uniformly continuous** if for each $\mathcal{V} \in \nu$ there is a $\mathcal{U} \in \mu$ such that whenever x and y are $\mathcal{U}$ close then $f(x)$ and $f(y)$ are $\mathcal{V}$ close (i.e., $|x - y| < \mathcal{U}$ implies $|f(x) - f(y)| < \mathcal{V}$). The **inverse image** $f^{-1}(\mathcal{V})$ of a uniform covering $\mathcal{V}$ is the set of all $f^{-1}(V)$ such that $V \in \mathcal{V}$. Then:

THEOREM 2.3 f is uniformly continuous if and only if for each uniform covering $\mathcal{V}$, $f^{-1}(\mathcal{V})$ is a uniform covering.

Proof: Assume $f:X \to Y$ is uniformly continuous where (X, μ) and (Y,ν) are uniform spaces. Let $\mathcal{V} \in \nu$. To show $f^{-1}(\mathcal{V}) \in \mu$ let $\mathcal{W} \in \nu$ such that $\mathcal{W}^* < \mathcal{V}$. Let $\mathcal{U} \in \mu$ such that $|x - y| < \mathcal{U}$ implies $|f(x) - f(y)| < \mathcal{W}$. Let $U \in \mathcal{U}$ and pick $p \in U$. For each $q \in U$, $|p - q| < \mathcal{U}$ implies $|f(p) - f(q)| < \mathcal{W}$ so $f(q) \in S(f(p),\mathcal{W})$. Since $\mathcal{W}^* < \mathcal{V}$ there is a $V \in \mathcal{V}$ such that $S(f(p),\mathcal{W}) \subset V$. Therefore $f(q) \in V$ which implies $q \in f^{-1}(V)$. Consequently $U \subset f^{-1}(V)$ so we conclude that $\mathcal{U} < f^{-1}(\mathcal{V})$ which implies $f^{-1}(\mathcal{V}) \in \mu$.

Conversely, assume $f^{-1}(\mathcal{V}) \in \mu$ for each $\mathcal{V} \in \nu$. Let $\mathcal{W} \in \nu$. Then there exists a $\mathcal{U} \in \mu$ such that $\mathcal{U} < f^{-1}(\mathcal{W})$. If $|x - y| < \mathcal{U}$ then there is a $U \in \mathcal{U}$ such that $x,y \in U \subset f^{-1}(W)$ for some $W \in \mathcal{W}$. Then $f(x), f(y) \in W$ which implies $|f(x) - f(y)| < \mathcal{W}$. Therefore f is uniformly continuous. ∎

COROLLARY 2.1 The composition of uniformly continuous functions is uniformly continuous.

THEOREM 2.4 Every uniformly continuous function is continuous.

Proof: Suppose $f:X \to Y$ is uniformly continuous. If V is a neighborhood of $f(x)$ then there is a uniform covering $\mathcal{V}$ such that $S(f(x),\mathcal{V}) \subset V$. But then $S(x,f^{-1}(\mathcal{V})) \subset f^{-1}(V)$ so $f^{-1}(V)$ is a neighborhood of x in X. Consequently f is continuous. ∎

Let $f:X \to Y$ be uniformly continuous and suppose there exists a uniformly continuous $g:Y \to X$ such that $g = f^{-1}$. Then f is one-to-one and onto and we say that f is a **uniform homeomorphism**. If $f:X \to Y$ is a uniform homeomorphism from the uniform space (X, μ) to the uniform space (Y,ν), then the uniform coverings of Y are precisely the coverings of the form $\mathcal{V} = \{f(U) \mid U \in \mathcal{U}\}$ for some $\mathcal{U} \in \mu$ and the uniform coverings of X are the coverings of the form $\mathcal{U} = \{f^{-1}(V) \mid V \in \mathcal{V}\}$ for some $\mathcal{V} \in \nu$.

THEOREM 2.5 For each uniform covering $\mathcal{U}$ of a uniform space X there is a uniformly continuous function f that maps X onto a metric space such that the inverse image of every set of diameter less than 1 is a subset of an element of $\mathcal{U}$.

Proof: Let $\{\mathcal{U}_n\}$ be a normal sequence of uniform coverings such that $\mathcal{U} = \mathcal{U}_1$. For each pair $x,y \in X$ put $\delta(x,y) = 2$ if y does not belong to $S(x, \mathcal{U})$; $\delta(x,y) = 0$ if $y \in S(x, \mathcal{U}_n)$ for each n; or $\delta(x,y) = 2^{-n}$ where n is the largest index such that $y \in S(x, \mathcal{U}_n)$. Then define

$$d^*(x,y) = inf\{\Sigma_{j=1}^{n}\delta(k_j, k_{j+1}) \mid \{k_j\} \text{ is a finite sequence with } x = k_1 \text{ and } y = k_n\}.$$

Now put $d(x,y) = min\{d^*(x,y),1\}$. Clearly $0 \le d(x,y) \le 1$ and $d(x,y) = d(y,x)$. To prove the triangle inequality let $x,y,z \in X$. Then $d(x,y) + d(y,z) \ge 1 \ge d(x,z)$ or $d(x,y) + d(y,z) = d^*(x,y) + d^*(y,z)$. Let

$$d^*(x,y) = inf\{\Sigma_{j=1}^{m}\delta(p_j, p_{j+1}) \mid \{p_j\} \text{ is a finite sequence with } x = p_1 \text{ and } y = p_m\}$$

$$d^*(y,z) = inf\{\Sigma_{j=1}^{n}\delta(q_j, q_{j+1}) \mid \{q_j\} \text{ is a finite sequence with } y = q_1 \text{ and } z = q_n\}$$

$$d^*(x,z) = inf\{\Sigma_{j=1}^{k}\delta(t_j, t_{j+1}) \mid \{t_j\} \text{ is a finite sequence with } x = t_1 \text{ and } z = t_k\}.$$

Any sums of the form $\Sigma_{j=1}^{m}\delta(p_j, p_{j+1})$ and $\Sigma_{j=1}^{n}\delta(q_j, q_{j+1})$ can be combined to form a sum $\Sigma_{j=1}^{m+n}\delta(t_j, t_{j+1})$ where $k = m + n$, $t_j = p_j$ for $j = 1 \ldots n$ and $t_j = q_{j-n}$ for $j = n+1 \ldots n+m$. Consequently $d^*(x,z)$ cannot be greater than $d^*(x,y) + d^*(y,z)$. But then $d(x, z) \le d(x,y) + d(y, z)$. Thus d is a pseudo-metric on X.

Define an equivalence relation $\sim$ on X by $x \sim y$ if and only if $d(x,y) = 0$. Then $M = X/\sim$ is a metric space and d induces a metric on M as shown in Section 1.5 which we will also denote here by d. As was shown there, if x^* and y^* are two members of M and x and y are representative members of x^* and y^* respectively then $d(x^*,y^*) = d(x,y)$ which justifies naming the metric on M with the same name as the pseudo-metric on X. Let $f:X \to M$ be the canonical projection. Then f is a continuous mapping of X onto M. To show f is uniformly continuous let $U \in \mathcal{U}_n$. If $x,y \in U$ then $d(x,y) \le 2^{-n}$. Therefore the inverse image of the uniform covering of M consisting of sets of diameter less than or equal to 2^{-n} is refined by $\mathcal{U}_n$. Consequently f is uniformly continuous.

It remains to show that the inverse image of a set of diameter less than 1 is contained in some element of $\mathcal{U}$. For this it suffices to show that whenever $d(x,y) < 1$ then $x,y \in U$ for some $U \in \mathcal{U}$. Suppose $d(x,y) < 2^{-n}$. Then there is a finite sum $\delta(p_1, p_2) + \ldots + \delta(p_{m-1}, p_m) < 2^{-n}$ where $x = p_1$ and $y = p_m$. If $m = 2$ then $\delta(x,y) < 2^{-n}$ which implies $\delta(x,y) \le 2^{-(n+1)}$ which in turn implies $x,y \in U_{n+1}$ for some $U_{n+1} \in \mathcal{U}_{n+1}$.

Let H be the collection of positive integers such that $m \in H$ if whenever $\delta(p_1, p_2) + \ldots + \delta(p_{m-1}, p_m) < 2^{-n}$ then $p_1, p_m \in U_{n+1}$ for some $U_{n+1} \in \mathcal{U}_{n+1}$. Let $m \in H$ and $\delta(p_1, p_2) + \ldots + \delta(p_m, p_{m+1}) < 2^{-n}$. Let i be the largest positive integer such that $\delta(p_1, p_2) + \ldots + \delta(p_i, p_{i+1}) < 2^{-(n+1)}$. If $i = m$ then $\delta(p_1, p_2) + \ldots + \delta(p_{m-1}, p_m) < 2^{-(n+1)}$ and $\delta(p_m, p_{m+1}) < 2^{-(n+1)}$ which implies that $p_1, p_m \in U_{n+2}$ and $p_m, p_{m+1} \in V_{n+2}$ for some $U_{n+2}, V_{n+2} \in \mathcal{U}_{n+2}$ since $m \in H$.

Consequently $p_1, p_{m+1} \in U_{n+1}$ for some $U_{n+1} \in \mathcal{U}_{n+1}$. Assume $i \neq m$ and let j be the least positive integer such that $\delta(p_{i+1}, p_{i+2}) + \ldots + \delta(p_j, p_{j+1}) < 2^{-(n+1)}$. Clearly we must have $\delta(p_{j+1}, p_{j+2}) + \ldots + \delta(p_m, p_{m+1}) < 2^{-(n+1)}$ if $j+1 \neq m+1$. But then $p_1, p_i \in U_{n+2}, p_i, p_j \in V_{n+2}$ and $p_j, p_{m+1} \in W_{n+2}$ for some $U_{n+2}, V_{n+2}, W_{n+2} \in \mathcal{U}_{n+2}$. Then $p_1, p_{m+1} \in Star(V_{n+2}, \mathcal{U}_{n+2}) \subset U_{n+1}$ for some $U_{n+1} \in \mathcal{U}_{n+1}$. Consequently $m + 1 \in$ H. We conclude H is the set of all positive integers greater than 1. Therefore, for each pair $x,y \in$ X with $d(x,y) < 1$, $x, y \in$ U for some U $\in \mathcal{U}$. ∎

Note that this same proof can be used to show that for each n, $S(x, 2^{-n}) \subset S(x, \mathcal{U}_n)$. Since it has already been demonstrated that for each n, $S(x, \mathcal{U}_n) \subset S(x, 2^{-n})$ we have that $S(x, \mathcal{U}_n) = S(x, 2^{-n})$. This will be useful in future proofs. The fact that X is a uniform space was only used in one place in the above theorem, namely, to show that the function f is uniformly continuous. Consequently we have the following result that will be used in the next chapter.

COROLLARY 2.2 For each normal sequence $\{\mathcal{U}_n\}$ of coverings in a completely regular space X there is a continuous function f that maps X onto a metric space such that the inverse image of every set of diameter less than 1 is a subset of an element of $\mathcal{U}_1$.

THEOREM 2.6 For each closed set F in a uniform space X and any point p not contained in F, there is a real valued uniformly continuous function f on X such that $f(F) = 0$ and $f(p) \geq 1$.

Proof: p is contained in X - F, which is open, so there exists a uniform covering $\mathcal{U}$ such that $S(p, \mathcal{U}) \subset$ X - F. By Theorem 2.5 there is a metric space (M, d) and a uniformly continuous function g:X → M such that whenever the diameter of N ⊂ M is less than 1, g^{-1}(N) ⊂ U for some U ∈ $\mathcal{U}$. Let $q \in$ F and suppose $d(g(p), g(q)) < 1$. Then there is a V ⊂ M of diameter less than 1 with $g(p), g(q) \in$ V. Let U ∈ $\mathcal{U}$ such that f^{-1}(V) ⊂ U which implies $p,q \in$ U which is impossible since $S(p, \mathcal{U})$ is contained in X - F. Therefore $d(g(p), g(q)) \geq 1$. Define f by $f(x) = d(g(x), g(F))$ for each $x \in$ X. Then $f(F) = 0$ and $f(p) \geq 1$.

It remains to show that f is uniformly continuous. For this define the function δ on M by $\delta(y) = d(y, g(F))$. Then $f = \delta$ © g. Since g is uniformly continuous it suffices to show that δ is uniformly continuous. For this we show that for each $m \in$ M and $\varepsilon > 0$, $\delta(S(m, \varepsilon)) \subset (r - \varepsilon, r + \varepsilon)$ for some real number r. Now

$$\delta(S(m, \varepsilon)) = \{d(y, g(F)) \mid y \in M \text{ and } d(m,y) < \varepsilon\}.$$

Put $r = d(m, g(F))$. If $d(y, g(F)) \in \delta(S(m, \varepsilon))$ then $d(m,y) + d(y, g(F)) \geq d(m,g(F))$. Hence, $d(y, g(F)) > r - \varepsilon$. Similarly, $d(y,m) + d(m, g(F)) \geq d(y, g(F))$ implies that $d(y, g(F)) < r + \varepsilon$. So $\delta(S(m, \varepsilon)) \subset (r - \varepsilon, r + \varepsilon)$ which concludes the proof. ∎

A useful feature of compact Hausdorff spaces is that they have only one uniformity that generates their topology. This fact has a variety of consequences that will be developed in later chapters.

THEOREM 2.7 A compact Hausdorff space has a unique uniformity.

Proof: Let μ and μ' be two uniformities for the compact Hausdorff space X. Let $\mathcal{U} \in \mu$. Then $\mathcal{U}$ has a finite subcovering say $\{U_1 \ldots U_n\}$. By Proposition 0.32 X is normal so by Lemma 1.3, $\{U_1 \ldots U_n\}$ is shrinkable to some closed covering $\{F_1 \ldots F_n\}$ such that $F_i \subset U_i$ for each $i = 1 \ldots n$. Let k be a positive integer with $k \leq n$. For each $p \in F_k$ there exists a $\mathcal{U}'_p \in \mu'$ with $S(p,\mathcal{U}'_p) \subset U_k$. Put $\mathcal{U}'_k = \{S(p,\mathcal{U}'_p) \mid p \in F_k\}$. By Proposition 0.31, F_k is compact so $\mathcal{U}'_k$ has a finite subcovering $\mathcal{W}'_k$. Since μ' is directed there is a $\mathcal{V}'_k \in \mu'$ such that $\mathcal{V}'_k <^* \mathcal{U}'_p$ for each $\mathcal{U}'_p$ such that $S(p,\mathcal{U}'_p) \in \mathcal{W}'_k$. Thus $S(p,\mathcal{V}'_k) \subset U_k$ for each $p \in F_k$.

Next choose $\mathcal{V}' \in \mu'$ such that $\mathcal{V}' <^* \mathcal{V}'_k$ for each $k = 1 \ldots n$. Let $x \in$ X. Then $x \in F_j$ for some positive integer $j \leq n$. Therefore $x \in S(x,\mathcal{V}') \subset S(x,\mathcal{V}'_j) \subset U_j$. Consequently $\mathcal{V}'$ refines $\{U_1 \ldots . U_n\}$. But then $\mathcal{U} \in \mu'$. We conclude that $\mu \subset \mu'$. A similar argument can be used to show that $\mu' \subset \mu$. Hence $\mu = \mu'$ so X has a unique uniformity. ∎

EXERCISES

1. Let $\mathcal{U}$ be an entourage uniformity for the space X and let $(Y,\mathcal{V})$ be another entourage uniform space. We will call a function f:X $\rightarrow$ Y **entourage uniformly continuous** if for each $V \in \mathcal{V}$ the set $\{(x,y) \mid (f(x), f(y)) \in V\}$ is a member of $\mathcal{U}$. Show that a function f:X $\rightarrow$ Y is entourage uniformly continuous if and only if it is uniformly continuous with respect to the covering uniformities $\mu = \{W_U \mid U \in \mathcal{U}\}$ and $\nu = \{W_V \mid V \in \mathcal{V}\}$ on X and Y respectively. [See Exercise 4, Section 2.1.]

2. Let f be a function from a set X into a uniform space (Y,ν) and let $\mathcal{V} \in \nu$. Then $f^{-1}(\mathcal{V}) = \{f^{-1}(\mathrm{V}) \mid \mathrm{V} \in \mathcal{V}\}$ is a covering of X. Put $f^{-1}(\nu) = \{f^{-1}(\mathcal{V}) \mid \mathcal{V} \in \nu\}$. Show that $f^{-1}(\nu)$ is the basis for the coarsest uniformity μ on X such that f is uniformly continuous with respect to μ and ν.

3. Let g:Y $\rightarrow$ X be a function from a uniform space (Y,ν) into a set X. Show there exists a finest uniformity μ on X such that g is uniformly continuous with respect to μ and ν, and that a basis for μ is the set $B = \{\mathcal{U} \mid \mathcal{U}$ is a covering of X such that $g^{-1}(\mathcal{U}) \in \nu\}$.

4. [A. Weil, 1937] For each subset S of a uniform space X, and every uniform covering $\mathcal{U}$, there exists a real-valued uniformly continuous function f on X such that f(S) = 1 and f(X - *Star*(S,$\mathcal{U}$)) = 0.

5. [S. Ginsburg and J. Isbell, 1959] For each countable uniform covering $\mathcal{U}$, there exists a uniformly continuous mapping into a separable metric space M such that $\mathcal{U}$ is refined by the inverse image of some uniform covering in M.

6. A covering $\{V_\alpha\}$ is said to be a **uniformly strict shrinking** of $\{U_\alpha\}$ if there exists a uniform covering $\mathcal{W}$ such that $Star(V_\alpha, \mathcal{W}) \subset U_\alpha$ for each α. Show that each uniform covering has a uniform uniformly strict shrinking (i.e., a uniformly strict shrinking that is itself a uniform covering).

2.3 Uniformizability and Complete Regularity

A topological space (X,τ) is said to be **uniformizable** if there exists a uniformity μ for X such that τ is the topology associated with μ. By Theorem 2.6 it follows that the uniformizable spaces are completely regular. The converse can also be shown (i.e., completely regular spaces are uniformizable) but first we need the following lemma.

LEMMA 2.1 For any family $\{f_\alpha\}$ of functions on a set X into uniform spaces (X_α, μ_α), there is a coarsest uniformity on X containing all inverse images of uniform coverings under these functions.

Proof: Let $\mathcal{U},\mathcal{V} \in \mu_\alpha$ for some α such that $\mathcal{U}^* < \mathcal{V}$. It is easily shown that $f_\alpha^{-1}(\mathcal{U}) <^* f_\alpha^{-1}(\mathcal{V})$. Consequently $f^{-1}(\mu_\alpha) = \{f_\alpha^{-1}(\mathcal{U}) | \ \mathcal{U} \in \mu_\alpha\}$ is a normal family and since unions of normal families are again normal families, $\mu = \cup_\alpha\{f_\alpha^{-1}(\mu_\alpha)\}$ is a normal family. Let ν be the collection of coverings of X that can be refined by finite intersections of members of μ. Then, as in the proof of Theorem 2.1, ν is a uniformity on X. Clearly it is the coarsest one containing μ. ■

The uniformity μ defined in Lemma 2.1 is called the **weak uniformity** induced by the family $\{f_\alpha\}$.

THEOREM 2.8 A space is uniformizable if and only if it is completely regular.

Proof: By Theorem 2.6 the uniformizable spaces are completely regular. Conversely, suppose (X,τ) is a completely regular space. Let $\{f_\alpha\}$ be the collection of real valued continuous functions on X. By the previous lemma, there is a coarsest uniformity c on X containing all the inverse images of uniform coverings under these functions. Let $\tau(c)$ denote the uniform topology. We will show $\tau(c) = \tau$.

To show $\tau \subset \tau(c)$ let U be open in X and suppose U does not belong to $\tau(c)$. Then there is a $p \in$ U such that $S(p, \mathcal{U})$ is not contained in U for each $\mathcal{U} \in c$. By Theorem 2.5 there is a uniformly continuous function f on X such that $f(p) \geq$

1 and $f(X - U) = 0$. Let $\mathcal{V} = \{S(r, 1/2) \mid r \text{ is a real number}\}$. Then $S(f(p),\mathcal{V})$ does not contain 0 which implies $f^{-1}(S(r, 1/2)) \subset U$ for each $V \in \mathcal{V}$ such that $f(p) \in V$. Therefore $S(p, f^{-1}(\mathcal{V})) \subset U$. Now there exists a $\mathcal{U} \in c$ such that $\mathcal{U} < f^{-1}(\mathcal{V})$ which implies $S(p, \mathcal{U}) \subset U$ so $U \in \tau(c)$.

Conversely, suppose N is a neighborhood of p in the uniform topology. Then $S(p, \mathcal{U}) \subset N$ for some $\mathcal{U} \in c$. By the previous lemma, there exist real valued continuous functions $f_1 \ldots f_n$ such that $f_1^{-1}(\mathcal{U}_1) \cap \ldots \cap f_n^{-1}(\mathcal{U}_n) <^* \mathcal{U}$ where $\mathcal{U}_i$ denotes the uniform covering of all spheres of radius less than ε_i for each $i = 1 \ldots n$. Let $\varepsilon = \min\{\varepsilon_i \mid i = 1 \ldots n\}$ and let $\mathcal{U}_\varepsilon$ denote a $\mathcal{U}_i$ that corresponds to this minimum ε_i. Then $f_1^{-1}(\mathcal{U}_\varepsilon) \cap \ldots \cap f_n^{-1}(\mathcal{U}_\varepsilon) <^* \mathcal{U}$. Hence

$$V = \cap\{f_i^{-1}((f_i(p) - \varepsilon/2, f_i(p) + \varepsilon/2)) \mid i = 1 \ldots n\} \subset U$$

for some $U \in \mathcal{U}$. But $f_i^{-1}((f_i(p) - \varepsilon/2, f_i(p) + \varepsilon/2))$ is open for each i so that V is open in X and $S(p,\mathcal{U}) \subset N$ which implies $U \subset N$ since $p \in f_i^{-1}((f_i(p) - \varepsilon/2, f_i(p) + \varepsilon/2))$ for each i. Thus $p \in V \subset N$ so N is a neighborhood of p with respect to the original topology. Therefore $\tau(c) \subset \tau$ which implies that $\tau(c) = \tau$. ∎

The uniformity c defined in the proof of Theorem 2.8 is called the **continuous function uniformity** on X. It can be shown that if we restrict our attention to the bounded real valued continuous functions, the coarsest uniformity c^* on X containing all the inverse images of uniform coverings under these functions also generates the topology τ of X [see Exercise 6]. This uniformity is called the **bounded continuous function uniformity** on X.

An important property of uniformities is that a basis can always be found that consists of open coverings and another can be found consisting of closed coverings. This is shown in the following theorem and its corollary.

THEOREM 2.9 Each uniform covering has a uniform refinement consisting of open (closed) sets.

Proof: Let (X, μ) be a uniform space and suppose $\mathcal{U} \in \mu$. By Theorem 2.5 there is a uniformly continuous function $f: X \to Y$ where Y is a metric space such that the inverse image of every set of diameter less than 1 is a subset of an element of $\mathcal{U}$. Let $\mathcal{V}$ be the covering of Y consisting of open spheres of radius 1/2 and let $Cl(\mathcal{V})$ be the covering of Y consisting of the closures of elements of $\mathcal{V}$. Then both $\mathcal{V}$ and $Cl(\mathcal{V})$ are uniform coverings of Y since $\mathcal{V} < Cl(\mathcal{V})$ and both $f^{-1}(\mathcal{V})$ and $f^{-1}(Cl(\mathcal{V}))$ refine $\mathcal{U}$.

The uniform continuity of f implies $f^{-1}(\mathcal{V})$ and $f^{-1}(Cl(\mathcal{V}))$ both belong to μ. Each member of $\mathcal{V}$ is open and each member of $Cl(\mathcal{V})$ is closed which implies each member of $f^{-1}(\mathcal{V})$ is open and each member of $f^{-1}(Cl(\mathcal{V}))$ is closed. Consequently, each uniform covering of X has a uniform refinement consisting of open (closed) sets. ∎

COROLLARY 2.3 Each uniformity has a basis consisting of open (closed) coverings.

Let A be a subset of a topological space X and let $Cl_X(A)$ and $Int_X(A)$ denote the closure of A and the interior of A respectively. A is said to be **regularly open** in X if the interior of the closure of A is identical with A, that is $Int_X(Cl_X(A)) = A$. A result that will be important later is that in a uniform space a basis for the uniformity can always be found that consists of regularly open sets.

THEOREM 2.10 Each uniform covering has a refinement consisting of regularly open sets.

Proof: Let (X, μ) be a uniform space and $\mathcal{U} \in \mu$. By Theorem 2.8 there exists a uniform refinement $\mathcal{V}$ of $\mathcal{U}$ consisting of closed sets and a uniform refinement $\mathcal{W}$ of $\mathcal{V}$ consisting of open sets. For each $W \in \mathcal{W}$ put

$$W^* = Int_X(Cl_X(W)) \subset Cl(W) \subset V$$

for some $V \in \mathcal{V}$. Then the covering $\mathcal{W}^* = \{W^* \mid W \in \mathcal{W}\}$ refines $\mathcal{V}$ and is refined by $\mathcal{W}$ which implies $\mathcal{W}^*$ is a uniform refinement of $\mathcal{U}$. Moreover, for each $W^* \in \mathcal{W}^*$ we have

$$Int_X(Cl_X(W^*)) = Int_X(Cl_X(Int_X(Cl_X(W)))) = Int_X(Cl_X(W)) = W^*$$

so that W^* is regularly open. Consequently, each uniform covering has a uniform refinement consisting of regularly open sets. ∎

COROLLARY 2.4 Each uniformity has a basis consisting of coverings of regularly open sets.

Given a uniformizable space, there may be many different uniformities that generate the given topology. The set of all such uniformities is called the **family of uniformities that generate** the given topology.

THEOREM 2.11 For each completely regular topology there is a finest uniformity that generates it. It consists of all coverings normal with respect to the family of all open coverings.

Proof: Let μ denote the family of all coverings refined by an open covering. By Theorem 2.2 there exists a finest uniformity λ contained in μ that consists of all coverings normal with respect to μ (i.e., $\mathcal{U} \in \lambda$ if and only if there exists a normal sequence $\{\mathcal{U}_n\}$ such that $\mathcal{U} = \mathcal{U}_1$ and for each n, $\mathcal{U}_n$ has an open refinement).

Next let ν be a uniformity that generates the topology and let $\mathcal{V} \in \nu$. Then there exists a normal sequence $\{\mathcal{V}_n\} \subset \nu$ such that $\mathcal{V} = \mathcal{V}_1$. By Theorem 2.8, for each n there exists an open covering $\mathcal{W}_n$ that refines $\mathcal{V}_n$. Thus $\mathcal{V} \in \lambda$. Consequently λ is finer than ν. It is easy to see that since ν generates the topology of the space, λ must generate it also. Therefore λ is the finest uniformity that generates the topology and λ consists of all coverings that are normal with respect to the family of all open coverings. ∎

The finest uniformity for a topological space is called the **universal** uniformity and is denoted by u. The open members of u are called simply **normal coverings.** A space equipped with the universal uniformity is sometimes called a **fine** space, and u is said to be *fine.*

THEOREM 2.12 Every continuous function from a fine space into another uniform space is uniformly continuous.

Proof: Let (X, u) and (Y, μ) be uniform spaces and assume f:X → Y is continuous. Let $\mathcal{U} \in \mu$. By Theorem 2.9 there is an open refinement $\mathcal{V}$ of $\mathcal{U}$. Since f is continuous, $f^{-1}(\mathcal{V})$ is an open covering of X. Since $\mathcal{V} \in \mu$ there exists a normal sequence $\{\mathcal{V}_n\}$ of uniform coverings of Y such that $\mathcal{V} = \mathcal{V}_1$. The inverse image of a star refinement is a star refinement. To show it for the sequence $\{\mathcal{V}_n\}$ note that for $V_{n+1} \in \mathcal{V}_{n+1}$

$$Star(f^{-1}(V_{n+1}), f^{-1}(\mathcal{V}_{n+1})) =$$

$$\cup\{f^{-1}(W_{n+1}) \mid W_{n+1} \in \mathcal{V}_{n+1} \text{ and } f^{-1}(V_{n+1}) \cap f^{-1}(W_{n+1}) \neq \emptyset\} =$$

$$\cup\{f^{-1}(W_{n+1}) \mid W_{n+1} \in \mathcal{V}_{n+1} \text{ and } V_{n+1} \cap W_{n+1} \neq \emptyset\} = f^{-1}[Star(V_{n+1}, \mathcal{V}_{n+1})].$$

Since the stars of $\mathcal{V}_{n+1}$ refine $\mathcal{V}_n$ it follows that the inverse images of stars of $\mathcal{V}_{n+1}$ refine the inverse images of $\mathcal{V}_n$. Therefore $Star(f^{-1}(V_{n+1}), f^{-1}(\{"V"\})) \subset f^{-1}(V_n)$ for some $V_n \in \mathcal{V}_n$. But then $f^{-1}(\mathcal{V}_{n+1}) <^* f^{-1}(\mathcal{V}_n)$. Consequently $\{f^{-1}(\mathcal{V}_n)\}$ is a normal sequence such that $f^{-1}(\mathcal{V}) = f^{-1}(\mathcal{V}_1)$. By Theorem 2.9. each $\mathcal{V}_n$ has an open refinement $\mathcal{W}_n$ and since f is continuous, $f^{-1}(\mathcal{W}_n)$ is an open refinement of $f^{-1}(\mathcal{V}_n)$. Therefore $f^{-1}(\mathcal{V})$ is normal with respect to the family of all open coverings of X. ∎

COROLLARY 2.5 Each continuous function on a compact uniform space into another uniform space is uniformly continuous.

Proof: By Theorem 2.7 a compact space has a unique uniformity, so a compact space is a fine space. By Theorem 2.12 every continuous function from a fine space into another uniform space is uniformly continuous. ∎

EXERCISES

1. Let $\mathcal{U}$ and $\mathcal{V}$ be coverings of the set X. Show that $\mathcal{U}$ and $\mathcal{V}$ have a coarsest common refinement. The coarsest common refinement is denoted by $\mathcal{U} \cap \mathcal{V}$.

2. Show that condition (1) in the definition of a uniformity can be replaced by the following condition: if $\mathcal{U}$ and $\mathcal{V}$ belong to μ then $\mathcal{U} \cap \mathcal{V}$ belongs to μ.

3. Let $\{f_\alpha\}$ be a family of functions from a set X onto uniform spaces $\{Y_\alpha\}$ that separates points (i.e., if p and q are distinct points of X then there is an α such that $f_\alpha(p) \neq f_\alpha(q)$). Show that the weak uniformity induced by $\{f_\alpha\}$ is a Hausdorff uniformity.

4. Let (Z,ν) be a Hausdorff uniform space, g a function from Z to the set X of Exercise 3, and assume $\{f_\alpha\}$ separates points. Show that a necessary and sufficient condition for g to be uniformly continuous into X in the weak uniformity induced by $\{f_\alpha\}$ is that the composition f_α © g:Z → X be uniformly continuous.

5. A uniformity on a space X is said to be a **metric uniformity** if there exists a metric d on the space such that the uniformity has a basis consisting of all coverings of the form $\{d(p, 2^{-n}) \mid p \in X\}$. Show that a uniformity is a metric uniformity if and only if it has a countable basis.

6. Let (X,τ) be a completely regular space. Let $\{f_\alpha\}$ be the collection of real valued bounded continuous functions on X and let c^* be the coarsest uniformity on X containing all the inverse images of uniform coverings under these functions. Show that c^* generates the topology of X.

2.4 Normal Coverings

There are a number of results about normal coverings that we will need in the following chapters. While these results do not deal with uniformities specifically, it seems best to cover these results now since many of the uniformities we will investigate later will be defined in terms of bases of certain normal coverings.

THEOREM 2.13 (J. W. Tukey, 1940) A space is normal if and only if each finite open covering is normal.

Proof: Assume each finite open covering is normal in the space X and let A and B be disjoint closed subsets of X. Then $\mathcal{W}$ = {X - A, X - B} is a finite open covering of X, and as such has an open refinement $\mathcal{U}$ with $\mathcal{U}^* < \mathcal{W}$. Suppose $Star(A, \mathcal{U}) \cap Star(B, \mathcal{U}) \neq \varnothing$. Then there exists open sets U,V ∈ $\mathcal{U}$ such that

$A \cap U \neq \varnothing$, $U \cap V \neq \varnothing$, and $V \cap B \neq \varnothing$. Since $\mathcal{U}^* < \mathcal{W}$, there exists a $W \in \mathcal{W}$ with $U \cup V \subset W$. But then $W \cap A \neq \varnothing$ and $W \cap B \neq \varnothing$ which is impossible since $\mathcal{W} = \{X - A, X - B\}$. Therefore, $Star(A, \mathcal{U}) \cap Star(B, \mathcal{U}) = \varnothing$ so X is normal.

Conversely, assume X is normal. We first prove that each binary open covering is normal. For this let $\mathcal{W} = \{U,V\}$ be a binary open covering of X (where $U \neq X \neq V$). Put $A = X - U$ and $B = X - V$. Then A and B are disjoint closed sets in X. By Urysohn's Lemma (Theorem 0.1), there exists a continuous function $f: X \to [0,1]$ such that $f(A) = 0$ and $f(B) = 1$. Define the pseudo-metric $d: X^2 \to [0,1]$ by $d(x,y) = |f(x) - f(y)|$. For each positive integer n, put $\mathcal{V}_n = \{S(x, 2^{-n}) \mid x \in X\}$. Then $\{\mathcal{V}_n\}$ is a normal sequence of open coverings. To show $\mathcal{V}_1 < \mathcal{W}$, let $S(x, 2^{-1}) \in \mathcal{V}_1$. Suppose $S(x, 2^{-1})$ is not a subset of U. Then $S(x, 2^{-1}) \cap A \neq \varnothing$ which implies there exists a $y \in A$ with $d(x,y) < 1/2$. Then $|f(x) - f(y)| < 1/2$ since $f(y) = 0$. But then $|f(x) - f(z)| > 1/2$ for each $z \in B$ since $f(z) = 1$. Therefore, Z does not belong to $S(x, 2^{-1})$ for each $z \in B$ which implies $S(x, 2^{-1}) \subset X - B = V$. Hence $\mathcal{V}_1 < \mathcal{W}$ so each binary open covering is normal.

It only remains to extend this result inductively to all finite open coverings. If each open covering of N or fewer members is normal for some positive integer N, let $\mathcal{U} = \{U_1 \ldots U_{N+1}\}$ be an open covering with $N+1$ members. Then $\mathcal{W} = \{U_1 \ldots U_{N-1}, U_N \cup U_{N+1}\}$ is an open covering with N members and $\mathcal{Z} = \{\cup_{i=1}^{N} U_i, U_{N+1}\}$ is an open covering with 2 members. Let $\{\mathcal{U}_n\}$ and $\{\mathcal{V}_n\}$ be normal sequences of open coverings such that $\mathcal{U}_1 < \mathcal{W}$ and $\mathcal{V}_1 < \mathcal{Z}$. Then $\{\mathcal{U}_n \cap \mathcal{V}_n\}$ is a normal sequence of open coverings such that $\mathcal{U}_1 \cap \mathcal{V}_1 < \mathcal{U}$. Consequently $\mathcal{U}$ is normal. We conclude that all finite open coverings are normal. ∎

A covering $\mathcal{U}$ of a space X is said to be **star-finite** if each $U \in \mathcal{U}$ meets at most finitely many members of $\mathcal{U}$. Tukey also gave a proof that a space is normal if and only if each star-finite open covering is normal. But in 1948, K. Morita pointed out an error in Tukey's proof in a paper titled *Star-Finite Coverings and the Star-Finite Property* that appeared in the first volume of Math. Japonicae (pp. 60-68). Morita's correct proof is developed below. We have previously introduced the concept of a finite intersection of coverings. If $\mathcal{U} = \{\mathcal{U}_\alpha\}$ is an arbitrary collection of coverings, we can define the intersection of $\mathcal{U}$ analogously as $\cap \mathcal{U}_\alpha = \{\cap U_\alpha \mid U_\alpha \in \mathcal{U}_\alpha \text{ for each } \alpha\}$. Under certain conditions, $\cap \mathcal{U}_\alpha$ may be a covering. If $\{U_\alpha\}$, $\alpha \in A$, is a covering and $\{V_\alpha\}$ is a refinement such that $V_\alpha \subset U_\alpha$ for each α then $\{V_\alpha\}$ is said to be a **precise** refinement of $\{U_\alpha\}$.

LEMMA 2.2 If $\mathcal{U} = \{U_\alpha\}$, $\alpha \in A$, is a star-finite open covering of a T_1 space X and $\{F_\alpha\}$ is a precise closed refinement, then the intersection $\mathcal{V}$ of all the binary coverings $\{U_\alpha, X - F_\alpha\}$ is an open Δ-refinement of $\mathcal{U}$ and $\mathcal{U} \cup \mathcal{V}$ is star-finite.

Proof: A member of $\mathcal{V}$ is of the form $V = [\cap_{\alpha\in B}U_\alpha]\cap[\cap_{\beta\in A-B}(X - F_\alpha)]$ for some $B \subset A$. If $V \neq \emptyset$, then B must be finite since $\mathcal{U}$ is star-finite. Also, $B \neq \emptyset$ since $\{F_\alpha\}$ covers X. Put $C = \{\beta \in A \mid \cap_{\alpha\in B}U_\alpha \cap U_\beta \neq \emptyset\}$. Then C is also finite. If β does not belong to C then $\cap_{\alpha\in B}U_\alpha\cap U_\beta = \emptyset$ so $\cap_{\alpha\in B}U_\alpha \subset X - F_\beta$, which implies $V = [\cap_{\alpha\in B}U_\alpha]\cap[\cap_{\beta\in C-B}(X - F_\beta)]$. Since C and B are finite, this shows V is open. To show $\mathcal{V}$ is a covering of X, let $p \in X$. Put $E = \{\alpha \in A \mid p \in U_\alpha\}$. Then $\cap_{\alpha\in E}U_\alpha \neq \emptyset$. If β does not belong to E then $p \in X - F_\beta$ so $p \in [\cap_{\alpha\in E}U_\alpha]\cap[\cap_{\beta\in A-E}(X - F_\beta)] \in \mathcal{V}$. Hence $\mathcal{V}$ is an open covering of X.

Clearly $\mathcal{V} < \mathcal{U}$ and $\mathcal{V} < \{U_\alpha, X - F_\alpha\}$ for each α. It remains to show that $\mathcal{V}$ is a Δ-refinement of $\mathcal{U}$ and that $\mathcal{U}\cup\mathcal{V}$ is star-finite. For this let $U_\gamma \in \mathcal{U}$ and $V \in \mathcal{V}$ such that $V \neq \emptyset$. Then we can write $V = [\cap_{\alpha\in B}U_\alpha]\cap[\cap_{\beta\in A-B}(X - F_\beta)]$ for some finite $B \subset A$. Let $D = \{\alpha \in A \mid U_\alpha\cap U_\gamma \neq \emptyset\}$. If $U_\gamma\cap V \neq \emptyset$ then $B \subset D$. Since D is a finite set, the number of sets $V \in \mathcal{V}$ which can intersect U_γ is also finite. Hence the number of members of $\mathcal{U}\cup\mathcal{V}$ that can intersect a member of $\mathcal{U}$ is finite. Since $\mathcal{V} < \mathcal{U}$ this implies $\mathcal{U}\cup\mathcal{V}$ is star-finite. Also, since $\mathcal{V} < \{U_\alpha, X - F_\alpha\}$ for each α, a member V of $\mathcal{V}$ is contained in either U_α or $X - F_\alpha$, for each α. If $V\cap F_\alpha \neq \emptyset$, then $V \subset G_\alpha$, so $Star(F_\alpha, \mathcal{V}) \subset U_\alpha$ for each α. If $p \in X$, then $p \in F_\beta$ for some $\beta \in A$. Therefore, $S(p,\mathcal{V}) \subset U_\beta$ so $\mathcal{V}$ is a Δ-refinement of $\mathcal{U}$. ∎

LEMMA 2.3 If $\mathcal{V}$ is an open Δ-refinement of an open covering $\mathcal{U} = \{U_\alpha\}$, $\alpha \in A$, of a space X, then the sets $F_\alpha = X - Star(X - U_\alpha, \mathcal{V})$ form a precise closed refinement of $\mathcal{U}$.

Proof: Clearly each F_α is closed and $F_\alpha \subset U_\alpha$ for each α. To show $\{F_\alpha\}$ covers X, let $p \in X$. If p does not belong to F_α for each α then $p \in Star(X-U_\alpha,\mathcal{V})$ for each α which implies $S(p,\mathcal{V})\cap(X - U_\alpha) \neq \emptyset$ for each α. But this contradicts the assumption that $\mathcal{V}$ is a Δ-refinement of $\mathcal{U}$. Consequently, $\{F_\alpha\}$ covers X. ∎

THEOREM 2.14 (K. Morita, 1948) A star-finite open covering $\mathcal{U} = \{U_\alpha\}$, $\alpha \in A$, admits a star-finite open Δ-refinement $\mathcal{V}$ such that $\mathcal{U}\cup\mathcal{V}$ is also star-finite if and only if there is a precise closed refinement $\mathcal{W}$ of $\mathcal{U}$.

The proof of Theorem 2.14 follows from Lemmas 2.2 and 2.3. If $\mathcal{U}$ is an open covering of X, we define $Star^n(F, \mathcal{U})$ for each positive integer n as follows: put $Star^1(F, \mathcal{U}) = Star(F, \mathcal{U})$ and $Star^n(F, \mathcal{U}) = Star(Star^{n-1}(F, \mathcal{U}))$ for each $n > 1$. If $\mathcal{U} = \{U_\alpha\}$, $\alpha \in A$, is a star-finite open covering of a T_1 space X, we can define an equivalence relation ~ on A by $\alpha \sim \beta$ if $U_\alpha \subset Star^n(U_\beta, \mathcal{U})$ for some positive integer n. Let D be the collection of equivalence classes with respect to ~. If $C \in D$, put $X_C = \cup_{\alpha\in C}U_\alpha$. Clearly $X = \cup_{C\in D}X_C$. Also, if C, C′ $\in D$ with $C \neq C'$, then $X_C\cap X_{C'} = \emptyset$ for otherwise, there exists some $p \in X_C\cap X_{C'}$ which implies $p \in U_\alpha\cap U_\beta$ for some $\alpha \in C$ and $\beta \in C'$ which in turn implies $\alpha \sim \beta$ which contradicts the assumption that $C \neq C'$. For each $C \in D$, X_C is obviously open. Also, $X - X_C = \cup_{\alpha\in A-C}U_\alpha$ is open, so X_C is closed. Therefore,

X_C is a component of X. Now $\{U_\alpha | \alpha \in C\}$ must be a countable open covering of X_C. We record these observations as:

THEOREM 2.15 If $\mathcal{U} = \{U_\alpha\}$, $\alpha \in A$, is a star-finite open covering of a T_1 space X, then A can be decomposed into a collection of equivalence classes D such that for each $C \in D$, $X_C = \cup_{\alpha \in C} U_\alpha$ is a component of X, $X_C \cap X_B = \varnothing$ if $C \neq B$ and $\{U_\alpha | \alpha \in C\}$ is countable.

THEOREM 2.16 (K. Morita, 1948) If $\mathcal{U} = \{U_i\}$ is a countable open covering of a T_1 space X and $\mathcal{V} = \{F_i\}$ is a precise closed refinement, and if for each i there exists a real valued continuous function f_i on X such that $f_i(F_i) = 0$ and $f_i(X - U_i) = 1$, then $\mathcal{U}$ admits a countable open star-finite Δ-refinement.

Proof: Using the continuous functions f_i, we can construct open sets U_i^n for $n = i, i+1, \ldots$ such that $F_i \subset U_i^n \subset Cl(U_i^n) \subset U_i^{n+1}$ for each $n = i, i+1, \ldots$ Then if we put $X_n = \cup_{i=1}^n U_i^n$ we have $X = \cup_{n=1}^\infty X_n$ and $Cl(X_n) \subset X_{n+1}$ for each positive integer n. Next put $H_n = X_n - Cl(X_{n-3})$ and $K_n = Cl(X_n) - X_{n-1}$ for each positive integer n, where we define $X_{-2} = X_{-1} = X_0 = \varnothing$. Then $K_n \subset H_{n+1}$, $\cup_{n=1}^\infty K_n = X$, and $H_m \cap H_n = \varnothing$ if $|m - n| \geq 3$. Since $K_n = \cup_{i=1}^n (K_n \cap Cl(U_i^n))$, the collection $\{K_n \cap Cl(U_i^n) | i = 1 \ldots n \text{ and } n = 1, 2, \ldots\}$ is a closed covering of X and hence $\{H_{n+1} \cap U_i^{n+1} | i = 1 \ldots n \text{ and } n = 1, 2, \ldots\}$ is an open covering of X. This latter covering is star-finite since $H_m \cap H_n = \varnothing$ for $|m - n| \geq 3$. Then if we construct the intersection $\mathcal{W}$ of all the binary coverings $\{H_{n+1} \cap U_i^{n+1}, X - (K_n \cap Cl(U_i^n))\}$ for $i = 1 \ldots n$ and each positive integer n, by Lemma 2.2, $\mathcal{W}$ is a countable open star-finite Δ-refinement of $\mathcal{U}$. ∎

THEOREM 2.17 (J. W. Tukey and K. Morita, 1948) A T_1 space X is normal if and only if every star-finite open covering is normal.

Proof: Assume X is normal and suppose $\mathcal{U} = \{U_\alpha\}$, $\alpha \in A$, is a star-finite open covering of X. By Theorem 2.15, A can be decomposed into a collection D of equivalence classes such that for each $C \in D$, $X_C = \cup_{\alpha \in C} U_\alpha$ is a component of X, $X_C \cap X_B = \varnothing$ if $C \neq B$, and $\{U_\alpha | \alpha \in C\}$ is countable for each $C \in D$. Since $\mathcal{U}$ is star-finite, $\{U_\alpha | \alpha \in C\}$ is point finite for each $C \in D$. Therefore, by Lemma 1.3, $\{U_\alpha | \alpha \in C\}$ is shrinkable to some covering $\{V_\alpha | \alpha \in C\}$ so $Cl(V_\alpha) \subset U_\alpha$ for each $\alpha \in C$. Consequently, there exists a closed covering $\Phi = \{F_\alpha | \alpha \in A\}$ with $F_\alpha \subset U_\alpha$ for each $\alpha \in A$ (simply put $F_\alpha = Cl(V_\alpha)$ for each $\alpha \in A$). Now, by Theorem 2.14, there exists a star-finite open Δ-refinement $\mathcal{V}$ of $\mathcal{U}$ such that $\mathcal{U} \cup \mathcal{V}$ is also star-finite. Then we can inductively construct a sequence $\{\mathcal{U}_n\}$ of open coverings of X such that $\mathcal{U}_{n+1}$ is a Δ-refinement of $\mathcal{U}_n$ for each n and $\mathcal{U}_1$ is a Δ-refinement of $\mathcal{U}$. Hence $\mathcal{U}$ is normal.

The proof that each star-finite open covering is normal implies X is normal is similar to the proof in Theorem 2.13 that each finite open covering is normal implies X is normal. ∎

Morita's paper also contained another theorem that will be needed in later chapters, so we present it here.

PROPOSITION 2.2 (K. Morita, 1948) A countable open covering in a normal space is normal if and only if it has a precise closed refinement.

Proof: Assume X is normal and suppose $\mathcal{U} = \{U_n\}$ is a countable open covering with a precise closed refinement $\Phi = \{F_n\}$. Since X is normal, there exists a continuous function f_n:X $\rightarrow$ [0, 1] such that $f_n(F_n) = 0$ and $f_n(X - U_n) = 1$ for each n. Then by Theorem 2.16, $\mathcal{U}$ admits a countable open star-finite Δ-refinement $\mathcal{V}$. By Theorem 2.17, $\mathcal{V}$ is normal, so $\mathcal{U}$ is normal. Conversely, suppose the countable open covering $\mathcal{U} = \{U_n\}$ is normal. Then $\mathcal{U}$ has an open Δ-refinement say $\mathcal{V}$. By Lemma 2.3, the sets $F_n = X - Star(X - U_n, \mathcal{V})$ form a precise closed refinement of $\mathcal{U}$. ∎

COROLLARY 2.6 A countable open covering of a normal space is normal if it is point finite.

Proof: Assume X is normal and suppose $\mathcal{U} = \{U_n\}$ is a countable open covering of X that is point finite. By Lemma 1.3, $\mathcal{U}$ is shrinkable to an open covering $\mathcal{V} = \{V_n\}$ such that $Cl(V_n) \subset U_n$ for each n. Then by Theorem 2.18, $\mathcal{U}$ is normal. ∎

PROPOSITION 2.3 A countable open covering of a normal space is normal if and only if it admits a countable star-finite open refinement.

Proof: Assume X is normal and $\mathcal{U} = \{U_n\}$ is a countable open covering of X that admits a countable star-finite open refinement $\mathcal{V}$. Then $\mathcal{V}$ is point finite, so by Corollary 2.6, $\mathcal{V}$ is normal. Hence $\mathcal{U}$ is normal. Conversely, suppose $\mathcal{U}$ is normal. Then $\mathcal{U}$ admits an open Δ-refinement say $\mathcal{V}$. By Lemma 2.3, the sets $F_n = X - Star(X-U_n, \mathcal{V})$ form a precise closed refinement of $\mathcal{U}$. Since X is normal, for each n there exists a continuous function f_n:X $\rightarrow$ [0, 1] such that $f_n(F_n) = 0$ and $f_n(X - U_n) = 1$. Then by Theorem 2.16, $\mathcal{U}$ admits a countable open star-finite Δ-refinement. ∎

Tukey also presented a theorem stating that a star-finite normal covering of a T_1 space has a star-finite Δ-refinement $\mathcal{V}$ such that $\mathcal{V}$ is also normal, and $\mathcal{U} \cup \mathcal{V}$ is star-finite, but the proof depended on the same erroneous theorem (Theorem 2.5) that invalidated his proof of our Theorem 2.17. Morita pointed out that this theorem also follows from his results that we have just recorded.

THEOREM 2.18 (J. W. Tukey and K. Morita, 1948) If $\mathcal{U}$ is a star-finite normal covering of a T_1 space X, then there exists a star-finite Δ-refinement $\mathcal{V}$ of $\mathcal{U}$ such that $\mathcal{V}$ is a normal covering and $\mathcal{U} \cup \mathcal{V}$ is also star-finite.

The proof of Theorem 2.18 depends on two lemmas whose proofs are left as exercises (Exercise 1 and 2) and on Lemmas 2.2 and 2.3.

EXERCISES

1. Show that if $\mathcal{U}$ is a normal covering of a T_1 space X, then the binary covering $\{U, Star(X - U, \mathcal{U})\}$ is normal for any open $U \subset X$.

2. Besides the assumption of Lemma 2.2, let us assume further that the binary coverings $\{U_\alpha, X - F_\alpha\}$ are all normal. Show that the covering $\mathcal{V}$ defined there is also normal.

3. Prove Theorem 2.18.

STAR REFINEMENTS OF COVERINGS

4. Assume $\mathcal{U}$ and $\mathcal{V}$ are coverings of a set X such that $\mathcal{V}^* < \mathcal{U}$. Show that there exists a covering $\mathcal{W}$ of X with $\mathcal{V} < \mathcal{W}^* < \mathcal{U}$ such that:

   (a) if $\mathcal{U}$ is finite, so is $\mathcal{W}$ [J. Tukey, 1940],
   (b) if $\mathcal{U}$ is point finite, so is $\mathcal{W}$ [J. Isbell, 1959],
   (c) if $\mathcal{U}$ is star finite, so is $\mathcal{W}$ [J. Tukey, 1940].

[Hint: Let $\mathcal{U} = \{U_\alpha\}$ and $\mathcal{V} = \{V_\beta\}$. Let γ be a subset of $\mathcal{U}$ such that the members of γ have a point in common, and let Γ be the family of all such γ. For each pair $\gamma,\delta \in \Gamma$ put $W_{\gamma\delta} = \cup\{V_\beta \mid V_\beta \subset U_\alpha$ for each $U_\alpha \in \gamma$ and $V_\beta^* \subset U_\alpha$ for each $U_\alpha \in \delta\}$. Let $\mathcal{W} = \{W_{\gamma\delta} \mid \gamma,\delta \in \Gamma\}$.]

5. [J. Isbell, 1959] Assume $\mathcal{U}$ and $\mathcal{V}$ are coverings of a set X such that $\mathcal{V}^{**} < \mathcal{U}$. Show that if $\mathcal{U}$ is countable, then there exists a countable covering $\mathcal{W}$ such that $\mathcal{V} < \mathcal{W}^* < \mathcal{U}$.

Chapter 3

TRANSFINITE SEQUENCES

3.1 Background

In the theory of metric spaces, sequences play a fundamental role. Recall that a function from one metric space to another is continuous if it preserves convergent sequences (Proposition 1.12) and that a metric space is compact if each sequence has a convergent subsequence (Theorem 1.10). Furthermore, it is possible to characterize the topology in metric spaces by means of convergent sequences (e.g., Proposition 1.10 and Corollary 1.6).

It was shown in Chapter 2 that for a much broader class of spaces (namely, the completely regular spaces) a structure called a uniformity exists that provides a uniform measure of nearness in much the same manner as a metric. One might hope that sequences would play a fundamental role in uniform spaces but unfortunately this is not the case. However, as we will see, this does not prohibit the existence of a theory of convergence in uniform spaces complete with the Cauchy concept and the existence of a suitable *completion* for each uniform space.

The basic problem with sequences is that they are countable. In a metric space each point p has a neighborhood base consisting of the sequence $\{S(p,2^{-n})\}$. It is the fact that *both* sequences and neighborhood bases in metric spaces have the same *order structure* that makes the theory of convergence of sequences so successful in metric spaces. In general, uniform spaces do not have countable neighborhood bases. In fact, as was seen in Chapter 2 (Exercise 6), if a unform space has a countable basis, then the uniformity is equivalent to a metric uniformity.

In this and the following chapters, different objects will be investigated as possible replacements for sequences. The simplest of these is the **transfinite sequence.** A transfinite sequence is simply a function ϕ from a limit ordinal γ into a space X. If for each $\alpha < \gamma$ we denote $\phi(\alpha)$ by x_α, we can use either the notation $\phi:\gamma \to X$ or $\{x_\alpha \mid \alpha < \gamma\}$ to denote the transfinite sequence. When the set γ is understood we can shorten the notation to $\{x_\alpha\}$.

Historically, it appeared to early researchers that the transfinite sequence was not general enough for characterizing topological properties and not much effort was expended in this area. Other objects, such as *nets* and *filters*, appeared to be more promising and, indeed, complete theories of convergence have been based on these concepts as we will see in the next chapter. However, it has recently been shown that all topological properties and certain uniform properties can be characterized in terms of transfinite sequences. This makes it worthwhile investigating how far the theory of transfinite sequences can be pushed in uniform spaces. Furthermore, as we shall see in later chapters, there are certain uniform structures for a Tychonoff space in which transfinite sequences play an important role. Also, we will see results in later chapters for which a simpler version was first proved for transfinite sequences.

3.2 Transfinite Sequences in Uniform Spaces

Let γ be a limit ordinal. A subset $R \neq \emptyset$ of γ is said to be **residual** in γ if whenever $r \in R$ then $\alpha \in R$ for each $\alpha > r$ in γ. A subset C of γ is said to be **cofinal** in γ if whenever $\alpha < \gamma$ there is a $c \in C$ such that $\alpha \leq c$. Let $\phi:\gamma \to X$ be a transfinite sequence and let $A \subset X$. Then ϕ is said to be **eventually** in A if there is a residual $R \subset \gamma$ with $\phi(R) \subset A$ and **frequently** in A if there is a cofinal $C \subset \gamma$ with $\phi(C) \subset A$. The transfinite sequence ϕ is said to **converge** to the point $p \in X$ if ϕ is eventually in each neighborhood of p and to **cluster** to p if it is frequently in each neighborhood of p.

If C is cofinal in γ, then the restriction $\phi|C$ denoted by ϕ_C is said to be a **subsequence** of ϕ. For each $\alpha < \gamma$ let $\phi(\alpha)$ be denoted by x_α. Then ϕ_C is often denoted by $\{x_\beta | \beta \in C\}$ or simply $\{x_\beta\}$ when there is no danger of confusing $\{x_\beta\}$ with $\{x_\alpha\}$. Let (X, μ) be a uniform space. The transfinite sequence $\{x_\alpha\}$ is said to be **Cauchy** if for each $\mathcal{U} \in \mu$, it is eventually in some $U \in \mathcal{U}$. It is said to be **cofinally Cauchy** if for each $\mathcal{V} \in \mu$, it is frequently in some $V \in \mathcal{V}$.

The first theorem we prove is a characterization of paracompactness in terms of transfinite sequences. It states that a space is paracompact if and only if each transfinite sequence that is cofinally Cauchy with respect to the finest uniformity clusters. To prove this theorem we need four lemmas which are, in fact, significant results in themselves. The first is due to C. H. Dowker (Canadian Journal of Mathematics, Volume 2, pp. 219-224, 1951). The following lemma is an equivalent formulation of that result due to F. Ishikawa (Proceedings of the Japan Academy, Volume 31, pp. 686-687, 1955). A space is said to be **countably paracompact** if each countable open covering has a locally finite open refinement. It is said to be **countably metacompact** if each countable open covering has a point finite open refinement. An ordered collection of sets $\{H_\alpha | \alpha < \gamma\}$ is said to be **ascending (descending)** if for each $\alpha < \beta, H_\alpha \subset H_\beta$ $(H_\beta \subset H_\alpha)$.

LEMMA 3.1 (F. Ishikawa, 1955) A space is countably paracompact if and only if each countable ascending open covering $\{U_n\}$ *has a countable ascending open refinement* $\{V_n\}$ *such that* $Cl(V_n) \subset U_n$ *for each n.*

Proof: Assume X is countably paracompact and $\{U_n\}$ is an ascending open covering of X. Then $\{U_n\}$ has a locally finite open refinement $\mathcal{W}$. For each W $\in \mathcal{W}$ let W* be the first U_n containing W and let $H_n = \cup\{W \mid W^* = U_n\}$. Then $H_n \subset U_n$ for each n and $\{H_n\}$ is a locally finite open refinement of $\{U_n\}$. Next, put

$$K_n = \cup\{H_j \mid j > n+1\} \text{ and } V_n = X - Cl(K_n)$$

for each n. Clearly $\{V_n\}$ is an ascending collection of open sets. To show $\{V_n\}$ is a covering of X let $x \in X$. Then there exists a neighborhood N of x that meets only finitely many H_n. Consequently there exists an m such that $N \cap (\cup\{H_j \mid j > m\}) = \emptyset$ which implies x does not belong to $Cl(\cup_{j>m} H_j)$ which in turn implies $x \in V_n$. Therefore $\{V_n\}$ covers X. Finally, to show $Cl(V_n) \subset U_n$ let $m = n + 1$ and note that $Cl(V_n)$ =

$$Cl[X - Cl(\cup_{j>m} H_j)] = Cl[X - (\cup_{j>m} Cl(H_j))] = Cl[\cap_{j>m}(X - Cl(H_j))]$$

since a locally finite collection is closure preserving (see Exercise 4) and by DeMorgan's Rule. Since each $X - H_j$ is closed,

$$Cl[\cap_{j>m}(X - Cl(H_j))] \subset \cap_{j>m}(X - H_j) = X - \cup_{j>m} H_j \subset \cup_{j<m} H_j \subset U_n.$$

Consequently $Cl(V_n) \subset U_n$ for each n.

Conversely assume that for each countable ascending open covering $\{U_n\}$ there is a countable ascending open refinement $\{V_n\}$ such that $Cl(V_n) \subset U_n$ for each n. Let $\{G_n\}$ be a countable open covering of X and for each n put $U_n = \cup\{G_j \mid j \leq n\}$. Then $\{U_n\}$ has a countable open refinement $\{V_n\}$ such that $Cl(V_n) \subset U_n$ for each n. For each n let $H_n = G_n - Cl(V_{n-1})$ where we consider $V_0 = \emptyset$. Then H_n is open and $H_n \subset G_n$ for each n. Furthermore,

$$\cup\{H_n\} \supset \cup\{G_n - U_{n-1}\} = \cup\{G_n\} = X \text{ (where } U_0 = \emptyset)$$

so that $\{H_n\}$ covers X. It remains to show that $\{H_n\}$ is locally finite. For this let $x \in X$ and pick m to be the first positive integer such that $x \in V_m$. Then V_m is a neighborhood of x such that $V_m \cap H_n = \emptyset$ for each $n \geq m+1$. Therefore $\{H_n\}$ is locally finite which implies X is countably paracompact. ∎

LEMMA 3.2 (F. Ishikawa, 1955) A space is countably metatcompact if and only if each countable ascending open covering $\{U_n\}$ has a countable closed refinement $\{F_n\}$ such that $F_n \subset G_n$ for each n.

The proof of Lemma 3.2 is analogous to the proof of Lemma 3.1 with the difference that instead of the neighborhoods N and V_m of x we need only consider the point x itself. The next lemma appeared in the Proceedings of the

American Mathematical Society in 1957 (Volume 8, pp. 822-828). A collection $\mathcal{U}$ of subsets of a topological space is said to be **closure preserving** if for every subcollection $\mathcal{V}$ of $\mathcal{U}$, the closure of the union is the union of the closures; i.e.,

$$Cl(\cup\{V \mid V \in \mathcal{V}\}) = \cup\{Cl(V) \mid V \in \mathcal{V}\}.$$

It is easily shown that a locally finite family is closure preserving.

LEMMA 3.3 (E. Michael, 1957) If an indexed family $\{U_\alpha\}$ is refined by a locally finite (closure preserving) family $\mathcal{V}$, then there exists a locally finite (closure preserving) indexed family $\{W_\alpha\}$ such that $W_\alpha \subset U_\alpha$ for each α and $\cup\{V \mid V \in \mathcal{V}\} = \cup\{W_\alpha\}$. Moreover, if each V in $\mathcal{V}$ is open (closed) then each W_α can be taken to be open (closed).

Proof: Assume $\{U_\alpha \mid \alpha \in A\}$ has a locally finite (closure preserving) refinement $\mathcal{V}$. For each $V \in \mathcal{V}$, pick α_V such that $V \subset U_{\alpha_V}$. For each α put

$$W_\alpha = \cup\{V \in \mathcal{V} \mid \alpha_V = \alpha\}.$$

Then $W_\alpha \subset U_\alpha$ for each α, $\{W_\alpha\}$ is locally finite (closure preserving), and $\cup\{V \mid V \in \mathcal{V}\} = \cup\{W_\alpha\}$. Moreover, if each V is open then each W_α is open. If each V is closed then since the family $\mathcal{V}$ is closure preserving, each W_α will be closed. ∎

The last lemma we need appeared in the Canadian Journal of Mathematics in 1967 (Volume 19, pp. 649-654). Let $\mathcal{U} = \{U_\alpha \mid \alpha < \gamma\}$ be a covering of a space X. If for each $\alpha, \beta \in \gamma$ such that $\alpha < \beta$ we have $U_\alpha \subset U_\beta$ then $\mathcal{U}$ is said to be a **well ordered covering** of X. X is said to be ***M*-paracompact** for some infinite cardinal M if each open covering of cardinality $\leq M$ has a locally finite open refinement.

LEMMA 3.4 (J. Mack, 1967) A space is paracompact if and only if each well ordered open covering has a locally finite open refinement.

Proof: Clearly, if a space is paracompact, each well ordered open covering has a locally finite open refinement. Conversely, assume each well ordered open covering has a locally finite open refinement. We use transfinite induction to prove X is M-paracompact for each infinite cardinal M. Let M be an infinite cardinal and suppose that for every infinite cardinal $\alpha < M$ that X is α-paracompact. Let $\{U_\alpha \mid \alpha < M\}$ be an open covering of X of cardinality M. For each $\beta < M$ put $V_\beta = \cup\{U_\alpha \mid \alpha \leq \beta\}$. Then $\mathcal{V} = \{V_\beta \mid \beta < M\}$ is a well ordered open covering of X and as such has a locally finite open refinement $\mathcal{W}$. By Lemma 3.3 we may assume $\mathcal{W} = \{W_\beta \mid \beta < M\}$ such that $W_\beta \subset V_\beta$ for each $\beta < M$. For each $\beta < M$ put

$$G_\beta = X - \cup\{Cl(W_\alpha) | \alpha \geq \beta\}.$$

Let $x \in X$. Then there exists a neighborhood N of x that meets only finitely many members of W. Consequently, there is a $\gamma < M$ such that $N \cap W_\alpha = \emptyset$ for each $\alpha \geq \gamma$ which implies x does not belong to $Cl(W_\alpha)$ for each $\alpha \geq \gamma$ which in turn implies $x \in G_\gamma$. Therefore $\Gamma = \{G_\beta | \beta < M\}$ is a well ordered open covering of X. Moreover,

$$Cl(G_\beta) \subset Cl[X - \cup\{W_\alpha | \alpha \geq \beta\}] = X - \cup\{W_\alpha | \alpha \geq \beta\} \subset \cup\{W_\alpha | \alpha < \beta\} \subset U_\beta.$$

Therefore $Cl(G_\beta) \subset V_\beta$ for each $\beta < M$. Since Γ is a well ordered open covering, Γ has a locally finite open refinement Φ. Consequently, Φ is a locally finite open refinement of V whose closures also refine V.

For each $C \in \Phi$ there exists an index β such that $Cl(C) \subset V_\beta$ which implies $\{U_\alpha | \alpha \leq \beta\}$ covers $Cl(C)$. By the induction hypothesis, X is β-paracompact. It is easily shown that $Cl(C)$ is β-paracompact (see Exercise 7). Therefore there exists a locally finite collection B_C in X that refines U and covers $Cl(C)$. Consequently, $B = \{B \cap C | B \in B_C \text{ and } C \in \Phi\}$ is a locally finite open refinement of U which implies X is M-paracompact. But then X is M-paracompact for each infinite cardinal M which implies X is paracompact. ■

The following theorem appeared in the 1980 paper titled *A note on transfinite sequences* (Fundamenta Mathematicae, Volume 106, pp. 213-226). It was first proved in 1969 and distributed in a preprint at the 1970 Pittsburg Topology Conference.

THEOREM 3.1 (N. Howes, 1969) A space is paracompact if and only if each transfinite sequence that is cofinally Cauchy with respect to the universal uniformity u clusters.

Proof: Assume X is paracompact and let $\{x_\alpha\}$ be a cofinally Cauchy transfinite sequence with respect to u. Suppose $\{x_\alpha\}$ does not cluster. Then for each $x \in X$ there is an open $U(x)$ containing x with $\{x_\alpha\}$ eventually in $X - U(x)$. Put $U = \{U(x) | x \in X\}$. By Theorem 1.1, U is normal so $U \in u$. But then $\{x_\alpha\}$ is frequently in some $U(y)$ so cannot be eventually in $X - U(y)$ which is a contradiction. Therefore $\{x_\alpha\}$ must cluster.

Conversely, assume each transfinite sequence that is cofinally Cauchy with respect to u clusters. By Lemma 3.4 it suffices to show that each well ordered open covering has a locally finite open refinement. Let $\{U_\alpha\}$ be a well ordered open covering of X. For each index α put $F_\alpha = X - U_\alpha$. Then $\cap F_\alpha = \emptyset$. Also, for each α let $<_\alpha$ be a well ordering of F_α and let $E = \{(F_\alpha, x) | x \in F_\alpha\}$. Define the well ordering $<$ on E by $(F_\alpha, x) < (F_\beta, y)$ if and only if $\alpha < \beta$ or $\alpha = \beta$ and $x <_\alpha y$. For each $(F_\alpha, x) \in E$ put $\psi(F_\alpha, x) = x$. Since E is order isomorphic with some limit ordinal γ, ψ is a transfinite sequence in X. Now ψ cannot cluster to any $p \in X$ for otherwise p would belong to each F_α which contradicts $\cap F_\alpha = \emptyset$. Denote ψ by $\{y_\beta\}$.

Let $\mathcal{U} \in u$. By Theorem 2.5, there is a uniformly continuous f_u:X $\rightarrow$ M where M is a metric space such that for each sphere $S(m, \varepsilon)$ of radius $\varepsilon < 1$ in M, $f_u^{-1}(S(m, \varepsilon)) \subset U$ for some $U \in u$. If $\{f_u(y_\beta)\}$ clusters to some $m \in M$, there is a cofinal $C \subset E$ with $\{f_u(y_\gamma) | \gamma \in C\} \subset S(m,1/2)$. Then

$$\{y_\gamma\} \subset f_u^{-1}(\{f_u(y_\gamma)\}) \subset f_u^{-1}(S(m,1/2)) \subset U.$$

Therefore, if $\{f_u(y_\beta)\}$ clusters in M for each $\mathcal{U} \in u$, then $\{y_\beta\}$ is cofinally Cauchy and therefore clusters in X which is a contradiction. Consequently, there must be a $\mathcal{U} \in u$ and a uniformly continuous function f:X $\rightarrow$ M for some metric space M such that for each sphere $S(m, \varepsilon)$ of radius $\varepsilon < 1$ in M, $f^{-1}(S(m,\varepsilon)) \subset U$ for some $U \in \mathcal{U}$, but $\{f(y_\beta)\}$ does not cluster in M.

Suppose $p \in \cap Cl[f(F_\alpha)]$. Let V be an open set containing p and let $(F_\beta, z) \in E$. Pick $\gamma > \beta$. Since $V \cap f(F_\gamma) \neq \emptyset$ there is an $x \in F_\gamma$ with $f(x) \in U$. If $\delta = (F_\gamma, x)$ then $y_\delta = x$ so $f(y_\delta) \in V$. But $(F_\beta, z) < \delta$ so $\{f(y_\beta)\}$ is frequently in each neighborhood of p which is a contradiction. Hence $\cap Cl[f(F_\alpha)] = \emptyset$. For each index α put $V_\alpha = M - Cl[f(F_\alpha)]$ and let $\mathcal{V} = \{V_\alpha\}$. Since M is metric and therefore paracompact, there is a locally finite open refinement $\mathcal{W}$ of $\mathcal{V}$ and hence $f^{-1}(\mathcal{W}) = \{f^{-1}(W) | W \in \mathcal{W}\}$ is locally finite in X. If $W \in \mathcal{W}$ there is a $V_\alpha \in \mathcal{V}$ with $W \subset V_\alpha = M - Cl[f(F_\alpha)]$ so that

$$f^{-1}(W) \subset f^{-1}(M - Cl(f(F_\alpha))) = X - f^{-1}(Cl(f(F_\alpha))) \subset X - F_\alpha = U_\alpha.$$

Therefore $f^{-1}(\mathcal{W})$ refines $\{U_\alpha\}$ so that X is paracompact. ∎

Note that the property that each transfinite sequence that is cofinally Cauchy with respect to u clusters is a *topological property* (i.e., one that is preserved under homeomorphisms) whereas this same property with respect to any other uniformity is a *uniform property* (i.e., one that is preserved under uniform homeomorphisms).

Let λ be the family of all countable normal coverings of a completely regular space X. In 1952 a paper appeared in the Osaka Mathematical Journal (Volume 4, pp. 23-40) titled *A Class of Topological Spaces* showing that the family e of all covering refined by members of λ is a uniformity for X. This e uniformity plays a fundamental role in *realcompact* spaces as will be seen in Chapter 7. Our present interest in the e uniformity is stated in Theorem 3.2.

LEMMA 3.5 (T. Shirota, 1952) The countable normal coverings of a completely regular space form a basis for a uniformity e that generates the topology.

Proof: Assume X is completely regular and $\mathcal{U} = \{U_n\}$ is a countable normal covering of X. We first show there is a countable normal covering $\mathcal{V}$ with $\mathcal{V}^* < \mathcal{U}$. Let $\mathcal{Z}$ be a normal covering with $\mathcal{Z}^* < \mathcal{U}$. By Lemma 2.3, the sets $F_n = X - Star(X - U_n, \mathcal{Z})$ form a precise closed refinement of $\mathcal{U}$. Then for each n there exists a continuous function $f_n: X \to [0,1]$ such that $f_n(F_n) = 1$ and $f_n(X - U_n) = 0$. To see this recall that by Corollary 2.2 there is a continuous function $f: X \to M$ where M is a metric space such that the inverse image of every set of diameter less than 1 is a subset of an element of $\mathcal{Z}$. Let d be the metric on M and for each $x \in X$ put $g(x) = d(f(F_n), f(x))$. By Proposition 1.4 g is continuous. Now put $g_n(x) = max\{g(x),1\}$ and $f_n(x) = 1 - g_n(x)$. Then f_n is the desired function. Next we define a continuous function $h: X \to I^{\omega}$ where

$$I^{\omega} = \Pi\{I_n \mid n = 1, 2, 3, \ldots\} \text{ and } I_n = [0,1]$$

for each index n. For this let $x \in X$ and put $h(x) = (f_1(x), f_2(x), f_3(x), \ldots)$. Then let $Y = h(X)$ and for each index n put $V_n = \{h(x) \in Y \mid f_n(x) > 0\}$. To show $\{V_n\}$ is an open covering of Y, note that if $x \in X$ there is an F_n containing x which implies $f_n(x) > 0$ and therefore $h(x) \in V_n$. To show each V_n is open note that $V_n = p_n^{-1}(0,1] \cap Y$ and $p_n^{-1}(0,1]$ is open in I^{ω} where p_n denotes the canonical projection of I^{ω} onto its nth coordinate subspace. Clearly $h^{-1}(V_n) \subset U_n$ for each n.

To complete the argument we use the fact that I^{ω} is a separable metric space. Clearly I^{ω} is separable since it is a countable product of separable spaces. That I^{ω} is a metric space follows from Lemma 1.5. For each index n let $W_n = p_n^{-1}(0,1]$. Then $V_n = W_n \cap Y$ and $\{W_n\}$ is an open covering of $I^{\omega} - \{0\}$ where 0 is the point of I^{ω} having all coordinates equal to zero. Since $I^{\omega} - \{0\}$ is open in I^{ω} it is also separable (it is left as an exercise to show this). Since $I^{\omega} - \{0\}$ is metric, by Theorem 1.2 it is fully normal. Therefore there are normal refinements $\mathcal{W}_1$ and $\mathcal{W}_2$ of $\{W_n\}$ with $\mathcal{W}_1^* < \mathcal{W}_2^* < \{W_n\}$. Let A be a countable dense subset of $I^{\omega} - \{0\}$. For each $a \in A$ pick $W(a) \in \mathcal{W}_1$ containing a. Then let $W_a \in \mathcal{W}_2$ such that $Star(W(a), \mathcal{W}_1) \subset W_a$.

To show $\{W_a \mid a \in A\}$ covers $I^{\omega} - \{0\}$ let $q \in I^{\omega} - \{0\}$ and let $W \in \mathcal{W}_1$ containing q. Then W contains some $a \in A$ which implies $W \subset Star(W(a), \mathcal{W}_1) \subset W_a$ and hence $q \in W_a$. Clearly $\{W_a\}$ is a countable normal star refinement of $\{W_n\}$. For each $a \in A$ put $V_a = W_a \cap Y$. Then $\{V_a\}$ is a countable normal refinement of $\{V_n\}$. Finally, put $\mathcal{V} = \{h^{-1}(V_a)\}$. Then $\mathcal{V}$ is the desired countable normal refinement of $\mathcal{U}$.

Next we show that the set λ of all countable normal coverings is a basis for a uniformity. From Theorem 2.1 we already know λ is a sub-basis for a uniformity. Therefore it only remains to show that if $\mathcal{U}, \mathcal{V} \in \lambda$ there is a $\mathcal{W} \in \lambda$ with $\mathcal{W} < \mathcal{U}$ and $\mathcal{W} < \mathcal{V}$. Since $\mathcal{U}$ and $\mathcal{V}$ are both countable so it $\mathcal{U} \cap \mathcal{V}$. If $\{\mathcal{U}_n\}$ and $\{\mathcal{V}_n\}$ are normal sequences of coverings such that $\mathcal{U} = \mathcal{U}_1$ and $\mathcal{V} = \mathcal{V}_1$ then $\mathcal{U}_{n+1} \cap \mathcal{V}_{n+1} <^* \mathcal{U}_n \cap \mathcal{V}_n$ for each positive integer n. Consequently $\mathcal{U} \cap \mathcal{V}$ is a countable normal covering. Put $\mathcal{W} = \mathcal{U} \cap \mathcal{V}$. Then $\mathcal{W}$ is the desired member of λ such that $\mathcal{W} < \mathcal{U}$ and $\mathcal{W} < \mathcal{V}$. Therefore λ is the basis for a uniformity which we denote by e.

It remains to show that e generates the topology of X. Rather than prove this directly we rely on a result (Lemma 3.7) that will be proved shortly. This lemma states that X admits a uniformity β that has a basis consisting of all the finite normal coverings. Clearly e is coarser than u and finer than β. But it has already been shown that u generates the topology of X and Lemma 3.7 will show that β also generates the topology of X. Consequently the topology generated by e must be coarser than the original topology and finer than the original topology. The only way for this to happen is if e generates the original topology. ∎

Another useful result from K. Morita's paper *Star-Finite Coverings and the Star-Finite Property*, referenced in Section 2.4 is the following:

LEMMA 3.6 (K. Morita, 1948) A regular Lindelöf space is paracompact.

Proof: Assume X is a regular Lindelöf space. We first show that X is normal. For this let H and K be disjoint closed sets in X. Then H and K are both Lindelöf. For each $p \in$ H let $U(p)$ be an open set containing p such that $Cl(U(p)) \cap$K $= \emptyset$ and for each $q \in$ K let $V(q)$ be an open set containing q such that $Cl(V(q)) \cap$H $= \emptyset$. Since H and K are both Lindelöf, we can find sequences $\{p_n\}$ and $\{q_n\}$ in H and K respectively such that $\{U(p_n)\}$ covers H and $\{V(q_n)\}$ covers K. For simplicity of notation, denote $U(p_n)$ by U_n and $V(q_n)$ by V_n for each n. Next let $A_1 = U_1$ and $B_1 = V_1 - Cl(U_1)$. For each positive integer $n > 1$ put $A_n = U_n - [\cup_{i=1}^{n-1} Cl(V_i)]$ and $B_n = V_n - [\cup_{i=1}^{n} Cl(U_i)]$. Then put A $= \cup_{n=1}^{\infty} A_n$ and B $= \cup_{n=1}^{\infty} B_n$. It is easily seen that H $\subset$ A and K $\subset$ B and that A and B are open sets.

To see that A$\cap$B $= \emptyset$, suppose $p \in$ A$\cap$B. Let j be the least positive integer such that $p \in A_j$ and let k be the least positive integer such that $p \in B_k$. Then $p \in U_j - [\cup_{i=1}^{j-1} Cl(V_i)]$ and $p \in V_k - [\cup_{i=1}^{k} Cl(U_i)]$. There are two cases to consider: first, $(j \le k)$ which implies $p \in U_j$ and p does not belong to U_j which is a contradiction, and second, $(j > k)$ which implies $p \in V_k$ and p does not belong to V_k which is also a contradiction. Hence A$\cap$B $= \emptyset$ so X is normal.

Next we show that each open covering has a star-finite refinement. For this let $\mathcal{U}$ be an open covering of X. For each $p \in$ X let $V(p)$ be an open neighborhood of p such that $Cl(V_p) \subset$ U for some U $\in \mathcal{U}$. Since X is Lindelöf, we can find a countable set of points $\{p_n\}$ such that X $= \cup_{n=1}^{\infty} V(p_n)$. For each n, let $U_n \in \mathcal{U}$ such that $Cl(V(p_n)) \subset U_n$. Since X is normal, for each n, there exists a continuous function f_n:X $\to$ [0,1] with $f_n(Cl(V(p_n))) = 0$ and $f_n(\text{X} - U_n) = 1$. Then by Theorem 2.16, $\{U_n\}$ admits an open star-finite refinement $\mathcal{V}$. But then $\mathcal{V}$ is a locally finite refinement of $\mathcal{U}$ so X is paracompact. ∎

THEOREM 3.2 (N. Howes, 1969) A completely regular space X is Lindelöf if and only if each transfinite sequence that is cofinally Cauchy with respect to the e uniformity clusters.

Proof: Assume X is Lindelöf and suppose $\{x_\alpha\}$ is a cofinally Cauchy transfinite sequence with respect to e that does not cluster. Then for each $x \in X$ there is an open $U(x)$ containing x such that $\{x_\alpha\}$ is eventually in X - $U(x)$. Let $\mathcal{U} = \{U(x) \mid x \in X\}$. Since X is Lindelöf $\mathcal{U}$ has a countable subcovering $\{U(x_i)\}$ and since regular Lindelöf spaces are paracompact, $\{U(x_i)\}$ is normal and therefore belongs to e. But then $\{x_\alpha\}$ is frequently in some $U(x_j)$ which is a contradiction. Therefore, $\{x_\alpha\}$ must cluster after all.

Conversely assume each transfinite sequence that is cofinally Cauchy with respect to e clusters. Then each transfinite sequence that is cofinally Cauchy with respect to u clusters, so by Theorem 3.1 X is paracompact and therefore countably metacompact. Next we show that a transfinite sequence $\{y_\beta \mid \beta < \gamma\}$ with no countable subsequence clusters. Let $\mathcal{U} \in e$. Then there is a countable normal covering $\{U_n\}$ that refines $\mathcal{U}$. For each index n put $E_n = \{\beta < \gamma \mid y_\beta \in U_n\}$ and let $E = \cup E_n$. Suppose E_n is not cofinal in γ for each n. Since $\{y_\beta\}$ has no countable subsequence, E is not cofinal in γ so there is a $\delta < \gamma$ such that y_δ is not contained in U_n for each n which is a contradiction. Consequently some E_m must be cofinal in γ so that $\{y_\beta\}$ is cofinally Cauchy with respect to e and hence clusters.

Finally, we show that a countably metacompact space in which each transfinite sequence with no countable subsequence clusters, is Lindelöf. Let κ be the least cardinal such that for some open covering $\mathcal{U} = \{U_\alpha \mid \alpha < \kappa\}$ has no countable subcovering. For each $\alpha < \kappa$ put $V_\alpha = \cup\{U_\beta \mid \beta \leq \alpha\}$ and let $F_\alpha = X - V_\alpha$. It is easily shown that each $F_\alpha \neq \varnothing$ so pick $x_\alpha \in F_\alpha$ for each index α. If $\{x_\alpha\}$ has no countable subsequence it clusters which implies $\cap F_\alpha = \varnothing$ which is impossible since $\cup V_\alpha = X$. Therefore assume the existence of a countable cofinal subset $\{\alpha_n\}$ of κ. But then $\{V_\alpha\}$ has a countable refinement $\{V_{\alpha_n}\}$ and hence by Lemma 3.2 there is a countable closed refinement $\{H_{\alpha_n}\}$ such that $H_{\alpha_n} \subset V_{\alpha_n}$ for each n. Now $\mathcal{U}_n = \{U_\beta \mid \beta \leq \alpha_n\}$ has cardinality less than κ and covers H_{α_n}. Since H_{α_n} is closed there is a countable subcollection of $\mathcal{U}_n$ that covers H_{α_n}. But since $\{H_{\alpha_n}\}$ is countable and covers X there must be a countable subcovering of $\mathcal{U}$. Therefore X is Lindelöf. ∎

COROLLARY 3.1 A regular, countably metacompact space is Lindelöf if and only if each transfinite sequence with no countable subsequence clusters.

Let λ be the family of all finite normal coverings of a space X. An important uniformity for X is the uniformity β consisting of all coverings refined by members of λ. This uniformity gives rise to the celebrated *Stone-Čech Compactification* that we will study in Chapter 6. Our present interest in the β uniformity will be evident in the statement of Theorem 3.3. To show β is a uniformity for X we proceed as follows:

LEMMA 3.7 Every finite normal covering has a finite normal star refinement.

Proof: Let $\mathcal{U}$ be a finite normal covering and $\mathcal{V}$ and $\mathcal{W}$ be normal coverings such that $\mathcal{V}^* < \mathcal{W}^* < \mathcal{U}$. For each $V, W \in \mathcal{V}$ put $V \sim W$ if

(1) V and W are contained in the same elements of $\mathcal{U}$ and
(2) $Star(V,\mathcal{V})$ and $Star(W,\mathcal{V})$ are contained in the same elements of $\mathcal{U}$.

Clearly ~ is an equivalence relation on $\mathcal{V}$. Since $\mathcal{U}$ is finite, there can be only finitely many equivalence classes. To see this, one can induct on the number of elements of $\mathcal{U}$. Clearly if $\mathcal{U}$ has only one member, there can be only finitely many equivalence classes. If it has already been established that when $\mathcal{U}$ has n members there are only finitely many equivalence classes and if we now assume $\mathcal{U}$ has n+1 members, say $U_1 \ldots U_{n+1}$, and let $\sim_n$ be the equivalence relation defined by (1) and (2) above with respect to $\{U_1, \ldots, U_n\}$ then there are finitely many equivalence classes $\{E_j\}$ for some $j = 1 \ldots k$ with respect to $\sim_n$. For each j, E_j is partitioned into two distinct sets

$$H_j = \{V \in E_j \mid V \text{ and } Star(V,\mathcal{V}) \subset U_{n+1}\} \text{ and } K_j = E_j - H_j.$$

But $\{H_j \mid j = 1 \ldots k\} \cup \{K_j \mid j = 1 \ldots k\}$ is merely the collection of equivalence classes with respect to ~ where ~ is defined by (1) and (2) above with respect to $\mathcal{U}$. Consequently, if $\mathcal{U}$ has n+1 members, there are only finitely many equivalence classes, which completes the induction argument.

Next, for each equivalence class E with respect to ~ let $Z(E) = \cup E$ and put $\mathcal{Z} = \{Z(E) \mid E$ is an equivalence class with respect to ~$\}$. Clearly $\mathcal{Z}$ is a finite normal covering since $\mathcal{V} < \mathcal{Z}$. We leave it to the reader to show that $\mathcal{Z}$ star refines $\mathcal{U}$. ∎

LEMMA 3.8 Each completely regular space admits a uniformity β that has a basis λ consisting of all finite normal coverings.

Proof: Let X be a completely regular space and u the universal uniformity on X. Then u consists of all coverings that are normal with respect to the family of all open coverings. So λ is the collection of all finite members of u. Let $\mathcal{U}$, $\mathcal{V}$ $\in$ λ. Then the covering $\mathcal{U} \wedge \mathcal{V}$ is finite. Moreover, $\mathcal{U} \wedge \mathcal{V} \in u$ by Exercise 2 of Section 2.3 so $\mathcal{U} \wedge \mathcal{V} \in$ λ. Hence λ satisfies condition (1) of the definition of a uniformity. By Lemma 3.7 $\mathcal{U}$ has a finite uniform star refinement $\mathcal{W} \in u$. By Theorem 2.9, $\mathcal{W}$ also has an open uniform refinement say $\mathcal{Z}$. Then

$$Z < Int(\mathcal{W}) = \{Int(\mathrm{W}) \mid \mathrm{W} \in \mathcal{W}\}$$

which implies $Int(\mathcal{W}) \in u$. But $Int(\mathcal{W})$ is finite since $\mathcal{W}$ is finite so $Int(\mathcal{W}) \in \lambda$. Clearly $Int(\mathcal{W}) < \mathcal{W}^* < \mathcal{U}$. Therefore λ satisfies condition (3) of the definition of a uniformity so λ is a basis for β.

It remains to show that β generates the topology of the space. For this let U be an open neighborhood of $p \in \mathrm{X}$. Then there is a uniformly continuous function $f{:}\mathrm{X} \to [0,1]$ such that $f(\mathrm{X} - \mathrm{U}) = 0$ and $f(p) = 1$. Let $\mathrm{V} = f^{-1}([0,1/2))$. Then $\{\mathrm{U,V}\}$ is a finite open covering of X. For each pair $x,y \in \mathrm{X}$ put $d(x,y) = |f(x) - f(y)|$. Clearly d is a pseudo-metric on X. For each positive integer n put $\mathcal{U}_n = \{S(p, 2^{-n}) \mid p \in \mathrm{X}\}$. As we have seen in previous proofs, $\mathcal{U}_{n+1}^* < \mathcal{U}_n$ for each n. Moreover, if $q \in \mathrm{X}$ such that $f(q) < 1/4$ then

$$S(q, \mathcal{U}_2) \subset \{p \in \mathrm{X} \mid |f(p) - f(q)| < 1/4\} \subset \{p \in \mathrm{X} \mid |f(p)| < 1/2\} \subset \mathrm{V}.$$

If $q \in \mathrm{X}$ such that $f(q) \geq 1/4$ then $S(q, \mathcal{U}_2) \subset \{p \in \mathrm{X} \mid |f(p)| > 0\} \subset \mathrm{U}$. Consequently $\mathcal{U}_2^* < \{\mathrm{U,V}\}$ and $\mathcal{U}_n$ is an open covering for each n. Therefore $\mathcal{W} = \{\mathrm{U,V}\}$ is a finite normal covering and $S(p,\mathcal{W}) = \mathrm{U}$. But then β generates the topology of X. ∎

THEOREM 3.3 (N. Howes, 1969) A space is compact if and only if each transfinite sequence that is cofinally Cauchy with respect to the β uniformity clusters.

Proof: Suppose X is compact and $\{x_\alpha\}$ is a transfinite sequence. For each index α put $M_\alpha = \{x_\beta \mid \beta > \alpha\}$ and let $U_\alpha = \mathrm{X} - Cl(M_\alpha)$. If $\{x_\alpha\}$ does not cluster $\{U_\alpha\}$ covers X and therefore has a finite subcovering $\{U_{\alpha_i}\}$. Let δ be an index such that $\alpha_i < \delta$ for each i. Then x_δ is not contained in U_{α_i} for each i which is a contradiction. Therefore every transfinite sequence in X clusters.

Conversely assume each transfinite sequence that is cofinally Cauchy with respect to β clusters and suppose X is not compact. Let γ be the least infinite cardinal such that there is an open covering $\{U_\alpha \mid \alpha < \gamma\}$ having no subcovering of smaller cardinality and for each $\alpha < \gamma$ put $V_\alpha = \cup\{U_\beta \mid \beta \leq \alpha\}$. Then $\{V_\alpha\}$ is a well ordered covering of X. For each $\alpha < \gamma$ pick $x_\alpha \in \mathrm{X} - V_\alpha$. Then the transfinite sequence $\{x_\alpha\}$ is cofinally Cauchy with respect to β. In fact, all transfinite sequences in X are cofinally Cauchy with respect to β for if $\mathcal{U} \in \beta$ there is a finite $\mathcal{V} \in \beta$ with $\mathcal{V} < \mathcal{U}$. Since $\mathcal{V}$ is finite, $\{x_\alpha\}$ must be frequently in some member of $\mathcal{V}$ and hence frequently in some member of $\mathcal{U}$.

According to our original assumption, $\{x_\alpha\}$ must cluster to some $p \in \mathrm{X}$. But then p cannot belong to V_α for each $\alpha < \gamma$ which is a contradiction since $\{V_\alpha\}$ covers X. Therefore $\{U_\alpha\}$ must have a subcovering of smaller cardinality. But if each infinite open covering has a subcovering of smaller cardinality, each infinite open covering must have a finite subcovering. Consequently X must be compact. ■

COROLLARY 3.2 A completely regular T_1 space is compact if and only if each transfinite sequence clusters.

EXERCISES

1. Show that X is countably paracompact if and only if for each countable descending chain of closed sets $\{F_n\}$ with $\cap F_n = \varnothing$ there exists a countable descending chain of open sets $\{G_n\}$ with $F_n \subset G_n$ for each n such that $\cap Cl(G_n) = \varnothing$.

2. Show that X is countably metacompact if and only if for each countable descending chain of closed sets $\{F_n\}$ with $\cap F_n = \varnothing$ there exists a countable descending chain of open sets $\{G_n\}$ with $F_n \subset G_n$ for each n such that $\cap G_n = \varnothing$.

3. Show that a regular countably metacompact space is Lindelöf if and only if each transfinite sequence with no countable subsequence clusters.

4. Show that a locally finite collection of sets is closure preserving (i.e., the closure of a union of members of a locally finite collection is the union of the closures).

5. Prove Lemma 3.2.

6. Mansfield (1957) defined a space to be **almost 2-fully normal** if for each open covering $\mathcal{U}$ there is an open refinement $\mathcal{V}$ such that if $p \in$ V and $q \in$ W for two members V and W of $\mathcal{V}$ with V$\cap$W = $\varnothing$, then there is a U $\in \mathcal{U}$ containing both p and q. We define a transfinite sequence ψ to be **cofinally Δ Cauchy** if for each open covering $\mathcal{U}$ of X there is a $p \in$ X such that ψ is frequently in $S(p, \mathcal{U})$. Show that an almost 2-fully normal T_1 space is paracomact if and only if each cofinally Δ Cauchy transfinite sequence clusters.

7. If X is M-paracompact for some infinite cardinal M and F is a closed subset of X, show that F is M-paracompact.

8. Prove Corollary 3.2.

9. Alexandrov and Urysohn (1929) introduced the concept of final compactness in the sense of complete accumulation points. A space is $[\alpha,\beta]$-compact in the sense of complete accumulation points, where α and β denote cardinals with $\alpha \leq \beta$, if every subset M of X whose cardinality is regular and lies in the interval

$[\alpha,\beta]$ has a point of complete accumulation; i.e., a point p such that if U is an open set containing p then the cardinality of $U \cap M$ is the same as the cardinality of M. A space is **finally compact in the sense of complete accumulation points** if it is $[\alpha,\beta]$-compact in the sense of complete accumulation points for all cardinals $\beta > \alpha$.

They then proved the following theorem: *A space is* $[\alpha,\beta]$*-compact in the sense of complete accumulation points if and only if every open covering* $\mathcal{U}$ *of* X *whose cardinality is regular and lies in the interval* $[\alpha,\beta]$ *has a subcovering* $\mathcal{U}^*$ *whose cardinality is less than the cardinality of* $\mathcal{U}$.

We define a space to be **linearly Lindelöf** if each well ordered open covering has a countable subcovering. Prove that the following properties are equivalent:

(1) linearly Lindelöf,
(2) final compactness in the sense of complete accumulation points,
(3) each transfinite sequence with no countable subsequence clusters.

10. RESEARCH PROBLEM

Miščenko (1962) exhibited a space that he called R^* that is completely regular, T_1, finally compact in the sense of complete accumulation points, but not Lindelöf. Later, M. Rudin (1971) showed that R^* is not normal.

Question: Does there exist a normal Hausdorff space that is linearly Lindelöf but not Lindelöf?

11. It was long a question as to whether or not the coverings of a given uniformity μ, the cardinalities of which are less than a given cardinal number κ form a basis for a uniformity μ_κ that generates the same topology. In case μ is the finest uniformity, Lemma 3.5 gives a positive answer when $\lambda = \omega$. The answer is now know to be positive in other cases:

[J. Isbell, 1964] if μ has a base consisting of point finite coverings,
[G. Vidossich, 1969] if μ has a base consisting of σ-point finite coverings,
[A. Kucia, 1973] if we assume the generalized continuum hypothesis (an axiom that is independent of the axioms of ZFC).

12. [J. Pelant, 1975] There exists a model of ZFC (due to J. E. Baumgartner) in which there exists a uniform space whose countable uniform coverings do not form the basis for a uniformity. Consequently, this question is independent of the axioms of ZFC.

3.3 Transfinite Sequences and Topologies

In this section, the theory of transfinite sequences is presented for arbitrary topological spaces. Although this section is independent of the concept of a uniformity, it is important to the development because it shows that as long as we only consider topological properties (as opposed to uniform properties), transfinite sequences are entirely adequate for characterizing these properties. It is only when uniform properties are considered that the transfinite sequences may be inadequate. We will give examples in later chapters where transfinite sequences cannot be used in the same way as nets and filters to characterize certain uniform properties. This, of course, does not rule out using transfinite sequences in some other way to characterize them.

Also, we will show that if we are careful in selecting which uniformities we use to generate a given topology, we can often characterize the uniform properties in which we are interested in terms of transfinite sequences. For instance, the class of *relatively fine* uniformities for a space, that were introduced in the author's 1994 paper in the journal Questions & Answers in General Topology (Vol. 12) titled *Relatively Fine Spaces*, is an example of such a class of uniformities. It is interesting to note that the u, e and β uniformities all belong to the class of relatively fine uniformities. There are several cases (for example Theorems 3.1 - 3.3) where the behavior of transfinite sequences with respect to a uniformity can be used to characterize topological properties. But this is not the same as being able to characterize uniform properties. At the present time, it is an open problem as to what extent transfinite sequences can be used to characterize uniform properties.

As previously mentioned, the success of the theory of convergent sequences in metric spaces is due to the fact that both sequences and neighborhood bases have the same order structure. In fact, in metric spaces, each point has a countable well ordered neighborhood base such that the well ordering is identical with the partial ordering of set inclusion. In more general topological spaces, the existence of a well ordered neighborhood base such that the well ordering is identical to the partial ordering of set inclusion is the exception rather than the rule.

Fortunately we can use the *Neighborhood Principle* (an equivalent form of the Axiom of Choice) to obtain a replacement for this well ordered (by inclusion) neighborhood base. This will enable us to characterize the topology of a space in terms of transfinite sequences. This of course means that every topological property (at least in theory) can be characterized in terms of the behavior of transfinite sequences.

The Ordering Lemma (N. Howes, 1968) If $(P,\leq)$ is a partially ordered set then there exists a well ordered cofinal subset $(C,<)$ such that the well ordering $<$ is compatible with $\leq$ on C.

Proof: Let (P, $\leq$) be a partially ordered set and let $<$ be a well ordering for P. For each $p \in$ P we show there exists a set C(p) having the properties:

(a) C(p) $\subset \{x \in \text{P} \mid x < p$ or $x = p\}$ and $p \in$ C(p) if and only if p is not $\leq q$ for each $q \in$ C(p).
(b) If $q < p$, then C(q) $\subset$ C(p).
(c) If $x \in$ C(p) and $y \in$ P with $x \leq y$, then $x < y$.

Let S be the subset of P consisting of members p of P for which such a C(p) does not exist. Either S = $\varnothing$, in which case C(p) exists for each $p \in$ P, or S has a first element with respect to $<$. Suppose S has a first element, say a. Put K(a) $= \cup\{\text{C}(p) \mid p < a\}$. There are two cases to consider:

Case 1: $a \leq b$ for some $b \in$ K(a).
Case 2: a is not $\leq b$ for each $b \in$ K(a).

In case 1 put C(a) = K(a) and in case 2 put C(a) = K(a)$\cup\{a\}$. In either case, C(a) satisfies properties (a) and (b). To show C(a) also satisfies property (c), suppose $x \in$ C(a) and $y \in$ P with $x \leq y$.

In case 1, $x \in$ K(a) which implies $x \in$ C(p) for some $p < a$. Then C(p) exists and satisfies property (c), so $x < y$. Therefore, C(a) satisfies property (c). In case 2, if $x \neq a$ then $x \in$ K(a) and the proof of case 1 that $x < y$ applies. Therefore, let $x = a$. Then $a \leq y$. Suppose $y < x$. Then $y < a$ which implies $y \in$ K(a) by the assumption of case 2. But $y \in$ K(a) implies y does not belong C(y) and since C(y) satisfies property (a), $y \leq z$ for some $z \in$ C(y) $\subset$ K(a). Then $a \leq y \leq z \in$ K(a) which contradicts the assumption of case 2. Therefore, $x < y$ so C(a) satisfies property (c). Hence a does not belong to S so S = $\varnothing$ and C(p) exists for each $p \in$ P.

Now put C $= \cup\{\text{C}(p) \mid p \in \text{P}\}$ and let $x \in$ C and $y \in$ P with $x \leq y$. $x \in$ C implies $x \in$ C(p) for some $p \in$ P and hence $x < y$, showing $<$ to be compatible with $\leq$ on C. It remains to show that C is cofinal in P. Let $y \in$ P and suppose y does not belong to C. Then y does not belong to C(y) which implies there is a $z \in$ C(y) with $y \leq z$. But C(y) $\subset$ C, so C is cofinal in P. Consequently the Well Ordering Principle implies the Ordering Lemma. ∎

The Neighborhood Principle (N. Howes, 1968) Each point in a topological space has a well ordered neighborhood base such that the well ordering is compatible with the partial ordering of set inclusion.

Proof: Let X be a space and p a point of X. Let B be any neighborhood base for p and notice that B is partially ordered by set inclusion (i.e., U,V $\in B$ with U $\subset$ V implies V $\leq$ U). By The Ordering Lemma there exists a well ordered cofinal subset $(C, <)$ of $(B, \leq)$ such that $<$ is compatible with $\leq$ on C. It can be seen that

C is the desired well ordered neighborhood base for *p* such that the well ordering is compatible with the partial ordering of set inclusion. ▪

The Neighborhood Principle (as well as the Ordering Principle) can be shown to be equivalent to the Axiom of Choice as follows: let X be a set and let τ be the finite complement topology on X (i.e., open sets are complements of finite sets). Let $(B,<)$ be a well ordered neighborhood base for some $p \in$ X. For each $y \in$ X distinct from p let $\Phi(y) =$ N such that N is the first neighborhood in B with respect to $<$ with N $\subset$ X - $\{y\}$. Then Φ:(X - $\{y\}$) $\rightarrow B$ and Φ^{-1}(N) contains at most finitely many members. For each A $\in B$ with $\Phi^{-1}(A) \neq \emptyset$ we can well order the finite set $\Phi^{-1}(A)$ (without the use of the Well Ordering Principle) by some well ordering $<_A$. Define the well ordering $\prec$ on X - $\{p\}$ as follows: if $x,y \in$ X - $\{p\}$ put $x \prec y$

(a) if $\Phi(x) < \Phi(y)$ or
(b) if $\Phi(x) = \Phi(y)$ and $x <_A y$ where A $= \Phi(x)$.

We conclude X can be well ordered so that the Neighborhood Principle implies the Well Ordering Principle and consequently the Axiom of Choice.

PROPOSITION 3.1 Let X be a topological space. Then U $\subset$ X is open if and only if no transfinite sequence in X - U clusters to a point of U.

Proof: Clearly if U is open, no transfinite sequence in X - U can cluster to a point of U. Conversely, assume no transfinite sequence in X - U clusters to a point of U and suppose U is not open. Then there is a $p \in$ U each neighborhood of which meets X - U. By The the Neighborhood Principle there exists a well ordered neighborhood base B of p such that the well ordering is compatible with the partial ordering of set inclusion. Without loss of generality, we may assume B is order isomorphic with some limit ordinal γ. Let $B = \{V_\alpha \mid \alpha < \gamma\}$.

For each $\alpha < \gamma$ pick $x_\alpha \in V_\alpha \cap$(X - U). Then the transfinite sequence $\{x_\alpha\}$ is contained in X - U. Let $\beta < \gamma$ and put C $= \{\alpha < \gamma \mid x_\alpha \in V_\beta\}$ and suppose C is not cofinal in γ. Then there exists a $\delta < \gamma$ such that $\xi < \delta$ for each $\xi \in$ C. Consequently x_α is not contained in V_β for each $\alpha \geq \delta$. Since $\{V_\alpha\}$ is a neighborhood base for p there is a $\kappa < \gamma$ such that $V_\kappa \subset V_\beta \cap V_\delta$. Since $<$ is compatible with the partial ordering of set inclusion, we must have $\delta \leq \kappa$. But $x_\kappa \in V_\kappa \subset V_\beta$ which is a contradiction since $\delta \leq \kappa$. Therefore C must be cofinal in γ. Consequently $\{x_\alpha\}$ is frequently in V_β and since V_β was chosen arbitrarily, $\{x_\alpha\}$ must be frequently in each member of $\{V_\alpha\}$. But then $\{x_\alpha\}$ clusters to p which contradicts our original assumption. Therefore U must be open. ▪

COROLLARY 3.3 Let X be a topological space. Then F $\subset$ X is closed if and only if a transfinite sequence in F can only cluster to a point of F.

In a metric space, a point is a limit point of a set A if and only if there is a sequence in A - $\{p\}$ that converges to p. A similar situation holds in arbitrary topological spaces with respect to transfinite sequences.

PROPOSITION 3.2 Let X be a topological space and let A be a subset of X. A point p is a limit point of A if and only if there is a transfinite sequence in A - $\{p\}$ that clusters to p.

Proof: If $\{x_\alpha\}$ is a transfinite sequence in A - $\{p\}$ that clusters to p then for each neighborhood U of p, $\{x_\alpha\}$ is frequently in U which implies $A \cap U \neq \emptyset$ so that p is a limit point of A.

Conversely assume p is a limit point of A and let $B = \{U_\beta \mid \beta < \gamma\}$ be a well ordered neighborhood base for p such that the well ordering is compatible with the partial ordering of set inclusion. For each $U_\beta \in B$ pick $x_\beta \in U_\beta \cap (A - \{p\})$. Then $\{x_\beta\}$ is a transfinite sequence in A - $\{p\}$. Next let $U_\alpha \in B$ and put $C = \{\beta < \gamma \mid x_\beta \in U_\alpha\}$. An argument similar to the one in Proposition 3.1 shows that C is cofinal in B and hence $\{x_\beta\}$ is frequently in U_α. Therefore $\{x_\beta\}$ is a transfinite sequence in A - $\{p\}$ that clusters to p. ■

In a metric space, no sequence can converge to two distinct points, whereas with transfinite sequences it is clear that they can cluster to two or more distinct points. Moreover, the property that no sequence can converge to two distinct points in a metric space is equivalent to the Hausdorff property. The question then arises: How can the Hausdorff property be characterized by the behavior of transfinite sequences when they do not have unique cluster points? Actually, the case for arbitrary spaces is not as different as it might first appear. A transfinite sequence $\{x_\alpha\}$ will be said to cluster to two points p and q **simultaneously** if for each pair of neighborhoods U and V of p and q respectively, $\{x_\alpha\}$ is frequently in $U \cap V$. Then we have the following:

PROPOSITION 3.3 A space X is Hausdorff if and only if no transfinite sequence can cluster to two distinct points simultaneously.

Proof: Clearly if X is Hausdorff no transfinite sequence can cluster to two distinct points simultaneously. Conversely, assume no transfinite sequence can cluster to two distinct points simultaneously and suppose X is not Hausdorff. Then there exists two distinct points p and q such that for each pair U,V of open sets, $U \cap V \neq \emptyset$. Let B be a neighborhood base for p and N a neighborhood base for q. Put $P = B \times N$. Define the partial ordering $\leq$ on P as follows: if $U_1, U_2 \in B$ and $V_1, V_2 \in N$ with $U_1 \subset U_2$ and $V_1 \subset V_2$ then $(U_2, V_2) \leq (U_1, V_1)$. By the Ordering Lemma there is a well ordered cofinal subset (E, <) of (P, $\leq$) such that < is compatible with $\leq$ on E. For each $(U,V) \in E$ pick $\psi(U,V)$ in $U \cap V$.

Then $\psi: E \rightarrow X$ is a transfinite sequence that can be shown to cluster to p and q simultaneously. For this let U and V be neighborhoods of p and q

respectively. Then (U,V) ∈ P and consequently there is a $(U_1,V_1) \in$ E with (U,V) $\le (U_1,V_1)$. But then $\psi(U_1,V_1) \in U_1 \cap V_1 \subset U \cap V$. Moreover, if we let C = $\{(A, B) \in E \mid \psi(A,B) \in U_1 \cap V_1\}$ then C is cofinal in E. To see this let (A,B) ∈ E. Then $U_2 = A \cap U_1$ and $V_2 = B \cap V_1$ are neighborhoods of p and q respectively. Moreover, $U_2 \subset A$ and $V_2 \subset B$ implies (A, B) $\le (U_2,V_2)$ and hence (A, B) $< (U_2,V_2)$. Also, $U_2 \subset U_1$ and $V_2 \subset V_1$ implies $U_2 \cap V_2 \subset U_1 \cap V_1$ so $\psi(U_2,V_2) \in U_1 \cap V_1$ and hence $(U_2,V_2) \in$ C. Therefore ψ is frequently in $(U_1,V_1) \subset$ (U,V) and consequently clusters simultaneously to p and q which contradicts our original assumption. Therefore X must be Hausdorff after all. ■

Notice the reliance on the notation ψ:E → X to denote a transfinite sequence in the above proof as opposed to the usual $\{x_\alpha\}$. This was because members of the set E were of the form (U,V) which would cause the *subscript notation* to take the form $\{x_{(U,V)}\}$. In what follows, when the subscript notation for transfinite sequences becomes complex we will resort to the *function notation.*

Let γ be a limit ordinal and for each $\alpha < \gamma$ let $\phi_\alpha:\gamma_\alpha \to X$ be a transfinite sequence. Consider the γ_α's to be formally disjoint and put $\Gamma = \cup\gamma_\alpha$. Define the well ordering < on Γ as follows: if $\delta,\kappa \in \Gamma$ then $\delta < \kappa$

(1) if $\alpha < \beta$ in γ and $\delta \in \gamma_\alpha$ and $\kappa \in \gamma_\beta$ or
(2) if δ and κ both belong to the same γ_α and $\delta < \kappa$ in γ_α.

The ordering < on Γ is usually called the **lexicographic** ordering of Γ. Next we define the transfinite sequence $\Sigma:\Gamma \to X$ called the **sum** of the transfinite sequences ϕ_α as follows:

$$\Sigma(\delta) = \phi_\alpha(\delta) \text{ where } \delta \in \gamma_\alpha.$$

With these definitions we are in a position to characterize the various topologies a space may have in terms of the behavior of classes of transfinite sequences. To motivate the definition of these classes we first prove the following:

PROPOSITION 3.4 Let X be a space, $\phi:\gamma \to X$ a transfinite sequence and for each $\alpha < \gamma$ let $\phi_\alpha:\gamma_\alpha \to X$ be transfinite sequences and Σ the sum of the ϕ_α's. Then

(1) If ϕ is constant then ϕ clusters.
(2) If $\gamma = A \cup B$ and ϕ clusters to p then either ϕ_A or ϕ_B exists and clusters to p. If A is residual in γ then ϕ_A must cluster to p.
(3) If for each α, $\phi_\alpha \subset \{\phi(\beta) \mid \alpha \le \beta\}$ and ϕ_α clusters to p then ϕ clusters to p.
(4) If for each α, ϕ_α clusters to $\phi(\alpha)$ and ϕ clusters to p then Σ clusters to p.

Proof: The validity of statement (1) follows immediately from the definition of clustering and statement (2) is easily verified. To establish (3) we note that for each neighborhood U of p and each $\alpha < \gamma$ there is some $\phi_\alpha(x) \in$ U for some $x \in \gamma_\alpha$. But $\phi_\alpha \subset \{\phi(\beta) \mid \alpha \leq \beta\}$ implies $\phi_\alpha(x) = \phi(\delta)$ for some $\delta \geq \alpha$. But then $\phi(\delta) \in$ U and hence ϕ is frequently in U. Therefore ϕ clusters to p.

To verify (4) let $\beta \in \cup\{\gamma_\alpha \mid \alpha < \gamma\}$ and let U be an open neighborhood of p. Then $\beta \in \gamma_\lambda$ for some $\lambda < \gamma$. Since ϕ clusters to p there is a $\delta > \lambda$ in γ with $\phi(\delta) \in$ U. But since ϕ_δ clusters to $\phi(\delta)$ there is a $\kappa < \gamma_\delta$ with $\phi_\delta(\kappa) \in$ U. Now $\phi_\delta(\kappa) = \Sigma(\kappa)$ so $\Sigma(\kappa) \in$ U and $\beta < \kappa$ in the lexicographic ordering on $\cup\{\gamma_\alpha \mid \alpha < \gamma\}$. Consequently Σ is frequently in U and hence clusters to p. ■

We are now ready to define the classes of transfinite sequences that were mentioned above. Let X be a set and C a class of ordered pairs (ψ, p) where ψ is a transfinite sequence in X and $p \in$ X. We call C a **cluster class** on X if C satisfies the four statements of Proposition 3.4; i.e., if ϕ clusters to p can be replaced by $(\phi, p) \in C$ in statements (1) - (4) of Proposition 3.4 and ϕ does not cluster to p can be replaced by (ϕ, p) does not belong to C. Then we have:

THEOREM 3.4 Let C be a cluster class on a set X and for each A $\subset$ X let Cl(A) be the set of all $x \in$ X with $(\phi, x) \in C$ and $\phi \subset$ A. Then "Cl" is a closure operator on X and $(\phi, x) \in C$ if and only if ϕ clusters to x relative to the topology associated with Cl.

Proof: We first show that *Cl* is a closure operator. First note that when *Cl* is applied to the empty set $\varnothing$, the result is $Cl(\varnothing) = \varnothing$. Next let A $\subset$ X. For each $a \in$ A let γ_a be a limit ordinal and define $\psi_a\colon\gamma_a \to$ X by $\psi_a(e) = a$ for each $e \in \gamma_a$. By statement (1) of Proposition 3.4 (ψ_a, a) belongs to C and hence $a \in Cl$(A) for each $a \in$ A. Thus A $\subset Cl$(A).

To show $Cl(Cl$(A)) = Cl(A) suppose $a \in Cl(Cl$(A)) which implies the existence of a transfinite sequence $\psi\colon\gamma \to Cl$(A) with $(\psi, a) \in C$. For each $e \in \gamma$, $\psi(e) \in Cl$(A) which implies there is a transfinite sequence $\psi_e\colon\gamma_e \to$ A such that $(\psi_e, \psi(e)) \in C$. But then the sum Σ of the transfinite sequences $\{\psi_e\}$ and the point a form a pair $(\Sigma, a) \in C$ by statement (4) of Proposition 3.4. Therefore $a \in Cl$(A) since $\Sigma \subset$ A. Consequently $Cl(Cl$(A)) $\subset Cl$(A) and hence $Cl(Cl$(A)) = Cl(A).

To show Cl(A$\cup$B) = Cl(A)$\cup Cl$(B) first note that $a \in Cl$(A) implies there is a pair $(\psi, a) \in C$ with $\psi \subset$ A and hence $\psi \subset$ A$\cup$B so $a \in Cl$(A$\cup$B). Similarly $a \in Cl$(B) implies $a \in Cl$(A$\cup$B) so Cl(A)$\cup Cl$(B) $\subset Cl$(A$\cup$B). Finally, $p \in Cl$(A$\cup$B) implies the existence of a pair $(\psi, p) \in C$ with $\psi \subset$ A$\cup$B. Let M = $\gamma \cap \psi^{-1}$(A) and N = $\gamma \cap \psi^{-1}$(B) where $\psi\colon\gamma \to$ X. Then γ = M$\cup$N and by statement (2) of Proposition 3.4, either (ψ_M, p) or $(\psi_N, p) \in C$. Since $\psi_M \subset$ A and $\psi_N \subset$ B either $p \in Cl$(A) or $p \in Cl$(B). Hence $p \in Cl$(A)$\cup Cl$(B). We conclude that Cl(A$\cup$B) = Cl(A)$\cup Cl$(B).

It remains to show that $(\psi, p) \in C$ if and only if ψ clusters to a relative to the topology associated with Cl which will now be referred to as τ. First suppose $(\psi, a) \in C$ but ψ does not cluster to a relative to τ. Then there is an open neighborhood U of a such that ψ is not frequently in U which implies a residual $R \subset \gamma$ with $\psi(R) \subset X - U$ where $\psi:\gamma \to X$. By statement (2) of Proposition 3.4, $(\psi_R, a) \in C$ which implies $a \in Cl(X - U)$. But this is a contradiction since U is open. Therefore ψ clusters to a relative to τ.

Conversely, suppose $\psi:\gamma \to X$ clusters to a. Then $a \in Cl(M_e)$ for each $e \in \gamma$ where $M_e = \{\psi(\beta) | e \leq \beta\}$. Therefore there is a transfinite sequence $\psi_e:\gamma_e \to M_e$ with $(\psi_e, a) \in C$ for each $e \in \gamma$. By statement (3) of Proposition 3.4 we must then have $(\psi, a) \in C$. ∎

Proposition 3.4 and Theorem 3.5 set up a one-to-one correspondence between the various topologies a set can have and the cluster classes on the set. It is clear from the definition of clustering that if C_1 and C_2 are two cluster classes and τ_1 and τ_2 are the associated topologies, that $C_1 \subset C_2$ if and only if $\tau_2 \subset \tau_1$. Moreover, it is interesting to note that if $(C_1 \cap C_2)$ denotes the smallest cluster class that is larger than each of C_1 and C_2 then $(C_1 \cap C_2)$ is the cluster class associated with $\tau_1 \cap \tau_2$.

We conclude this chapter with a characterization of continuous functions in terms of transfinite sequences. Various other interesting mappings (e.g., open mappings, closed mappings, quotient mappings, etc.) can be characterized in a similar manner but this will be left for the exercises.

PROPOSITION 3.5 Let $f:X \to Y$ be a function from a space X into a space Y. Then f is continuous if and only if for each transfinite sequence $\{x_\alpha\}$ in X that clusters to some $p \in X$, $\{f(x_\alpha)\}$ clusters to f(p).

Proof: Assume f is continuous and suppose $\{x_\alpha\}$ is a transfinite sequence in X that clusters to some point $p \in X$. Let U be a neighborhood of $f(p)$ in Y and pick a neighborhood V of p such that $f(V) \subset U$. Then $\{x_\alpha\}$ frequently in V implies $\{f(x_\alpha)\}$ is frequently in $f(V) \subset U$, so $\{f(x_\alpha)\}$ clusters to $f(p)$.

Conversely, assume that for each transfinite sequence $\{x_\alpha\}$ in X that clusters to some point p in X, $\{f(x_\alpha)\}$ clusters to $f(p)$. Suppose f is not continuous. Then there is a $p \in X$ and a neighborhood U of $f(p)$ such that $f(V)$ is not contained in U for each neighborhood V of p. By Theorem 3.4 there exists a well ordered neighborhood base $\{V_\alpha | \alpha < \gamma\}$ of p such that $<$ is compatible with the partial ordering of set inclusion. For each $\alpha \in A$ pick $x_\alpha \in V_\alpha$ such that $f(x_\alpha)$ is not contained in U. Then $\{x_\alpha\}$ clusters to p but $\{f(x_\alpha)\}$ does not cluster to $f(p)$ which is a contradiction. We conclude that f must be continuous. ∎

EXERCISES

1. Show that X is T_1 if and only if for each pair of distinct points in X there are two transfinite sequences clustering to the two points respectively but neither clustering to the other point.

2. A function is said to be **open** if the image of each open set is again an open set. Show that an onto function f is open if and only if for each transfinite sequence $\{y_\beta\}$ in Y that clusters to some $p \in$ Y and for each $q \in f^{-1}(p)$ there is a transfinite sequence $\{x_\alpha\}$ in $\cup\{f^{-1}(y_\beta)\}$ that clusters to q.

3. A function is said to be **closed** if the image of each closed set is a closed set. Show that a function f is closed if and only if whenever a transfinite sequence $\{y_\beta\}$ clusters to some $p \in$ Y there is a transfinite sequence $\{x_\alpha\}$ that is a subset of $f^{-1}(\{y_\beta\})$ that clusters to $f^{-1}(p)$.

4. A function is said to be a **quotient** if whenever the inverse image of a set is open, then the set itself must be open. Show that a function f is a quotient if and only if for each transfinite sequence $\{y_\beta\}$ clustering to some $p \in$ Y there is a transfinite sequence $\{x_\alpha\}$ contained in the inverse image of $\{y_\beta\}$ under f that is frequently in each open inverse image of an open set in Y that contains $f^{-1}(p)$.

5. Show that Corollary 3.2 still holds when the assumption of complete regularity is removed from the hypothesis; i.e., show that any space is compact if and only if each transfinite sequence clusters.

Chapter 4

COMPLETENESS, COFINAL COMPLETENESS AND UNIFORM PARACOMPACTNESS

4.1 Introduction

In 1915, A paper by E. H. Moore appeared in the Proceedings of the National Academy of Science U.S.A. titled *Definition of limit in general integral analysis*. This study of unordered summability of sequences led to a theory of convergence by Moore and H. L. Smith titled *A general theory of limits* which appeared in the American Journal of Mathematics in 1922. In 1937, G. Birkhoff applied the Moore-Smith theory to general topology in an article titled *Moore-Smith convergence in general topology*, which appeared in the Annals of Mathematics, No. 38, pp. 39-56. In 1940, J. W. Tukey made extensive use of the theory in his monograph titled *Convergence and uniformity in topology* published in the Annals of Mathematics Studies series. Tukey worked with objects that were generalizations of sequences that he referred to as **phalanxes.** They were a special case of the objects that are usually called **nets** today.

An equivalent theory of convergence using objects called **filters** emerged in the thirties from the Bourbaki group in France. The theory of filters is the convergence theory of choice for many topologers. There are many things to recommend it but it is also very awkward to use in certain situations. For instance, in the treatment of hyperspaces (Chapter 5), the objects of the hyperspace are subsets of the original space. Filters, which are themselves collections of subsets of a space, become awkward to construct in the hyperspace because they are collections of collections of subsets.

For some proofs, filters carry along too much baggage. In proofs about the completions of uniform spaces it is sometimes desirable to pick a convergence object from the original space that is arbitrarily close to a convergence object in the completion. Then conclusions are made about the convergence object in the original space and these conclusions are then shown to hold for the convergence object in the completion due to its proximity to the object in the original space. A natural way to attempt this with filters is to restrict each subset in the completion filter to the original space, make conclusions about the restricted filter, then take the closures of the members of the restricted filter to get back to the completion. The problem here is that taking the closure picks up too many points. Since the original filter in the completion is not always obtained, it may not be possible to draw conclusions about the original filter in the completion.

Surely with time, one should be able to find filter type proofs for the theorems we will be presenting in the sequel, but this will not be our viewpoint. Instead, in some of the areas where the filter type proofs are especially nice, the recasting of the results into filter terminology will be suggested as exercises.

4.2 Nets

A non-void set D is said to be **directed** by the binary relation $\leq$ provided that the following conditions hold:

(1) if m,n and $d \in \mathrm{D}$ with $m \leq n$ and $n \leq d$ then $m \leq d$,
(2) $d \leq d$ for each d in D and
(3) if $m,n \in \mathrm{D}$ then there exists a d in D with $m \leq d$ and $n \leq d$.

Clearly a directed set is a particular type of partially ordered set. Condition (3) in the definition of a directed set is what sets the directed set apart from an ordinary partially ordered set. Also, it should be noticed that well ordered sets are necessarily directed sets. Some useful examples of directed sets that are not necessarily well ordered are given below.

Let X be a topological space and let B be a local neighborhood base for a point $p \in \mathrm{X}$. Define the relation $\leq$ on B as follows: $\mathrm{U} \leq \mathrm{V}$ if $\mathrm{U},\mathrm{V} \in B$ and $\mathrm{V} \subset \mathrm{U}$. Clearly $(B, \leq)$ satisfies conditions (1) and (2) above so $(B, \leq)$ is a partially ordered set. To show $(B, \leq)$ also satisfies (3) assume $\mathrm{U},\mathrm{V} \in B$. Then $\mathrm{W} = \mathrm{U} \cap \mathrm{V}$ belongs to B and $\mathrm{W} \subset \mathrm{U}$ and $\mathrm{W} \subset \mathrm{V}$. Thus $\mathrm{U} \leq \mathrm{W}$ and $\mathrm{V} \leq \mathrm{W}$, so $(B, \leq)$ satisfies (3).

Another directed set that we will employ frequently is the uniformity itself. Let (X, μ) be a uniform space. The relations $<$ (refinement) and $<^*$ (star refinement) have already been defined for μ (see Section 2.1). Clearly $(\mu, <)$ and $(\mu, <^*)$ satisfy conditions (1) and (2) defined above so that both are partially ordered sets. That $<$ also satisfies condition (3) above follows from condition (1) of Section 2.1.

At this point we are careful to remind the reader that we consider these orderings as *"proceeding toward the left"* in the sense that in proving condition (3) for $<^*$ we will show that if $\mathcal{U},\mathcal{V} \in \mu$ then there exists a $\mathcal{W} \in \mu$ such that $\mathcal{W}^* < \mathcal{U}$ and $\mathcal{W}^* < \mathcal{V}$. By (1) of Section 2.1 there exists a $\mathcal{Z} \in \mu$ such that $\mathcal{Z} < \mathcal{U}$ and $\mathcal{Z} < \mathcal{V}$. By (3) there is a $\mathcal{W}$ in μ with $\mathcal{W} <^* \mathcal{Z}$. Therefore $\mathcal{W} <^* \mathcal{U}$ and $\mathcal{W} <^* \mathcal{V}$. Consequently both $(\mu, <)$ and $(\mu, <^*)$ are directed sets.

A subset $\mathrm{R} \neq \varnothing$ of a directed set $(\mathrm{D}, \leq)$ is said to be **residual** if whenever $m,n \in \mathrm{D}$ with $m \in \mathrm{R}$ and $m \leq n$ then $n \in \mathrm{R}$. If $\mathrm{C} \subset \mathrm{D}$ such that whenever $m \in \mathrm{D}$ there is an $n \in \mathrm{C}$ with $m \leq n$ we say C is **cofinal** in D. A **net** is a function ψ:D

$\to$ X from a directed set D into a space X. In the event D is well ordered then ψ is simply a transfinite sequence with which we are already familiar. For each $\alpha \in$ D let $x_\alpha = \psi(\alpha)$. We will often identify a net ψ with its range $\psi(D) = \{x_\alpha \mid \alpha \in D\}$. If the set D is understood we simply write $\{x_\alpha\}$. We say a net $\{x_\alpha\}$ is **frequently** in U $\subset$ X if there is a cofinal C $\subset$ D with $\{x_\beta \mid \beta \in C\} \subset U$. We say $\{x_\alpha\}$ is **eventually** in U if there is a residual R $\subset$ D with $\{x_\gamma \mid \gamma \in R\} \subset U$. $\{x_\alpha\}$ is said to **converge** to a point p in X if it is eventually in each neighborhood of p and to **cluster** to p if it is frequently in each neighborhood of p.

PROPOSITION 4.1 A subset U of a space X is open if and only if no net in X - U converges to a point of U.

Proof: Assume U is open in X and let $\{x_\alpha\}$ be a net that converges to some $p \in$ U. From the definition of convergence $\{x_\alpha\}$ must eventually be in U. But then $\{x_\alpha\}$ cannot lie entirely in X - U. Conversely, assume no net in X - U converges to a point of U and suppose U is not open. Then there is a $p \in$ U every neighborhood of which meets X - U. Let D be a local basis for p and let $\leq$ be the partial ordering of set inclusion on D. We have already seen that D is directed with respect to this ordering. For each V $\in$ D pick $x_V \in V \cap (X - U)$. Then $\{x_V\}$ is a net in X - U. It is easily shown that $\{x_V\}$ converges to p. Indeed, let W $\in$ D and put R(W) = $\{V \in D \mid V \subset W\}$. Then R(W) is residual in D and $x_V \in$ W for each V $\in$ R(W). But this is a contradiction since $\{x_V\}$ lies in X - U. We conclude that U must be open. ∎

PROPOSITION 4.2 A point p of a space X belongs to the closure of E $\subset$ X if and only if there is a net in E that converges to p.

Proof: Assume $p \in Cl(E)$ and let D be a local basis for p. Then D is directed by the partial ordering of set inclusion. For each $V \in$ D pick $x_V \in V \cap E$. Then $\{v_V\}$ is a net in E. An argument similar to the one in Proposition 4.1 shows that $\{x_V\}$ converges to p. Conversely, assume $\{x_\alpha\}$ is a net in E converging to a point $p \in$ X. Let U be an open set containing p. Then $\{x_\alpha\}$ is eventually in U which implies $U \cap E \neq \varnothing$. Thus $p \in Cl(E)$. ∎

PROPOSITION 4.3 A point p of a space X is a limit point of E $\subset$ X if and only if there is a net in E - $\{p\}$ that converges to p.

Proof: Let p be a limit point of E $\subset$ X and let D be a local basis for p. For each V $\in$ D pick $v_V \in V \cap (E - \{p\})$. Then $\{x_V\}$ is a net in E - $\{p\}$ converging to p. Conversely, if $\{x_\alpha\}$ is a net in E - $\{p\}$ converging to $p \in$ X then for each open set U containing p, $\{x_\alpha\}$ is eventually in U. But then $U \cap (E - \{p\}) \neq \varnothing$ so p is a limit point of E. ∎

PROPOSITION 4.4 A space X is Hausdorff if and only if each net in X converges to at most one point.

Proof: Let X be a Hausdorff space and assume p and q are distinct points of X. Since X is Hausdorff, there are disjoint open sets U and V such that $p \in U$ and $q \in V$. If $\{x_\alpha\}$ is a net in X that converges to p then $\{x_\alpha\}$ is eventually in U so it cannot eventually be in V. Thus $\{x_\alpha\}$ cannot converge to q. We conclude that a net can converge to at most one point.

Conversely, assume each net in X can converge to at most one point and suppose X is not Hausdorff. Then there exists two distinct points p and q such that for each pair of neighborhoods U of p and V of q, $U \cap V \neq \emptyset$. Let $B(p)$ and $B(q)$ be local bases for p and q respectively and put $D = B(p) \times B(q)$. Define $\leq$ on D as follows: for each $a, c \in D$ where $a = (A, B)$ and $c = (U, V)$, put $a \leq c$ if $U \subset A$ and $V \subset B$. Then $(D, \leq)$ is a directed set. For each $c = (U,V) \in D$ pick $x_c \in U \cap V$. Then $\{x_c\}$ is a net in X. To show that $\{x_c\}$ converges to both p and q let $U \in B(p)$ and $V \in B(q)$. Put $d = (U,V)$ and let $R(d) = \{a \in D \mid d \leq a\}$. Then $R(d)$ is residual in D. If $a \in R(d)$ then $a = (A, B)$ for some $A \in B(p)$ and $B \in B(q)$ such that $A \subset U$ and $B \subset V$. Moreover, $x_a \in A \cap B \subset U \cap V$. Therefore, for each $a \in R(d)$, $x_a \in U$ and $x_a \in V$, so $\{x_a\}$ is eventually in both U and V. We conclude $\{x_a\}$ converges to both p and q which is a contradiction. Hence X must be Hausdorff. ∎

A net $\phi: E \to X$ is called a **subnet** of the net $\psi: D \to X$ if there exists a function $\lambda: E \to D$ such that:

(1) $\phi = \psi \circledcirc \lambda$ and
(2) for each $d \in D$ there is an $e \in E$ such that if $e \leq c$ then $d \leq \lambda(c)$.

Such a function λ is called a **cofinal function.** It is easily shown that if C is a cofinal subset of D that the identity function $i: C \to D$ is a cofinal function and hence a net $\psi: D \to X$ restricted to a cofinal subset C is a subnet of ψ. However, in the theory of convergence of nets, we will see that such subnets are usually not very useful. Normally we will be interested in subnets whose domains are not subdomains of the domain of the original net.

It should be noted that a subnet is a more complex object than a subsequence of a transfinite sequence and that a subsequence of a transfinite sequence is a subnet when the transfinite sequence is considered as a net. The principle advantage of nets is that one works with convergence rather than clustering (compare Proposition 4.4 with Proposition 3.3).

PROPOSITION 4.5 If B is a family of subsets of the space X with the property that the intersection of two members of B contains a member of B and if the net $\{x_\alpha\}$ is frequently in each member of B then $\{x_\alpha\}$ has a subnet that is eventually in each member of B.

Proof: Let $\{x_\alpha | \alpha \in A\}$ be the net in the hypothesis above. Since the intersection of any two members of B contains a member of B, B is directed by set inclusion. Then

$$E = \{(\alpha, F) \in A \times B | x_\alpha \in F\}$$

is directed by the ordering $\leq$ defined as follows: if (α_1, F_1) and $(\alpha_2, F_2) \in E$ then $(\alpha_2, F_2) \leq (\alpha_1, F_1)$ if $\alpha_2 \leq \alpha_1$ and $F_1 \subset F_2$. Now define a function $\lambda: E \to A$ by $\lambda(\alpha, F) = \alpha$ for each element (α, F) of E. Clearly λ is cofinal so the net $\{y_e | e \in E\}$ defined by $y_e = x_{\lambda(e)}$ for each $e \in E$ is a subnet of $\{x_\alpha\}$.

Next let $H \in B$. Since $\{x_\alpha\}$ is frequently in H, there exists a $\beta \in A$ such that $x_\beta \in H$. Hence $e' = (\beta, H) \in E$. Let $e = (\alpha, F)$ be an arbitrary member of E such that $e' \leq e$. Then $\beta \leq \alpha$ and $F \subset H$ so

$$y_e = x_{\lambda(\alpha, F)} = x_\alpha \in F \subset H.$$

Thus $\{x_\alpha\}$ is eventually in H. ∎

PROPOSITION 4.6 A net clusters to a point if and only if it has a subnet that converges to the point.

Proof: Let p be a cluster point of the net $\{x_\alpha | \alpha \in A\}$ and let B be a local basis for p. Then the intersection of any two members of B contains a member of B and $\{x_\alpha\}$ is frequently in each member of B. Hence by Proposition 4.5, there exists a subnet $\{y_\beta\}$ of $\{x_\alpha\}$ that is eventually in each member of B. Hence $\{y_\beta\}$ converges to p.

Conversely, assume that the net $\{x_\alpha\}$ has a subnet $\{y_\beta\}$ that converges to p. Let D be the domain of $\{y_\beta\}$ and λ the cofinal function from D to A that defines $\{y_\beta\}$. Let U be a neighborhood of p and let $\delta \in A$. Since λ is cofinal there is an element $\varepsilon \in D$ such that $\lambda(\beta) \geq \delta$ for each $\beta \geq \varepsilon$. Since $\{y_\beta\}$ converges to p there is an element $\varepsilon' \geq \varepsilon$ with $y_{\varepsilon'} \in U$. Now let $\delta' = \lambda(\varepsilon')$. Then we have $\delta' \geq \delta$ and $x_{\delta'} = x_{\lambda(\varepsilon')} = y_{\varepsilon'} \in U$. Hence $\{x_\alpha\}$ clusters to p. ∎

PROPOSITION 4.7 A space is compact if and only if each net has a convergent subnet.

Proof: By Proposition 4.6 it suffices to show that a space is compact if and only if each net clusters. For this let $x = \{x_\alpha | \alpha \in A\}$ be a net in the compact space X. For each $\alpha \in A$ let M_α be set set of all x_β such that $\beta \geq \alpha$. Since A is directed by $\geq$, $\{M_\alpha | \alpha \in A\}$ has the finite intersection property so the family $\{Cl(M_\alpha) | \alpha \in A\}$ also has the finite intersection property. Since X is compact, by Proposition 0.35, there is a $p \in X$ which belongs to $Cl(M_\alpha)$ for each $\alpha \in A$. We will show that x clusters to p. For this let U be a neighborhood of p and let $\alpha \in A$. Since $p \in Cl(M_\alpha)$, it follows that $M_\alpha \cap U \neq \varnothing$. Hence there is a $\beta \in A$

with $\beta \geq \alpha$ and $x_\beta \in$ U. This implies x is frequently in U. We conclude x clusters to p.

Conversely, assume every net in X clusters. Let F be a family of closed sets in X with the finite intersection property. Let G denote the family of finite intersections of members of F. Then G also has the finite intersection property. Since $F \subset G$, it suffices to show that the intersection of all members of G is nonempty to show that X is compact. Now G is directed by set inclusion $\subset$. For each F $\in G$ pick $x_F \in$ F. Then the assignment F $\rightarrow x_F$ defines a net $x = \{x_F\}$ in X. By hypothesis, x clusters to some $p \in$ X. Let H, K $\in G$ such that K $\subset$ H. Then $x_K \in$ K $\subset$ H, so x is eventually in the closed set H. By Propositions 4.1 and 4.6, $p \in$ H. Hence $p \in$ H for each H $\in G$. We conclude that X is compact. ∎

PROPOSITION 4.8 A function f:X → Y is continuous at p ∈ X if and only if for each net $\{x_\alpha\}$ in X converging to p, the net $\{f(x_\alpha)\}$ in Y converges to f(p).

Proof: Suppose f:X → Y is continuous at p and let V be a neighborhood of $f(p)$ in Y. By definition of continuity at p, there is a neighborhood U of p in X such that f(U) $\subset$ V. Since $\{x_\alpha\}$ converges to p, there is a residual R in the domain of $\{x_\alpha\}$ such that $x_\beta \in$ U for each $\beta \in$ R. Thus $f(x_\beta) \in f$(U) $\subset$ V for each $\beta \in$ R, so $\{f(x_\alpha)\}$ converges to $f(p)$.

Conversely, assume f is not continuous at p. Then there is an open neighborhood V of $f(p)$ in Y such that every neighborhood of p meets f^{-1}(Y-V). Let B be a local basis for p. Then B is directed by set inclusion. For each U $\in B$ pick x_U in U$\cap f^{-1}$(Y - V). The assignment U $\rightarrow x_U$ defines a net $\{x_U\}$ in X that converges to p. Now the composition $\{f(x_U)\}$ is a net in Y - V. Since V is open and $f(p) \in$ V, $\{f(x_U)\}$ cannot converge to $f(p)$. ∎

In the last chapter, cluster classes of transfinite sequences were discussed and it was shown that the various topologies a space can have correspond to the various cluster classes of transfinite sequences in the space. A similar situation holds for nets, only now it is possible to use convergence instead of clustering. In order to define the concept of a *convergence class*, it is first necessary to have a result on *iterated limits*. Iterated limits are usually first encountered in calculus where the limits are the limits of ordinary sequences. Iterated limits of nets can be defined in an analogous manner as in what follows.

If (D,$<_D$) and (E,$<_E$) are directed sets, then the Cartesian product D × E is directed by < where < is defined as follows: $(d,e) < (a,b)$ if $d <_D a$ and $e <_E b$. The ordering < is then called the **product ordering**. If we have a family of directed sets (D_α,$<_\alpha$) for each $\alpha \in$ A, the Cartesian product ΠD_α is the set of all functions f on A such that $f_\alpha = f(\alpha)$ is a member of D_α for each $\alpha \in$ A. The product ordering on ΠD_α is then defined as follows: if d and $e \in \Pi D_\alpha$ then d <

e if $d_\alpha <_\alpha e_\alpha$ for each $\alpha \in$ A. It is shown that the Cartesian product under the product ordering is indeed a directed set.

Next, consider the case where D is a directed set and for each $\delta \in$ D, E_δ is another directed set. Let $\Sigma = \{(\delta,\varepsilon) \mid \delta \in$ D and $\varepsilon \in E_\delta\}$. Consider a function $\psi:\Sigma \to$ X where X is a topological space. Then for each $\delta \in$ D it is possible that the net $\psi_\delta = \{\psi(\delta,\varepsilon) \mid \varepsilon \in E_\delta\}$ converges to some point $p_\delta \in$ X. We also denote p_δ by $lim_\alpha\psi(\delta,\alpha)$ and say that the limit of ψ_δ exists and equals p_δ. Furthermore, it is possible that the net $\phi = \{p_\delta \mid \delta \in$ D$\}$ also converges to some point $p \in$ X. In this case we say that p is the **iterated limit** $lim_\delta lim_\alpha\psi(\delta,\alpha)$ of ψ with respect to δ and α. Let P = D $\times \Pi\{E_\delta \mid \delta \in$ D$\}$.

THEOREM 4.1 Let D be a directed set and for each $\delta \in D$ let E_δ be another directed set. Let Σ and P be defined as above. Then there exists a net $\lambda:P \to \Sigma$ such that for any function $f:\Sigma \to X$ for which the iterated limit $lim_\delta lim_\alpha f(\delta,\alpha)$ exists, f © λ converges to the iterated limit.

Proof: Define λ:P $\to \Sigma$ as follows: for each $(\delta, d) \in$ P put $\lambda(\delta, d) = (\delta, d(\delta))$. Suppose $lim_\delta lim_\alpha f(\delta,\alpha) = q$ and U is an open neighborhood of q. We must find a member (δ, d) of P such that if $(\delta, d) < (\beta, g)$ then f © $\lambda(\beta, g) \in$ U. Pick $\gamma \in D$ such that $lim_\beta f(\beta,\nu) \in$ U for each β following γ and then, for each such β pick $d(\beta) \in E_\beta$ such that $f(\beta,\nu) \in$ U for all ν following $d(\beta)$ in E_β. If β is a member of D which does not follow γ let $d(\beta)$ be an arbitrary member of E_β. If $(\beta, g) > (\gamma, d)$, then $\beta > \gamma$, hence $lim_\beta f(\beta,\nu) \in$ U, and since $g(\beta) > d(\beta)$ we have f © $\lambda(\beta,g) = f(\beta, g(\beta)) \in$ U. We conclude that f © λ converges to q. ∎

Let C be a collection of pairs (ϕ, p) of nets ϕ in the space X and points $p \in$ X. We say C is a **convergence class** for X if it satisfies the following conditions:

(1) if ϕ is a constant net such that $\phi(\alpha) = p$ for each α then $(\phi, p) \in C$,
(2) if $(\phi, p) \in C$ then so does (ψ, p) for each subnet ψ of ϕ,
(3) if (ϕ, p) does not belong to C then there is a subnet ψ of ϕ such that (ξ, p) does not belong to C for each subnet ξ of ψ and
(4) let D be a directed set and for each $\delta \in$ D let E_δ be another directed set. Let Σ and P and λ be defined as in Theorem 4.1. Then for each function $f:\Sigma \to$ X for which $lim_\delta lim_\alpha f(\delta,\alpha)$ exists, $(f$ © $\lambda, p) \in C$.

THEOREM 4.2 Let C be a convergence class for a set X and for each $A \subset X$ let Cl(A) be the set of all $p \in X$ with $(\psi, p) \in C$ and $\psi \subset A$. Then "Cl" is a closure operator on X and $(\psi, p) \in C$ if and only if ψ converges to p relative to the topology associated with Cl.

The proof of Theorem 4.2 is left as an exercise (see Exercise 2). A net in a space X is said to be a **universal net** if for each A $\subset$ X the net is eventually in A or eventually in X - A. Universal nets have the property that if they are frequently in a set that they must eventually be in the set. Consequently, universal nets converge to each of their cluster points.

PROPOSITION 4.9 Every net has a universal subnet.

Proof: Let $x = \{x_\alpha \mid \alpha \in D\}$ be a net in X and let $\Phi = \{F \subset X \mid x$ is eventually in $F\}$. Clearly if A, B $\in \Phi$ then $A \cap B \neq \emptyset$. Consequently Φ has the two properties:

(1) x is frequently in each member of Φ and
(2) Φ has the finite intersection property.

Let S be the set of all collections of sets in X that contain Φ and have these two properties. Then S is ordered by set inclusion $\subset$. If $\{\Phi_\alpha\}$ is a chain in S with respect to $\subset$ then it is easily shown that $\cup\Phi_\alpha$ also has properties (1) and (2) above. By Zorn's Lemma (Theorem 0.1.(4)), there is a maximal collection Ω containing Φ and having properties (1) and (2).

Let $A \subset X$ and assume A does not belong to Ω. Then either x is eventually in X - A or there is a $B \in \Omega$ such that $A \cap B = \emptyset$. If x is eventually in X - A then $X - A \in \Phi \subset \Omega$. Therefore suppose there is a $B \in \Omega$ such that $A \cap B = \emptyset$. Then x must be frequently in X - A or else it could not be frequently in B and $B \subset X - A$. Therefore, we can append X - A to Ω to get a larger collection Ω' that also has properties (1) and (2). But since Ω is maximal with respect to properties (1) and (2) above, $\Omega' = \Omega$ so $(X - A) \in \Omega$.

Since x is frequently in each member of Ω and Ω has the finite intersection property, by Proposition 4.5 there is a subnet y of x that is eventually in each member of Ω. Let $A \subset X$. If $A \in \Omega$ then y is eventually in A. If A is not contained in Ω then, as shown above, $(X - A) \in \Omega$ so y is eventually in X - A. Consequently y is a universal subnet of x. ∎

THEOREM 4.3 A space is compact if and only if each universal net converges.

Proof: Let X be compact and let let x be a universal net in X. By Propositions 4.6 and 4.7, x must cluster. But a universal net that clusters must also converge. Hence x converges. Therefore compactness implies that each universal net converges. Conversely, if every universal net in X converges, then by Proposition 4.9, every net has a universal subnet so every net in X has a convergent subnet. Then by Proposition 4.7, X must be compact. ∎

Let (X, μ) be a uniform space. A net $\{x_\alpha\}$ in X is said to be **Cauchy** if it is eventually in some sphere of radius $\mathcal{U}$ for each $\mathcal{U} \in \mu$. $\{x_\alpha\}$ is said to be

cofinally Cauchy if it is frequently in some sphere of radius $\mathcal{U}$ for each $\mathcal{U} \in \mu$. The proofs of the following two propositions are left as exercises (Exercise 3).

PROPOSITION 4.10 A net $\{x_\alpha\}$ in a uniform space (X, μ) is Cauchy if and only if for each $\mathcal{U} \in \mu$ there is a $U \in \mathcal{U}$ such that $\{x_\alpha\}$ is eventually in U.

PROPOSITION 4.11 A net $\{x_\alpha\}$ in (X, μ) is cofinally Cauchy if and only if for each $\mathcal{U} \in \mu$ there is a $U \in \mathcal{U}$ such that $\{x_\alpha\}$ is frequently in U.

EXERCISES

1. Let $\{x_\alpha \mid \alpha \in D\}$ be a net in the pseudo-metric space (X, d). Show that $\{x_\alpha\}$ converges to $p \in X$ if and only if the net $\{y_\alpha \mid \alpha \in D\}$, defined by $y_\alpha = d(p, x_\alpha)$ for each $\alpha \in D$, converges to zero.

2. Prove Theorem 4.2 (see Theorem 3.4).

3. Prove Propositions 4.10 and 4.11.

4. Let $\{x_\alpha \mid \alpha \in D\}$ be a net in **R** (the reals). If $\alpha < \beta$ in D implies $x_\alpha \geq X_\beta$ ($x_\alpha \leq x_\beta$), then $\{x_\alpha\}$ is said to be **monotone increasing (decreasing)**. Show that a monotone increasing net that is bounded above, or a monotone decreasing net that is bounded below converges.

FILTERS

5. A **filter** on a set X is a collection F of subsets of X satisfying the following properties: (1) F is closed under finite intersections, (2) the empty set does not belong to F, and (3) every subset of X containing a member of F belongs to F. Show that the following three examples of filters satisfy properties (1) through (3) above.

The Neighborhood Filter: Let X be a space and let $p \in X$. Let F denote the neighborhood base at p.

The Filter Associated with a Net: Let $\{x_\alpha \mid \alpha \in D\}$ be a net. Put $F = \{F \subset X \mid x_\beta \in F$ for each $\beta \in R$ for some residual $R \subset D\}$.

The Intersection Filter: Let $\{F_\alpha \mid \alpha \in A\}$ be a non-empty family of filters on X. Then put $F = \cap F_\alpha$.

6. A filter is said to **converge** to a point $p \in X$ if each neighborhood of p belongs to F. F is said to **cluster** to p if every neighborhood of p meets every member of F. A **filter base** is a collection B of subsets of X satisfying

properties (1) and (2) in the definition of a filter above. Clearly, the collection F of all subsets of X containing a member of B is a filter called the filter **generated** by B. The filter base B is said to converge to $p \in$ X if the filter generated by B converges to p and to cluster to p if the filter generated by B clusters to p. Show the following:

(a) $U \subset X$ is open if and only if no filter base in X - U converges to a point of U.
(b) $p \in Cl(E)$ if and only if there is a filter base in E that converges to p.
(c) p is a limit point of $E \subset X$ if and only if there is a filter base in E - $\{p\}$ that converges to p.

7. Show that a space is Hausdorff if and only if each filter converges to at most one point.

8. Filters can be compared in the following way: Let F and G be two filters on X. F is said to be **finer** than G and G is said to be **coarser** than F if $G \subset F$. In addition if $F \neq G$ then F is said to be **strictly finer** than G and G is **strictly coarser** than F. Two filters are said to be **comparable** if one is finer than the other. Show the following:

(a) If a filter F clusters to a point $p \in$ X then there is a filter G finer than F that converges to p.
(b) A space is compact if and only if each filter clusters.
(c) A space is compact if and only if each filter is contained in a convergent filter.

4.3 Completeness, Cofinal Completeness and Uniform Paracompactness

We have already encountered the concept of completeness in metric spaces (Chapter 1, Section 6). Just as with metric spaces, completeness plays a fundamental role in the theory of uniform spaces. The concept of completeness in metric spaces generalizes in a natural way to uniform spaces. The next three propositions show that Cauchy nets and cofinally Cauchy nets behave in a manner analogous to Cauchy sequences and cofinally Cauchy sequences in metric spaces. Their proofs are also left as exercises (Exercises 1 and 2).

PROPOSITION 4.12 A convergent net is Cauchy.

PROPOSITION 4.13 A net that clusters is cofinally Cauchy.

PROPOSITION 4.14 If a Cauchy net clusters to p then it also converges to p.

The following theorem is the net version of Theorems 3.1, 3.2 and 3.3. It appeared in the 1971 paper titled *On Completeness* (Pacific Journal of Mathematics, Volume 38, Number 2, pp. 431-440). The proof is similar to the proofs of Theorems 3.1, 3.2 and 3.3, so we leave it as an exercise (Exercise 6).

THEOREM 4.4 (N. Howes, 1971) A space is paracompact (resp. Lindelöf or compact) if and only if each net that is cofinally Cauchy with respect to the u (resp. e or β) uniformity clusters.

A uniform space is said to be **complete** if each Cauchy net converges. It is said to be **cofinally complete** if each cofinally Cauchy net clusters.

COROLLARY 4.1 A cofinally complete uniform space is complete.

The Lebesgue property for metric spaces also generalizes in a natural way to uniform spaces. We say (X, μ) has the **Lebesgue property** if for each open covering $\mathcal{V}$ of X, there is a $\mathcal{U} \in \mu$ such that $\mathcal{V}$ can be refined by the covering consisting of spheres of radius $\mathcal{U}$. An equivalent characterization of the Lebesgue property is the following:

PROPOSITION 4.15 (X, μ) has the Lebesgue property if and only if each open covering of X has a refinement in μ (i.e., μ is fine and X is paracompact).

Proof: Assume each open covering $\mathcal{V}$ of X has a refinement say $\mathcal{U}$ in μ. Let $\mathcal{W} <^* \mathcal{U}$. Then the spheres of radius $\mathcal{W}$ also refine $\mathcal{V}$ so (X, μ) has the Lebesgue property. Conversely, if X has the Lebesgue property and $\mathcal{V}$ is an open covering of X, pick $\mathcal{U} \in \mu$ such that the spheres of radius $\mathcal{U}$ refine $\mathcal{V}$. Clearly then $\mathcal{U}$ refines $\mathcal{V}$. ∎

PROPOSITION 4.16 A compact Hausdorff uniform space has the Lebesgue property.

Proof: Let (X, μ) be a compact Hausdorff uniform space. Then by Theorem 2.7 and Lemma 3.7 we have μ = β is fine. Since X is paracompact, it has the Lebesgue property by Proposition 4.15. ∎

PROPOSITION 4.17 A uniform space with the Lebesgue property is cofinally complete.

Proof: Let (X, μ) be a uniform space with the Lebesgue property. Suppose (X,μ) is not cofinally complete. Then there exists a cofinally Cauchy net $\{x_\alpha\}$ that does not cluster. For each $p \in$ X let U(p) be an open set containing p such that $\{x_\alpha\}$ is eventually in X - U(p) and put $\mathcal{U} = \{U(p) \mid p \in X\}$. Since (X, μ) has the Lebesgue property, by Proposition 4.15, $\mathcal{U}$ has a refinement $\mathcal{V} \in \mu$. Since

$\{x_\alpha\}$ is cofinally complete it is frequently in V for some V $\in$ $\mathcal{V}$. But V $\subset$ U(p) for some $p \in$ X so $\{x_\alpha\}$ cannot be eventually in X - U(p) which is a contradiction. Thus (X, μ) is cofinally complete. ■

A uniform space (X, μ) is said to be **precompact** or **totally bounded** if each uniform covering has a finite subcovering. Clearly each precompact metric space is a precompact uniform space.

THEOREM 4.5 A uniform space is precompact if and only if each net has a Cauchy subnet.

Proof: Assume each net in the uniform space (X, μ) has a Cauchy subnet and suppose X is not precompact. Then there is a $\mathcal{U} \in \mu$ such that $\{S(x, \mathcal{U}) | x \in X\}$ has no finite subcovering. Pick $p_1, p_2 \in$ X such that p_2 does not belong to $S(p_1, \mathcal{U})$. Next assume $\{p_1 \ldots p_n\}$ has been defined such that for each pair i,j with $i < j \leq n$ we have p_j does not belong to $S(p_i, \mathcal{U})$. Since $\cup\{S(p_1, \mathcal{U}) \ldots S(p_n, \mathcal{U})\} \neq$ X we can choose $p_{n+1} \in$ X with p_{n+1} not belonging to $S(p_i, \mathcal{U})$ for each $i = 1 \ldots n$. Consequently it is possible (by induction) to construct a sequence $\{p_n\} \subset$ X such that for each pair of positive integers i,j with i , j we have p_j does not belong to $S(p_i, \mathcal{U})$.

Let $x = \{x_\alpha | \alpha \in D\}$ be such a Cauchy subnet of the sequence $\{p_n\}$ for some cofinal function λ:D $\rightarrow$ N such that $x_\alpha = p_{\lambda(\alpha)}$ for each $\alpha \in$ D. Let $\mathcal{V} \in \mu$ with $\mathcal{V} <^* \mathcal{U}$. Since x is Cauchy there is a residual R $\subset$ D and a $p_0 \in$ X such that $p_{\lambda(\alpha)} \in S(p_0,\mathcal{V})$ for each $\alpha \in$ R. Since λ is a cofinal function, λ(R) is cofinal in N which implies $\{p_n\}$ is frequently in $S(p_0,\mathcal{V})$. Pick positive integers k,m with $k < m$ such that $p_k, p_m \in S(p_0,\mathcal{V})$. Then $S(p_k,\mathcal{V}) \cup S(p_0,\mathcal{V}) \subset S(p_k, \mathcal{U})$. But then $p_m \in S(p_k, \mathcal{U})$ which is a contradiction. We conclude (X, μ) is precompact.

Conversely, assume (X, μ) is precompact and let $x = \{x_\alpha\}$, $\alpha \in$ D, be a net in X. If $\mathcal{U} \in \mu$, then $\mathcal{U}$ has a finite subcovering, say $\{U_1 \ldots U_N\}$. Since $\{x_\alpha\}$ cannot be eventually in X - U_j for $j = 1 \ldots$ N, it must be frequently in one of them. Therefore, every net in (X, μ) is cofinally Cauchy. By Proposition 4.9, $\{x_\alpha\}$ has a universal subnet $\{x_{\lambda(\beta)}\}$ where λ:B $\rightarrow$ D is a cofinal function from a directed set B into D. Let $\mathcal{V} \in \mu$. Since $\{x_{\lambda(\beta)}\}$ is cofinally Cauchy, there exists a V $\in$ $\mathcal{V}$ such that $\{x_{\lambda(\beta)}\}$ is frequently in V. But then $\{x_{\lambda(\beta)}\}$ must eventually be in V, so $\{x_{\lambda(\beta)}\}$ is Cauchy. ■

PROPOSITION 4.18 A closed subspace of a complete uniform space is complete.

PROPOSITION 4.19 A uniform space is compact if and only if it is complete and precompact.

The proofs of the two propositions above are essentially the same as the proofs of Propositions 1.23 through 1.25 with sequences being replaced by nets. They are left as an exercise (Exercise 5).

In a 1977 paper titled *A note on uniform paracompactness* that appeared in the Proceedings of the American Mathematical Society (Volume 62, Number 2, pp. 359-362), M. Rice introduced the concept of **uniform paracompactness** which is defined as the property that every open covering has a uniformly locally finite open refinement. A refinement $\mathcal{V}$ of a covering $\mathcal{U}$ is said to be **uniformly locally finite** if there exists a uniform covering $\mathcal{W}$ such that each member of $\mathcal{W}$ meets only finitely many members of $\mathcal{V}$.

THEOREM 4.6 A uniform space is cofinally complete if and only if it is uniformly paracompact.

Proof: We will prove that a uniform space (X, μ) is cofinally complete if and only if each directed open covering of X is uniform and then use the equivalent form of uniform paracompactness given in Exercise 7(b) to finish the argument. To prove the sufficiency, assume each directed open covering is uniform and suppose (X, μ) is not cofinally complete. Then there exists a cofinally Cauchy net $\psi: D \to X$ that does not cluster. For each $\alpha \in D$ put $H_\alpha = \{\psi(\delta) \mid \delta \geq \alpha\}$ and let $F_\alpha = Cl(H_\alpha)$. Then $\cap F_\alpha = \emptyset$ since ψ does not cluster. For each $\alpha \in D$ put $U_\alpha = X - F_\alpha$. Then $\mathcal{U} = \{U_\alpha\}$ is a directed open covering of X, so $\mathcal{U} \in \mu$. But for each $\alpha \in D$, ψ is eventually in $X - U_\alpha$ which is a contradiction. Therefore, (X, μ) is cofinally complete.

Conversely, assume (X, μ) is cofinally complete. Let $\mathcal{U} = \{U_\alpha \mid \alpha \in D\}$ be a directed open covering of X where $(D, <)$ is a directed set. If $X = U_\alpha$ for some $\alpha \in D$ then $\mathcal{U} \in \mu$ so assume $F_\alpha = X - U_\alpha \neq \emptyset$ for each α. Put $E = \{(F_\alpha, x) \mid \alpha \in D \text{ and } x \in F_\alpha\}$ and for each α let $<_\alpha$ be a well ordering of F_α. Then $(E, <)$ is a directed set where $<$ is defined on E by $(F_\alpha, x) < (F_\beta, y)$ if $\alpha < \beta$ in D or $\alpha = \beta$ and $x <_\alpha y$. For each $(F_\alpha, x) \in E$ put $\psi(F_\alpha, x) = x$. Then $\psi: E \to X$ is a net in X. Since $\mathcal{U}$ is a covering of X, ψ cannot cluster, so ψ cannot be cofinally Cauchy. Therefore, there exists a $\mathcal{V} \in \mu$ such that if $V \in \mathcal{V}$, there exists a $\delta \in D$ with $\psi(F_\alpha, x) \in X - V$ for each $\alpha > \delta$. But $F_\delta = \{\psi(F_\delta, x) \mid x \in F_\delta\}$, so $V \subset U_\delta$. Therefore, $\mathcal{V} < \mathcal{U}$, so $\mathcal{U} \in \mu$. ∎

EXERCISES

1. Prove Propositions 4.12 and 4.13.

2. Prove Proposition 4.14.

3. Let (X, μ) be a uniform space and let $\mathcal{U} \in \mu$. $A \subset X$ is said to be **$\mathcal{U}$-small** if $A \subset U$ for some $U \in \mathcal{U}$. A is **$\mathcal{U}$-large** if its complement is $\mathcal{U}$-small. A

collection H of proper subsets of X is **heavy** if for each $\mathcal{U} \in \mu$ there is an H $\in H$ that is $\mathcal{U}$-large. Show that X is complete if and only if each heavy covering has a finite subcovering.

4. A collection $\mathcal{U}$ of proper subsets of X is said to be **bound** to another collection $\mathcal{V}$ if for each finite subcollection $\mathcal{W}$ of $\mathcal{U}$ that does not cover X, neither does $\mathcal{W} \cup \{V\}$ for each V $\neq$ X in $\mathcal{V}$. A collection that is bound to a heavy collection is called **heavily bound**. Show that a uniform space is cofinally complete if and only if each heavily bound open covering has a finite subcovering.

5. Prove Propositions 4.18 and 4.19.

6. Prove Theorem 4.4.

UNIFORMLY PARACOMPACT SPACES

7. [M. Rice, 1977] The following statements are equivalent:
   (a) X is uniformly paracompact.
   (b) Each directed open covering is uniform.
   (c) If $\mathcal{U}$ is an open covering of X, there exists a uniform covering $\mathcal{V}$ such that $\mathcal{U} | V$ has a finite subcovering for each V $\in \mathcal{V}$.

8. [M. Rice, 1977] A locally compact uniform space is uniformly paracompact if and only if it is uniformly locally compact. A uniform space (X, μ) is said to be **uniformly locally compact** if there exists a $\mathcal{U} \in \mu$ consisting of compact sets.

UNIFORMLY PARACOMPACT METRIC SPACES

9. [M. Rice, 1977] The collection of points of a uniformly paracompact metric space that admit no compact neighborhood is compact.

10. [A. Hohti, 1981] A necessary and sufficient condition for a metric space to be uniformly paracompact is that there exists a compact K $\subset$ X with X - *Star*(K,ε) is uniformly locally compact for each $\varepsilon > 0$.

11. [A. Hohti, 1981] If (X, d) and (Y, δ) are uniformly paracompact metric spaces, then (X $\times$ Y, $d \times \delta$) is uniformly paracompact if and only if one of the following conditions hold:

   (a) *either* (X, d) or (Y, δ) is compact *or*
   (b) *both* (X, d) and (Y, δ) are locally compact.

12. [A. Hohti, 1981] Let α be an infinite ordinal. Put $H(\alpha) = ([0,1] \times \alpha)/E$ where

xEy if and only if $x = y$ or $p_1(x) = 0 = p_1(y)$. Define a metric d on $H(\alpha)$ by $d(x,y) = |p_1(x) - p_1(y)|$ if $p_2(x) = p_2(y)$ or $d(x, y) = p_1(x) + p_1(y)$ otherwise. Then $(H(\alpha), d)$ is the *hedgehog* metric space with α spines. $H(\alpha)$ is a uniformly paracompact metric space that is not locally compact. It is therefore a counter example to a statement by P. Fletcher and W. Lindgren in 1978 that the product of a uniformly locally compact space with a C-complete (uniformly paracompact) space is C-complete.

4.4 The Completion of a Uniform Space

As seen in Chapter 2 (Section 2.1, Exercise 1), every metric space is a uniform space. Our first proposition below states that a complete metric space is also complete when considered as a uniform space. This may not seem surprising at first glance, but it might if we rephrase it in the following manner: In a metric space, the convergence of all Cauchy sequences forces the convergence of all Cauchy nets (of all cardinalities). The proof of this proposition is left as an exercise (Exercise 1).

PROPOSITION 4.20 A complete metric space is also complete when considered as a uniform space.

Our next result will show that Proposition 4.20 can be generalized to uniform spaces; i.e., that the convergence of all Cauchy nets on a certain ordered set forces the convergence of all Cauchy nets. We will then use these specialized nets to construct a completion of the uniform space from equivalence classes of these specialized nets in a manner similar to the construction of the metric completion in Chapter 1. Our first step will be to define these specialized nets. For this let (X, μ) be a uniform space and let ν be a cofinal subset of μ of least cardinality. Then ν is a basis for μ. Next, well order ν by some ordering $\prec$ such that the cardinality of each initial interval is less than the cardinality of ν. From the proof of the Ordering Lemma, it can be seen that there exists a cofinal subset λ of ν (with respect to the ordering $<^*$) such that the well ordering $\prec$ is compatible with $<^*$ on λ (i.e., $\mathcal{U} <^* \mathcal{V}$ implies that $\mathcal{V} \prec \mathcal{U}$).

Then λ is a well ordered basis for μ of least cardinality such that the well ordering $\prec$ is compatible with the directed ordering $<^*$ and the cardinality of each initial interval (with respect to $\prec$) is less than the cardinality of λ. We call λ a **fundamental basis** for μ and a net $\{x_\alpha | \alpha \in \lambda\} \subset X$ a **fundamental net** with respect to λ. Let $\{x_\alpha | \alpha \in D\}$ be a net in X and let $(E, <)$ be a directed set. A function $\zeta: E \to D$ is called a **compatible** function if whenever $\alpha, \beta \in E$ with $\alpha < \beta$ then $\zeta(\beta)$ is not strictly less than $\zeta(\alpha)$ in D. A net $\{y_\beta | \beta \in E\} \subset X$ where $y_\beta = x_{\zeta(\beta)}$ for each $\beta \in E$ is said to be **contained** in the net $\{x_\alpha\}$. Note that a cofinal function is a compatible function so a subnet of a net is contained in the net. Also note that a compatible function need not be cofinal so that a net $\{y_\beta\}$ contained in a net $\{x_\alpha\}$ need not *span* the net $\{x_\alpha\}$ as a subnet would.

THEOREM 4.7 Each Cauchy net ϕ contains a fundamental Cauchy net ψ such that ϕ converges to a point p if and only if ψ converges to p.

Proof: (Construction of ψ) Let $\phi = \{x_\alpha | \alpha \in D\}$ where D is ordered by $\leq$. Well order D by $<$ such that the cardinality of each initial interval is less than the cardinality of D. Then there is a cofinal $E \subset D$ with respect to $\leq$ such that $<$ is compatible with $\leq$ on E. Now $\phi_E = \{x_\alpha | \alpha \in E\}$ is a Cauchy subnet of ϕ. For each $a \in \lambda$ there is a $U_a \in a$ and a residual $R_a \subset E$ with respect to $\leq$ such that $\phi_E(R_a) \subset U_a$. Put $V_a = Star(U_a, a)$ and let $\mathcal{U} = \{U_a\}$ and $\mathcal{V} = \{V_a\}$. Then $\mathcal{U}$ has the finite intersection property and for each $U_a \in \mathcal{U}$, if U is another member of a with ϕ_E eventually in U then $U \subset V_a$. Also, $\mathcal{V}$ is directed by set inclusion; i.e., if $a,b \in \lambda$ there is a $c \in \lambda$ with $V_c \subset V_a \cap V_b$. For each $a \in \lambda$ define $E_a = \{\delta \in E | \phi_E(\delta) \in V_a\}$ and for each $b \in \lambda$ let $\zeta(b)$ be the first element of E_b (with respect to $<$). Then $\zeta: \lambda \to E$ is a compatible function and $\psi = \phi_E © \zeta$ is a fundamental net contained in ϕ. Let $a \in \lambda$ and pick $b \in \lambda$ with $b <^* a$. Then $V_b \subset W$ for some $W \in a$ so $\psi(b) = x_{\zeta(b)} \in W$. Also, if $c \in \lambda$ such that $c <^* b$ then $V_c \subset V_b$. Thus $\psi(c) \in W$. Put $R = \{c \in \lambda | c <^* a\}$. Then R is residual in λ and $\psi(R) \subset W$ so ψ is Cauchy.

(Proof ϕ converges if ψ converges) Assume ψ converges to $p \in X$ and N is a neighborhood of p. Then there are $a,b \in \lambda$ with $a <^* b$ and $Star(p,b) \subset N$. Also, there is a $c \in \lambda$ with $c <^* a$ and $\psi(c) \in Star(p,a)$. Then p and $\psi(c)$ both belong to some $W_a \in a$. Also $\psi(c) \in V_c$ and $V_c \subset V_a$. Hence $Star(p,a) \subset Star(W_a, a) \subset W_b$ for some $W_b \in b$ and $V_a \subset Star(W_a, a) \subset W_b$. Now $\phi(E_c) = \phi_E(E_c) \subset V_c \subset V_a$ so $\phi(R_c) \subset V_a$. Since R_c is residual in E with respect to $\leq$, it is cofinal in D with respect to $\leq$, so ϕ is frequently in $V_a \subset W_b \subset Star(p,b) \subset N$. Since ϕ is Cauchy, ϕ converges to p.

(Proof ψ converges if ϕ converges) Assume ϕ converges to p and N is a neighborhood of p. Then there are $a,b \in \lambda$ with $a <^* b$ and $Star(p,b) \subset N$. Also, there is a residual $R \subset D$ with respect to $\leq$ with $\phi(R) \subset W_a$ for some $W_a \in a$ such that $p \in W_a$. Then $R \cap E$ is residual in E and $\phi_E(R \cap E) \subset W_a$. Since R_a is residual in E so is $R \cap R_a$. Thus there is a $\delta \in R \cap R_a$ with $x_\delta = \phi_E(\delta) \in W_a$. But $x_\delta \in V_a$, so $W_a \cap V_a \neq \emptyset$. Also, $\psi(a) = x_{\zeta(a)}$ implies $\psi(a) \in \phi_E(E_a) \subset V_a$. Moreover, if $c \in \lambda$ with $c <^* a$ then $V_c \subset V_a$ so $\psi(c) \in V_a$. Thus ψ is eventually in V_a. Now $a <^* b$ implies there is a $W_b \in b$ with $W_a \cup V_a \subset W_b$ so $p \in W_b$ which implies $V_a \subset Star(p,b) \subset N$. Hence ψ converges to p. ∎

Until now, we have avoided the definition of a subspace of a uniform space. Subspaces are one of the *fundamental constructions* to be dealt with in the next chapter. The approach so far has been to avoid these constructions (i.e., building new uniform spaces from old ones) as long as possible in order to examine the behavior of the generalizations of sequences and convergence (from metric spaces) and some of their topological consequences in uniform spaces as soon as possible.

Strictly speaking, the completion of a uniform space can be introduced without introducing the concept of a subspace, but the most useful way of thinking of the completion is, that it is a (perhaps) *larger* uniform space that contains the original uniform space as a subspace. Consequently, we now give a simple definition of a uniform subspace, but will revisit the concept in the next chapter in a more formal setting.

A subset A of a uniform space (X, μ) can be given a uniform structure that is derived from the uniformity μ in the following way: for each $\mathcal{U} \in \mu$ put $\mathcal{U}_A = \{U \cap A \mid U \in \mathcal{U}\}$. Then let $\mu_A = \{\mathcal{U}_A \mid \mathcal{U} \in \mu\}$. It is easily shown that μ_A is a uniformity on A. μ_A is called the uniformity **induced** on A by the uniformity μ. μ_A is also said to be the uniformity of μ **relativized** to A. It should be noted that the uniformity μ_A causes the identity mapping i_A:A → X to be uniformly continuous.

Our next task is to construct a completion for (X, μ) from the fundamental Cauchy nets. For this let Σ be the set of all fundamental Cauchy nets in X and define an equivalence relation ~ on Σ by $\phi \sim \psi$ if for each $\mathcal{U} \in \lambda$ there is a $U \in \mathcal{U}$ such that φ and ψ are both eventually in U. Let X′ be the set of all equivalence classes with respect to ~ and for each $\mathcal{U} \in \mu$ put $\mathcal{U}^\wedge = \{U^\wedge \mid U \in \mathcal{U}\}$ where

$$U^\wedge = \{\phi' \in X' \mid \phi \text{ is eventually in } U \text{ for each } \phi \in \phi'\}.$$

Then $\mu^\wedge = \{\mathcal{U}^\wedge \mid \mathcal{U} \in \mu\}$ is a basis for a uniformity for X′. Let μ′ be the uniformity generated by $\mu^\wedge$. Then (X′, μ′) is a uniform space. Note that if $x \in X$ there is a unique $\phi' \in X'$ that consists of all fundamental Cauchy nets that converge to *x*. For each $x \in X$ let *i*(*x*) denote the unique equivalence class φ′ whose members converge to *x*. Then *i*:X → X′ is well defined and since X is Hausdorff, *i* is one-to-one. Let $\mathcal{U} \in \mu$ and $U \in \mathcal{U}$. It is easily shown that $i(U) = U^\wedge \cap i(X)$. Define $i(\mathcal{U}) = \{i(U) \mid U \in \mathcal{U}\}$ for each $\mathcal{U} \in \mu$. Then for each $\mathcal{U}' \in \mu'$ there is a $\mathcal{U}^\wedge \in \mu^\wedge$ that refines $\mathcal{U}'$ so $\mathcal{U}^\wedge \cap i(X) = \{U^\wedge \cap i(X) \mid U \in \mathcal{U}\} = i(\mathcal{U})$ refines $\mathcal{U}' \cap i(X)$. But then $\mathcal{U}$ refines $i^{-1}(\mathcal{U} \cap i(X))$ so $i^{-1}(\mathcal{U}' \cap i(X)) \in \mu$. Consequently *i*(X) is a uniform subspace. It is easily shown that *i*(X) is dense in X′.

To show X′ is Hausdorff let φ′ and ψ′ be distinct elements of X′. Then there exists $\phi \in \phi'$ and $\psi \in \psi'$ such that φ is not equivalent to ψ. Hence there is a $\mathcal{U} \in \mu$ such that either φ or ψ is not eventually in U for each $U \in \mathcal{U}$. Note this implies that if $\mathcal{V} \in \mu$ with $\mathcal{V} < \mathcal{U}$ then either φ or ψ is not eventually in V for each $V \in \mathcal{V}$. Pick $\mathcal{W} \in \mu$ with $\mathcal{W}^{**} < \mathcal{U}$. Since both φ and ψ are Cauchy there are $W_1, W_2 \in \mathcal{W}$ with φ eventually in W_1 and ψ eventually in W_2. Now $\mathcal{W}^* < \mathcal{U}$ implies $W_1 \cap W_2 = \varnothing$ so $\hat{W_1} \cap \hat{W_2} = \varnothing$.

Let $\theta \in \phi'$. There is a $W \in \mathcal{W}$ and residual $P,Q \subset \lambda$ with $\theta(P)$, $\phi(Q) \subset W$. Also, there is a residual $R \subset \lambda$ with $\phi(R) \subset W_1$. Then $S = P \cap Q \cap R$ is residual in λ, $\theta(S)$ and $\phi(S) \subset W$ and $\phi(S) \subset W_1$, so there is a $V_1 \in \mathcal{W}^*$ with $W \subset V_1$. Thus $\phi' \in \hat{V_1}$. Similarly, there is a $V_2 \in \mathcal{W}^*$ with $\psi' \in \hat{V_2}$. $\mathcal{W}^{**} < \mathcal{U}$ implies $V_1 \cap V_2 = \varnothing$ so $\hat{V_1} \cap \hat{V_2} = \varnothing$. Since $\hat{V_1}$ and $\hat{V_2}$ are neighborhoods of ϕ' and ψ' respectively in X′, X′ is Hausdorff.

It remains to show $(X', \mu^{\prime})$ is complete. Let $\{\phi_u' \mid \mathcal{U} \in \lambda\}$ be a fundamental net in $i(X)$. Then for each $\mathcal{U} \in \lambda$, there is an $x_u \in X$ with $\phi_u' = i(x_u)$. Put $\psi(\mathcal{U}) = x_u$. Then $\psi:\lambda \to X$ is a fundamental net. It is easily seen that ψ is Cauchy. Let U′ be an open set in X′ containing ψ'. Then there is a $\mathcal{U}\hat{} \in \mu\hat{}$ with $\psi' \in Star(\psi', \mathcal{U}\hat{}) \subset U'$ so there is a $V\hat{} \in \mathcal{U}\hat{}$ with $\psi' \in V\hat{} \subset U'$. Then $\psi = \{x_u\}$ is eventually in U so $\phi_u' \in U'$. Thus $\{\phi_u'\}$ converges to ψ'.

Next let $\{\phi_u' \mid \mathcal{U} \in \lambda\}$ be a fundamental Cauchy net in X′. For each $a \in \lambda$ there is a residual $R_a \subset \lambda$ with $\{\phi_u' \mid \mathcal{U} \in R_a\} \subset \hat{U_a}$ for some $\hat{U_a} \in a\hat{}$. Let $\hat{V_a} = Star(\hat{U_a}, a\hat{})$ and put $\mathcal{U}\hat{} = \{\hat{U_a}\}$ and $\mathcal{V}\hat{} = \{\hat{V_a}\}$. Just as in the proof of the preceding theorem, for each $a \in \lambda$, if $U\hat{} \in a\hat{}$ and $\{\phi_u'\}$ is eventually in $U\hat{}$, then $U\hat{} \cap \hat{U_a} \neq \varnothing$ so $U\hat{} \subset \hat{V_a}$, and if $a,b \in \lambda$ with $a <^* b$ then $\hat{V_a} \subset \hat{V_b}$.

Since $i(X)$ is dense in X′, for each $a \in \lambda$ there is an $x_a \in X$ with $i(x_a) \in \hat{V_a}$. Then $\{i(x_a)\}$ is a fundamental net in $i(X)$. It is easily shown that $\{i(x_a)\}$ is Cauchy and consequently converges to some $p' \in X'$. Let N be a neighborhood of p' and $b \in \lambda$ with $Star(p', b\hat{}) \subset N$. Pick $a \in \lambda$ with $a <^* b$. Since $\{i(x_a)\}$ converges to p' there is a $c \in \lambda$ with $c <^* a$ and $i(x_c) \in Star(p', a\hat{})$. Thus p' and $i(x_c)$ both belong to some $W\hat{} \in a\hat{}$. Now $i(x_c) \in \hat{V_c}$ and $\{\phi_u'\}$ is eventually in $\hat{U_c} \subset \hat{V_c}$. Hence there is a $\hat{W_a} \in a\hat{}$ with $\hat{V_c} \subset \hat{W_a}$ so $\hat{W_a} \cap \hat{U_a} \neq \varnothing$. Since $i(x_c) \in W\hat{} \cap \hat{W_a}$ we have $W\hat{} \subset Star(\hat{W_a}, a\hat{}) \subset \hat{W_b}$ for some $\hat{W_b} \in b\hat{}$. Since $\hat{U_a} \subset \hat{W_b}$, $\{\phi_u'\}$ is eventually in $\hat{W_b} \subset Star(p', b\hat{}) \subset N$ and we conclude $\{\phi_u'\}$ converges to p'. Then by the above theorem, (X', μ') is complete.

(X', μ') is called the completion of (X, μ). Since the function i is one-to-one and uniformly continuous we often identify X with the dense uniform subspace $i(X)$ of X′. In this case, $i(x) = x$ for each $x \in X$; i.e., the function i is the identity mapping on X. Using this identification, we notice that for each $\mathcal{U} \in \mu$ there is a $\mathcal{U}' \in \mu'$ such that $\mathcal{U} = i^{-1}(\mathcal{U}')$, so $\mathcal{U} = \mathcal{U}' \cap X = \{U' \cap X \mid U' \in \mathcal{U}'\}$. Also, if $\mathcal{V}' \in \mu'$ then $\mathcal{V}' \cap X \in \mu$. We record these results as:

THEOREM 4.8 Every uniform space has a completion; i.e., if (X, μ) is a uniform space, there is a complete uniform space (X', μ') and a one-to-one uniformly continuous function $i:X \to X'$ such that $i(X)$ is a dense uniform subspace of X′.

Theorem 4.8 corresponds to Theorem 1.15 for metric spaces. Theorems 1.16 and 1.17 can also be generalized to uniform spaces as follows:

THEOREM 4.9 If f is a uniformly continuous function on a subset A of a uniform space X into a complete uniform space Y then f has a unique uniformly continuous extension f' to Cl(A).

Proof: Let μ be the uniformity on X and ν the uniformity on Y. For each $x \in$ A let $\{x_\alpha\}$ be the constant fundamental Cauchy net ($x_\alpha = x$ for each α) that converges to x and for each $x \in Cl(A) - A$ pick a fundamental Cauchy net $\{x_\alpha\} \subset A$ that converges to x. Let $\mathcal{V} \in \nu$. Since f is uniformly continuous, there exists a $\mathcal{U} \in \mu$ such that whenever $a,b \in A$ with $a,b \in U$ for some $U \in \mathcal{U}$, then $f(a), f(b) \in V$ for some $V \in \mathcal{V}$. But then the net $\{f(x_\alpha)\}$ is Cauchy in Y for each $x \in Cl(A)$. Since Y is complete, $\{f(x_\alpha)\}$ converges to some $x' \in Y$. Define $f': Cl(A) \to Y$ by $f'(x) = x'$ for each $x \in Cl(A)$. Clearly, $f'(a) = f(a)$ for each $a \in A$ so f' is and extension of f.

To show f' is well defined, let $x \in Cl(A)$ and let $\{y_\alpha\}$ be another fundamental Cauchy net converging to x. Suppose $x' \neq y'$. Then there exists a $\mathcal{V} \in \nu$ with $S(x',\mathcal{V}) \cap S(y',\mathcal{V}) = \emptyset$. By the uniform continuity of f, there exists a $\mathcal{U} \in \mu$ such that $a,b \in U \in \mathcal{U}$ implies $f(a), f(b) \in V$ for some $V \in \mathcal{V}$. Since both $\{x_\alpha\}$ and $\{y_\alpha\}$ converge to x, there exists a β such that $\alpha > \beta$ implies $x_\alpha, y_\alpha \in U$ for some $U \in \mathcal{U}$ which in turn implies $f(x_\alpha), f(y_\alpha) \in V$ for some $V \in \mathcal{V}$. But this is impossible since $\{f(x_\alpha)\}$ is eventually in $S(x',\mathcal{V})$ and $\{f(y_\alpha)\}$ is eventually in $S(y',\mathcal{V})$. Hence $x' = y'$ so f' is well defined.

To show f' is uniformly continuous on $Cl(A)$, let $\mathcal{W} \in \nu$ and pick $\mathcal{V} \in \nu$ with $\mathcal{V}^* < \mathcal{W}$. Since f is uniformly continuous on A, there exists a $\mathcal{U} \in \mu$ such that whenever $a,b \in A$ with $a,b \in U$ for some $U \in \mathcal{U}$, then $f(a), f(b) \in V$ for some $V \in \mathcal{V}$. Let $\mathcal{Z} \in \mu$ with $\mathcal{Z}^* < \mathcal{U}$. Let $x,y \in Cl(A)$ such that $x,y \in Z$ for some $Z \in \mathcal{Z}$. Since $\{x_\alpha\}$ converges to x, $\{y_\alpha\}$ converges to y, $\{f(x_\alpha)\}$ converges to x', and $\{f(y_\alpha)\}$ converges to y', it is possible to choose a β such that $\alpha > \beta$ implies $x_\alpha, x \in Z_1$, and $y_\alpha, y \in Z_2$ for some $Z_1, Z_2 \in \mathcal{Z}$, and $x', f(x_\alpha) \in V_1$ and $y', f(y_\alpha) \in V_2$ for some $V_1, V_2 \in \mathcal{V}$. Then $x_\alpha, y_\alpha \in Star(Z,\mathcal{Z}) \subset U$ for some $U \in \mathcal{U}$. Hence $f(x_\alpha), f(y_\alpha) \in V$ for some $V \in \mathcal{V}$. But then $x', y' \in Star(V,\mathcal{V}) \subset W$ for some $W \in \mathcal{W}$. Therefore, $x,y \in Z$ which implies $f'(x), f'(y) \in W$, so f' is uniformly continuous.

To show f' is unique, suppose F is another uniformly continuous extension of f from A to $Cl(A)$. Let $x \in Cl(A) - A$. Then $\{x_\alpha\}$ converges to x. Since both f' and F are continuous, it follows that $\{f'(x_\alpha)\}$ converges to $f'(x)$ and $\{F(x_\alpha)\}$ converges to $F(x)$. But $\{x_\alpha\} \subset A$ implies $\{f'(x_\alpha)\} = \{f(x_\alpha)\} = \{F(x_\alpha)\}$. Since a net in a Hausdorff space has (at most) a unique limit, $f'(x) = F(x)$, so f' is unique. ■

THEOREM 4.10 For each uniform space (X, μ), its completion (X', μ') is unique; i.e., if $(X^\wedge, \mu^\wedge)$ is another completion of (X, μ) then there is a uniform homeomorphism $h: X' \to X^\wedge$ that keeps each point of X fixed.

Proof: Let j:X → X^ denote the uniform homeomorphism that maps X into the completion X^. By Theorem 4.9 there exists a unique uniformly continuous extension j':X′ → X^. Since j' is an extension of j we have $j = j' \circledcirc i$ where i:X → X′ denotes the uniform homeomorphism that maps X into its completion X′. Similarly, there is a unique uniformly continuous extension i^:X^ → X′ of i. Then $i = i\hat{} \circledcirc j$.

For each $x' \in$ X′ put $g'(x') = i\hat{}(j'(x'))$. Then $g' = j' \circledcirc i\hat{}$ so g':X′ → X^ is uniformly continuous. If $x \in$ X then $g'(x) = i\hat{}(j'(x)) = i\hat{}(j(x)) = i(x) = x$ so g' is an extension of the identity map i:X → X′. But by Theorem 4.9, this extension is unique so $g' = i'$ (the identity map on X′). Hence $j' \circledcirc i\hat{} = i'$ so $i\hat{} = (j')^{-1}$. But then X′ and X^ are uniformly homeomorphic. ■

EXERCISES

1. Prove Proposition 4.20.

2. Show that the completion of a uniform space is compact if and only if the uniformity is precompact.

3. Show that a complete subspace of a Hausdorff uniform space is closed.

4. [K. Morita, 1951] Let μ* be the uniformity of μX. For each $\mathcal{U} = \{U_\alpha\}$ in μ, put $\mathcal{U}' = \{U'_\alpha\}$ where $U'_\alpha = \mu X - Cl_{\mu X}(X - U_\alpha)$ for each α. Then $\{\mathcal{U}' \mid \mathcal{U} \in \mu\}$ is a basis for μ*.

CAUCHY FILTERS AND WEAKLY CAUCHY FILTERS

A filter F in a uniform space (X, μ) is said to be **Cauchy** if for each $\mathcal{U} \in \mu$, there exists an F ∈ F such that F ⊂ U for some U ∈ $\mathcal{U}$. It is said to be **weakly Cauchy** if for each $\mathcal{U} \in \mu$, there exists a U ∈ $\mathcal{U}$ with U∩F ≠ ∅ for each F ∈ F.

5. Show that (X, μ) is complete if and only if each Cauchy filter in X converges.

6. [N. Howes, 1971] (X, μ) is cofinally complete if and only if each weakly Cauchy filter in (X, u) clusters.

7. [H. Corson, 1958] X is paracompact if and only if each weakly Cauchy filter in (X, u) clusters.

4.5 The Cofinal Completion or Uniform Paracompactification

Since each uniform space has a unique completion, it is natural to define a uniform space (X', μ') to be a **cofinal completion** of the uniform space (X, μ) if (X', μ') is cofinally complete and (X, μ) is uniformly homeomorphic to a dense uniform subspace of (X', μ') and to ask: "When does (X, μ) have a cofinal completion, and if a cofinal completion exists, is it unique?" Providing answers to these questions is the objective of this section.

It turns out these same techniques can be used to provide answers to a variety of other questions asked by K. Morita and H. Tamano. In the late fifties, Tamano considered the following question: "What is a necessary and sufficient condition for a uniform space to have a paracompact completion?" Although he did not arrive at a solution, he obtained elegant characterizations of completeness, paracompactness and the structure of the completion by means of a concept called the *radical* of a uniform space. These results were published in the Journal of the Mathematical Society of Japan in 1960 (Volume 12, No. 1, pp. 104-117) under the title *Some Properties of the Stone-Čech Compactification.* We will analyze these results in Chapter 6.

In 1970, K. Morita presented a paper titled *Topological completions and M-spaces* at an international Topology Conference held at the University of Pittsburgh. In Section 7 of that paper, five unsolved problems were listed including a special case of Tamano's question mentioned above and the question: "What is a necessary and sufficient condition for a Tychonoff space to have a Lindelöf topological completion?" By a **topological completion** we mean the completion with respect to the finest uniformity.

In this section and in later chapters, we provide answers to all these questions. All of the solutions are in terms of the cofinally Cauchy nets and their behavior with respect to various uniformities. Since some of these questions ask if the completion of a uniform space has a certain topological property, one might be interested in knowing if there are solutions in terms of topological properties, or if topological properties exist that are only necessary or only sufficient. Since the literature is rich with characterizations of paracompactness and the Lindelöf property, one might expect a variety of solutions to these problems. However, at the present time, this area is largely unexplored.

We define a uniform space to be **preparacompact** if each cofinally Cauchy net has a Cauchy subnet. Recall that a uniform space is precompact if *every* net has a Cauchy subnet. Consequently, preparacompactness is a generalization of precompactness. To continue the parallel, recall that complete precompact spaces are compact and the completion of a precompact space is compact. We will see that complete preparacompact spaces are paracompact and that the completion of a preparacompact space is paracompact.

A uniform space will be called **countably bounded** if each uniform covering has a countable subcovering. If (X, μ) is a uniform space and (X^*, μ^*) its completion, let u^* denote the universal uniformity for X^* and let ν be the uniformity induced on X by u^*. The ν uniformity is called the uniformity derived from u^* or simply the **derived** uniformity. A directed set D is said to be **ω directed** or **countably directed** if for each countable $\{d_i\} \subset D$ there is a $d \in D$ such that $d_i \leq d$ for each i. A net $\{x_\alpha | \alpha \in D\}$ is ω **directed** if D is an ω directed set.

LEMMA 4.1 A completely regular space is Lindelöf if and only if each ω directed net clusters.

Proof: Assume that each ω directed net clusters in the completely regular space X. Suppose X is not Lindelöf. Then there is a covering $\mathcal{U}$ of X having no countable subcovering. For each countable $\mathcal{V} \subset \mathcal{U}$ put $G(\mathcal{V}) = \cup\{U | U \in \mathcal{V}\}$ and $F(\mathcal{V}) = X - G(\mathcal{V})$. Let S be the set of all countable subsets of $\mathcal{U}$ and for each $\mathcal{V} \in S$ pick $x_\nu \in F(\mathcal{V})$. The assignment $\mathcal{V} \to x_\nu$ defines an ω directed net $\{x_\nu | \mathcal{V} \in S\}$ in X that clusters to some $p \in X$ where S is directed by set inclusion. Let $U \in \mathcal{U}$ such that $p \in U$. Then $\{x_\nu\}$ is eventually in $F(\{U\})$ and hence cannot be frequently in U which is a contradiction. Consequently X is Lindelöf.

Conversely assume X is Lindelöf and let $\{x_\alpha | \alpha \in D\}$ be an ω directed net in X. Let $\mathcal{U}$ be a member of the e uniformity of X. Then there is a countable subcovering $\mathcal{V} = \{V_i\}$ such that $\mathcal{V} \in e$ and $\mathcal{V}$ refines $\mathcal{U}$. Suppose $\{x_\alpha\}$ is not frequently in some member of $\mathcal{V}$ and put $D_i = \{\delta \in D | x_\delta \in V_i\}$. Then there exists a $\delta_i \in D$ such that $\delta \leq \delta_i$ for each $\delta \in D_i$ or else $\{x_\delta | \delta \in D_i\}$ would be frequently in V_i. Since $\{x_\alpha\}$ is ω directed there is a $\delta \in D$ such that $\delta_i \leq \delta$ for each i. Now $x_\delta \in V_j$ for some j since $\mathcal{V}$ covers X. Hence $\delta \in D_j$ which implies $\delta < \delta_j$. But $\delta_j \leq \delta$ which is a contradiction. Therefore x_α must frequently be in some member of $\mathcal{V}$. Since $\mathcal{V}$ was chosen arbitrarily, $\{x_\alpha\}$ is cofinally Cauchy with respect to e and therefore clusters by Theorem 4.4. ∎

The following theorem appeared in an article titled *On completeness* in the Pacific Journal of Mathematics in 1971 (Volume 38, pp. 431-440).

THEOREM 4.11 (N. Howes, 1970) Let (X, μ) be a uniform space and ν the derived uniformity. Then:

(1) (X, μ) has a paracompact completion if and only if (X,ν) is preparacompact and

(2) (X, μ) has a Lindelöf completion if and only if (X,ν) is countably bounded and preparacompact, and

(3) (X, μ) has a compact completion if and only if (X,ν) is precompact.

Proof of (1): Let (X', μ') be the completion of (X, μ) and let u' be the universal uniformity for X′. Then (X,ν) is a dense uniform subspace of (X', u'). Assume (X,ν) is preparacompact and that $\psi: D \to X'$ is a cofinally Cauchy net with respect to u'. Since (X', μ') is complete, so is (X', u'). Let $E = D \times u'$ and define $\leq$ on E by $(d, \mathcal{U}') \leq (e, \mathcal{V}')$ if $d \leq e$ and $\mathcal{V}' <^* \mathcal{U}'$. For each $(d, \mathcal{U}') \in E$ put $\theta(d,\mathcal{U}') = a$ for some $a \in X$ such that a and $\psi(d)$ both belong to some $U' \in \mathcal{U}'$. Then the correspondence $(d, \mathcal{U}') \to \theta(d, \mathcal{U}')$ defines a net $\theta: E \to X$.

Let $\mathcal{U}' \in u'$ and pick $\mathcal{V}' \in u'$ with $\mathcal{V}' <^* \mathcal{U}'$. Since ψ is cofinally Cauchy there is a cofinal $C \subset D$ with $\psi(C) \subset V'$ for some $V' \in \mathcal{V}'$. Put

$$A = \{(d,\mathcal{W}') \mid d \in C \text{ and } \mathcal{W}' <^* \mathcal{V}'\}.$$

Then A is cofinal in E. Let $(d,\mathcal{W}') \in A$. Then $\theta(d,\mathcal{W}') = y \in X$ such that y and $\psi(d)$ both belong to some $W' \in \mathcal{W}'$. Since $(d,\mathcal{W}') \in A$, $d \in C$ which puts $\psi(d)$ in V'. Consequently we have:

$$y \in Star(V',\mathcal{W}') \subset Star(V',\mathcal{V}') \subset U'$$

for some $U' \in \mathcal{U}'$. Therefore $\theta(A) \subset U'$ which implies θ is cofinally Cauchy in (X', u'). But $\theta(E) \subset X$ implies θ is cofinally Cauchy in (X,ν). Consequently θ has a Cauchy subnet ξ. But then ξ converges to some $x' \in X'$. Therefore θ clusters to x'. It remains to show that ψ clusters to x'. For this let O be an open set containing x'. Then there is a $\mathcal{U}' \in u'$ such that $x' \in Star(x',\mathcal{U}') \subset O$ where the members of $\mathcal{U}'$ are open sets. Pick $\mathcal{V}' \in u'$ such that $\mathcal{V}' <^* \mathcal{U}'$. Let S be cofinal in E such that $\theta(S) \subset V'$ for some $V' \in \mathcal{V}'$ containing x'. Put

$$D(S) = \{d \in D \mid (d,\mathcal{W}') \in S \text{ for some } \mathcal{W}' <^* \mathcal{V}'\}.$$

Then D(S) is cofinal in D. For each $d \in D(S)$, $\psi(d)$ and $\theta(d,\mathcal{W}')$ are contained in some $W' \in \mathcal{W}'$ for some $(d,\mathcal{W}') \in S$ where $\mathcal{W}' <^* \mathcal{V}'$ which implies $\psi(d)$ and $\theta(d,\mathcal{W}')$ are contained in some $V_1 \in \mathcal{V}'$ for each $(d,\mathcal{W}') \in S$. But $(d,\mathcal{W}') \in S$ implies $\theta(d,\mathcal{W}') \in V'$. Hence

$$\psi(d) \in V' \cup V_1 \subset Star(V',\mathcal{V}') \subset U'$$

for some $U' \in \mathcal{U}'$. But $x' \in V'$ implies $x' \in U'$. Then $U' \subset O$ since $Star(x',\mathcal{U}') \subset O$. Consequently $\psi(d) \in O$ for each $d \in D(S)$ which implies ψ clusters to x'. Therefore each cofinally Cauchy net in X′ clusters which implies (X', u') is cofinally complete. But then X′ is paracompact by Theorem 4.4.

Conversely suppose X′ is paracompact. By Theorem 4.4 (X', u') is cofinally complete. Let ψ be a cofinally Cauchy net with respect to ν. Since (X,ν) is a uniform subspace of (X', u'), we know that ψ is cofinally Cauchy in (X', u'). Also since (X', u') is cofinally complete, ψ clusters to some $p \in X'$. But then ψ has a subnet θ that converges to p. Then θ is Cauchy in (X', u'). But $\theta \subset X$ and therefore θ is Cauchy in (X,ν). Consequently, each cofinally Cauchy net in (X,ν) has a Cauchy subnet so that (X,ν) is preparacompact.

Proof of (2): Assume first that X′ is Lindelöf. Then X′ is paracompact and hence (X′, u') is cofinally complete. But then (X,ν) is preparacompact as was shown in part (1). Next let $\mathcal{V} \in \nu$. Then $\mathcal{V}$ has a uniform refinement $\mathcal{U}$ consisting of closed sets. For each U ∈ $\mathcal{U}$ put U′ = $Cl_{X'}$(U) and let $\mathcal{U}'$ = {U′ | U′ ∈ $\mathcal{U}$}. Then $\mathcal{U} \in u'$ and hence has a countable subcovering say $\{U_i'\}$. Then $\{U_i\}$ covers X. In fact, if $p \in$ X then $p \in$ X′ which implies $p \in U_j'$ for some positive integer j. Hence $p \in Cl_{X'}(U_j)$. Let O be open in X such that $p \in$ O. Then O = O′∩X for some O′ that is open in X′. Now $p \in$ O′ which implies $O' \cap U_j \neq \emptyset$ since $p \in Cl_{X'}(U_j)$. Then there is a $t \in O' \cap U_j$. $t \in U_j$ implies $t \in$ X so we have $(O' \cap X) \cap U_j \neq \emptyset$. Hence $O \cap U_j \neq \emptyset$ so that $p \in Cl_X(U_j) = U_j$. Since $\{U_i\}$ covers X there exists some $\{V_i\} \subset \mathcal{V}$ such that $\cup V_i$ = X. Consequently X is countably bounded.

Conversely assume (X,ν) is countably bounded and preparacompact. (X′,μ′) is paracompact by part (1) so that (X′, u') is cofinally complete by Theorem 4.4. Let $\mathcal{U}' \in u'$. Since X′ is paracompact there exists a locally finite open refinement $\mathcal{V}' = \{V_\beta' \mid \beta \in B\}$. Since X is normal we can shrink $\mathcal{V}'$ to an open covering $\mathcal{W}' = \{W_\beta' \mid \beta \in B\}$ such that $Cl_{X'}(W_\beta') \subset V_\beta'$ for each β ∈ B. Then $\mathcal{W}'$ is also locally finite. Since $\mathcal{W}'$ is an open covering in a paracompact space X′, it must be a member of the universal uniformity u'. But then $\mathcal{W}$ = {W′∩X | W′ ∈ $\mathcal{W}'$} belongs to ν which implies there is a countable subcovering $\{W_{\beta_i}\} \subset \mathcal{W}$. Since $\mathcal{W}'$ is locally finite in X′, $\{W_{\beta_i}\}$ is locally finite in X′. Hence $\cup_{i=1}^{\infty} Cl_{X'}(W_{\beta_i}) = Cl_{X'}(\cup_{i=1}^{\infty} W_{\beta_i}) = Cl_{X'}(X) = X'$ since X is dense in X′. Therefore $\{Cl_{X'}(W_{\beta_i})\}$ covers X and

$$Cl_{X'}(W_{\beta_i}) \subset Cl_{X'}(W_{\beta_i}') \subset V_{\beta_i}'$$

for each positive integer i. Hence $\{V_{\beta_i}'\}$ covers X′. But since $\mathcal{V}'$ refines $\mathcal{U}'$ there exists a countable subcovering $\{U_{\beta_i}'\} \subset \mathcal{U}'$ that covers X′. Therefore (X′,u') is also countably bounded. We are now in a position to show that X′ is Lindelöf. We will use the fact that (X′, u') is cofinally complete and countably bounded to show that each ω-directed net in X′ clusters. We then invoke Lemma 4.1 to obtain the desired result. Let ψ:D → X′ be an ω directed net and let $\mathcal{U}' \in u'$. Then $\mathcal{U}'$ has a countable subcovering say $\{U_i'\}$. Put $D_i = \{d \in D \mid \psi(d) \in U_i'\}$ for each i and suppose D_i is not cofinal in D for each i. Then there exists a $d_i \in$ D for each i such that $d \le d_i$ for each $d \in D_i$. Since ψ is ω directed there is a $d_0 \in$ D such that $d_i \le d_0$ for each i. Since $\{U_i'\}$ covers X′, $\psi(d_0) \in U_j'$ for some positive integer j which implies $d_0 \in D_j$ which in turn implies D_j is cofinal in D since $D = \cup_{i=1}^{\infty} D_i$. Therefore ψ is frequently in U_j. Hence ψ is cofinally Cauchy and consequently must cluster. Thus X′ is Lindelöf by Lemma 4.1.

Proof of (3): By Problem 3 of Section 4.4, (X, μ) has a compact completion (X′, μ′) if and only if μ is precompact. But if ν is precompact then μ ⊂ ν implies μ is precompact which in turn implies X′ is compact. Conversely, if

X' is compact then $\mu' = u'$ which implies $\mu = \nu$ and μ is precompact which implies ν is precompact. ∎

COROLLARY 4.2 Let u be the universal uniformity for a completely regular T_1 space X. Then:

(1) (X, u) has a paracompact completion if and only if it is preparacompact,

(2) (X, u) has a Lindelöf completion if and only if it is countably bounded and preparacompact.

COROLLARY 4.3 The completion of a preparacompact uniform space is cofinally complete.

COROLLARY 4.4 A countably bounded cofinally complete uniform space is Lindelöf.

COROLLARY 4.5 A paracompact space is Lindelöf if and only if it is countably bounded with respect to the universal uniformity.

In view of the existence of a unique completion for each uniform space, it is natural to ask when a uniform space has a cofinal completion, and if a cofinal completion exists, when is it unique?

THEOREM 4.12. A uniform space has a cofinal completion if and only if it is preparacompact. In this case it is unique and identical to the ordinary completion (i.e., the ordinary completion is cofinally complete).

Proof: (Sufficiency) Suppose (X, μ) is preparacompact and (X', μ') is its completion. It needs to be shown that (X', μ') is cofinally complete. In the sufficiency part of the proof of Theorem 4.11.(1), it was shown that if ν is the uniformity *derived* from μ rather than the uniformity μ and if (X,ν) is preparacompact then (X', μ') is cofinally complete. The proof that (X', μ') is cofinally complete based on the assumption that (X, μ) is preparacompact is similar so it will not be included here.

(Necessity) Assume (X, μ) has a cofinal completion (X', μ'). Let $\{x_\alpha \mid \alpha \in D\}$ be a cofinally Cauchy net in (X, μ). Then $\{x_\alpha\}$ clusters to some $p \in X'$ so $\{x_\alpha\}$ has a subnet $\{y_\beta\}$ that converges to p. But then $\{y_\beta\}$ is Cauchy in (X',μ'). Since $\{y_\beta\}$ lies in X it is Cauchy in (X, μ). Consequently, every cofinally Cauchy net in (X, μ) has a Cauchy subnet so (X, μ) is preparacompact. Since (X', μ') is cofinally complete it is also complete. But then (X', μ') is a completion of (X, μ). Since the completion of a uniform space is unique this means the cofinal completion (when it exists) is identical to the completion. ∎

EXERCISES

1. Show that a completely regular T_1 space is paracompact if and only if it is complete and preparacompact with respect to the universal uniformity.

2. Show that a uniform space (X, μ) is countably bounded if and only if each ω directed net is cofinally Cauchy.

3. Let μ be a uniformity that generates the topology of X and let $\mathcal{U}$ be an open covering of X. We call $p \in$ X a **residue point** for $\mathcal{U}$ with respect to $\mathcal{V} \in \mu$ if there does not exist a U $\in$ $\mathcal{U}$ with $Star(p, \mathcal{V}) \subset$ U. Let $H_{\mathcal{V}}$ be the set of residue points with respect to $\mathcal{V}$. Put $F_{\mathcal{V}} = Cl(H_{\mathcal{V}})$ and $G_{\mathcal{V}} =$ X - $F_{\mathcal{V}}$. Then $\mathcal{W} = \{G_{\mathcal{V}} \mid \mathcal{V} \in \mu\}$ is called the **residue covering** derived from $\mathcal{U}$. Show that $\mathcal{W}$ is an open covering of X.

4. Show that a completely regular T_1 space is paracompact if and only if each residue open covering with respect to the universal uniformity has a finite subcovering.

5. Show that a completely regular T_1 space is paracompact if and only if each heavily bound open covering (Exercise 5, Section 4.3) with respect to the universal uniformity has a finite subcovering.

6. Show that in a completely regular T_1 space the following are equivalent:

   (a) the Lindelöf property,
   (b) each residue open covering with respect to e has a finite subcovering,
   (c) each heavily bound open covering with respect to e has a finite subcovering.

7. A topological space is said to be **entirely normal** if the collection of all neighborhoods of the diagonal in X × X forms an entourage uniformity (Exercise 4, Section 2.1). Show that entire normality and almost-2-fully normal (Exercise 6, Section 3.2) are equivalent.

8. A net in a topological space X will be called **cofinally Δ Cauchy** if for each open covering $\mathcal{U}$ of X there is a $p \in$ X such that the net is frequently in $S(p, \mathcal{U})$. X is **cofinally Δ complete** if each cofinally Δ Cauchy net clusters. Show that an entirely normal space is paracompact if and only if it is cofinally Δ complete.

9. Show that a metacompact space is cofinally Δ complete.

10. Show that entire normality implies collectionwise normality.

11. Show that a metacompact space is paracompact if and only if it is collectionwise normal.

12. RESEARCH PROBLEM

We would like to have a characterization of preparacompactness in terms of coverings, perhaps analogous to the characterization of precompactness in terms of coverings. The following result is in this direction, but something better is needed. A collection Σ of subsets of X is said to be a **directed collection** if for each $A,B \in \Sigma$, $A \cup B \in \Sigma$.

THEOREM A uniform space (X, μ) is preparacompact if and only if each heavily bound collection having no finite subcollection that covers X is contained in a heavy directed collection.

See Exercises 3 and 4 of Section 4.3.

Chapter 5

FUNDAMENTAL CONSTRUCTIONS

5.1 Introduction

In this chapter we consider some important constructions of uniform spaces from other uniform spaces. Our first concern will be to consider the so called *classical constructions* that are studied for most spaces and algebraic structures that arise in the study of mathematics, namely subspaces, sums, products and quotients. Our approach will be to derive these constructions from a few fundamental concepts. These fundamental concepts take the form of *limits* of collections of uniformities. We will make these concepts precise in the next section.

The reason for waiting until now to introduce the fundamental constructions is that many of the interesting results about these constructions involve concepts related to completeness. For example, it can be shown that the product of complete uniform spaces is a complete uniform space. Without these other concepts, it is difficult to appreciate the utility of these constructions. We opted instead to first develop the theory of completeness in uniform spaces and then see how it applies to these constructions.

After the classical constructions, we proceed to some other constructions (some of which also apply to a variety of other structures such as topological spaces). We should remark that we have already seen one such construction; namely, the completion of a uniform space. Of great interest to us will be the concepts of the *hyperspace* of a uniform space, the *inverse limit* of a directed family of uniform spaces, the *weak completion* of a uniform space and the *spectrum* of a uniform space. If the hyperspace is complete, the original space is said to be *supercomplete*. Like cofinal completeness, supercompleteness turns out to be a strong form of completeness. In fact, it was recently discovered that supercompleteness is a property that lies between completeness and cofinal completeness. This result will allow us to strengthen several results from Chapter 4.

The concept of supercompleteness was introduced by S. Ginsburg and J. Isbell in 1954 in an abstract titled *Rings of convergent functions* that appeared in the Bulletin of the American Mathematical Society, Volume 60, page 259. But the definition of supercompleteness in the simple form mentioned above is due to Isbell in a paper that appeared in the Pacific Journal of Mathematics in 1962 titled *Supercomplete spaces* (Volume 12, pp. 287-290).

The concept of the inverse limit of a family of uniform spaces leads to the concept of the spectrum of a uniform space, and the spectral analysis of uniform spaces whose spectra exist. The study of uniform spaces by means of their spectra has been pursued by B. Pasynkov (1963) and K. Morita (1970). Also, the concept of the inverse limit of a directed family of uniform spaces is needed to express the weak completion of a space with respect to a uniformity introduced by K. Morita in 1970.

Finally, we will introduce the *locally fine coreflection* and the *subfine coreflection* of a given uniform space. These were introduced by Ginsberg and Isbell in 1959 and shown to be equivalent by J. Pelant in 1987. This new construction has a variety of uses. In the last section of this chapter we will introduce the concepts of categories and functors. This material is optional as far as being needed in later chapters. It is included for two reasons. First, much of the literature of uniform spaces after Isbell makes use of the category theoretic vocabulary, and second, category theory, like set theory, is a tool that is often helpful in the study of uniform spaces. The term "coreflection" refers to a category theoretic concept, and it will be shown that the locally fine coreflection satisfies this notion. Thereafter, categories and functors will only be used in the exercises.

5.2 Limit Uniformities

Given a collection $\{\tau_\alpha\}$ of topologies for a set X, there exists a finest topology that is coarser than each τ_α called the **infimum** topology on X with respect to $\{\tau_\alpha\}$ and denoted $inf_\alpha\tau_\alpha$. Similarly, there exists a coarsest topology that is finer than each τ_α called the **supremum** topology on X with respect to $\{\tau_\alpha\}$ and denoted $sup_\alpha\tau_\alpha$. To see this, note that $\cap\tau_\alpha$ is a topology for X that is coarser than each τ_α. If τ is another topology for X that is coarser than each τ_α, then $\tau \subset \cap\tau_\alpha$ so $\cap\tau_\alpha$ is the finest topology for X that is coarser than each τ_α (i.e., $inf_\alpha\tau_\alpha = \cap\tau_\alpha$). Also note that if Σ is the collection of topologies that are finer than each τ_α, then $\Sigma \neq \varnothing$ since the discrete topology belongs to Σ. Consequently, $\cap\Sigma$ is a topology for X. For each α, $\tau_\alpha \subset \sigma$ for each $\sigma \in \Sigma$. Hence $\tau_\alpha \subset \cap\Sigma$ for each α so $\cap\Sigma$ is finer than each τ_α. If τ is another topology finer than each τ_α, then $\tau \in \Sigma$ which implies $\cap\Sigma \subset \tau$. Therefore, $\cap\Sigma$ is the coarsest topology that is finer than each τ_α (i.e., $sup_\alpha\tau_\alpha = \cap\Sigma$). It can be shown (Exercise 1) that a sub-base for $sup_\alpha\tau_\alpha$ is the set $\cup\tau_\alpha$.

The concepts of *infimum* and *supremum* uniformities are analogous to the topological concepts. If $\{\mu_\alpha\}$ is a collection of uniformities for a set X, there is a finest uniformity that is coarser than each μ_α called the **infimum** uniformity and denoted $inf_\alpha\mu_\alpha$. There also exists a coarsest uniformity that is finer than each μ_α called the **supremum** uniformity and denoted $sup_\alpha\mu_\alpha$. By Theorem

2.1, $\cup\mu_\alpha$ is a sub-basis for a uniformity μ on X which implies $\{\cap_{i=1}^n \mathcal{U}_{\alpha_i} \mid \mathcal{U}_{\alpha_i} \in \mu_{\alpha_i}$ for some $i = 1 \ldots n\}$ is a basis for μ. If μ' is another uniformity for X that is finer than each μ_α then μ' must contain each $\cap_{i=1}^n \mathcal{U}_{\alpha_i}$ so $\mu \subset \mu'$. Hence $\mu = sup_\alpha\mu_\alpha$. Similarly, if $\Sigma = \{\sigma_\gamma \mid \gamma \in G\}$ is the collection of all uniformities that are coarser than each μ_α then $\Sigma \neq \varnothing$ since $\{X\} \in \Sigma$. Put $\nu = sup_\gamma\sigma_\gamma$. If $\mathcal{V} \in \nu$ then by the definition of ν, there exists $\sigma_1 \ldots \sigma_n \in \Sigma$ and $\mathcal{W}_i \in \sigma_i$ for each $i = 1 \ldots n$ such that $\cap_{i=1}^n \mathcal{W}_i < \mathcal{V}$. For each α, $\sigma_i \subset \mu_\alpha$ for each $i = 1 \ldots n$ so $\mathcal{W}_i \in \mu_\alpha$ for each $i = 1 \ldots n$ which implies $\cap_{i=1}^n \mathcal{W}_i \in \mu_\alpha$. Hence $\mathcal{V} \in \mu_\alpha$ so $\mu \subset \mu_\alpha$ for each α. So $\nu \in \Sigma$ and by the definition of ν, ν is finer than any other member of Σ so ν is the finest uniformity that is coarser than each μ_α. Therefore $\nu = inf_\alpha\mu_\alpha$.

PROPOSITION 5.1 If $\{\mu_\alpha\}$ is a collection of uniformities for a set X and $\{\tau_\alpha\}$ is the corresponding collection of topologies generated by the μ_α's, then the topology generated by $sup_\alpha\mu_\alpha$ is $sup_\alpha\tau_\alpha$.

Proof: Let τ be the topology generated by $sup_\alpha\mu_\alpha$. Since $sup_\alpha\mu_\alpha$ is finer than each μ_α, τ is finer than each τ_α so $sup_\alpha\tau_\alpha \subset \tau$. If $U \in \tau$ then for each $p \in U$ there exists a basic covering $\cap_{i=1}^n \mathcal{U}_{\alpha_i} \in sup_\alpha\mu_\alpha$ such that $S(p, \cap_{i=1}^n \mathcal{U}_{\alpha_i}) \subset U$. Now $S(p, \mathcal{U}_{\alpha_i}) \in sup_\alpha\tau_\alpha$ for each $i = 1 \ldots n$ since $sup_\alpha\tau_\alpha$ is finer than τ_{α_i} for each $i = 1 \ldots n$. But then $\cap_{i=1}^n S(p, \mathcal{U}_{\alpha_i}) \in sup_\alpha\tau_\alpha$. If $x \in \cap_{i=1}^n S(p, \mathcal{U}_{\alpha_i})$, then $x,p \in U_{\alpha_i}$ for some $U_{\alpha_i} \in \mathcal{U}_{\alpha_i}$ for each $i = 1 \ldots n$. But then $x,p \in \cap_{i=1}^n U_{\alpha_i} \in \cap_{i=1}^n \mathcal{U}_{\alpha_i}$ which implies $x \in S(p, \cap_{i=1}^n \mathcal{U}_{\alpha_i})$. Therefore

$$\cap_{i=1}^n S(p, \mathcal{U}_{\alpha_i}) \subset S(p, \cap_{i=1}^n \mathcal{U}_{\alpha_i}) \subset U$$

which implies $U \in sup_\alpha\tau_\alpha$. Hence $\tau \subset sup_\alpha\tau_\alpha$ which implies $\tau = sup_\alpha\tau_\alpha$. ∎

Unfortunately, a proposition similar to 5.1 does not hold for $inf_\alpha\mu_\alpha$. Whereas $\cup\mu_\alpha$ and $\cup\tau_\alpha$ are sub-bases for the uniformity $sup_\alpha\mu_\alpha$ and the topology $sup_\alpha\tau_\alpha$ respectively, $\cap\mu_\alpha$ is not necessarily even a uniformity while $\cap\tau_\alpha = inf_\alpha\tau_\alpha$. The reason for this is explored in the exercises at the end of the section.

Let f:X → Y be a function from a set X into a topological space Y. Then f determines a coarsest topology τ_f on X such that f is continuous. A basis for the open sets in τ_f is the collection $B = \{f^{-1}(U) \mid U$ is open in Y$\}$. If $F = \{f_\alpha: X \to Y_\alpha \mid \alpha \in A\}$ is a family of functions from X into topological spaces Y_α, we define the topology τ_F to be $sup_\alpha\tau_{f_\alpha}$. τ_F is called the **projective limit** topology for the collection F. It can be shown (Exercise 3) that a sub-basis for τ_F is the set $S = \{f_\alpha^{-1}(U_\alpha) \mid U_\alpha$ is open in Y_α for some $\alpha\}$ and that τ_F is the coarsest topology on X such that each f_α in F is continuous.

Similarly, if g:Y → X is a function from a topological space Y onto a set X, there is a finest topology τ_g on X such that g is continuous. A basis for the open sets in τ_g is the collection $B = \{U \subset X \mid g^{-1}(U)$ is open in $Y\}$. If $G = \{g_\alpha : Y_\alpha \to X \mid \alpha \in A\}$ is a family of functions from topological spaces Y_α onto X, we define the topology τ_G to be $inf_\alpha \tau_{g_\alpha}$. τ_G is called the **inductive limit** topology for the collection G. It can be shown (Exercise 4) that a basis for τ_G is the set $B = \{U \subset X \mid g_\alpha^{-1}(U)$ is open in Y_α for each $\alpha \in A\}$ and that τ_G is the finest topology on X such that each $g_\alpha \in G$ is continuous.

The concepts of projective and inductive limits are also relevant to collections of uniformities. The only difference is that now we are interested in making collections of functions uniformly continuous rather than simply continuous. If f:X $\to$ Y is a function from a set X into a uniform space (Y, ν), then by Exercise 2 of Section 2.2, f determines a coarsest uniformity μ_f on X such that f is uniformly continuous. A basis for μ_f is $f^{-1}(\nu) = \{f^{-1}(\mathcal{V}) \mid \mathcal{V} \in \nu\}$. Let $F = \{f_\alpha : X \to (Y_\alpha, \nu_\alpha) \mid \alpha \in A\}$ be a collection of functions from X into uniform spaces (Y_α, ν_α). Define μ_F to be $sup_\alpha \mu_{f_\alpha}$. μ_F is called the **projective limit** uniformity on X. It can be shown (Exercise 5) that a sub-basis for μ_F is the set $S = \{f_\alpha^{-1}(\mathcal{V}_\alpha) \mid \mathcal{V}_\alpha \in \nu_\alpha$ for some $\alpha\}$ and that μ_F is the coarsest uniformity on X such that each f_α in F is uniformly continuous.

Similarly, if g:Y $\to$ X is a function from a uniform space (Y,ν) into X, by Exercise 3 of Section 2.2, g determines a finest uniformity μ_g on X such that g is uniformly continuous. A basis for μ_g is the set $B = \{\mathcal{U} \mid \mathcal{U}$ is a covering of X such that $g^{-1}(\mathcal{U}) \in \nu\}$. Let $G = \{g_\alpha : Y_\alpha \to X \mid \alpha \in A\}$ be a family of functions from uniform spaces (Y_α, ν_α) onto X. We define the uniformity μ_G to be $inf_\alpha \mu_{g_\alpha}$ and call μ_G the **inductive limit** uniformity on X. It can be shown (Exercise 5) that a basis for μ_G is the set $\{\mathcal{U} \mid \mathcal{U}$ is a covering of X and $g_\alpha^{-1}(\mathcal{U}) \in \nu_\alpha$ for each $\alpha\}$ and that μ_G is the finest uniformity on X such that each g_α in G is uniformly continuous.

PROPOSITION 5.2 The topology generated by the projective limit uniformity is the projective limit topology.

The proof follows immediately from Proposition 5.1 and the definitions of projective limit topology and projective limit uniformity.

EXERCISES

1. Let $\{\tau_\alpha\}$ be a collection of topologies for a set X. Show that the set $S = \cup\tau_\alpha$ is a sub-basis for $sup_\alpha \tau_\alpha$.

2. Let $F = \{f_\alpha : X \to Y_\alpha \mid \alpha \in A\}$ be a family of functions from a set X into topological spaces Y_α and let τ_F be the projective limit topology with respect to F. Show that the set $S = \{f_\alpha^{-1}(U_\alpha) \mid U_\alpha$ is open in Y_α for some $\alpha\}$ is a sub-basis

for τ_F and that τ_F is the coarsest topology on X such that each f_α in F is continuous.

3. Let $G = \{g_\alpha : Y_\alpha \to X \mid \alpha \in A\}$ be a family of functions from topological spaces Y_α onto a set X and let τ_G be the inductive limit topology with respect to G. Show that $B = \{U \subset X \mid g_\alpha^{-1}(U)$ is open in Y_α for each $\alpha \in A\}$ is a basis for τ_G and that τ_G is the finest topology on X such that each g_α in G is continuous.

4. Let $F = \{f_\alpha : X \to (Y_\alpha, \nu_\alpha) \mid \alpha \in A\}$ be a family of functions from a set X into uniform spaces (Y_α, ν_α) and let μ_F be the projective limit uniformity with respect to F. Show that the set $S = \{f_\alpha^{-1}(\mathcal{V}_\alpha) \mid \mathcal{V}_\alpha \in \nu_\alpha$ for some $\alpha\}$ is a subbasis for μ_F and that μ_F is the coarsest uniformity on X such that each f_α in F is uniformly continuous.

5. Let $G = \{g_\alpha : Y_\alpha \to X \mid \alpha \in A\}$ be a family of functions from uniform spaces (Y_α, ν_α) onto a set X and let μ_G be the inductive limit uniformity with respect to G. Show that the set $B = \{\mathcal{U} \mid \mathcal{U}$ is a covering of X and $g_\alpha^{-1}(\mathcal{U}) \in \nu_\alpha$ for each $\alpha\}$ is a basis for μ_G and that μ_G is the finest uniformity on X such that each g_α in G is uniformly continuous.

5.3 Subspaces, Sums, Products and Quotients

In this section, the classical constructions will be examined as special cases of inductive and projective limit uniformities. It will be seen that subspaces and product spaces are special cases of projective limit uniformities, while sum and quotient spaces are examples of inductive limit uniformities. If (X, μ) is a uniform space and $Y \subset X$, define $i : Y \to X$ by $i(y) = y$ for each $y \in Y$. Let μ_i denote the projective limit uniformity on Y determined by the single function i. It is left as an exercise (Exercise 1) to show that the uniformity μ_i is identical to the subspace uniformity introduced in Section 4.4.

Given a collection $F = \{X_\alpha \mid \alpha \in A\}$ of uniform spaces with uniformities μ_α for each α, we define the **uniform product space** (P, π) to be the Cartesian product set $P = \Pi X_\alpha$ (introduced in the Forward) together with the projective limit uniformity π determined by the collection $\{p_\alpha \mid \alpha \in A\}$ of all canonical projections $p_\alpha : P \to X_\alpha$. Recall that the members of P are functions $f : A \to \cup X_\alpha$ such that $f(\alpha) \in X_\alpha$ for each α. The mappings p_α are defined by $p_\alpha(f) = f(\alpha)$ for each $f \in P$. When dealing with uniform product spaces such as $P = \Pi X_\alpha$, it is customary to refer to the individual uniform spaces X_α from which the product space is constructed as the **coordinate** (uniform) **spaces**.

PROPOSITION 5.3 The topology associated with the uniform product space is the topology of the topological product of the coordinate spaces (considered as topological spaces).

Proof: Let τ_α be the topology associated with μ_α for each α and let τ denote the product topology on $P = \Pi X_\alpha$. It will suffice to show that for each sub-basic U $\in \tau$ containing $x \in P$, there exists a $\mathcal{V} \in \pi$ with $S(x,\mathcal{V}) \subset U$ and for each $S(x,\mathcal{W})$ where $\mathcal{W} \in \pi$, there exists a basic $V \in \tau$ with $x \in V \subset S(x,\mathcal{W})$. If $U \in \tau$ is a sub-basic open set then $U = \{f \in P \mid f(\beta) \in U_\beta\}$ for some open $U_\beta \subset X_\beta$. Since $x \in U$, $p_\beta(x) = x(\beta) \in U_\beta$ which implies there exists some $\mathcal{V}_\beta \in \mu_\beta$ with $S(x(\beta),\mathcal{V}_\beta) \subset U_\beta$. Since π is the projective limit uniformity determined by the collection $\{p_\alpha\}$ of canonical projection, by Exercise 5 of Section 5.2, $p_\beta^{-1}(\mathcal{V}_\beta)$ is a sub-basic member of π. Let $f \in S(x, p_\beta^{-1}(\mathcal{V}_\beta)) = \{p_\beta^{-1}(V_\beta) \mid x(\beta) \in V_\beta \in \mathcal{V}_\beta\}$. Then $f \in p_\beta^{-1}(V_\beta)$ for some $V_\beta \in \mathcal{V}_\beta$ that contains $x(\beta)$. Hence $f(\beta) \in V_\beta \subset S(x(\beta),\mathcal{V}_\beta) \subset U_\beta$ so $f \in U$. Therefore, $S(x, p_\beta^{-1}(\mathcal{V}_\beta)) \subset U$ so τ is coarser than the uniform product topology on P.

Conversely, suppose $\mathcal{W}$ is a basic member of π. Since π is the projective limit uniformity determined by the collection $\{p_\alpha\}$ of canonical projections, by Exercise 5 of Section 5.2, there exists a finite collection $\mathcal{W}_{\alpha_1} \ldots \mathcal{W}_{\alpha_n}$ of coverings in $\mu_{\alpha_1} \ldots \mu_{\alpha_n}$ respectively such that $\mathcal{W} = \cap_{i=1}^n p_{\alpha_i}^{-1}(\mathcal{W}_{\alpha_i})$. For each $i = 1 \ldots n$ choose $V_{\alpha_i} \in \mathcal{W}_{\alpha_i}$ with $x(\alpha_i) \in V_{\alpha_i}$. Then $x \in V = \cap_{i=1}^n V_{\alpha_i} = \cap_{i=1}^n p_{\alpha_i}^{-1}(V_{\alpha_i})$ and V is a basic member of τ. If $f \in V$ then both x and f belong to $\cap_{i=1}^n p_{\alpha_i}^{-1}(V_{\alpha_i})$. But $S(x,\mathcal{W}) = \{f \in P \mid x, f \in \cap_{i=1}^n p_{\alpha_i}^{-1}(W_{\alpha_i})$ for some $W_{\alpha_i} \in \mathcal{W}_{\alpha_i}$ for each $i = 1 \ldots n\}$. Hence $x \in V \subset S(x,\mathcal{W})$ so τ is finer than the uniform product topology on P. Therefore, τ is the uniform product topology on P. ∎

Quotient uniformities are an important example of inductive limit uniformities. If (X, μ) is a uniform space and R is an equivalence relation on X, let $q{:}X \to X/R$ be the canonical projection of X onto the quotient set X/R (introduced in Chapter 0). If we then give $Q = X/R$ the inductive limit uniformity μ_q determined by the single function q, then (Q, μ_q) is called the **quotient uniform space** with respect to R. It is left as an exercise (Exercise 5) to show that the topology associated with μ_q is the topology of the quotient (topological) space X/R, also introduced in the Introduction.

A word of caution is in order here. The space Q may not be Hausdorff (see Exercise 7). To insure that the quotient of a Hausdorff uniform space is again Hausdorff, we need the concept of a *uniform quotient* (see Exercise 8). It is of interest to observe that every onto mapping $f{:}X \to Y$ from X to a set Y determines an equivalence relation R_f on X by taking $(x,y) \in R_f$ if $f(x) = f(y)$. Moreover, the canonical projection $q{:}X \to X/R_f$ determines a one-to-one function $g{:}X/R_f \to Y$ such that $f = g © q$. g is the function defined by $g(x^*) = f(x)$ where x^* is the equivalence class containing x.

If Y is also a uniform space with uniformity ν, then by Proposition 0.20, f is continuous if and only if g is continuous. It is easily seen that this continuity condition also holds for uniform continuity; i.e., f is uniformly continuous if and only if g is uniformly continuous. In fact, if g is uniformly continuous, then f must be uniformly continuous since it is the composition of the uniformly continuous function g and the canonical projection q (which is uniformly continuous by definition). Conversely, if f is uniformly continuous then for each $\mathcal{V} \in \nu$, $f^{-1}(\mathcal{V}) \in \mu$ implies that $q[f^{-1}(\mathcal{V})] \in \mu_q$ (the quotient uniformity on X/R_f). We will show that $g^{-1}(\mathcal{V}) = q[f^{-1}(\mathcal{V})]$ thereby showing g to be uniformly continuous. For this let $V \in \mathcal{V}$ and suppose $x^* \in g^{-1}(V)$. Then $g(x^*) = f(x) \in V$. Let $y \in f^{-1}[f(x)]$. Then $f(y) = f(x)$ which implies $y \in x^*$ so $q(y) = x^*$. Thus $x^* \in q(f^{-1}(V))$ so $g^{-1}(V) \subset q(f^{-1}(V))$. Next suppose $z^* \in q(f^{-1}(V))$. Then $z \in f^{-1}(V)$ which implies $f(z) \in V$. But $g(z^*) = f(z)$ so $z^* \in g^{-1}(V)$. Therefore $q(f^{-1}(V)) \subset g^{-1}(V)$ which implies $g^{-1}(V) = q(f^{-1}(V))$.

In Section 5 of Chapter 1, we introduced the concept of the metric space associated with a pseudo-metric space (X, d) by defining the equivalence relation ~ in X by $x \sim y$ if $d(x,y) = 0$ and defining a metric ρ on X/~ by $\rho(x^*,y^*) = d(p^{-1}(x^*), p^{-1}(y^*))$ where p:X → X/~ was the canonical projection and x^* and y^* were the equivalence classes with respect to ~ containing x and y respectively. A similar construction can be done for uniform spaces; i.e., if (X,μ) is a non-Hausdorff uniform space, it is possible to construct a Hausdorff uniform space from X. For this, define R_μ on X by $(x,y) \in R_\mu$ if $x \in S(y,\mathcal{U})$ for each $\mathcal{U} \in \mu$. Then let q:X → X/R_μ be the canonical projection of X onto the quotient set X/R_μ and let μ_q be the quotient uniformity determined by q. Then (X/R_μ, μ_q) is a uniform space called the **Hausdorff uniform space associated with** (X, μ).

PROPOSITION 5.4 The Hausdorff uniform space associated with (X,μ) determines a Hausdorff topology and the canonical projection q:X → X/R_μ is both an open and closed mapping.

Proof: Let x^* and y^* be distinct members of X/R_μ. Then y does not belong to x^* which implies y does not belong to $S(x,\mathcal{U})$ for some $\mathcal{U} \in \mu$. By Exercise 6 of Section 5.2, $B = \{\mathcal{V} \mid \mathcal{V}$ is a covering of X/R_μ and $q^{-1}(\mathcal{V}) \in \mu\}$ is a basis for μ_q. Now $q(\mathcal{U}) = \{q(U) \mid U \in \mathcal{U}\}$ covers X/R_μ and $\mathcal{U} < q^{-1}[q(\mathcal{U})]$ so $q(\mathcal{U}) \in \mu_q$. If $y^* \in S(x^*, q(\mathcal{U}))$ then both x^* and $y^* \in q(U)$ for some $U \in \mathcal{U}$ which implies $x,y \in U$ so $y \in S(x,\mathcal{U})$ which is a contradiction. Hence y^* does not belong to $S(x^*,q(\mathcal{U}))$. Let $\mathcal{W} \in \mu_q$ such that $\mathcal{W}^* < q(\mathcal{U})$. Then $S(x^*,\mathcal{W}) \cap S(y^*,\mathcal{W}) = \emptyset$ so (X/R_μ, μ_q) determines a Hausdorff topology.

To show that the canonical projection q is both an open and closed mapping, first let U be an open set in X and suppose $x^* \in q(U)$. Then $q(x) = q(y)$ for some $y \in U$. Therefore, $x \in y^*$. Let $\mathcal{V} \in \mu$ such that $S(y,\mathcal{V}) \subset U$. Then $x \in S(y,\mathcal{V})$ which implies $x^* \in S(y^*, q(\mathcal{V})) \subset q(U)$. Hence $q(U)$ is open in X/R_μ so q is an open mapping. If F is closed in X and if x^* does not belong to $q(F)$ then $x \in F$ which implies there exists a $\mathcal{W} \in \mu$ such that $S(x,\mathcal{W}) \cap Star(F,\mathcal{W}) = \emptyset$. Let $\mathcal{V} \in \mu$ with $\mathcal{V}^* < \mathcal{W}$. Suppose $S(x^*, q(\mathcal{V})) \cap q(F) \neq \emptyset$. Then there exists some $y^* \in$ X/R_μ such that $x^*,y^* \in q(V)$ for some $V \in \mathcal{V}$ and such that $y^* \in q(F)$. $y^* \in q(F)$ implies that there exists some $z \in F$ with $q(z) = q(y)$ which in turn implies $y \in z^*$, so $y \in S(z,\mathcal{V})$. Moreover, $x^*,y^* \in q(V)$ implies there are r, s

$\in \mathcal{V}$ with $q(r) = q(x)$ and $q(s) = q(y)$ so $y \in S(s,\mathcal{V})$ and $x \in S(r,\mathcal{V})$. Consequently, there exist V_1, V_2, and $V_3 \in \mathcal{V}$ such that $x,r \in V_1, s,y \in V_2$, and $y,z \in V_3$. But then there are W_1 and $W_2 \in \mathcal{W}$ with $V_1 \cup V_2 \subset W_1$ and $V_2 \cup V_3 \subset W_2$. Now $x,s \in W_1$ and $s,z \in W_2$ implies $W_1 \cap W_2 \neq \emptyset, W_1 \subset S(x,\mathcal{W})$ and $W_2 \subset Star(F,\mathcal{W})$. Hence $S(x,\mathcal{W}) \cap Star(F,\mathcal{W}) \neq \emptyset$ which is a contradiction. Therefore, $S(x^*, q(\mathcal{V})) \cap q(F) = \emptyset$. Thus $q(F)$ must be closed so q is also a closed mapping. ∎

Our last example of an inductive limit uniformity is the *uniform sum* of a collection $\{(X_\alpha, \mu_\alpha)\}$ of uniform spaces. We define a new uniform space ΣX_α by first defining the points of ΣX_α to be the ordered pairs (x, α) where $x \in X_\alpha$. The mappings $i_\alpha : X_\alpha \to \Sigma X_\alpha$ defined by $i_\alpha(x) = (x, \alpha)$ are called the **canonical injections**. The uniformity σ of ΣX_α is defined to be the inductive limit uniformity determined by the collection $\{i_\alpha\}$ of canonical injections. The uniform space $(\Sigma X_\alpha, \sigma)$ is called the **uniform sum** of the uniform spaces X_α. The proof of the following proposition is left as an exercise (Exercise 6).

PROPOSITION 5.5 The topology associated with the uniform sum of a collection of uniform spaces is the topology of the disjoint topological sum of these spaces considered as topological spaces.

The canonical injections essentially transfer the uniformities μ_α from the spaces X_α onto the disjoint "pieces" $i_\alpha(X_\alpha)$ of ΣX_α and since σ is the inductive limit uniformity determined by the i_α's, σ is the finest uniformity on ΣX_α that makes all the i_α's uniformly continuous.

EXERCISES

1. Show that if (X, μ) is a uniform space and $Y \subset X$, that the projective limit uniformity μ_i on Y determined by the function $i: Y \to X$ such that $i(y) = y$ for each $y \in Y$ is identical with the subspace uniformity on Y introduced in Section 4.4.

2. Show that a net in a product space converges to a point p if and only if its projection in each coordinate subspace converges to the projection of p.

3. Show that a net in a product uniform space is Cauchy if and only if the projection of the net in each coordinate subspace is Cauchy.

4. Show that the product ΠX_α of complete uniform spaces (X_α, μ_α) is complete.

5. Let (X, μ) be a uniform space and let R be an equivalence relation on X. Show that the topology associated with the quotient uniform space with respect

to R is the topology of the quotient (topological) space X/R introduced in the Introduction.

6. Prove Proposition 5.5.

7. Show that if (X, μ) is a non-normal uniform space with the finest uniformity u, that there exists an equivalence relation R on X such that X/R is not Hausdorff.

UNIFORM QUOTIENTS

A **uniform relation** in a uniform space (X, μ) is an equivalence relation R on X such that for each pair of non-equivalent points $x,y \in$ X, there exists a sequence $\{\mathcal{U}_n\} \subset \mu$ satisfying:

(1) If xRx' and yRy', then x' and y' are in no common member of $\mathcal{U}_1$.
(2) If $U_{n+1} \in \mathcal{U}_{n+1}$, there exists a $U_n \in \mathcal{U}_n$ such that if $p \in U_{n+1}$ and pRq, then $S(q, \mathcal{U}_{n+1}) \subset U_n$.

A **uniform quotient** is a quotient uniform space X/R where the equivalence relation R is a uniform relation.

8. Show that a uniform quotient of a Hausdorff uniform space is Hausdorff.

9. Show that the canonical projection q:X → X/R of a uniform quotient is uniformly continuous.

UNIFORMLY LOCALLY COMPACT SPACES

A uniform space is said to be **uniformly locally compact** if it has a uniform covering consisting of compact sets. For each countable ordinal α put $X_\alpha = \alpha + 1$ and let τ_α be the order topology on X_α. Then X_α is compact, so X_α has a unique uniformity μ_α consisting of all open coverings. Let (Σ, μ) be the sum of the collection $\{(X_\alpha, \mu_\alpha) | \alpha < \omega_1\}$. Let R be the equivalence relation on Σ defined by $x \sim y$ if and only if x and y belong to the same X_α. Let Y = ω_1 with the order topology.

10. [S. Ginsburg and J. Isbell, 1959] Y has a unique uniformity with a basis consisting of all finite open coverings. R is a uniform relation on (Σ, μ). Σ/R = Y. Let q:Σ → Y be the canonical projection. There exists a $\mathcal{U} \in \mu$ such that $q(\mathcal{U}) = \{q(U) | U \in \mathcal{U}\}$ is not a uniform covering in Y. Σ is complete, but $q(\Sigma)$ = Y is not complete. Σ is uniformly locally compact.

5.4 Hyperspaces

Let (X, μ) be a uniform space and let X′ denote the set of all non-void closed subsets of X. If H, K ∈ X′ with H ⊂ *Star* (*K*, $\mathcal{U}$) and K ⊂ *Star* (*H*, $\mathcal{U}$) for some $\mathcal{U}$ ∈ μ, then H and K are said to be **$\mathcal{U}$-close**, denoted by |H - K| < $\mathcal{U}$. Note that this relationship is reflexive, i.e., |H - K| < $\mathcal{U}$ implies |K - H| < $\mathcal{U}$. If $\mathcal{U}$ ∈ μ and F ∈ X′, put *B*(*F*,$\mathcal{U}$) = {K ∈ X′ | |F - K| < $\mathcal{U}$} and let $\mathcal{U}'$ = {*B*(*F*,$\mathcal{U}$) | F ∈ X′}. Then define μ* = {$\mathcal{U}'$ | $\mathcal{U}$ ∈ μ}. We want to show that μ* is the basis for a uniformity μ′ on X′ but first we establish some useful lemmas.

LEMMA 5.1 If $\mathcal{U}$* < $\mathcal{V}$, |*H - F*| < $\mathcal{U}$, *and* |*K - F*| < $\mathcal{U}$, *then* |*H - K*| < $\mathcal{V}$.

Proof: |H - F| < $\mathcal{U}$ implies H ⊂ *Star* (*F*, $\mathcal{U}$) and F ⊂ *Star* (*H*, $\mathcal{U}$). |K - F| < $\mathcal{U}$ implies K ⊂ *Star* (*F*, $\mathcal{U}$) and F ⊂ *Star* (*K*, $\mathcal{U}$). Therefore H ⊂ *Star* (*Star* (*K*, $\mathcal{U}$), $\mathcal{U}$) ⊂ *Star* (*K*, $\mathcal{V}$). Similarly K ⊂ *Star* (*H*, $\mathcal{V}$) so |H - K| < $\mathcal{V}$. ∎

LEMMA 5.2 If $\mathcal{U}$* < $\mathcal{V}$ *and F* ∈ *X′ then S*(*F*,$\mathcal{U}'$) ⊂ *B*(*F*,$\mathcal{V}$) ⊂ *S*(*F*,$\mathcal{V}'$).

Proof: Let H ∈ *S*(*F*,$\mathcal{U}'$). Then H ∈ *B*(*K*,$\mathcal{U}$) for some K ∈ X′ such that F ∈ *B*(*K*,$\mathcal{U}$). Therefore, H ⊂ *Star* (*K*, $\mathcal{U}$), K ⊂ *Star* (*H*, $\mathcal{U}$), F ⊂ *Star* (*K*, $\mathcal{U}$), and K ⊂ *Star* (*F*, $\mathcal{U}$), so H ⊂ *Star* (*Star* (*F*, $\mathcal{U}$),$\mathcal{U}$) ⊂ *Star* (*F*, $\mathcal{V}$). Similarly, F ⊂ *Star* (*H*, $\mathcal{V}$) which implies H ∈ *B*(*F*,$\mathcal{V}$). Therefore *S*(*F*,$\mathcal{U}'$) ⊂ *B*(*F*,$\mathcal{V}$). *B*(*F*,$\mathcal{V}$) ⊂ *S*(*F*,$\mathcal{V}'$) follows from the definition of *S*(*F*,$\mathcal{V}'$). ∎

PROPOSITION 5.6 μ* *is a basis for a uniformity* μ′ *on X′*.

Proof: We need to establish (1) and (3) of the definition of a uniform space. We can do this simultaneously by showing that if $\mathcal{V}'$,$\mathcal{W}'$ ∈ μ*, there exists a $\mathcal{U}'$ ∈ μ* with $\mathcal{U}'$ <* $\mathcal{V}'$∩$\mathcal{W}'$. For this let $\mathcal{U}$ ∈ μ such that $\mathcal{U}$* < $\mathcal{V}$∩$\mathcal{W}$. Let U′ ⊂ $\mathcal{U}'$. Then U′ = *B*(*E*,$\mathcal{U}$) for some E ∈ X′. Let F ∈ *Star* (*U′*,$\mathcal{U}'$). Then F ∈ *B*(*K*,$\mathcal{U}$) for some K ∈ X′ such that K ∈ *B*(*E*,$\mathcal{U}$) which implies |F - K| < $\mathcal{U}$ and |K - E| < $\mathcal{U}$. By Lemma 5.1, |F - E| < $\mathcal{V}$ so F ∈ *B*(*E*,$\mathcal{V}$) ∈ $\mathcal{V}'$. Therefore, *Star* (*U′*,$\mathcal{U}'$) ⊂ *B*(*E*,$\mathcal{V}$) so $\mathcal{U}'$ <* $\mathcal{V}'$. Similarly, $\mathcal{U}'$ <* $\mathcal{W}'$ so $\mathcal{U}'$ <* $\mathcal{V}'$∩$\mathcal{W}'$. Thus μ* is a basis for a uniformity μ′ on X′. ∎

The uniform space (X′, μ′) is called the **hyperspace** of (X, μ). Hyperspaces of topological spaces have also been studied. The original notion of a hyperspace is due to F. Hausdorff (see *Mengenlehre*, 3rd edition, Springer, Berlin, 1927). Hausdorff defined a metric on the set of non-empty closed and bounded subsets of a given metric space. L. Vietoris generalized the concept to topological spaces (see *Bereiche Zweiter Ordnung*, Monatshefte fur Mathematik und Physik, Volume 33, 1923, pp. 49-62). N. Bourbaki introduced the uniformity for the hyperspace of non-void closed subsets of a uniform space in terms of an entourage uniformity (*Topologie General*, Paris, Hermann, 1940).

In this section we translate Bourbaki's approach in terms of covering uniformities.

PROPOSITION 5.7 The hyperspace of a uniform space is Hausdorff.

Proof: Let (X, μ) be a uniform space and (X', μ') its hyperspace. Choose H, K $\in X'$ with $H \neq K$. Then either H contains a point not in K or K contains a point not in H. Without loss of generality we may assume there is an $x \in H$ such that x does not belong to K. Theorem 2.6 can be used to show there exists $\mathcal{U},\mathcal{V} \in \mu$ with $\mathcal{U}^* < \mathcal{V}$ such that $S(x,\mathcal{V}) \cap Star(K,\mathcal{V}) = \emptyset$. Suppose $K \in B(H,\mathcal{U})$. Then $K \subset Star(H,\mathcal{U})$ and $H \subset Star(K,\mathcal{U})$ which implies $x \in Star(K,\mathcal{U})$ which is a contradiction. Therefore, K does not belong to $B(H,\mathcal{U})$ so X′ is T_1. Since X′ is regular, it is also Hausdorff. ∎

LEMMA 5.3 A net $\{F_\alpha\}$ in X′ converges (clusters) to some $F \in X'$, if and only if for each $\mathcal{V} \in \mu$, $\{F_\alpha\}$ is eventually (frequently) in $B(F,\mathcal{V})$.

Proof: Let $\mathcal{V} \in \mu$ and choose $\mathcal{U} \in \mu$ with $\mathcal{U}^* < \mathcal{V}$. If $\{F_\alpha\}$ converges to F then $\{F_\alpha\}$ is eventually (frequently) in $S(F,\mathcal{U}')$. But by Lemma 5.2, $S(F,\mathcal{U}') \subset B(F,\mathcal{V})$. Conversely, if for each $\mathcal{V} \in \mu$, $\{F_\alpha\}$ is eventually (frequently) in $B(F,\mathcal{V})$, then it is eventually in $S(F,\mathcal{V}')$ by Lemma 5.2, so $\{F_\alpha\}$ converges to F. ∎

LEMMA 5.4 $\{F_\alpha\}$ is Cauchy in X′ if and only if for each $\mathcal{V} \in \mu$, there exists a β such that for each $\alpha \geq \beta$, $F_\alpha \in B(F_\beta,\mathcal{V})$.

Proof: Let $\mathcal{U} \in \mu$ with $\mathcal{U}^* < \mathcal{V}$. Since $\{F_\alpha\}$ is Cauchy, there exists a β such that for some $H \in X'$, $F_\alpha \in B(H,\mathcal{U})$ for each $\alpha \geq \beta$. Then for each $\alpha \geq \beta$, $F_\alpha \subset Star(H,\mathcal{U})$ and $H \subset Star(F_\alpha,\mathcal{U})$. Also, $F_\beta \subset Star(H,\mathcal{U})$ and $H \subset Star(F_\beta,\mathcal{U})$. Therefore, $F_\alpha \subset Star(Star(F_\beta,\mathcal{U}),\mathcal{U}) \subset Star(F_\beta,\mathcal{V})$. Similarly, $F_\beta \subset Star(F_\alpha,\mathcal{V})$ so $F_\alpha \in B(F_\beta,\mathcal{V})$ for each $\alpha \geq \beta$. The converse it obvious since $B(F_\beta,\mathcal{V}) \in \mathcal{V}'$. ∎

Let $\{F_\alpha\}$ be a net in X′. A point $p \in X$, each of whose neighborhoods meets cofinally many of the F_α's is said to be a **cluster point** of $\{F_\alpha\}$ (in X).

PROPOSITION 5.8 The set K of cluster points of a net $\{F_\alpha\}$ in X′ is closed. If $\{F_\alpha\}$ converges in X′, it converges to K.

Proof: Let p be a limit point of K and let U be a neighborhood of p. Then U contains a point $k \in K$. Since k is a cluster point of $\{F_\alpha\}$, U frequently meets $\{F_\alpha\}$ which implies $p \in K$. Hence K is closed. Next assume $\{F_\alpha\}$ converges to $F \in X'$. Suppose x does not belong to F. Then, as pointed out in Proposition 5.7, there is a $\mathcal{U} \in \mu$ with $S(x,\mathcal{U}) \cap Star(F,\mathcal{U}) = \emptyset$. By Lemma 5.3, $\{F_\alpha\}$ is eventually in $B(F,\mathcal{U})$. Therefore, there is a β with $F_\alpha \in B(F,\mathcal{U})$ for each $\alpha \geq \beta$ which implies $F_\alpha \subset Star(F,\mathcal{U})$ for each $\alpha \geq \beta$. Hence x is not a cluster point of

F. Consequently, $F \subset K$ and there can be no points of K that are not in F, so F = K. ∎

Our first undertaking with hyperspaces will be to characterize supercompleteness (see Section 5.1 for the definition) in terms of the behavior of a certain class of nets in the original space X. In 1988, B. Burdick informed the author that he had shown that cofinal completeness implies supercompleteness and wondered if the reverse implication held. The author pointed out that real Hilbert space (Chapter 1) can be shown to be a supercomplete metric space that is not cofinally complete (see Exercise 1). Thereafter, Burdick discovered the following characterization of supercompleteness that yields the implication: cofinal completeness => supercompleteness as a corollary.

Burdick defines a net $\{x_\alpha\}$, $\alpha \in A$, in a uniform space (X, μ) to be **almost Cauchy** if for each $\mathcal{U} \in \mu$ there exists a collection C of cofinal subsets of A such that for each $K \in C$, $\{x_\gamma \mid \gamma \in K\} \subset U$ for some $U \in \mathcal{U}$, and such that $\cup C$ is residual in A. Clearly, Cauchy nets are almost Cauchy and almost Cauchy nets are cofinally Cauchy. His characterization of supercompleteness is that a uniform space is supercomplete if and only if each almost Cauchy net clusters.

Burdick's proof depends on a theorem of Isbell on *partially convergent functions*. Isbell's development of partially convergent functions will be given in a later section. For now we present an alternate constructive proof directly from the definitions of supercompleteness and almost Cauchy nets. This constructive proof has the advantage of making clear just how the clustering of almost Cauchy nets is related to supercompleteness. The proof is rather long, so we decompose it into two lemmas plus the main constructions. We define a net $\{F_\alpha\}$ in X′ to be **proper** if $\{F_\alpha\}$ is ordered by set inclusion ($\alpha \leq \beta$ if and only if $F_\beta \subset F_\alpha$) and if $F_\alpha \subset F \in X'$ for some α then $F \in \{F_\alpha\}$.

LEMMA 5.5 A proper net $\{F_\alpha\}$ in X′ is Cauchy if and only if for each $\mathcal{V} \in \mu$ there is a β with $F_\beta \subset Star(F_\alpha,\mathcal{V})$ for each α.

Proof: Assume $\{F_\alpha\}$ is a proper Cauchy net in X′. Let $\mathcal{V} \in \mu$. Pick $\mathcal{U} \in \mu$ with $\mathcal{U}^{**} < \mathcal{V}$. By Lemma 5.4, there is a γ such that for each $\alpha \geq \gamma$, $F_\alpha \in B(F_\gamma,\mathcal{U})$. Now $Cl(Star(F_\gamma,\mathcal{U})) = F_\beta$ for some β since $\{F_\alpha\}$ is proper. Let $\alpha \geq \gamma$. Then $F_\alpha \in B(F_\gamma,\mathcal{U})$ which implies $F_\gamma \subset Star(F_\alpha,\mathcal{U})$ so $Star(F_\gamma,\mathcal{U}) \subset Star(F_\alpha,\mathcal{U}^*)$, which implies $F_\beta \subset Star(F_\alpha,\mathcal{V})$. If $\alpha < \gamma$ then $F_\gamma \subset F_\alpha$ so $F_\beta \subset Star(F_\gamma,\mathcal{V}) \subset Star(F_\alpha,\mathcal{V})$. Consequently, there is a β with $F_\beta \subset Star(F_\alpha,\mathcal{V})$ for each α.

Conversely, assume that for each $\mathcal{V} \in \mu$ there is a β with $F_\beta \subset Star(F_\alpha,\mathcal{V})$ for each α. If $\alpha \geq \beta$ then $F_\beta \subset Star(F_\alpha,\mathcal{V})$ and $F_\alpha \subset F_\beta \subset Star(F_\beta,\mathcal{V})$. Therefore $F_\alpha \in B(F_\beta,\mathcal{V})$ for each $\alpha \geq \beta$ so $\{F_\alpha\}$ is Cauchy. ∎

LEMMA 5.6 (X, μ) is supercomplete if and only if each proper Cauchy net in X′ converges.

Proof: Let $\{F_\alpha\}$ be a Cauchy net in X′. For each α put $H_\alpha = \cup\{F_\beta \mid \beta \geq \alpha\}$. Let $\{S_\gamma\}$ denote the collection of elements of X′ containing some H_α where $\{S_\gamma\}$ is directed by set inclusion ($\gamma < \delta$ if and only if $S_\delta \subset S_\gamma$). Clearly $\{S_\gamma\}$ is proper. Since $\{F_\alpha\}$ is Cauchy, so is $\{S_\gamma\}$. To see this, let $\mathcal{V} \in \mu$ and pick $\mathcal{U} \in \mu$ with $\mathcal{U}^* < \mathcal{V}$. By Lemma 5.4, there exists a β such that $F_\alpha \in B(F_\beta,\mathcal{U})$ for each $\alpha \geq \beta$. Therefore, $F_\alpha \subset Star(F_\beta,\mathcal{U})$ for each $\alpha \geq \beta$ which implies $Cl(H_\beta) \subset Star(F_\beta,\mathcal{V})$. Now $Cl(H_\beta) = S_\delta$ for some $S_\delta \in \{S_\gamma\}$. For each $\gamma \geq \delta$, $S_\gamma \subset S_\delta$ which implies $S_\gamma \subset Star(F_\beta,\mathcal{V})$. Also, for each $\gamma \geq \delta$, there exists a $\lambda \geq \beta$ with $H_\lambda \subset S_\gamma$ which implies $F_\lambda \subset S_\gamma$ and $F_\beta \subset Star(F_\lambda,\mathcal{U})$ so $F_\beta \subset Star(S_\gamma,\mathcal{U})$ which implies $S_\gamma \in B(F_\beta,\mathcal{V})$ for each $\gamma \geq \delta$. Therefore, $\{S_\gamma\}$ is a proper Cauchy net.

Then by hypothesis, $\{S_\gamma\}$ converges to its set of cluster points K. Let $\mathcal{V} \in \mu$. Pick $\mathcal{U} \in \mu$ with $\mathcal{U}^{***} < \mathcal{V}$. By Lemma 5.3, $\{S_\gamma\}$ is eventually in $B(K,\mathcal{U})$ so there exists a δ with $S_\gamma \in B(K,\mathcal{U})$ for each $\gamma \geq \delta$. Now there exists a β with $Cl(H_\beta) \subset S_\delta$. Since $\{F_\alpha\}$ is Cauchy, there exists a ξ with $F_\alpha \in B(F_\xi,\mathcal{U})$ for each $\alpha \geq \xi$. Pick λ such that $\lambda \geq \beta$ and $\lambda \geq \xi$. Then $Cl(H_\lambda) \subset Cl(H_\beta) \subset S_\delta$ so $Cl(H_\lambda) = S_\sigma$ for some $\sigma \geq \delta$. Hence $Cl(H_\lambda) \in B(K,\mathcal{U})$. Since $\lambda \geq \xi$, $F_\alpha \in B(F_\xi,\mathcal{U})$ for each $\alpha \geq \lambda$ which implies $F_\alpha \subset Star(F_\xi,\mathcal{U})$ and $F_\xi \subset Star(F_\alpha,\mathcal{U})$ for each $\alpha \geq \lambda$. Thus $Cl(H_\lambda) \subset Star(F_\xi,\mathcal{U}^*)$ and $F_\xi \subset Star(Cl(H_\lambda),\mathcal{U}^*)$. Hence $Cl(H_\lambda) \in B(F_\xi,\mathcal{U}^*)$. Consequently, we have

$$|F_\alpha - F_\xi| < \mathcal{U}, |F_\xi - Cl(H_\lambda)| < \mathcal{U}^*, \text{ and } |Cl(H_\lambda) - K| < \mathcal{U} \text{ for each } \alpha \geq \lambda.$$

Then multiple applications of Lemma 5.1 yields $|F_\alpha - K| < \mathcal{V}$ for each $\alpha \geq \lambda$ so $\{F_\alpha\}$ converges to K. Hence X is supercomplete. ▪

Burdick's theorem appeared in an article titled *A note on completeness of hyperspaces* published in the Proceedings of the Fifth Northeast Topology Conference (1991) by Marcel-Dekker. The following proof is due to the author.

THEOREM 5.1 (B. Burdick, 1991) A uniform space (X, μ) is supercomplete if and only if each almost Cauchy net clusters.

Proof: Assume each almost Cauchy net clusters and suppose (X, μ) is not supercomplete. Then by Proposition 5.8 and Lemma 5.6, there exists a proper Cauchy net $\{F_\alpha\}$, $\alpha \in A$, in X′ that does not cluster to its set of cluster points K. There are two cases to consider here. First, K might be the empty set, in which case K would not belong to X′ and therefore could not be a cluster point of $\{F_\alpha\}$. On the other hand, K may not be empty. This is the case we will assume first. Later we will see that an argument similar to the one we will use here can be used to show that under the hypothesis of this theorem, K cannot be the empty set.

If $K \neq \emptyset$, there exists a $\mathcal{U} \in \mu$ such that $\{F_\alpha\}$ is eventually outside $B(K,\mathcal{U})$ which implies there exists a β with F_α not contained in $B(K,\mathcal{U})$ for each $\alpha \geq \beta$. Let $p \in K$. Then there exists a cofinal $C \subset A$ with $p \in Star(F_\gamma,\mathcal{U})$ for each $\gamma \in C$. Pick $\delta \in C$ with $\delta > \beta$. Then $F_\delta \subset F_\beta$ which implies $p \in Star(F_\beta,\mathcal{U})$. Therefore, $K \subset Star(F_\beta,\mathcal{U})$ which implies F_β is not contained in $Star(K,\mathcal{U})$ or else $F_\beta \in B(K,\mathcal{U})$ which is a contradiction.

Similarly, F_α is not contained in $Star(K,\mathcal{U})$ for each $\alpha \geq \beta$. Suppose γ is not greater than or equal to β. Then there is a δ such that $\gamma \leq \delta$ and $\beta \leq \delta$ which implies F_δ is not contained in $Star(K,\mathcal{U})$ and $F_\delta \subset F_\gamma$. Hence F_γ is not contained in $Star(K,\mathcal{U})$. Consequently, F_α is not contained in $Star(K,\mathcal{U})$ for each $\alpha \in A$. Put $U = Star(K,\mathcal{U})$ and for each $\alpha \in A$ let $H_\alpha = F_\alpha - U$. Then $\{H_\alpha\}$ is a Cauchy net in X′ that is directed by set inclusion. Let $\{G_\gamma\}, \gamma \in B$, be the collection of elements of X′, directed by set inclusion, that contain a member of $\{H_\alpha\}$. Then $\{G_\gamma\}$ is a proper Cauchy net in X′.

Let $D = \{(x,\gamma) \in X \times B \mid x \in G_\gamma\}$ and define $\leq$ on D by $(x,\gamma) \leq (y,\delta)$ if $\gamma \leq \delta$. For each $(x,\gamma) \in D$ put $\psi(x,\gamma) = x$. Then $\psi: D \to X$ is a net which we will show is almost Cauchy. For this let $\mathcal{W} \in \mu$. Pick $\mathcal{U} \in \mu$ with $\mathcal{U}^* < \mathcal{W}$. Since $\{G_\gamma\}$ is stable, there exists a λ with $G_\gamma \in B(G_\lambda,\mathcal{U})$ for each $\gamma \geq \lambda$. Pick $x \in Star(G_\lambda,\mathcal{U})$. Then some $W \in \mathcal{W}$ contains $Star(x,\mathcal{U})$. Let $\mathcal{V} = \{W \in \mathcal{W} \mid W$ contains $Star(x,\mathcal{U})$ for some $x \in G_\lambda\}$. Let $R = \{(y,\beta) \in D \mid \beta \geq \lambda\}$. Then R is residual in D. Let $(y,\beta) \in R$ which implies $\beta \geq \lambda$ which in turn implies $G_\beta \subset Star(G_\lambda,\mathcal{U})$ so $y \in Star(G_\lambda,\mathcal{U})$ which implies $y \in S(x,\mathcal{U})$ for some $x \in G_\lambda$. Then there exists a $W \in \mathcal{V}$ containing y. Therefore, $\psi(y,\beta) = y \in W$. Let $C_W = \{(y,\beta) \in R \mid \psi(y,\beta) \in W$ for some $W \in \mathcal{V}\}$. Then $\cup\{C_W \mid W \in \mathcal{V}\} = R$.

So $\{C_W \mid W \in \mathcal{V}\}$ is a family of subsets of D whose union is residual in D such that $\psi(C_W) \subset W \in \mathcal{V}$. It remains to show that C_W is cofinal in D for each $W \in \mathcal{V}$. For this let $W_0 \in \mathcal{V}$ and $(y,\beta) \in D$. Then there exists an $x_0 \in G_\lambda$ with $S(x_0,\mathcal{U}) \subset W_0$. Pick $(z,\delta) \in R$ with $(x_0,\lambda) \leq (z,\delta)$ and $(y,\beta) \leq (z,\delta)$. Then $G_\delta \subset G_\lambda$ which implies $z \in G_\lambda$. But $(z,\delta) \in R$ implies $\delta \geq \lambda$ which in turn implies $G_\delta \in B(G_\lambda,\mathcal{U})$, so $G_\lambda \subset Star(G_\delta,\mathcal{U})$. Therefore, $x_0 \in S(s,\mathcal{U})$ for some $s \in G_\delta$ which implies $s \in S(x_0,\mathcal{U})$. Then $(x_0,\lambda) \leq (s,\delta)$, $(y,\beta) \leq (s,\delta)$, and $\psi(s,\delta) \in S(x_0,\mathcal{U}) \subset W_0$. Hence C_{W_0} is cofinal in D so ψ is almost Cauchy.

By hypothesis, ψ clusters to some $p \in X$. Let V be a neighborhood of p. Then there exists a cofinal $C \subset D$ with $\psi(C) \subset V$. Let $\alpha \in A$. Then $F_\alpha = G_\beta$ for some $\beta \in B$. Pick $x \in G_\beta$. Then there exists $(y,\delta) \in C$ with $(x,\beta) \leq (y,\delta)$ and $\psi(y,\delta) \in V$. Therefore, $\beta \leq \delta$ which implies $G_\delta \subset G_\beta$ and $\psi(y,\delta) = y \in G_\delta \subset G_\beta \subset F_\alpha$ so $F_\alpha \cap V \neq \emptyset$. Hence p is a cluster point of $\{F_\alpha\}$ so $p \in K$. But $\psi \subset X - U$ which is closed, so $p \in X - U$ which is a contradiction.

Next we have the case $K = \emptyset$. In this case, let D in the argument above be the set $\{(x,\gamma) \in X \times A \mid x \in F_\alpha\}$. Define $\leq$ on D as before and define $\psi: D \to X$ by $\psi(x,\alpha) = x$ for each $(x,\alpha) \in D$. Then just as in the argument above, we can show that ψ is almost Cauchy and hence clusters to some $p \in X$. But then as the argument above shows, $p \in K$, so $K \neq \emptyset$. Consequently, (X, μ) must be supercomplete.

Conversely, assume X is supercomplete. Let $\{x_\alpha\}$, $\alpha \in A$, be an almost Cauchy net in X. For each α put $H_\alpha = \{x_\beta \mid \beta \geq \alpha\}$ and $F_\alpha = Cl(H_\alpha)$. Then $\{F_\alpha\}$ is a net in X′. Let $\mathcal{V} \in \mu$ and pick $\mathcal{U} \in \mu$ with $\mathcal{U}^* < \mathcal{V}$. Since $\{x_\alpha\}$ is almost Cauchy, there exists a collection $\{C_\gamma\}$, $\gamma \in B$, of cofinal subsets of A and a collection $\{U_\gamma\} \subset \mathcal{U}$ such that $\cup_\gamma C_\gamma$ is residual in A and $\{x_\beta \mid \beta \in C_\gamma\} \subset U_\gamma$ for each γ. Pick $\alpha_0 \in A$ such that $\beta \in \cup_\gamma C_\gamma$ for each $\beta \geq \alpha_0$. Then $H_{\alpha_0} \subset \cup_\gamma C_\gamma$. For each γ pick $V_\gamma \in \mathcal{V}$ such that $Star(U_\gamma, \mathcal{U}) \subset V_\gamma$. We want to show $F_{\alpha_0} \subset \cup_\gamma U_\gamma$. Let $y \in F_{\alpha_0}$. If $y \in H_{\alpha_0}$, clearly $y \in \cup_\gamma V_\gamma$. Suppose y is a limit point of H_{α_0}. Let $U \in \mathcal{U}$ be a neighborhood of y. Then there exists an $x_\beta \in H_{\alpha_0}$ with $x_\beta \in U$. Now $x_\beta \in H_{\alpha_0}$ implies $x_\beta \in U_\gamma$ for some γ so $U \subset V_\gamma$. Hence $F_{\alpha_0} \subset \cup_\gamma V_\gamma$.

Let $\alpha_1 \in H_{\alpha_0}$. Then $F_{\alpha_1} \subset F_{\alpha_0}$ which implies $F_{\alpha_1} \subset Star(F_{\alpha_0}, \mathcal{V})$. Pick $\gamma \in B$. Let $\beta \in C_\gamma$ such that $\beta > \alpha_1$. Then $x_\beta \in U_\gamma$ and $x_\beta \in F_{\alpha_1}$ which implies $x_\beta \in V_\gamma \cap F_{\alpha_1}$ so $\cup V_\gamma \subset Star(F_{\alpha_1}, \mathcal{V})$. But $F_{\alpha_0} \subset \cup_\gamma V_\gamma$ so $F_{\alpha_0} \subset Star(F_{\alpha_1}, \mathcal{V})$. Hence $F_{\alpha_1} \in B(F_{\alpha_0}, \mathcal{V})$ for each $\alpha_1 \in H_{\alpha_0}$ so $\{F_\alpha\}$ is Cauchy. Since X′ is complete, $\{F_\alpha\}$ converges to its set of cluster points K. Let $p \in K$. Let W be a neighborhood of p in X. Then there exists a cofinal $C \subset A$ with $F_\alpha \cap W \neq \emptyset$ for each $\alpha \in C$. If $y \in F_\alpha \cap W$ then $y \in H_\alpha$ or y is a limit point of H_α. In either case, W contains some x_β with $\beta \geq \alpha$. For each $\alpha \in C$ pick $\kappa(\alpha) \geq \alpha$ such that $x_{\kappa(\alpha)} \in F_\alpha \cap W$. Then $\{x_{\kappa(\alpha)} \mid \alpha \in C\} \subset W$ is cofinal in A. Hence, $\{x_\alpha\}$ clusters to p. ∎

COROLLARY 5.1 Cofinal completeness implies supercompleteness which in turn implies completeness.

COROLLARY 5.2 (J. Isbell, 1962) If X is paracompact, then it is supercomplete with respect to u.

Notice that our proof that supercompleteness implies each almost Cauchy net clusters only relies on the fact that there exists a cluster point of $\{F_\alpha\}$. Consequently, the existence of a cluster point for each proper Cauchy net in X′ is equivalent to supercompleteness. Furthermore, if p is a cluster point for the proper Cauchy net $F = \{F_\alpha\}$ and $\mathcal{U} \in \mu$, then for each $U \in \mathcal{U}$ containing p, $U \cap F_\beta \neq \emptyset$ for cofinally many F_β's in F. Let γ be any index. Then there exists a $\delta \geq \gamma$ with $U \cap F_\delta \neq \emptyset$ and since F is directed by inclusion we have $U \cap F_\gamma \neq \emptyset$. Therefore, $S(p, \mathcal{U}) \cap F_\alpha \neq \emptyset$ for each $F_\alpha \in F$. Thus we can pick an $x_\mathcal{U} \in S(p, \mathcal{U}) \cap F_\alpha$ for each $\mathcal{U} \in \mu$. Then $\{x_\mathcal{U}\}$ converges to p which implies p is a limit point of F_α for each α. Therefore, $p \in F_\alpha$ for each α since each F_α is closed. Hence $\cap F_\alpha \neq \emptyset$. Similarly, if $\cap F_\alpha \neq \emptyset$ then any $p \in \cap F_\alpha$ is a cluster point of F. We record these observations as

PROPOSITION 5.9 If (X, μ) is a uniform space then the following statements are equivalent:

(1) (X, μ) is supercomplete,

(2) Each proper Cauchy net in X' has a cluster point and

(3) For each proper Cauchy net F in X', $\cap F \neq \emptyset$.

EXERCISES

1. A function f:X → Y from a uniform space (X, μ) into a uniform space (Y,ν) determines a function f':X' → Y' where X' and Y' are the hyperspaces of X and Y respectively. f' is defined by $f'(A) = Cl_Y(f[A])$ for each A ∈ X'. f' is called the **hyperfunction** of f. Show that f' is uniformly continuous if and only if f is uniformly continuous.

THE HYPERSPACE OF A METRIC SPACE

2. The **Hausdorff distance** h between two (closed) sets A and B in a metric space (X, d) is defined as the maximum of $sup\{d(a, B) \mid a \in A\}$ and $sup\{d(A,b) \mid b \in B\}$ where $d(x, S) = inf\{d(x,y) \mid y \in S\}$ for any subset S. Show

(a) h is a metric on HX (the hyperspace of X) that generates the topology of HX so that the hyperspace of a metric space is again a metric space.

(b) If (X, d) is complete then (HX, h) is complete.

THE HYPERSPACE OF A COMPACT SPACE

3. Show that the hyperspace of a compact space is compact.

4. Show that a discrete space of the power of the continuum, with the uniformity determined by all countable coverings, is complete but not supercomplete.

5. Show that real Hilbert space is supercomplete but not cofinally complete.

6. Define the **limit inferior** of a net $\{F_\alpha\}$ in X' to be the set $inf\{F_\alpha\} = \{x \in X \mid$ for each neighborhood U of x, $\{F_\alpha\}$ eventually meets U$\}$. Similarly, define the **limit superior** to be the set $sup\{F_\alpha\} = \{x \in X \mid$ for each neighborhood U of x, $\{F_\alpha\}$ frequently meets U$\}$. If K is the set of cluster points of $\{F_\alpha\}$ then

(a) $sup\{F_\alpha\} = K = inf\{F_\alpha\}$

(b) (X, μ) is supercomplete if and only if whenever $\{F_\alpha\}$ is Cauchy then for each $\mathcal{U} \in \mu$ there is a β with $F_\alpha \subset Star(K, \mathcal{U})$ for each $\alpha \geq \beta$.

7. Use the results of Exercise 8 to show that if X is paracompact, then it is supercomplete with respect to u, without reference to cofinally Cauchy or almost Cauchy nets (i.e., do not appeal to Corollary 5.1).

5.5 Inverse Limits and Spectra

We begin our discussion of inverse limits of topological and uniform spaces with a special case that will motivate the concept in general. Let $\{X_n\}$ be a sequence of topological spaces where n ranges over the non-negative integers and suppose that for each positive integer n, there is a continuous function $f_n{:}X_n \to X_{n-1}$. The sequence of spaces and mappings $\{X_n, f_n\}$ is known as an **inverse limit sequence.** Inverse limit sequences are often represented by diagrams like the one below:

$$\xrightarrow{\;} \ldots \xrightarrow{f_n} X_n \xrightarrow{f_{n-1}} X_{n-1} \to \ldots \xrightarrow{f_2} X_1 \xrightarrow{f_1} X_0$$

If $m < n$, then the composition mapping $f_n^m = f_{m+1}$ © f_{m+2} © . . . © f_n is a continuous mapping from X_n to X_m. Consider the sequence of points $\{x_n\}$ such that for each n, $x_n \in X_n$ and $x_n = f_{n+1}(x_{n+1})$. Then $\{x_n\}$ can be identified with a point of the product space $\Pi_{n=0}^{\infty} X_n$ by means of the function $g{:}J \to \cup X_n$, where J denotes the non-negative integers, defined by $g(n) = x_n$. By means of this identification, the set Y of all such sequences can be considered to be a subset of $\Pi_{n=0}^{\infty} X_n$. Then Y, equipped with the subspace topology, is called the **inverse limit space** of the sequence $\{X_n, f_n\}$. Y is denoted by $lim_{\leftarrow} X_n$ or by X_∞. The functions f_n are sometimes called **bonding maps.**

LEMMA 5.7 If $\{X_n, f_n\}$ is an inverse limit sequence with onto bonding maps and for some countable set $\{a_n\}$ of positive integers there is a set $\{x_{a_n}\}$ such that $x_{a_n} \in X_{a_n}$ for each n and such that if $m < n$, then $f_{a_n}^{a_m}(x_{a_n}) = x_{a_m}$, then there exists a point in $lim_{\leftarrow} X_n$ whose coordinate in X_{a_n} is x_{a_n} for each n.

Proof: First assume $\{a_n\}$ is infinite. For each positive integer m, there exists a least positive integer a_n such that $a_n \geq m$. If $a_n = m$ put $x_m = x_{a_n}$, otherwise let $x_m = f_m^{a_n}(x_{a_n})$. Then $\{x_m\}$ is a point of $lim_{\leftarrow} X_n$ such that $x_{a_n} \in X_{a_n}$ for each n.

Next, assume $\{a_n\}$ is finite. Then there exists a greatest a_n, say a_j. If $m < a_j$ put $x_m = f_m^{a_j}(x_{a_j})$. For each $m \geq a_j$, assume x_m has already been defined. Since f_{m+1} is onto, pick $x_{m+1} \in X_{m+1}$ such that $f_{m+1}(x_{m+1}) = x_m$. Then by induction we can complete the sequence $\{x_m\}$ such that $\{x_m\} \in lim_{\leftarrow} X_n$ and for each n, $x_{a_n} \in X_{a_n}$. ■

Lemma 5.7 illustrates the fundamental property of inverse limit spaces, that for any coordinate x_k, all elements of $lim_{\leftarrow} X_n$ having that coordinate have all other coordinates x_m with $m < k$ determined by the inverse limit sequence, whereas there may be some room for choice of the coordinates x_m with $m > k$.

If the bonding maps are not onto, $lim_{\leftarrow} X_n$ may be the empty set. An example of this occurs when the X_n are all countable discrete spaces, say $X_n = \{x_m^n\}$ and the bonding maps $f_n:X_n \to X_{n-1}$ are of the form $f_n(x_m^n) = x_{m+1}^{n-1}$. Clearly $\{X_n, f_n\}$ is an inverse limit sequence, but if we begin with a point x_m^0, it is only possible to pick the first m coordinates before we reach x_1^m and there does not exist an x_j^{m+1} such that $f_{m+1}(x_j^{m+1}) = x_1^m$. Consequently, there do not exist any sequences $\{x_n\}$ such that $x_n \in X_n$ and $f_n(x_n) = x_{n-1}$ for each $n > 1$. Therefore, $lim_{\leftarrow} X_n = \emptyset$. Under certain conditions, we can assure that $lim_{\leftarrow} X_n \neq \emptyset$. For instance,

THEOREM 5.2 If each space X_n in the inverse limit sequence $\{X_n, f_n\}$ is a compact Hausdorff space then $lim_{\leftarrow} X_n \neq \emptyset$.

Proof: For each positive integer n let $Y_n \subset \Pi X_n$ be defined by $\{y_i\} \in Y_n$ if for each $j < n$, $y_{j-1} = f_j(y_j)$. Then $lim_{\leftarrow} X_n = \cap Y_n$. We will show that for each n, Y_n is closed. Suppose $p \in \Pi X_n - Y_n$. Then for some $j < n$ we have $f_{j+1}(p_{j+1}) \neq p_j$. Since X_j is Hausdorff, there exists disjoint open sets U_j and V_j containing p_j and $f_{j+1}(p_{j+1})$ respectively. Put $V_{j+1} = f_{j+1}^{-1}(V_j)$ and let U_p be a basic open set in ΠX_n containing p and having U_j and V_{j+1} as its jth and $j{+}1^{st}$ factors respectively. Then no point of Y_n lies in U_p since if $q = \{q_n\} \in U_p$, then $q_{j+1} \in V_{j+1}$ which implies $q_j \in V_j$ so q does not belong to U_p. Therefore, Y_n is closed. Since $\{Y_n\}$ is a decreasing chain of closed sets in the compact space ΠX_n, $\cap Y_n \neq \emptyset$ so $lim_{\leftarrow} X_n \neq \emptyset$.∎

There are many applications of inverse limit spaces of inverse limit sequences in topology. What we are interested in here is extending the concept to uniform spaces, and extending it in a more general setting. For this first notice that by changing the definition of an inverse limit sequence $\{X_n, f_n\}$ so that the X_n are now uniform spaces and the bonding maps are uniformly continuous, we get an inverse limit uniform space $lim_{\leftarrow} X_n$ that is a uniform subspace of the uniform product space ΠX_n. This follows from the parallel between product topological spaces and product uniform spaces [see Section 5.3].

But what we have in mind is something more general than this. Let (D,<) be a directed set and suppose that for each $\alpha \in D$, X_α is a topological space. Further suppose that whenever $\alpha, \beta \in D$ with $\alpha \leq \beta$, there is a continuous function $f_\beta^\alpha: X_\beta \to X_\alpha$, and that these functions satisfy the following rules:

(1) f_α^α is the identity function for each $\alpha \in D$
(2) $f_\beta^\alpha \circledcirc f_\gamma^\beta = f_\gamma^\alpha$ whenever $\alpha < \beta < \gamma$.

Let $Y = \{X_\alpha \mid \alpha \in D\}$ and $F = \{f_\beta^\alpha \mid \alpha, \beta \in D \text{ with } \alpha \leq \beta\}$. Then the pair $\{Y, F\}$

is called an **inverse limit system.** Since the set of non-negative integers forms a directed set, it is clear that an inverse limit sequence is a special case of an inverse limit system. To define the inverse limit space of an inverse limit system, we let $\{x_\alpha\}$ denote a net in $\cup X_\alpha$ such that $x_\alpha \in X_\alpha$ for each $\alpha \in$ D and such that if $\alpha < \beta$ in D then $f^\alpha_\beta(x_\beta) = x_\alpha$. Then $\{x_\alpha\}$ can be identified with a point p of ΠX_α having coordinates $p_\alpha = x_\alpha$. The collection of all such nets, under this identification, is a subspace of ΠX_α that we denote by $lim_{\leftarrow} X_\alpha$ and call the **inverse limit space** of the system $\{Y, F\}$.

LEMMA 5.8 $lim_{\leftarrow} X_\alpha$ *of the inverse limit system* {Y, F} *is a closed subspace of the product space* ΠX_α.

The proofs of Lemma 5.8 and of the following Theorem are left as exercises (Exercises 1 and 2).

THEOREM 5.3 The inverse limit space of an inverse limit system of compact Hausdorff spaces is a compact Hausdorff space, and if each space of the inverse limit system is non-empty, then the inverse limit space is non-empty.

Again, if in the definition of the inverse limit system of $\{Y, F\}$, we require the $X_\alpha \in Y$ to be uniform spaces and the *bonding maps* in F to be uniformly continuous functions, we get an inverse limit uniform space $lim_{\leftarrow} X_\alpha$ that is a uniform subspace of the uniform product space ΠX_α. We now give an important example of an inverse limit uniform space that we will use later on.

Let (X,ν) be a uniform space and let $\{\Phi_\lambda \mid \lambda \in \Lambda\}$ denote the collection of all normal sequences of open uniform coverings. For each λ let $\Phi_\lambda = \{\mathcal{U}^\lambda_n \mid n = 1, 2, \ldots\}$ where $\mathcal{U}^\lambda_n \in \nu$ and $\mathcal{U}^\lambda_n <^* \mathcal{U}^\lambda_{n-1}$ for $n = 2, 3, \ldots$ We let X_λ denote the topological space obtained from X by taking $\{S(x, \mathcal{U}^\lambda_n) \mid n = 1, 2, \ldots\}$ as a basis for the neighborhoods of x at each point $x \in$ X. It is easily verified that X_λ has the topology of the pseudo-metric space (X, d_λ) as constructed in the proof of Theorem 2.5 from a normal sequence of uniform coverings and that $S(x, \mathcal{U}^\lambda_n) = S(x, 2^{-n})$ with respect to d_λ. Next, let X/Φ_λ be the quotient space obtained from the equivalence relation ~ defined by setting $x \sim y$ if $y \in S(x, \mathcal{U}^\lambda_n)$ for each n. Again, it is easily verified that X/Φ_λ has the topology of the metric space M in the proof of Theorem 2.5 obtained by taking the quotient space of the pseudo-metric space (X, d_λ) with respect to the equivalence relation ~ defined by $x \sim y$ if and only if $d_\lambda(x,y) = 0$. Hence for each $\lambda \in \Lambda$, X_λ is a pseudo-metrizable space and X/Φ_λ is a metrizable space.

If for each $\lambda \in \Lambda$ we denote the identity map on X, viewed as a mapping from X onto X_λ, by i_λ and if we denote the quotient mapping from X_λ onto X/Φ_λ by ϕ^*_λ then the composition mapping $\phi_\lambda = \phi^*_\lambda \circledcirc i_\lambda$:X $\rightarrow$ X/Φ_λ is continuous. In order to construct an inverse limit system from the family $\{X/\Phi_\lambda \mid \lambda \in \Lambda\}$, we first introduce a partial ordering < on Λ as follows: let $\lambda, \mu \in \Lambda$ such that for each positive integer n, there exists a positive integer k such that $\mathcal{U}_k^\mu < \mathcal{U}_n^\lambda$, then we put $\lambda < \mu$. Then $(\Lambda, <)$ is countably directed. To see this let $\{\lambda_n\}$ be a countable set of members of Λ. For each positive integer k let $\mathcal{U}_k^\mu = \mathcal{U}_1^\lambda \cap \ldots \cap \mathcal{U}_k^\lambda$ and put $\Phi_\mu = \{\mathcal{U}_n^\mu \mid n = 1, 2, \ldots\}$. Then clearly $\mu > \lambda_n$ for each positive integer n.

Now suppose $\lambda < \mu$. Then if $U \subset X$ is open in X_λ, it must also be open in X_μ. Now $y \in \cap_{n=1}^\infty S(x, \mathcal{U}_n^\mu)$ implies that $y \in \cap_{n=1}^\infty Star(x, \mathcal{U}_n^\lambda)$ for any pair of points $x,y \in X$. Consequently, there exists a continuous mapping $\phi_\mu^\lambda: X/\Phi_\mu \to X/\Phi_\lambda$ defined by mapping an equivalence class in X/Φ_μ onto the equivalence class of X/Φ_λ to which a member of the equivalence class in X/Φ_μ belongs. Then we have the following commutative diagram:

$$\begin{array}{ccccc} X & \xrightarrow{i_\mu} & X_\mu & \xrightarrow{\phi_\mu^*} & X/\Phi_\mu \\ \| & & \downarrow i_\mu^\lambda & & \downarrow \phi_\mu^\lambda \\ X & \xrightarrow{i_\lambda} & X_\lambda & \xrightarrow{\phi_\lambda^*} & X/\Phi_\lambda \end{array}$$

where i_μ^λ denotes the identity map on X considered as a mapping from X_μ onto X_λ. It is left as an exercise [Exercise 3] to show that ϕ_μ^λ is uniformly continuous with respect to the metric uniformities on X/Φ_μ and X/Φ_λ respectively. Then $\{X/\Phi_\lambda, \phi_\mu^\lambda\}$ is an inverse limit system of metrizable spaces. Let $\nu(X) = lim_{\leftarrow} X/\Phi_\lambda$ and let π_λ denote the canonical projection of $\nu(X)$ onto X/Φ_λ. Then the uniformity of $\nu(X)$ has a basis consisting of the coverings

$$\{\pi_\lambda^{-1}(\phi_\lambda(\mathcal{U}_n^\lambda)) \mid \lambda \in \Lambda, n = 1, 2, \ldots\}.$$

$\nu(X)$ is called the **weak completion** of X with respect to ν. A uniform space is said to be **weakly complete** if each ω-directed Cauchy net clusters. We will show that (X,ν) is uniformly homeomorphic to a dense uniform subspace of $\nu(X)$ and that $\nu(X)$ is weakly complete. The weak completion of a uniform space with respect to a uniformity was introduced by K. Morita in 1970 in a paper titled *Topological completions and M-spaces* published in Sci. Rep. Tokyo Kyoiku Daigaku 10, No. 271, pp. 271-288.

To see that $\nu(X)$ is weakly complete, let $\psi: D \to \nu(X)$ be an ω-directed Cauchy net. Then for each $\lambda \in \Lambda$, $\pi_\lambda © \psi: D \to X/\Phi_\lambda$ is an ω-directed Cauchy net in the metric space X/Φ_λ. The notion of weak completeness is different from the notion of completeness, because every metric space is weakly complete with respect to its metric uniformity. We prove it for X/Φ_λ by showing that $\psi_\lambda = \pi_\lambda © \psi$ clusters in X/Φ_λ. For this let n be a positive integer and let R_n be residual in D such that $\psi_\lambda(R_n) \subset S(x_n, 2^{-n})$ for some $x_n \in X/\Phi_\lambda$. Then $\cap_{n=1}^\infty \psi_\lambda(R_n) \neq \emptyset$, so pick $y \in \cap_{n=1}^\infty \psi_\lambda(R_n)$. Suppose $\cap_{\delta \in D} Cl(\psi_\lambda(R_\delta)) = \emptyset$ where $R_\delta = \{\gamma \in D \mid \delta \leq \gamma\}$. Then there exists an $\alpha \in D$ such that y does not belong to $Cl(\psi_\lambda(R_\alpha))$. Let m be the least positive integer such that $S(y,2^{-m}) \cap Star(Cl(\psi_\lambda(R_\alpha)), 2^{-m}) = \emptyset$. Now for each $n > m$, $y \in \psi_\lambda(R_n)$ which implies $\psi_\lambda(R_n) \subset S(y, 2^{-m})$ so $\psi_\lambda(R_n) \cap \psi_\lambda(R_\alpha) = \emptyset$ which is a contradiction. Hence $\cap_{\delta \in D} Cl(\psi_\lambda(R_\delta)) \neq \emptyset$ which implies ψ_λ clusters in X/Φ_λ. Since ψ_λ is Cauchy in X/Φ_λ, we see that ψ_λ converges to some $y_\lambda \in X/\Phi_\lambda$.

Now $\{y_\lambda\}$ defines a point in $\Pi_\lambda X/\Phi_\lambda$ and since ψ_λ converges to y_λ for each λ, we see (by Exercise 2 of Section 5.3) that ψ must converge to $\{y_\lambda\}$ in $\Pi_\lambda X/\Phi_\lambda$. But since $\psi(D) \subset \nu(X) = lim_{\leftarrow} X/\Phi_\lambda$ and since $lim_{\leftarrow} X/\Phi_\lambda$ is closed in $\Pi_\lambda X/\Phi_\lambda$ by Lemma 5.8, we have that ψ clusters in $\nu(X)$. Consequently, $\nu(X)$ is weakly complete.

To see that (X,ν) is uniformly homeomorphic to a dense subspace of $\nu(X)$, notice that for any $x \in X$, $\{\phi_\lambda(x) \mid \lambda \in X\}$ defines a point of $\nu(X)$ [see commutative diagram above to verify $\phi_\mu^\lambda(\phi_\mu(x)) = \phi_\lambda(x)$ for each pair $\lambda, \mu \in \Lambda$ with $\lambda < \mu$]. Define $\phi: X \to \nu(X)$ by $\phi(x) = \{\phi_\lambda(x)\}$ for each $x \in X$. It is easily shown [see Exercise 4] that ϕ is uniformly continuous. To see that ϕ is one-to-one, notice that if $x,y \in X$ such that $x \neq y$, then $\phi_\lambda(x) \neq \phi_\lambda(y)$ for some $\lambda \in \Lambda$. To see that ϕ is a uniform homeomorphism, we show that ϕ^{-1} is uniformly continuous by showing that $\pi_\lambda^{-1}(\phi_\lambda(\mathcal{U}_{n+1}^\lambda)) < \phi(\mathcal{U}_n^\lambda)$ for each $\lambda \in \Lambda$ and positive integer n.

For this, suppose $x^*, y^* \in \phi(X)$ such that $x^*, y^* \in \pi_\lambda^{-1}(\phi_\lambda(U_{n+1}))$ for some $U_{n+1} \in \mathcal{U}_{n+1}^\lambda$. Then $x^* = \{\phi_\alpha(x) \mid \alpha \in \Lambda\}$ for some $x \in X$ and $y^* = \{\phi_\alpha(y)\}$ for some $y \in X$, and $\pi_\lambda(x^*), \pi_\lambda(y^*) \in \phi_\lambda(U_{n+1})$. Therefore, $\phi_\lambda(x), \phi_\lambda(y) \in \phi_\lambda(U_{n+1})$. Then there exists $x', y' \in U_{n+1}$ such that $\phi_\lambda(x') = \phi_\lambda(x)$ and $\phi_\lambda(y') = \phi_\lambda(y)$ which implies that $d_\lambda(x, x') = 0$ and $d_\lambda(y, y') = 0$. Consequently, there exists V_{n+1} and $W_{n+1} \in \mathcal{U}_n^\lambda$ such that $x, x' \in V_{n+1}$ and $y, y' \in W_{n+1}$. Then there exists a $U_n \in \mathcal{U}_n^\lambda$ such that $U_{n+1} \cup V_{n+1} \cup W_{n+1} \subset U_n$. Hence $x,y \in U_n$ so $\phi(x), \phi(y) \in \phi(U_n)$. But $\phi(x) = \{\phi_\alpha(x)\} = x^*$ and $\phi(y) = \{\phi_\alpha(y)\} = y^*$ so $x^*, y^* \in \phi(U_n)$. Therefore, $\pi_\lambda^{-1}(\phi_\lambda(\mathcal{U}_{n+1}^\lambda)) < \phi(\mathcal{U}_n^\lambda)$ for each $\lambda \in \Lambda$ and positive integer n.

To see that $\phi(X)$ is dense in $\nu(X)$, suppose there exists a $y \in \nu(X)$ such that y does not belong to $Cl(\phi(X))$. Then there exists a $\lambda \in \Lambda$ such that $S(y, \pi_\lambda^{-1}(\phi_\lambda(\mathcal{U}_n^\lambda))) \cap Cl(\phi(X)) = \emptyset$. Then $S(\pi_\lambda(y), \phi_\lambda(\mathcal{U}_n^\lambda)) \cap \pi_\lambda(Cl(\phi(X))) = \emptyset$. But $\pi_\lambda(Cl(\phi(X))) = X/\Phi_\lambda$ and $S(\pi_\lambda(y), \phi_\lambda(\mathcal{U}_n^\lambda)) \subset X/\Phi_\lambda$ so we have a contradiction. Therefore, $\phi(X)$ is dense in $\nu(X)$. Consequently, it makes sense to call $\nu(X)$ a weak completion of X. But, a weak completion need not be unique. As already pointed out, every metric space is weakly complete with respect to its metric uniformity. If we start with a metric space that is not complete, we see that both X and the completion of X are distinct weak completions of X. It is important to recognize that Morita's definition of *the* weak completion with respect to a uniformity is only *defined* to be *the* weak completion with respect to the uniformity. It does not enjoy this distinction in virtue of its inherent uniqueness like the ordinary completion does.

THEOREM 5.4 (K. Morita, 1970) The mapping $\phi: X \to \nu(X)$ is onto if and only if X is weakly complete with respect to ν.

Proof: We first show that if X is weakly complete with respect to ν then ϕ is onto. For this, let $y \in \nu(X)$. Put $\Psi = \{\phi_\lambda^{-1}(\pi_\lambda(y)) \mid \lambda \in \Lambda\}$. Let $\{\lambda_n\}$ be a countable collection of indicies in Λ. Since $(\Lambda, <)$ is countably directed, there exists a $\gamma \in \Lambda$ such that $\lambda_n < \gamma$ for each positive integer n. Therefore, for each n

$$\phi_\gamma^{-1}(\pi_\gamma(y)) = \{x \in X \mid x \in \pi_\gamma(y)\} \subset \{x \in X \mid x \in \pi_{\lambda_n}(y)\} = \phi_{\lambda_n}^{-1}(\pi_{\lambda_n}(y)).$$

Consequently, $\phi_\gamma^{-1}(\pi_\gamma(y)) \subset \Pi_{n=1}^\infty \phi_{\lambda_n}^{-1}(\pi_{\lambda_n}(y))$ so Ψ has the countable intersection property. Also, if $\mathcal{U} \in \nu$, let $\Phi_\lambda = \{\mathcal{U}_n\}$ be a normal sequence of open members of ν such that $\mathcal{U}_1 < \mathcal{U}$. Pick $x \in \pi_\lambda(y)$ and let $U_2 \in \mathcal{U}_2$ such that $x \in U_2$. If $z \in \phi_\lambda^{-1}(\pi_\lambda(y))$ then $z \in S(x, \mathcal{U}_2) \subset U_1$ for some $U_1 \in \mathcal{U}_1$ Therefore, $\phi_\lambda^{-1}(\pi_\lambda(y)) \subset U_1 \subset \mathrm{U}$ for some $\mathrm{U} \in \mathcal{U}$. So Ψ contains a $\mathcal{U}$-small member for each $\mathcal{U} \in \nu$. Finally, notice that each member of Ψ is closed since for each $\phi_\lambda^{-1}(\pi_\lambda(y)) \in \Psi$, $\pi_\lambda(y)$ is a point of the metric space X/Φ_λ and hence a closed set.

Let Σ be the collection of countable intersections of members of Ψ and let $D = \{(A, x) \mid x \in A \in \Sigma\}$. Define $<$ on D by $(B,y) < (A, x)$ if $A \subset B$. Then $(D,<)$ is a countably directed set. For each $(A, x) \in D$ put $\psi(A, x) = x$. Then the correspondence $(A, x) \to \psi(A, x)$ defines an ω-directed net $\psi: D \to X$. Since Ψ contains a $\mathcal{U}$-small member for each $\mathcal{U} \in \nu$, the net ψ is Cauchy. Since X is weakly complete with respect to ν, ψ converges to some $p \in X$. But then $p \in Cl(W)$ for each $W \in \Psi$. Since each member of Ψ is closed, there exists a point $x \in \cap_\lambda\{\phi_\lambda^{-1}(\pi_\lambda(y)) \mid \lambda \in \Lambda\}$ which implies $\phi_\lambda(x) = \pi_\lambda(y)$ for each $\lambda \in \Lambda$. Therefore, $\phi(x) = \{\phi_\lambda(x) \mid \lambda \in \Lambda\} \in \nu(X)$ and $\pi_\lambda(\phi(x)) = \phi_\lambda(x)$ for each $\lambda \in \Lambda$. But $\pi_\lambda(y) = \phi_\lambda(x)$ for each $\lambda \in \Lambda$, so $y = \phi(x)$. Therefore, ϕ is onto.

Conversely, assume $\psi: D \to X$ is an ω-directed Cauchy net in X with respect to ν. Then for each $\lambda \in \Lambda$ and each positive integer n, there exists a residual $R(\lambda, n)$ such that $\psi(R(\lambda, n)) \subset U_n^\lambda$ for some $U_n^\lambda \in \mathcal{U}_n^\lambda$. Now $\cap_{n=1}^\infty R(\lambda,n) \neq \varnothing$ so $\cap_{n=1}^\infty \phi_\lambda(\psi(R(\lambda, n))) \neq \varnothing$ in X/Φ_λ. Suppose there are two distinct points x_λ and y_λ in $\cap_{n=1}^\infty \phi_\lambda(\psi(R(\lambda, n)))$. Since $\{\phi_\lambda(\mathcal{U}_n^\lambda) \mid n = 1, 2, 3 \ldots\}$ determines the topology of X/Φ_λ, there exists some positive integer m such that $S(x_\lambda, \phi_\lambda(\mathcal{U}_m^\lambda)) \cap S(y_\lambda, \phi_\lambda(\mathcal{U}_m^\lambda)) = \varnothing$. Now $x_\lambda \in \phi_\lambda(\psi(R(\lambda, n))) \subset \phi_\lambda(U_n^\lambda) \subset S(x_\lambda, \phi_\lambda(\mathcal{U}_m^\lambda))$ which implies y_λ does not belong to $\phi_\lambda(\psi(R(\lambda, n)))$ which is a contradiction. Hence there exists exactly one point in $\cap_{n=1}^\infty \phi_\lambda(\psi(R(\lambda, n)))$ which we denote by y_λ.

For each pair of positive integers m,n put $P(m,n) = R(\mu, m) \cap R(\lambda, n)$. Clearly, $Q = \cap_{m,n=1}^\infty P(m,n)$ is residual in D. Also, $Q \subset \cap_{m=1}^\infty R(\mu, m)$ and $Q \subset \cap_{n=1}^\infty R(\lambda,n)$. Therefore, $\cap_{m,n=1}^\infty \phi_\lambda(\psi(P(m,n))) = \{y_\lambda\}$ and $\cap_{m,n=1}^\infty \phi_\mu(\psi(P(m,n))) = \{y_\mu\}$. Let $z \in \cap_{m,n=1}^\infty \psi(P(m,n))$. Then $\phi_\mu(z) = y_\mu$ so z is a representative member of the equivalence class y_μ in X. But $\phi_\lambda(z) = y_\lambda$ so $\phi_\mu^\lambda(y_\mu) = y_\lambda$.

Therefore, the point $y = \{y_\lambda \mid \lambda \in \Lambda\}$ lies in $\nu(X)$. We want to show that the net $\phi \circledcirc \psi: D \to \nu(X)$ converges to y so that if ϕ is onto then ψ converges in X (by Proposition 4.8) and hence X is weakly complete with respect to ν. Let $S(y, \pi_\lambda^{-1}(\phi_\lambda(\mathcal{U}_n^\lambda)))$ be a basic open neighborhood of y in $\nu(X)$. Now $\phi_\lambda(\psi(R(\lambda,n))) \subset S(y_\lambda, \phi_\lambda(\mathcal{U}_n^\lambda))$ so

$$\pi_\lambda^{-1}(\phi_\lambda(\psi(R(\lambda, n)))) \subset \pi_\lambda^{-1}(S(y_\lambda, \phi_\lambda(\mathcal{U}_n^\lambda))) \subset S(y, \pi_\lambda^{-1}(\phi_\lambda(\mathcal{U}_n^\lambda))).$$

By the definition of $\phi: X \to \nu(X)$, it is clear that for each $x \in X$, $\phi(x) \in \pi_\lambda^{-1}(\phi_\lambda(x))$ so $\phi \circledcirc \psi\,(R(\lambda, n)) \subset S(y, \pi_\lambda^{-1}(\phi_\lambda(\mathcal{U}_n^\lambda)))$. Hence, $\phi \circledcirc \psi$ converges to y in $\nu(X)$. ▪

Theorem 5.4 shows that if we start with a weakly complete uniform space (X,ν) then the weak completion $\nu(X)$ of X with respect to ν is simply X. In this case, the inverse limit system $\{X/\Phi_\lambda, \phi_\mu^\lambda\}$ is called the **spectrum** of X with respect to ν. In any case, $\{X/\Phi_\lambda, \phi_\mu^\lambda\}$ is called the **spectrum** of $\nu(X)$. Although the concepts of the weak completion with respect to ν and the ordinary completion with respect to ν are different in general, there are cases for which both concepts coincide, as the following theorem shows. The equivalence of (1), (3) and (4) in the following theorem is due to B. Pasynkov [*ω-mappings and inverse spectra*, Soviet Math. Dokl., Volume 4, pp. 706-709, 1963].

THEOREM 5.5 (K. Morita, 1970 and B. Pasynkov, 1963) The following statements are equivalent:

(1) X is complete with respect to u

(2) X is weakly complete with respect to u

(3) X is the inverse limit of an inverse system of metric spaces

(4) X is homeomorphic to a closed subset of a product of metric spaces.

Proof: Clearly (1) → (2) and (3) → (4). By Theorem 5.4, (2) → (3). Since metric spaces are topologically complete, the assumption (4) implies X is homeomorphic to a closed subset of a product of topologically complete spaces. But then the product space is topologically complete. Since X is closed in the product space, it is complete with respect to the finest uniformity for the product space relativized to X. Consequently, X is complete with respect to u. ▪

EXERCISES

1. Prove Lemma 5.8.

2. Let (X,ν) be a uniform space and let $\{\Phi_\lambda \mid \lambda \in \Lambda\}$ be the collection of all normal sequences of open uniform coverings. For each $\lambda \in \Lambda$ let X_λ and X/Φ_λ be the pseudo-metric space and the metric space derived from X with respect to

Φ_λ as constructed in this section. For each pair $\lambda, \mu \in \Lambda$ with $\lambda < \mu$, let ϕ_μ^λ be the bonding map from X/Φ_μ into X/Φ_λ. Show that ϕ_μ^λ is uniformly continuous with respect to the metric uniformities on X/Φ_μ and X/Φ_λ.

3. Let (X,ν) be a uniform space and let $\nu(X)$ be the weak completion with respect to ν. Let $\phi:X \to \nu(X)$ be the function defined in this section by $\phi(x) = \{\phi_\lambda(x) \mid \lambda \in \Lambda\}$ for each $x \in X$. Show that ϕ is uniformly continuous.

4. Let $\{Y, F\}$ be an inverse system of uniform spaces where $Y = \{X_\alpha \mid \alpha \in D\}$ and $F = \{f_\beta^\alpha \mid \alpha, \beta \in D \text{ with } \alpha < \beta\}$. Let C be a cofinal subset of D. Put $Z = \{X_\alpha \mid \alpha \in C\}$ and $G = \{f_\beta^\alpha \mid \alpha, \beta \in C \text{ with } \alpha < \beta\}$. Show that the inverse limit of $\{Z,G\}$ is homeomorphic with the inverse limit of $\{Y,F\}$.

5.6 The Locally Fine Coreflection

A uniform space is said to have some property *P uniformly locally* if there is a uniform covering such that each member of this covering has property *P*. For instance, a uniform space (X, μ) is said to be **uniformly locally compact** if there exists a $\mathcal{U} \in \mu$ such that each $U \in \mathcal{U}$ is compact. There are a variety of such properties *P* that have been studied as uniformly local properties. Similarly, a collection $\{D_\alpha\}$ of subsets of X is said to be **uniformly discrete** if there exists a $\mathcal{U} \in \mu$ such that for each pair α,β, $Star(D_\alpha,\mathcal{U}) \cap Star(D_\beta,\mathcal{U}) = \emptyset$.

In a paper titled *Some operators on uniform spaces* (Transactions of the American Mathematical Society, Volume 93, pp. 145-168), published in 1959, S. Ginsburg and J. Isbell introduced the notion of *locally fine* uniform spaces, and the concept of the *locally fine coreflection* (a finer uniformity) associated with a given uniformity. To define these concepts, they made use of the notion of a uniformly locally uniform covering of a uniform space (X, μ). A covering $\mathcal{V}$ of X is said to be **uniformly locally uniform** if there exists a $\mathcal{U} \in \mu$ such that for each $U \in \mathcal{U}$, the covering $\{U \cap V \mid V \in \mathcal{V}\}$ of U is a uniform covering (in the induced subspace uniformity on U). This means that if $\mathcal{U} = \{U_\alpha\}$ then for each fixed α, we may replace $\mathcal{V}$ with a uniform covering $\mathcal{V}_\alpha = \{V_\beta^\alpha\}$ such that $\{U_\alpha \cap V_\beta^\alpha\} = \mathcal{V}_\alpha \mid U_\alpha$, the *trace* of $\mathcal{V}_\alpha$ on U_α ($\mathcal{V}_\alpha$ restricted to U_α).

A uniform space (X, μ) is then defined to be **locally fine** if each uniformly locally uniform covering belongs to μ. It is easily seen that (X, μ) is locally fine if and only if it is closed under what Ginsburg and Isbell call the *staggered intersection* operation $\{U_\alpha \cap V_\beta^\alpha\}$ where $\{U_\alpha\} \in \mu$ and $\{V_\beta^\alpha\}$ is a uniform covering of U_α for each α. Next, they constructed a new uniformity $\lambda(\mu)$ for X, that is the coarsest uniformity for X finer than μ, that is locally fine. The uniform space $(X, \lambda(\mu))$ is called the **locally fine coreflection** of X (the meaning of the term *coreflection* will be explained in the next section – it is not necessary for understanding the uniformity $\lambda(\mu)$).

To define $\lambda(\mu)$ we proceed via transfinite induction using the concept of the *derivative* of a uniformity. Let $\mu^0 = \mu$ and define the **derivative** μ^1 of μ to be the family of all coverings which have refinements of the form $\{U_\alpha \cap V_\beta^\alpha\}$, where $\{U_\alpha\} \in \mu$ and for each fixed α, $\{V_\beta^\alpha\} \in \mu | U_\alpha$. It is easily shown that μ^1 is a filter of coverings with respect to refinement, and that $\mu^0 \subset \mu^1$. Then, for each ordinal number α, we can define μ^α by the inductive rules $\mu^{\alpha+1} = (\mu^\alpha)^1$ for all ordinals α and for limit ordinals β by $\mu^\beta = \cup\{\mu^\alpha | \alpha < \beta\}$. Clearly, there must exist a *last derivative* $\mu^\kappa = \mu^{\kappa+1}$. We put $\lambda(\mu) = \mu^\kappa$. It is easily shown that $\lambda(\mu)$ is a filter of coverings with respect to refinement. It is considerably more difficult to show that $\lambda(\mu)$ is a uniformity for X. The proof given by Ginsburg and Isbell (1959) and the revised proof given by Isbell in his book (1964) are not entirely correct. The proof given below incorporates the idea of the lemma they tried to prove prior to the theorem in order to avoid some of the problems with their proofs.

THEOREM 5.6 For a complete metric space (X, μ) where μ is the metric uniformity, $\lambda(\mu) = u$.

Proof: By Stone's Theorem (Chapter 1), u consists of all coverings that have open refinements. Each member of μ has an open refinement. This property is preserved under the operation of taking the derivative. Consequently, $\lambda(\mu) \subset u$.

To show the converse, let $\mathcal{V} = \{V_\beta\}$ be an open covering of X. For each positive integer n, let $\mathcal{U}_n = \{U_\alpha^n\}$ be a uniform covering of X consisting of the spheres of radius 2^{-n}. Let P be the set of all indices (n, α) of the sets U_α^n in the covering $\mathcal{U}_n$ and $<$ be the partial ordering on P defined by: $(n+1, \beta) < (n, \alpha)$ if $U_\beta^{n+1} \cap U_\alpha^n \neq \varnothing$, and $(n+k, \gamma) < (n,\alpha)$ if there exists a chain $(n+k, \gamma) < (n+k-1,\delta) < \ldots < (n, \alpha)$. Let S be the set of all (n, α) with $Star(U_\alpha^n, \mathcal{U}_n) \subset V_\beta$ for some β. Since (X, μ) is complete and each U_α^n is of radius 2^{-n}, each infinite descending chain in P converges to some limit point x. Since some ε-sphere of x is contained in some V_β, only finitely many of the U_α^n in the chain can have indices in the chain P - S.

We now define an ordinal valued function δ on P - S as follows: if every predecessor of p is in S then $\delta(p) = 0$. If $\delta(q)$ is defined for each predecessor q of p in P - S, then $\delta(p) = \alpha$ where α is the least ordinal greater than all these $\delta(q)$. First, we observe that this defines δ on all of P - S, for otherwise there would exist some $p_1 \in P - S$ on which δ is not defined which implies there is some predecessor $p_2 < p_1$ in P - S for which δ is not defined, and so on, so we get an infinite descending chain $\ldots < p_n < \ldots < p_2 < p_1$ of elements of P - S for which δ is not defined which is a contradiction.

Next, we show that for each $(n,\alpha) \in P - S$, $\{U_\alpha^n \cap U_\gamma^k | (k,\gamma) \in S\} \in \lambda(\mu) | U_\alpha^n$. For this we induct on the values of δ. To get the induction started (and to prove another result we will need later), let p either belong to S or have $\delta(p) = 0$. In either case, each predecessor of p is in S. Let $Q_0 = \{s \in S | s$ is an immediate predecessor of p}, and for each non-negative integer n let $Q_{n+1} = \{s \in S \mid s$ is an immediate predecessor of some $q \in Q_n\}$. Then $\cup Q_n$ is the set of all predecessors of p in S. Now let $\mathcal{U}_0^p = \{U_p \cap U_s \mid s \in Q_0\} \in \mu \mid U_p = \mu^0$ on U_p. Then $\mathcal{U}_1^p = \{U_p \cap U_s \mid s \in \cup_{i \le 1} Q_i\} \in \mu^1$ on U_p and $\mathcal{U}_0^p \subset \mathcal{U}_1^p$. Continuing this process by induction, we obtain a sequence of coverings $\{\mathcal{U}_n^p\} = \{U_p \cap U_s \mid s \in \cup_{i \le n} Q_i\}$ with $\mathcal{U}_n^p \subset \mathcal{U}_{n+1}^p$ and $\mathcal{U}_n^p \in \mu^n$ for each non-negative integer n. Consequently, $\mathcal{U}_\infty^p = \{U_p \cap U_s \mid s \in \cup Q_n\} \in \lambda(\mu) \mid U_p$. Now $p = (n,\alpha)$ for some non-negative integer n and some index α. Then for each positive integer m such that $m < n$, if (m, γ) does not precede (n, α) then $U_\alpha^n \cap U_\gamma^m = \varnothing$. Hence if $s \in S$ with s not in $\cup Q_n$ and $U_p \cap U_s \neq \varnothing$ then $s = (k,\beta)$ for some non-negative integer k with $k \le n$ and some index β. But then $U_\beta^k \in \mathcal{U}_k$. Since there are only finitely many $\mathcal{U}_k$ with $k \le n$ we see that $\{U_p \cap U_s \mid s \in S\} \in \lambda(\mu) \mid U_p$.

To keep the induction going, we assume $\{U_p \cap U_s \mid s \in S\} \in \lambda(\mu) \mid U_p$ for each $p \in P$ with $\delta(p) < \kappa$ for some ordinal κ. Let $q \in P - S$ with $\delta(q) = \kappa$. We maintain that $\{U_q \cap U_s \mid s \in S\} \in \lambda(\mu) \mid U_q$. To see this, note that $\{U_q \cap U_s \mid s \in S\}$ is refined by $\{(U_q \cap U_r) \cap U_s \mid r$ is an immediate predecessor of q and $s \in S\}$, that $\{U_q \cap U_r \mid r$ is an immediate predecessor of $q\} \in \mu \mid U_q$ and for each $r < q$, $\{U_r \cap U_s \mid s \in S\} \in \lambda(\mu) \mid U_r$ by our induction assumption. To see this last assertion note that if $r < q$ and $r \in P - S$ then $\delta(r) < \delta(s)$, whereas if $r \in S$, then all predecessors of r are in S, which is one of the cases we have already proved. Consequently, $\{U_q \cap U_s \mid s \in S\} \in \lambda(\mu) \mid U_q$, so $\{U_p \cap U_s \mid s \in S\} \in \lambda(\mu) \mid U_p$ for each $p \in P - S$.

Now $\{U_p \mid p$ is a maximal element in $P\} \in \mu$ since if p is maximal, then $p = (1,\alpha)$ for some index α. For each maximal p, either $p \in S$ or $p \in P - S$. If $p \in P - S$ then $\{U_p \cap U_s \mid s \in S\} \in \lambda(\mu) \mid U_p$. If $p \in S$, then each predecessor of p is in S so $\{U_p \cap U_s \mid s \in S\} \in \lambda(\mu) \mid U_p$. Therefore, $\mathcal{U} = \{U_s \mid s \in S\} \in \lambda(\mu)$ by definition of $\lambda(\mu)$. But clearly, $\mathcal{U} < \mathcal{V}$, so $\mathcal{V} \in \lambda(\mu)$. Hence $u \subset \lambda(\mu)$, so $\lambda(\mu) = u$. ■

PROPOSITION 5.10 For each $\mathcal{U} \in \mu$, where (X, μ) is a uniform space, there exists a normal sequence $\{\mathcal{U}_n\}$ of locally finite open members of $\lambda(\mu)$ with $\mathcal{U}_1 < \mathcal{U}$.

Proof: If $\mathcal{U} \in \mu$, there exists a uniformly continuous $f{:}X \to M$ where M is a metric space, and a uniform covering $\mathcal{V}$ of M with $f^{-1}(\mathcal{V}) < \mathcal{U}$ (Theorem 2.5). Let M′ be the completion of M. Then $f{:}X \to M'$ is uniformly continuous and $f^{-1}(\mathcal{V}') < \mathcal{U}$ where $\mathcal{V}'$ is the Morita extension of $\mathcal{V}$ to M′ (see Exercise 4, Section 4.4). Since M′ is paracompact, there exists a normal sequence $\{\mathcal{V}_n'\}$ of locally finite open coverings of M′ with $\mathcal{V}_1' < \mathcal{V}'$. By Theorem 5.6, each $\mathcal{V}_n' \in \lambda(\nu)$ where ν is the uniformity of M′. Suppose $\mathcal{V}_n \in \nu^\alpha$ for some ordinal α. By a straightforward induction argument, we can show that $\mathcal{U}_n = f^{-1}(\mathcal{V}_n^*) \in \mu^\alpha$. Hence $\mathcal{U}_n \in \lambda(\mu)$ for each n. Clearly $\mathcal{U}_n$ is locally finite for each n and $\{\mathcal{U}_n\}$ is a normal sequence of open, locally finite members of $\lambda(\mu)$ with $\mathcal{U}_1 < \mathcal{U}$. ■

PROPOSITION 5.11 For a uniform space (X, μ), λ(μ) is a uniformity for X.

Proof: We will prove inductively that for each $\mathcal{U} \in \lambda(\mu)$, there exists a normal sequence $\{\mathcal{U}_n\}$ of open, locally finite members of $\lambda(\mu)$ with $\mathcal{U}_1 < \mathcal{U}$. By Proposition 5.10, it is true for members of μ. Assume α is the least ordinal for which the assertion fails for some $\mathcal{U} \in \mu^\alpha$. Now $\mathcal{U} = \{V_\beta \cap W_\gamma^\beta\}$ where $\{V_\beta\} \in \mu^\kappa$ for some $\kappa < \alpha$ and for each index β, $\{W_\gamma^\beta\} \in \mu^{\kappa(\beta)}$ for some $\kappa(\beta) < \alpha$. Hence $\{V_\beta\}$ has an open locally finite refinement $\{H_\delta\} \in \lambda(\mu)$ with an open locally finite star refinement $\{K_\theta\} \in \lambda(\mu)$. For each δ choose a β(δ) with $H_\delta \subset V_{\beta(\delta)}$. Since $\{H_\delta\}$ is locally finite, each K_θ is contained in only finitely many of the H_δ, say $H_{\delta_1} \ldots H_{\delta_n}$. Put $\mathcal{W}_\theta = \{W_\gamma^{\beta(\delta_1)}\} \cap \ldots \cap \{W_\gamma^{\beta(\delta_n)}\}$. Since each μ^α is a filter, it can be seen from the induction hypothesis that $\mathcal{W}_\theta$ has an open, locally finite, star refinement $\{Z_\xi^\theta\}$ that is normal with respect to the collection of open, locally finite members of λ(μ). But then, $\{K_\theta \cap Z_\xi^\theta\}$ is an open, locally finite, star refinement of $\mathcal{U}$ that can be shown to be normal with respect to the open, locally finite members of λ(μ). Therefore, the assertion does not fail after all, so λ(μ) is a uniformity. ∎

COROLLARY 5.3 The uniformity λ(μ) has a uniformly locally finite basis.

THEOREM 5.7 For a uniform space (X, μ), λ(μ) is the coarsest locally fine uniformity finer than μ. Moreover, each locally fine uniformity has a basis of uniformly locally finite coverings.

Proof: By construction, λ(μ) is locally fine, and finer than μ. If ν is another locally fine uniformity finer than μ, then it is finer than μ^1. By a straightforward induction argument, ν is finer than μ^α for each ordinal α and hence finer than λ(μ). Therefore, λ(μ) is the coarsest locally fine uniformity finer than μ. Finally, if μ is locally fine, then λ(μ) = μ, so each $\mathcal{U} \in \mu$ has an open uniformly locally finite refinement. Therefore, each locally fine uniformity has a uniformly locally finite basis. ∎

Let P be a partially ordered set. $R \subset P$ is **residual** in P if for each $r \in R$, $s \in R$ for each $s \in P$ with $r < s$. Note that the restriction $R \neq \varnothing$ has been dropped from the corresponding definition for directed sets. A function $\psi: P \to X$ is called a **partial net** (denoted p-net) in X. A p-net ψ in a uniform space (X,ν) is **partially Cauchy** if for each $\{U_\alpha\} \in \nu$, there exists a family $\{R_\alpha\}$ of residual subsets of P such that for each α, $\psi(R_\alpha) \subset U_\alpha$ and $\cup R_\alpha$ is cofinal in P. The definition of a cofinal subset of P is the same as the definition for directed sets. We extend this definition by allowing ν to be μ^α for some uniformity μ and ordinal α. In this case, we say ψ is **partially Cauchy with respect to the** α^{th} **derivative of** μ. ψ is said to **partially converge** to some $x \in X$ if for each neighborhood U of x, $\psi^{-1}(U)$ contains a non-empty residual subset of P. If P is a directed set, the definitions of partially Cauchy and partially convergent are the same as Cauchy and convergent respectively.

LEMMA 5.9 If $\psi:P \to (X, \mu)$ is partially Cauchy, so is $\psi:P \to (X,\lambda(\mu))$.

Proof: Suppose $\psi:P \to (X, \lambda(\mu))$ is not partially Cauchy. Then there exists a least ordinal α for which $\psi:P \to (X, \mu^\alpha)$ is not partially Cauchy. α must be a non-limit ordinal, say $\beta + 1$, for otherwise μ^α is just the union of its predecessors, for which ψ is partially Cauchy. Let $\{U_\gamma \cap V_\delta^\gamma\} \in \mu^\alpha$. Then $\{U_\gamma\} \in \mu^\beta$ and for each γ, $\{V_\delta^\gamma\} \in \mu^\beta$. For $p \in P$, there exists a successor q, all of whose successors are in one of the U_γ. For this γ, q has a successor r, all of whose successors are in one of the V_δ^γ. Hence there exists a residual $R(\gamma,\delta) \neq \varnothing$ with $\psi(R(\gamma,\delta)) \subset U_\gamma \cap V_\delta^\gamma$ such that $p \leq r$ for each $r \in R(\gamma,\delta)$. But then there exists a cofinal union of residual sets, each mapped by ψ into one of the $U_\gamma \cap V_\delta^\gamma$. For each $U_\gamma \cap V_\delta^\gamma$ not containing one of these residual images of ψ, put $R(\gamma,\delta) = \varnothing$. Then ψ is partially Cauchy with respect to μ^α which is a contradiction. Therefore, $\psi:P \to (X, \lambda(\mu))$ is partially Cauchy. ∎

The proof of the following theorem and the terminology used in the proof is somewhat different from Isbell's original proof, but his central ideas are preserved. This different route frees us from having to introduce additional terminology and the concept of a *stable Cauchy filter*, and from proving some additional lemmas.

THEOREM 5.8 (J. Isbell, 1962) For a uniform space (X, μ), the following are equivalent:

(1) (X, μ) is supercomplete.

(2) X is paracompact and $\lambda(\mu)$ is fine.

(3) Each partially Cauchy p-net in (X, μ) partially converges.

Proof: (1) → (2) Condition (2) implies each open covering belongs to $\lambda(\mu)$. Suppose this is not the case, i.e., suppose the open covering $\mathcal{U}$ is not $\lambda(\mu)$-uniform. Let (X', μ') be the hyperspace of (X, μ). We construct a proper Cauchy net in X′ that has no cluster point. By Proposition 5.10 this is a contradiction. For this let B consist of those closed sets $F \subset X$ with the trace of $\mathcal{U}$ on some uniform neighborhood of X - F being $\lambda(\mu)$-uniform. Let $F, K \in B$. Then $F \cap K$ is closed.

$F, K \in B$ implies there exists $\mathcal{V}, \mathcal{W} \in \mu$ with $\mathcal{U}$ being $\lambda(\mu)$-uniform on $Star(X - F,\mathcal{V})$ and $\mathcal{U}$ being $\lambda(\mu)$-uniform on $Star(X - K,\mathcal{W})$. Let $\mathcal{Z} = \mathcal{V} \cap \mathcal{W}$. Then $\mathcal{Z} \in \mu$ and $\mathcal{U}$ is $\lambda(\mu)$-uniform on both $Star(X - F, \mathcal{Z})$ and $Star(X - K, \mathcal{Z})$. If $F \cap K = \varnothing$ then $F \subset X - K$ so $\mathcal{U}$ is $\lambda(\mu)$-uniform on $Star(F, \mathcal{Z})$. Therefore, for each $Z \in \mathcal{Z}$, $\mathcal{U}|Z \in \lambda(\mu)|Z$ so $\mathcal{U} \in \lambda(\mu)$ which contradicts our assumption, so $F \cap K \neq \varnothing$. Also, since $\mathcal{U}$ is $\lambda(\mu)$-uniform on both $Star(X\text{-}F,\mathcal{Z})$ and $Star(X\text{-}K,\mathcal{Z})$, $\mathcal{U}$ is $\lambda(\mu)$-uniform on $Star(X\text{-}F,\mathcal{Z}) \cup Star(X\text{-}K,\mathcal{Z})$ which contains $Star(X\text{-}F \cap K,\mathcal{Z})$. Consequently, $F \cap K \in B$.

Let $\mathcal{D}$ be the collection of closed sets in X that contain a member of $\mathcal{B}$ and define $\leq$ on $\mathcal{D}$ by $F \leq K$ if $K \subset F$. Then $(\mathcal{D}, \leq)$ is a directed set. For each $F \in \mathcal{D}$ put $\psi(F) = F$. Then $\psi:\mathcal{D} \to X'$ is a proper net in X′. We need to show that ψ is also Cauchy. By Lemma 5.5, we need to show that for each $\mathcal{V} \in \mu$ there exists an $F \in \mathcal{D}$ with $\psi(F) \subset Star(\psi(K),\mathcal{V})$ for each $K \in \mathcal{D}$. For this let $\mathcal{W} \in \mu$ with $\mathcal{W}^* < \mathcal{V}$ and put $F = Cl(\{W \in \mathcal{W} \,|\, \mathcal{U}$ is not $\lambda(\mu)$-uniform on W$\})$. Then $F \in \mathcal{B}$ since the union of the $W \in \mathcal{W}$ on which $\mathcal{U}$ is $\lambda(\mu)$-uniform is a uniform neighborhood of X - F containing $A = Star(X - F,\mathcal{W})$. To see this let $W \in \mathcal{W}$ with $W \cap (X - F) = \emptyset$. Then W is not a member of $\mathcal{W}$ on which $\mathcal{U}$ is $\lambda(\mu)$-uniform. Hence $W \subset A$.

For each $K \in \mathcal{D}$, there exists $H \in \mathcal{B}$ with $H \subset K$. $Star(H,\mathcal{W})$ contains a dense subset of F, for if $x \in \cup\{W \in \mathcal{W} \,|\, \mathcal{U}$ is not $\lambda(\mu)$-uniform on W$\}$ then $x \in W_0 \in \mathcal{W}$ on which $\mathcal{U}$ is not $\lambda(\mu)$-uniform which implies W_0 is not contained in X - H (since $H \in \mathcal{B}$) so $W_0 \subset Star(H,\mathcal{W})$ which implies $x \in Star(H,\mathcal{W})$. But then $F \subset Star(H,\mathcal{W}^*) \subset Star(K,\mathcal{V})$. Since $\psi(F) = F$ and $\psi(K) = K$ we have $\psi(F) \subset Star(\psi(K),\mathcal{V})$ for each $K \in \mathcal{D}$ so ψ is Cauchy.

So $\psi:\mathcal{D} \to X'$ is a proper Cauchy net. By Proposition 5.10, ψ has a cluster point $p \in X$. Then there exists a $U \in \mathcal{U}$ with $p \in U$. Pick $\mathcal{V} \in \mu$ with $S(p,\mathcal{V}^*) \subset U$ and put $F = X - S(p,\mathcal{V})$. Then F is closed and $Star(X - F,\mathcal{V}) \subset S(p,\mathcal{V}^*) \subset U$ so $\mathcal{U}$ is $\lambda(\mu)$-uniform on $Star(X - F,\mathcal{V})$ since $U | S(p,\mathcal{V}^*) = X | S(p,\mathcal{V}^*)$. Hence $F \in \mathcal{D}$ and $S(p,\mathcal{V}) \cap F = \emptyset$. Then for each $K \in \mathcal{D}$ with $K \geq F$, $K \cap S(p,\mathcal{V}) = \emptyset$ so p is not a cluster point of ψ which is a contradiction. Therefore, (1) → (2).

(2) → (3). By Lemma 5.9 we need only show that a partially Cauchy partial net in $(X, \lambda(\mu))$ partially converges. Assume this is not the case, i.e., that there exists a partially Cauchy partial net $\psi:P \to (X, \lambda(\mu))$ that does not partially converge. Then there exists an open covering $\mathcal{U}$ of X such that $\psi^{-1}(U)$ contains no non-empty residual subset of P for each $U \in \mathcal{U}$. By (2), $\lambda(\mu) = u$ and since X is paracompact, $\mathcal{U} \in u$. Hence ψ is not partially Cauchy which is a contradiction. Therefore, (2) → (3).

(3) → (1) Let $\psi:\mathcal{D} \to X'$ be a proper Cauchy net in the hyperspace (X', μ') and suppose ψ has no cluster point. Let $\mathcal{P}$ be the collection of subsets S of X that are uniform neighborhoods of sets that meet each $\psi(d)$ for $d \in \mathcal{D}$. Then $\mathcal{P}$ is partially ordered by set inclusion. For each $P \in \mathcal{P}$ pick $f(P) \in P$. Then $f:\mathcal{P} \to X$ is a partial net. Let $x \in X$. Since x is not a cluster point of ψ, there exists a $\mathcal{U} \in \mu$ and $\delta \in \mathcal{D}$ with $S(x, \mathcal{U}^*) \cap \psi(\delta) = \emptyset$. Then $Q = Star(\psi(\delta), \mathcal{U}) \in \mathcal{P}$ and $S(x,\mathcal{U}) \cap Q = \emptyset$. Suppose f partially converges to x. Then there exists a residual $\mathcal{R} \subset \mathcal{P}$ with $f(\mathcal{R}) \subset S(x,\mathcal{U})$. Let $R \in \mathcal{R}$. Now $R = Star(T,\mathcal{V})$ for some $\mathcal{V} \in \mu$ and some T that meets $\psi(d)$ for each $d \in \mathcal{D}$. Put $A = \psi(\delta) \cap T$ and let $\mathcal{W} = \mathcal{U} \cap \mathcal{V} \in \mu$. Then $B = Star(A,\mathcal{W}) \in \mathcal{P}$ and $B \subset Q \cap R$. Therefore, $R \leq B$ and $f(B) \in Q$ which implies $f(B)$ does not belong to $S(x, \mathcal{U})$ which is a contradiction. Hence f cannot partially converge to any $x \in X$.

Let $\{U_\alpha\} \in \mu$. If S $\in P$, there exists a $\mathcal{V} \in \mu$ with $Star(\mathrm{T},\mathcal{V}^*) \subset \mathrm{S}$ and $\mathcal{V}^{**} < \{U_\alpha\}$. By Lemma 5.5, there exists a $\delta \in D$ with $\psi(\delta) \subset Star(\psi(d),\mathcal{V})$ for each $d \in D$. Let $x \in \psi(\delta) \cap \mathrm{T}$. Some U_γ contains $\mathrm{R} = S(x,\mathcal{V}^*) \subset Star(\mathrm{T},\mathcal{V}^*) \subset \mathrm{S}$. Since $x \in \psi(\delta)$, each $\psi(d)$ meets $S(x,\mathcal{V})$. Hence $\mathrm{R} \in P$. $\mathrm{R} \subset \mathrm{S}$ implies $\mathrm{S} \leq \mathrm{R}$ in P and all successors of R in P are mapped into U_γ by f. Therefore, $f{:}P \to \mathrm{X}$ is partially Cauchy and hence must partially converge by (3). But this is a contradiction, so each proper Cauchy net in (X', μ') clusters. By Proposition 5.10, (X, μ) is supercomplete. ∎

In the sense expressed in Theorem 5.8, the uniformity $\lambda(\mu)$ characterizes supercompleteness in paracompact uniform spaces (X, μ). In a yet unpublished paper, B. Burdick shows that there exists another uniformity $\theta(\mu)$ that characterizes cofinal completeness in the same way. We now develop a useful alternate construction of $\lambda(\mu)$. We denote this new construction by $l(\mu)$ for the time being. $l(\mu)$ is called the *subfine coreflection* of μ. To define $l(\mu)$, we first need to define what a subfine uniform space is. A uniform space is said to be **subfine** if it is a subspace of a fine uniform space. Subfine spaces are a generalization of fine spaces whose definition is much more intuitive than the definition of the locally fine spaces. At the time Isbell published his book in 1964, it was an open problem whether locally fine spaces are subfine. In a 1987 paper titled *Locally fine uniformities and normal covers* (Czechoslovak Math. Jour. 37(112), pp. 181-187), J. Pelant solved the problem with an affirmative answer. Pelant's complex proof, together with other proofs he references would take up more space than we can allot to this subject. The interested reader can find his proof and the necessary references in the citation above.

In what follows, it will first be shown that every uniform space (X, μ) has a coarsest subfine uniformity $l(\mu)$ finer than μ, and that by Pelant's Theorem, $l(\mu) = \lambda(\mu)$. For the construction of $l(\mu)$, we need the concept of an *injective* uniform space. A uniform space Y is **injective** if whenever A is a uniform subspace of X, every uniformly continuous function $f{:}\mathrm{A} \to \mathrm{Y}$ can be extended to a uniformly continuous function $F{:}\mathrm{X} \to \mathrm{Y}$.

THEOREM 5.9 The closed interval [0,1] *is an injective space.*

Proof: Let $i{:}\mathrm{A} \to \mathrm{X}$ be a uniform imbedding and $f{:}\mathrm{A} \to \mathrm{I} = [0,1]$ a uniformly continuous function. Let μ be the uniformity of X and μ_β the collection of all coverings of X refined by finite members of μ. It is left as an exercise to show that μ_β is a uniformity for X that is coarser than μ (Exercise 6). Then the identity function $j_X{:}(\mathrm{X}, \mu) \to (\mathrm{X}, \mu_\beta)$ is uniformly continuous. Let $\pi{:}\mathrm{X} \to \mu_\beta\mathrm{X}$ be the uniform imbedding of (X, μ_β) into its completion $\mu_\beta\mathrm{X}$.

Now $\mu_\beta | \mathrm{A}$ is a uniformity on A coarser than μ so the identity function $j_A{:}(\mathrm{A}, \mu) \to (\mathrm{A}, \mu_\beta)$ is uniformly continuous and $Cl_{\mu_\beta X}(\mathrm{A})$ is the completion of (A, μ_β) since completions are unique. So $\mu_\beta\mathrm{A}$ is a closed subspace of $\mu_\beta\mathrm{X}$. Since μ_β is a precompact uniformity, by Proposition 4.19, $\mu_\beta\mathrm{X}$ is compact.

Since I is precompact, f:A $\rightarrow$ I is also uniform continuous with respect to $\mu_\beta \mid A$. By Theorem 4.9, f can be extended to a uniformly continuous function $f^\beta:\mu_\beta A \rightarrow I$. Then since $\mu_\beta A$ is closed in $\mu_\beta X$ and $\mu_\beta X$ is normal by Proposition 0.32, we can apply Tietze's Extension Theorem (0.2) to get a continuous extension $g:\mu_\beta X \rightarrow I$. Since both I and $\mu_\beta X$ are compact, their unique uniformities are fine, so g is uniformly continuous. Define F:X $\rightarrow$ I by $F = g \circledcirc \pi \circledcirc j_X$. Then for each $a \in A$, $F(a) = g(\pi(j_A(a))) = g(a) = f^\beta(a) = f(a)$, so F is an extension of f. F is uniformly continuous since it is the composition of uniformly continuous functions. Hence [0,1] is an injective space. ∎

*COROLLARY 5.4 Any closed interval $I \subset$ **R** is an injective space.*

PROPOSITION 5.12 Each product of injective spaces is injective.

Proof: If Y = ΠY_α is the product of injective uniform spaces Y_α and the uniform subspace A $\subset$ X is mapped uniformly continuously into Y, then the functions $f_\alpha = p_\alpha \circledcirc f$:A $\rightarrow Y_\alpha$ (where p_α is the αth coordinate projection) is a uniformly continuous function, so it can be extended to a uniformly continuous function g_α:X $\rightarrow Y_\alpha$. Then the function g:X $\rightarrow$ Y defined by $g(x) = \{g_\alpha(x)\}$ extends f:A $\rightarrow$ Y, since $f(a) = \{f_\alpha(a)\}$ for each $a \in A$. To see that g is uniformly continuous, let $\mathcal{U}$ be a uniform covering of ΠY_α. Then there exists a finite collection $\alpha_1 \ldots \alpha_n$ of indices and uniform coverings $\mathcal{V}_{\alpha_1} \ldots \mathcal{V}_{\alpha_n}$ of $Y_{\alpha_1} \ldots Y_{\alpha_n}$ respectively such that $p_{\alpha_1}^{-1}(\mathcal{V}_{\alpha_1}) \cap \ldots \cap p_{\alpha_n}^{-1}(\mathcal{V}_{\alpha_n}) < \mathcal{U}$. Then

$$g^{-1}(p_{\alpha_1}^{-1}(\mathcal{V}_{\alpha_1}) \cap \ldots \cap p_{\alpha_n}^{-1}(\mathcal{V}_{\alpha_n})) = g^{-1}(p_{\alpha_1}^{-1}(\mathcal{V}_{\alpha_1})) \cap \ldots \cap g^{-1}(p_{\alpha_n}^{-1}(\mathcal{V}_{\alpha_n})) =$$

$$[p_{\alpha_1} \circledcirc g]^{-1}(\mathcal{V}_{\alpha_1}) \cap \ldots \cap [p_{\alpha_n}^{-1} \circledcirc g]^{-1}(\mathcal{V}_{\alpha_n}) = g_{\alpha_1}^{-1}(\mathcal{V}_{\alpha_1}) \cap \ldots \cap g_{\alpha_n}^{-1}(\mathcal{V}_{\alpha_n})$$

which is uniformly continuous in X since $g_{\alpha_i}^{-1}(\mathcal{V}_{\alpha_i})$ is uniformly continuous for each $i = 1 \ldots n$. Therefore, $g^{-1}(\mathcal{U})$ is a uniform covering of X so g is uniformly continuous. Consequently, Y is an injective uniform space. ∎

A uniform covering $\mathcal{U}$ of a uniform space X is said to be **realized** by a uniformly continuous function f:X $\rightarrow$ Y if for some uniform covering $\mathcal{V}$ of Y we have $f^{-1}(\mathcal{V}) < \mathcal{U}$.

PROPOSITION 5.13 If a family of uniformly continuous functions $f_\alpha:X \rightarrow Y_\alpha$ realizes every uniform covering of X, then the mapping $f:X \rightarrow \Pi Y_\alpha$ defined by $f(x) = \{f_\alpha(x)\}$ for each $x \in X$, is an imbedding.

Proof: Clearly f is uniformly continuous by an argument similar to the one used to prove that g is uniformly continuous in Proposition 5.12 above. To show that f is an imbedding, we need to show that for each uniform covering $\mathcal{U}$ of X, $f(\mathcal{U})$ is a uniform covering of ΠY_α. To see this, note that there exists an index γ for $\mathcal{U}$ and a uniform covering $\mathcal{V}$ of Y_γ such that $f_\gamma^{-1}(\mathcal{V}) < \mathcal{U}$ implies $(p_\gamma \circledcirc f)^{-1}(\mathcal{V}) <$

$\mathcal{U}$ which in turn implies $f^{-1}(p_\gamma^{-1}(V)) < \mathcal{U}$. But $p_\gamma^{-1}(V)$ is a sub-basic uniform covering of ΠY_α and since f is uniformly continuous $f^{-1}(p_\gamma^{-1}(V))$ is a uniform covering of X. Hence $f(\mathcal{U}) = p_\gamma^{-1}(V)$ which is a uniform covering of ΠY_α. ∎

THEOREM 5.10 Every uniform space can be uniformly imbedded in a product of metric spaces.

Proof: By Theorem 2.5, for each uniform covering $\mathcal{U}_\alpha$ of a uniform space X, there exists a uniformly continuous function f_α that maps X onto a metric space M_α such that the inverse image of each set of diameter < 1 is a subset of an element of $\mathcal{U}_\alpha$. Therefore, $\mathcal{U}_\alpha$ is realized by f_α. Then by Proposition 5.13, the mapping f:X → ΠM_α defined by $f(x) = \{f_\alpha(x)\}$ for each $x \in$ X is an imbedding. ∎

For a set S, the metric space l_∞(S) of all bounded real valued functions on S with the distance function $d(f, g) = sup\{|f(x) - g(x)| \mid x \in X\}$, is called the **$l_\infty$-space** of S.

THEOREM 5.11 Each metric space can be uniformly imbedded in the unit sphere of an l_∞-space(S) for some set S.

Proof: Let (M, d) be a metric space. For each pair $x,y \in$ M put $\delta(x,y) = min\{d(x,y),1\}$. Clearly (M, δ) is uniformly homeomorphic with (M, d). For each $x \in$ M, let x^* be the function on M defined by $x^*(y) = \delta(x,y)$. Then $d(x^*,y^*) = sup\{|\delta(x,z) - \delta(y,z)| \mid z \in M\}$. For any $z \in$ M, the largest $\delta(x,z)$ can be is $\delta(x, y) + \delta(y, z)$ because of the triangle inequality. Therefore, $sup\{|\delta(x, z) - \delta(y, z)| \mid z \in M\} \leq sup\{|\delta(x,y) + \delta(y, z) - \delta(y, z)| \mid z \in M\} = \delta(x,y)$ so for each pair $x,y \in$ M, $d(x^*,y^*) \leq \delta(x,y)$. Conversely, $|x^*(y) - y^*(y)| = |\delta(x,y) - 0| = \delta(x,y)$ so $sup\{|x^*(z) - y^*(z)| \mid z \in M\} = \delta(x,y)$. Hence, for each pair $x,y \in$ M, $d(x^*,y^*) = \delta(x,y)$, so (M, δ) is uniformly homeomorphic with f(M, δ) ⊂ l_∞(M) where f:M → l_∞(M) is the uniform imbedding defined by $f(x) = x^*$ for each $x \in$ M. Since (M, d) is uniformly homeomorphic with (M, δ), we have (M, d) is uniformly homeomorphic with f(M) ⊂ l_∞(M). Since $|x^*(m)| \leq 1$ for each $m \in$ M, (M, d) can be uniformly imbedded in the unit sphere of an l_∞-space. ∎

PROPOSITION 5.14 The unit sphere of an l_∞-space is injective.

Proof: Let S_1(X) denote the unit sphere of the l_∞-space l_∞(X). Let Z be a uniform space and f:Z → S_1(X) a uniformly continuous function. We first show that f is uniformly continuous if and only if the function g:Z × X → [-1,1] is uniformly continuous where g is defined by $g(z, x) = [f(z)](x)$ and X is considered to be uniformly discrete. For this, suppose f is uniformly continuous and $\mathcal{V}$ is a uniform covering of [-1,1]. Then there exists a uniform covering $\mathcal{U}$ of Z such that f maps any two points that are $\mathcal{U}$-close in Z into two functions that are $\mathcal{V}$-close in S_1(X). Then the product $\mathcal{U} \times \mathcal{W}$ where $\mathcal{W} = \{\{x\} \mid x \in X\}$

is a uniform covering of Z × X. Moreover, $\mathcal{U} \times \mathcal{W}$ is finer than $g(\mathcal{V}^*)$. To see this note that $g(\mathcal{V}^*) = \{g(Star(V,\mathcal{V})) \mid V \in \mathcal{V}\}$. For each $V \in \mathcal{V}$, $g(Star(V,\mathcal{V})) = \{(z, x) \in Z \times X \mid [f(z)](x) \in Star(V,\mathcal{V})\}$. Now if (z_1, x) and $(z_2, x) \in U \times \{x\}$ for some $U \times \{x\} \in \mathcal{U} \times \mathcal{W}$, then z_1 and z_2 are $\mathcal{U}$-close in Z which implies $f(z_1)$ and $f(z_2)$ are $\mathcal{V}$-close in $S_1(X)$, so $[f(z_1)](x)$ and $[f(z_2)](x)$ are $\mathcal{V}$-close in [-1,1]. Hence $[f(z_1)](x)$, $[f(z_2)](x) \in V_0$ for some $V_0 \in \mathcal{V}$, so $g(U \times \{x\}) \subset Star(V_0, \mathcal{V})$. Therefore, $\mathcal{U} \times \mathcal{W}$ is finer than $g(\mathcal{V}^*)$ so g is uniformly continuous.

Conversely, suppose g is uniformly continuous. Then there exists a uniform covering $\mathcal{U}$ of Z and a uniform covering $\mathcal{W}$ of X such that $\mathcal{U} \times \mathcal{W} < g(\mathcal{V})$. Without loss of generality we may assume $\mathcal{W} = \{\{x\} \mid x \in X\}$. Pick $x_0 \in X$. If $z_1, z_2 \in U$ for some $U \in \mathcal{U}$, then $(z_1, x_0), (z_2, x_0) \in U \times \{x_0\} \subset g(V)$ for some $V \in \mathcal{V}$. But $g(V) = \{(z, x) \in Z \times X \mid [f(z)](x) \in V\}$ so $[f(z_1)](x)$, $[f(z_2)](x) \in V$. Since this holds for each x_0 in X, $f(z_1)$ and $f(z_2)$ are two functions in $S_1(X)$ that are $\mathcal{V}$-close. Hence f is uniformly continuous.

To show that $S_1(X)$ is an injective uniform space, let A be a uniform subspace of Y and let f:A → $S_1(X)$ be a uniformly continuous function. Then by the above result, g:A × X → [-1,1] is uniformly continuous where $g(a, x) = [f(a)](x)$. g can be extended to a uniformly continuous function G:Y × X → [-1,1] (this follows from Theorem 4.9 and Exercise 4 of Section 2.2). But then F:Y → $S_1(X)$ is uniformly continuous where for each $y \in Y$, $F(y)$ is the function in $S_1(X)$ defined by $[F(y)](x) = G(y, x)$. Clearly, F is an extension of f, so $S_1(X)$ is an injective space. ∎

THEOREM 5.12 Each uniform space can be uniformly imbedded in an injective uniform space.

Proof: Let X be a uniform space. By Theorem 5.10, X can be imbedded in a product ΠM_α of metric spaces. By Theorem 5.11, each M_α can be imbedded in the unit sphere $S_1(M_\alpha)$ of the l_∞-space $l_\infty(M_\alpha)$. Consequently, X can be imbedded in $\Pi S_1(M_\alpha)$. By Proposition 5.14, $S_1(M_\alpha)$ is an injective uniform space for each α. Then by Proposition 5.12, $\Pi S_1(M_\alpha)$ is an injective uniform space, so X can be imbedded in an injective uniform space. ∎

For a uniform space (X, μ), let (Y, ν) be an injective uniform space in which (X, μ) can be imbedded. Let u be the finest uniformity on Y and put $l(\mu) = u \mid X$. Clearly (X, $l(\mu)$) is a subfine uniform space. The next lemma shows that $l(\mu)$ is well defined. $l(\mu)$ is called the **subfine coreflection** of μ.

LEMMA 5.10 If (X, μ) is uniformly imbedded in an injective uniform space (Z, ζ), and if u is the fine uniformity on Z, then (X, l(μ)) and (X, m(μ)) are uniformly homeomorphic where m(μ) = u | X.

Proof: Let f:A → (X, μ) be uniformly continuous where A is a subspace of a fine uniform space B. Then f has an extension g:B → (Y, ν) since Y is injective and (X, μ) is imbedded in (Y, ν). Since B is a fine space, g is uniformly continuous into (Y, u) where u is the fine uniformity on Y. But then the restriction of g to A is uniformly continuous, so f:A → (X, l(μ)) is uniformly continuous. Therefore, any uniformly continuous function f from a subfine space into (X, μ) is also uniformly continuous into (X, l(μ)).

The identity function i:(X, m(μ)) → (X, μ) is uniformly continuous, so by the above result, i:(X, m(μ)) → (X, l(μ)) is uniformly continuous. If (Y, ν) were replaced with (Z, ζ) in the above argument, we would obtain the result that any uniformly continuous function f from a subfine space into (X, μ) is also uniformly continuous into (X, m(μ)). Hence i^{-1}:(X, l(μ)) → (X, m(μ)) is uniformly continuous, so (X, m(μ)) is uniformly homeomorphic with (X, l(μ)). ▪

COROLLARY 5.5 The subfine coreflection of an injective space or a complete metric space is fine.

Proof: The injective case is clear from the proof of Lemma 5.10. If (X, μ) is a complete metric space, then by Theorem 5.11 and Proposition 5.14, (X, μ) is a closed subspace of an injective metric space (Y, ν). Since X is closed in Y, each open covering of X together with the set Y - X is an open covering of Y. Hence the fine uniformity on Y induces the fine uniformity on X. ▪

LEMMA 5.11 If (X, μ) is subfine then it is locally fine.

Proof: Assume (X, μ) is subfine. It suffices to show that λ(μ) = μ. For this let (X, μ) be uniformly imbedded in some fine space (Y, ν). Then μ = ν|X. Then as shown in the proof of Lemma 5.10, the identity function i:(X, μ) → (X, l(μ)) is uniformly continuous. Clearly i^{-1}:(X, l(μ)) → (X, μ) is uniformly continuous since l(μ) is finer than μ. Therefore, (X, μ) is uniformly homeomorphic with (X,l(μ)) so μ = l(μ). ▪

LEMMA 5.12 l(μ) is the coarsest subfine uniformity for X finer than μ. In particular, μ is subfine if and only if l(μ) = μ.

Proof: Suppose there exists a subfine uniformity θ for X such that μ ⊂ θ ⊂ l(μ). Then i:(X, θ) → (X, μ) is uniformly continuous, so by the proof of Lemma 5.10, i:(X, θ) → (X, l(μ)) is uniformly continuous. Clearly i^{-1}:(X, l(μ)) → (X, θ) is uniformly continuous since l(μ) is finer than θ. Therefore, (X, l(μ)) is uniformly homeomorphic with (X, θ), so l(μ) is the coarsest subfine uniformity for X finer than μ. ▪

LEMMA 5.13 A point finite uniform covering has a uniformly locally finite uniform shrinking. A σ-disjoint uniform covering has a σ-discrete uniform shrinking.

Proof: Let $\{U_\alpha\}$ be a point finite uniform covering. By Exercise 6 of Section 2.2, there exists a uniform covering $\{V_\alpha\}$ that is a uniformly strict shrinking of $\{U_\alpha\}$. Let $\mathcal{W}$ be a uniform covering such that for each α, $Star(V_\alpha, \mathcal{W}) \subset U_\alpha$. Let $p \in X$ and pick $W \in \mathcal{W}$ with $p \in W$. If $W \cap V_\beta \neq \emptyset$ then $p \in U_\beta$. Consequently, W can meet at most finitely many of the V_α's since p can be in at most finitely many of the U_α's. Therefore, $\{V_\alpha\}$ is uniformly locally finite.

If $\{U_\alpha\}$ is a σ-disjoint uniform covering, there exists countably many subcollections $\{U_\beta^n\} \subset \{U_\alpha\}$ such that each $\{U_\beta^n\}$ is a disjoint collection. Since for each α, $Star(V_\alpha, \mathcal{W}) \subset U_\alpha$, the collection $\{V_\beta^n\}$ is uniformly discrete. Hence $\{V_\alpha\}$ is a σ-uniformly discrete uniform shrinking of $\{U_\alpha\}$. ∎

LEMMA 5.14 Each uniform covering of a subfine uniform space has a locally finite uniform refinement and a σ-disjoint uniform refinement.

Proof: Let (X, μ) be a subfine uniform space. Then there exists a fine uniform space (Y, μ^*) such that (X, μ) is a subspace of (Y, μ^*). Let $\mathcal{U} \in \mu$. There exists a $\mathcal{U}^* \in \mu^*$ with $\mathcal{U} = \mathcal{U}^* \cap X$. By Theorem 2.5, there exists a uniformly continuous $f: X \to (M, d)$ where d is a metric for M, such that for each $x \in X$, $f(S(x, \mathcal{U}^*)) = S(f(x), 1/2)$. Since M is paracompact, the covering $\mathcal{V} = \{S(m, 1/2) \mid m \in M\}$ has a locally finite refinement $\mathcal{W}$ that belongs to the finest uniformity for M. Since f is continuous, $f^{-1}(\mathcal{W}) \in \mu^*$ because μ^* is fine. Then $f^{-1}(\mathcal{W}) \mid X$ is a locally finite member or μ that refines $\mathcal{U}$.

Similarly, $\mathcal{V}$ has a σ-disjoint refinement $\mathcal{Z}$. Hence $f^{-1}(\mathcal{Z})$ is a σ-disjoint open refinement of $\mathcal{U}^*$ that is itself uniform, so $f^{-1}(\mathcal{Z}) \mid X$ is a σ-disjoint uniform refinement of $\mathcal{U}$. ∎

THEOREM 5.13 Each subfine uniform space has a basis of uniformly locally finite coverings and a basis of σ-uniformly discrete coverings.

The proof of Theorem 5.13 follows from the preceding lemmas. A uniformity is said to be **point finite** if each member has a uniform point finite refinement.

THEOREM 5.14 If μ is a point finite uniformity, so is the derivative μ^1.

Proof: Let $\{U_\alpha \cap V_\beta^\alpha\} \in \mu^1$. Let $\mathcal{U} = \{U_\alpha\}$ and $\mathcal{V}^\alpha = \{V_\beta^\alpha\}$ for each α. Without loss of generality we may assume $\mathcal{U}$ is point finite. Choose a point finite $A = \{A_\gamma\} \in \mu$ with $A <^* \mathcal{U}$ and for each α, a $\mathcal{W}^\alpha <^* \mathcal{V}^\alpha$ in μ. Each A_γ is contained in at most finitely many of the U_α, say $U_{\alpha_1} \ldots U_{\alpha_n}$. Put $G^\gamma = \mathcal{W}^{\alpha_1} \cap \ldots \cap \mathcal{W}^{\alpha_n}$ and let $Z^\gamma = \{Z_\delta^\gamma\}$ be a point finite uniform refinement of G^γ. Then $\{A_\gamma \cap Z_\delta^\gamma\} \in \mu^1$ is point finite.

To show that $\{A_\gamma \cap Z_\delta^\gamma\} <^* \{U_\alpha \cap V_\beta^\alpha\}$, note that for any pair γ, δ we can choose U_α containing $Star(A_\gamma, A)$ and a V_β^α containing $Star(Z_\beta^\gamma, \mathcal{W}^\alpha)$. If $A_\theta \cap Z_\zeta^\theta$

meets $A_\gamma \cap Z^\gamma_\delta$ then $A_\theta \subset Star(A_\gamma, A) \subset U_\alpha$ which implies $Z^\theta < G^\theta < \mathcal{W}^\alpha$, so $Z^\theta_\xi \subset Star(Z^\gamma_\beta, \mathcal{W}^\alpha) \subset V^\alpha_\beta$. Hence $A_\theta \cap Z^\theta_\xi \subset A_\gamma \cap Z^\gamma_\delta$. We conclude that $\{A_\gamma \cap Z^\gamma_\delta\} <^* \{U_\alpha \cap V^\alpha_\beta\}$ so μ^1 is also a point finite uniformity. ∎

COROLLARY 5.6 Each subfine uniform space is locally fine.

Proof: Let (X, μ) be a subfine space and (Y, μ*) a fine space such that (X, μ) is a uniform subspace of (Y, μ*). From the definition it is clear that (X, μ^1) is a uniform subspace of (X, μ^{*1}). By Theorem 5.13 μ and μ* are point finite, so by Theorem 5.14, μ^1 and μ^{*1} are also. Since μ* is fine, $\mu^{*1} = \mu^*$ which implies $\mu^1 = \mu$. A straightforward induction on α yields $\mu^\alpha = \mu$ for each α, so $\lambda(\mu) = \mu$. ∎

EXERCISES

DERIVATIVES OF UNIFORMITIES

It took over 20 years to settle the question of whether the αth derivative μ^α of a uniformity μ is a uniformity or not. Ginsburg and Isbell announced a proof in 1955 that the answer is affirmative. In their 1959 paper referenced earlier in this section, they retracted their announcement and left the question open.

1. [J. Isbell, 1964] If μ is a uniformity that has a basis consisting of point finite coverings, then μ^1 (and hence any μ^α) is a uniformity.

2. [J. Pelant, 1975] A uniformity μ for X has a basis consisting of point finite coverings if and only if the first derivative of the product uniformity $\Pi^\kappa_{\alpha=1}\mu_\alpha$, where each $\mu_\alpha = \mu$, of the product space $(X, \mu)^\kappa$ is a uniformity for each cardinal κ.

3. [J. Pelant, 1975] There exists a uniform space with no basis consisting of point finite coverings. Consequently, by Exercise 2, there exists a uniformity with a derivative that is not a uniformity. (Note: The existence of a uniform space without a basis consisting of point finite coverings was also an open problem from Isbell's book.)

INJECTIVE SPACES

4. Let X and Y be uniform spaces and let *U*(X,Y) denote the collection of all uniformly continuous functions from X into Y. For an imbedding $i{:}A \to X$, define the function $F{:}U(X,Y) \to U(A,Y)$ by $F(h) = h \circledcirc i$ for each $h \in U(X,Y)$. Show that Y is injective if and only if *F* is onto.

5. Show that if a uniform space Y has the property that whenever A is a closed

uniform subspace of X, every uniformly continuous function f:A $\rightarrow$ Y can be extended to a uniformly continuous function F:X $\rightarrow$ Y, then Y is complete and hence injective. [Hint: Suppose Y is not complete and Z is its completion. Put A = Y and construct X as follows: Let κ be a regular cardinal. $\kappa + 1$ is a compact Hausdorff space in the order topology. A continuous function f on κ or $\kappa + 1$ into a metric space is finally (residually) constant. Since Y is imbeddable in a product of metric spaces, κ can be chosen large enough to make each continuous function on κ finally constant. Also, κ should be greater than the cardinality of any subset of Y having a limit point in Z - Y. Let S $\subset$ Y have a limit point in Z - Y. Put X = [Y $\times$ (κ + 1) - κ]$\cup$[S $\times$ (κ + 1)]$\cup$[$p \times \kappa$]. Then X $\subset$ Z $\times$ (κ + 1). Identify Y with Y $\times$ (κ + 1) - κ so that Y $\subset$ X. It can be shown that Y is closed in X and any continuous extension over S $\times$ (κ + 1) of the identity i:Y $\rightarrow$ Y must agree with coordinate projection on a uniform neighborhood of S $\times$ [(κ + 1) - κ] and cannot be extended over {p} $\times$ κ.]

6. Let (X, μ) be a uniform space and let $p(\mu)$ be the collection of all coverings of X refined by a finite member of μ. Show that $p(\mu)$ is a uniformity for X that is coarser than μ.

7. Show that a uniform retract of an injective uniform space is also injective. A **retraction** is a mapping r such that r © $r = r$. A **uniform retract** (Y, ν) of a uniform space (X, μ) is a uniform subspace such that there exists a uniformly continuous function r:X $\rightarrow$ Y such that r is a retraction.

5.7 Categories and Functors

In this section we introduce the elements of *category theory*. None of the following sections or chapters will depend on a knowledge of this subject, so this section can be skipped without sacrificing the accessibility to subsequent sections. However, there will be exercises at the end of future sections that will involve an elementary understanding of category theory. The approach here is to keep the development of category theory separate from our development of uniform spaces to allow the reader the choice of pursuing this area or not. Some researchers in the theory of uniform spaces tend to combine categorical notions with uniform space concepts in their publications. Consequently, a knowledge of category theory is helpful in reading some of the literature in the theory of uniform spaces.

On the other hand, the reliance on categorical notions and terminology by these uniform space researchers has often been criticized as making this field almost inaccessible to anyone but the specialist. In future sections and chapters, all uniform concepts will be introduced without reliance on categorical terminology or concepts, to allow as immediate access to the ideas as we know how. However, in order to illustrate the current usage of category theory as a tool in the study of uniform spaces (just as set theory is used as a tool), we will present a selection of problems that involve categorical notions that should suffice to prepare the reader to access that branch of uniform space theory that uses category theory.

For those who have studied even elementary set theory, it is understood that the *class* of all sets cannot itself be a set. The assumption that it is leads to a statement that is simultaneously true and false, as first noted by B. Russell. This is the famous *Russell Paradox*. However, it is still useful to retain the notion of a class of all sets, just it is useful to have the notion of the class of all uniform spaces or the class of all topological spaces.

By a **concrete category** C, we mean a class O of sets (referred to as **objects** of C), and for each ordered pair of objects (X, Y), a *set* of functions $f{:}X \to Y$ (denoted by $M(X, Y)$ and called the **morphisms** of C with domain X and range Y) such that

(1) The identity function on each object X is a morphism of C with domain X and range X.
(2) Each composition of morphisms of C is a morphism of C.

The class O of sets may be larger than any cardinal number, and will be in all examples given in this book.

By a **covariant functor** $F{:}C \to D$ of two concrete categories C and D, we mean a pair of functions F_O (called the **object function**) and F_M (called the **morphism function**) such that F_O assigns to each object X in C an object $F_O(X)$ in D, and F_M assigns to each morphism $f{:}X \to Y$ of C a morphism $F_M(f){:}F_O(X) \to F_O(Y)$ such that

(3) For each identity morphism 1_X of C, $F_M(1_X) = 1_{F_O(X)}$.
(4) For each composed morphism $g \circledcirc f$ of C, $F_M(g \circledcirc f) = F_M(g) \circledcirc F_M(f)$.

Since F_O only applies to objects and F_M only applies to morphisms, it is customary to denote both F_O and F_M simply by F as no confusion is likely to arise. This shorter notation is useful for denoting the *composition of covariant functors* $F{:}C \to D$ and $G{:}D \to E$ by $G \circledcirc F(X) = G(F(X))$ and $G \circledcirc F(f) = G(F(f))$.

An **isomorphism** is a covariant functor $F{:}C \to D$ such that there exists a covariant functor $F^{-1}{:}D \to C$ with both $F^{-1} \circledcirc F$ and $F \circledcirc F^{-1}$ being the identity functors on C and D respectively. If the functor $F{:}C \to D$ is an isomorphism, the concrete categories C and D are said to be **isomorphic** and both C and D are said to be **instances** of the same **abstract category**. A **categorical property** is a property P that if possessed by a concrete category C, is preserved by isomorphisms (i.e., if D is a concrete category isomorphic with C, then D also has property P). Similarly, we can define categorical concepts and definitions.

For instance, the notion of a mapping $f{:}X \to Y$ having an inverse $f^{-1}{:}Y \to X$ is a *categorical concept* in the following sense. If we define $f{:}X \to Y$ has an inverse $f^{-1}{:}Y \to X$ to mean that there exists a mapping $f^{-1}{:}Y \to X$ such that $f^{-1} © f = 1_X$ and $f © f^{-1} = 1_Y$, then if X and Y belong to the concrete category C such that $F{:}C \to D$ is an isomorphism, then it is easily seen from the definitions that the morphism $F(f){:}F(X) \to F(Y)$ also has an inverse.

On the other hand, the notion that the morphism $f{:}X \to Y$ is one-to-one is not a categorical notion. However, being one-to-one implies an important categorical notion that is closely related to being one-to-one. In fact, in uniform spaces, being one-to-one is equivalent to this notion. A morphism $f{:}X \to Y$ of a concrete category C is said to be a **monomorphism** if for each pair of morphisms $g{:}Z \to X$ and $h{:}Z \to X$ we have $f © g = f © h$ implies $g = h$. Clearly if f is one-to-one it is a monomorphism. On the other hand, if f is not one-to-one then $f(x) = f(y)$ for some $x \neq y$ in X. But then the constant functions $g{:}Z \to \{x\}$ and $h{:}Z \to \{y\}$ are not equal, yet $f © g = f © h$. Hence f is not a monomorphism. Therefore, f is a monomorphism if and only if it is one-to-one. It is the fact that constant functions are morphisms in the category of uniform spaces that makes monomorphisms equivalent to one-to-one mappings.

Similarly to this *left cancellation* property that defines the monomorphisms, the *right cancellation* property defines the **epimorphisms.** f is said to be an epimorphism if for each pair of morphisms $g{:}Y \to Z$ and $h{:}Y \to Z$, we have $g © f = h © f$ implies $g = h$. Epimorphisms are closely related to the concept of being *onto*. Indeed, an onto morphism is clearly an epimorphism. But being an epimorphism is a more general property, for if $f(X)$ is merely dense in Y, it suffices to be an epimorphism. To see this, note that if $g © f = h © f$ and $y \in Y$, we can pick a net ϕ in $f(X)$ converging to y. Since g and h are morphisms (uniformly continuous functions in the category of uniform spaces), $g © \phi$ converges to $g(y)$ and $h © \phi$ converges to $h(y)$. But $g © f = h © f$ and $\phi \subset f(X)$ implies $g © \phi = h © \phi$. Since nets converge to unique limits in uniform spaces, we have $g(y) = h(y)$. Therefore, f is an epimorphism.

A morphism $f{:}X \to X$ is called a **retraction** if $f^2 = f © f = f$. If a retraction is either an epimorphism or a monomorphism, it is an identity morphism.

By a **contravariant functor** $F{:}C \to D$ of two concrete categories C and D, we mean a functor F that assigns to each object X of C an object $F(X)$, but to each morphism $f{:}X \to Y$ of C, a morphism *in the opposite direction* $F(f){:}F(Y) \to F(X)$ of D to C that satisfies property (3) above, but instead of satisfying property (4), it satisfies

(4′) For each composed morphism $g © f$ of C, $F_M(g © f) = F_M(f) © F_M(g)$.

Observe that the composition of two contravariant functors is a covariant functor. Also, functors of different *variances* can be composed with the resultant functor being a contravariant functor.

A **duality** is a contravariant functor $F:C \to D$ such that there exists a contravariant functor $F^{-1}:D \to C$ with both $F^{-1} © F$ and $F © F^{-1}$ being identity functors. If C and D are concrete categories, and $F:C \to D$ is a duality, then every concept, definition, or theorem about C gives rise to a *dual* concept, definition or theorem about D. To illustrate what is meant by dual concepts and definitions, we consider the concepts of *sums* and *products* of objects of an arbitrary concrete category C. As we shall see, these concepts are dual concepts. We have already considered the concepts of sums and products of uniform spaces. We now give *categorical definitions* for sums and products of objects in arbitrary concrete categories in such a way that the ordinary definitions of sums and products of uniform spaces are equivalent to the categorical definitions in the category of uniform spaces.

Let $\{X_\alpha\}$ be a collection of objects in the concrete category C and let $\{i_\alpha\}$ be a collection of morphisms such that $i_\alpha:X_\alpha \to \Sigma$ for some $\Sigma \in C$. Then Σ is called the **sum** of the objects $\{X_\alpha\}$ if the following conditions hold:

(5) If $f:\Sigma \to Y$ and $g:\Sigma \to Y$ are distinct morphisms, then for some index α, the morphisms $f © i_\alpha$ and $g © i_\alpha$ are distinct.
(6) For each family of morphisms $f_\alpha:X_\alpha \to Y$, there exists a morphism $f:\Sigma \to Y$ such that $f © i_\alpha = f_\alpha$ for each α.

Similarly, an object $\Pi \in C$ is said to be a **product** of the objects $\{X_\alpha\}$ if there is a collection $\{p_\alpha\}$ of morphisms such that $p_\alpha:\Pi \to X_\alpha$ for each α, if the following conditions hold:

(7) If $f:Z \to \Pi$ and $g:Z \to \Pi$ are distinct morphisms, then for some index α, the morphism $p_\alpha © f$ and $p_\alpha © g$ are distinct.
(8) For each family of morphisms $f_\alpha:Z \to X_\alpha$, there exists a morphism $f:Z \to \Pi$ such that $p_\alpha © f = f_\alpha$ for each α.

To show that the categorical definitions of sum and product are dual, we must show that if Σ is the sum of $\{X_\alpha\}$ in C, then $\Pi = F(\Sigma)$ is the product of $\{F(X_\alpha)\}$ in D and vice-versa. For this, assume Σ is the sum of $\{X_\alpha\}$ in C and put $\Pi = F(\Sigma)$. Suppose $F(f)$ and $F(g)$ are distinct morphisms such that $F(f):F(Y) \to \Pi$ and $F(g):F(Y) \to \Pi$. Then $f:\Sigma \to Y$ and $g:\Sigma \to Y$. Since $F^{-1} © F$ is the identity, $F^{-1} © F(f) = f$ and $[F^{-1} © F](g) = g$. Therefore, $f = g$ implies $[F^{-1} © F](f) = [F^{-1} © F](g)$, so $F(f) = F(g)$ which is a contradiction. Therefore, f and g are distinct. Since Σ is the sum of $\{X_\alpha\}$, there exists an α with $f © i_\alpha \neq g © i_\alpha$. Then by an argument similar to the above, $F(f © i_\alpha) \neq F(g © i_\alpha)$ which implies $F(i_\alpha) © F(f) \neq F(i_\alpha) © F(g)$. For each α, put $p_\alpha = F(i_\alpha)$ and let $Z = F(Y)$. Then if $F(f):Z \to \Pi$ and $F(g):Z \to \Pi$ are distinct morphisms, for some index α, the morphisms $p_\alpha \circledcirc F(f)$ and $p_\alpha \circledcirc F(g)$ are distinct. Roughly speaking, condition (5) in C forces condition (7) in D.

Next, suppose $F(f_\alpha){:}Z \to F(X_\alpha)$ is a family of morphisms in D. Then $f_\alpha{:}X_\alpha \to Y$ is a family of morphisms in C, so by (6), there exists a morphism $f{:}\Sigma \to Y$ with $f \circledcirc i_\alpha = f_\alpha$ for each α. But then $F(f){:}Z \to \Pi$ such that $p_\alpha \circledcirc F(f) = F(f_\alpha)$ for each α. Again, roughly speaking, condition (6) in C implies condition (8) in D. Hence $\Pi = F(\Sigma)$ is the product of $\{F(X_\alpha)\}$ in D.

The proof that if Π is the product of $\{X_\alpha\}$ in D, then $\Sigma = F^{-1}(\Pi)$ is the sum of $\{F^{-1}(X_\alpha)\}$ is similar. This establishes the *duality* of the concepts of sum and product in arbitrary concrete categories. To illustrate what is meant by *dual theorems* we will show that the *categorical theorem*: sums are unique (in C), gives rise to the dual (categorical) theorem: products are unique (in D). To see this, let Σ be the sum of $\{X_\alpha\}$ in C. Then $\Pi = F(\Sigma)$ is the product of $\{F(X_\alpha)\}$ in D as we have already seen.

If Π' is another product of $\{F(X_\alpha)\}$, then $\Sigma' = F^{-1}(\Pi')$ is another sum of $\{X_\alpha\}$ so there exists a uniform homeomorphism $i{:}\Sigma' \to \Sigma$ (since sums are unique in C). Therefore, $F(i){:}\Pi \to \Pi'$ and since $i^{-1} \circledcirc i = 1_{\Sigma'}$ and $i \circledcirc i^{-1} = 1_\Sigma$, we have:

$$F(i^{-1} \circledcirc i) = F(i) \circledcirc F(i^{-1}) = F(1_{\Sigma'}) = 1_{F(\Sigma')}.$$

Similarly, $F(i^{-1}) \circledcirc F(i) = 1_{F(\Sigma)}$, so $F(i)$ is a uniform homeomorphism of $\Pi = F(\Sigma)$ onto $\Pi' = F(\Sigma')$. Therefore, products are unique in D. The theorem that products are unique in C implies sums are unique in D is proved similarly. This establishes the duality of these two theorems.

The *principle of duality* says that any categorical concept, definition or theorem gives rise to a dual concept, definition or theorem.

It is also of interest to give categorical definitions of the concepts of *subspace* and *quotient space*. For the concept of a subspace, we recall that for a uniform space X, a subspace can be considered to be a uniform space S such that there exists a one-to-one uniformly continuous function $i{:}S \to X$ such that the uniformity of X relativized to $i(S)$ is identical with the identification uniformity of $i(S)$. By a **uniform imbedding** of S into X, we mean a uniform homeomorphism of S onto a subspace S' of X. Clearly, if S is a subspace of X, the mapping $i{:}S \to X$ above is a uniform imbedding.

Then a one-to-one mapping $i{:}S \to X$ is a uniform imbedding if and only if whenever $i = g \circledcirc f$ where g and f are uniformly continuous and f is one-to-one and onto, then f is a uniform homeomorphism. To see this first assume $i{:}S \to X$ is a uniform imbedding and $i = g \circledcirc f$ where g and f are uniformly continuous and f is one-to-one and onto. Let ν be the uniformity of S and μ the uniformity of X restricted to $i(S)$. For $V \in \nu$, $i(V) = \{i(V) \mid V \in \mathcal{V}\} \in \mu$ since i is a uniform imbedding. Let λ be the uniformity of $S' = f(S)$. Since g is uniformly continuous, $g^{-1}(i(V)) \in \lambda$. But $f = g^{-1} \circledcirc i$ because i is one-to-one and $g = i \circledcirc f^{-1}$ which implies g is one-to-one. Hence $f(V) \in \lambda$, so f is a uniform homeomorphism.

Conversely, if the one-to-one mapping $i = g \circledcirc f$ where g and f are uniformly continuous and f is one-to-one and onto implies f is a uniform homeomorphism, then if $V \in \nu$, $f(V) \in \lambda$. Assume $i(V)$ does not belong to μ. Then $g(g^{-1}(i(V)))$ is not in μ since g is one-to-one. But $f(V) \in \lambda$ implies $g^{-1}(i(V)) \in \lambda$ since $f = g^{-1} \circledcirc i$. Therefore, the identification uniformity μ^* of g is strictly finer than μ. Let λ' be the uniformity induced on S' by μ with respect to g. Then λ' is strictly coarser than $\lambda = g^{-1}(\mu^*)$. Hence f is uniformly continuous with respect to ν and λ', but (S,ν) and (S', λ') are not homeomorphic which is a contradiction. Therefore, $i(V) \in \mu$ so i is a uniform imbedding.

Next, we use the characterization of uniform imbeddings to motivate the categorical definition of imbeddings. A monomorphism $i:S \to X$ in a concrete category C is said to be an **imbedding** if whenever $i = g \circledcirc f$ where g and f are morphisms, and f is one-to-one and onto, then f is an isomorphism. This means that if $f:S \to S'$ then S' has the same structure as S, so that no structure can be inserted on S that is *between* the structure of S and that of $i(S)$. In the case of uniform spaces, this means that the uniformity of S is the one induced by the function i. It is also clear that in any concrete category, two imbeddings with the same image have isomorphic domains.

The dual categorical concept of an imbedding is that of a *quotient* morphism. An epimorphism $q:X \to Q$ in a concrete category D is said to be a **quotient** if whenever $q = d \circledcirc h$ where d and h are morphisms and d is one-to-one and onto, then d is an isomorphism. To show that in the category of uniform spaces, a uniform quotient mapping satisfies this categorical definition, let (X, μ) be a uniform space and $q:X \to Y$ a uniform quotient mapping. Then q is onto and Y has the identification uniformity with respect to μ generated by q. Let ν be the uniformity of Y, then $V \in \nu$ if and only if $f^{-1}(V) \in \mu$.

Let $q = d \circledcirc h$ where d and h are uniformly continuous and d is one-to-one and onto. Let λ be the uniformity of $Q' = h(X)$. If $W \in \lambda$, then $h^{-1}(W) \in \mu$. Since q is a uniform quotient, $q(h^{-1}(W)) \in \nu$. But then $d(h(h^{-1}(W))) = d(W) \in \nu$, so d is a uniform homeomorphism. Consequently, in the category of uniform spaces, uniform quotient mappings satisfy the categorical definition of a quotient morphism.

A **subcategory** S of a concrete category A is a category such that each object of S is an object of A and each morphism of S is a morphism of A. S is said to be a **full subcategory** of A if every morphism of A whose domain and range are both objects of S is a morphism of S. A full subcategory R of a category A is said to be a **reflection** of A if for each object $X \in A$, there is an object $X^* \in R$ and a morphism $r:X \to X^*$ (called the **reflection morphism**) such that each morphism $f:X \to Y$ where $Y \in R$ can be factored uniquely over X^* by r

(i.e., $f = g © r$ for a unique $g:X^* \to Y$). The space X^* is called the **reflection of** X in R or simply the R-**reflection** of X.

There are many interesting reflections of the category of all uniform spaces. We have already encountered some of them unknowingly. For instance, the class of all complete uniform spaces C is a reflection of the category U of all uniform spaces. For each uniform space $X \in U$, the completion $X^* \in C$ is the C-reflection of X and the natural imbedding $i:X \to X^*$ is the reflection morphism such that any morphism $f:X \to Y \in C$ factors uniquely over X^* by i.

For any reflection R of a category A, an R-reflection of an object $X \in A$ is unique up to isomorphism. To see this, note that if X^* and X^+ are both R-reflections of X, then there exists reflection morphisms $r^*:X \to X^*$ and $r^+:X \to X^+$ such that any morphism $f:X \to Y \in R$ factor uniquely over both X^* and X^+ by r^* and r^+ respectively. Therefore, $r^* = g^+ © r^+$ and $r^+ = g^* © r^*$ for some unique g^* and g^+. Now $g^*:X \to X^+$ and $g^+:X^+ \to X^*$. Moreover, $r^+ = g^* © (g^+ © r^+) = (g^* © g^+) © r^+$ so $(g^* © g^+) = 1_{X^+}$. Similarly, $g^+ © g^* = 1_{X^*}$. Therefore, $g^+ = g^{*-1}$ so X^* and X^+ are isomorphic (in the class of uniform spaces this means uniformly homeomorphic).

The fact that an R-reflection is unique allows us to make the simplifying assumption that if $X \in R$, the only R-reflection of X is X itself and the only reflection morphism $r:X \to X$ is the identity morphism. This allows us to state the following fundamental theorem about reflective categories.

THEOREM 5.15 If R is a reflection of a concrete category A, $X, Y \in A$ and X^*, Y^* are their R-reflections respectively, then any morphism $f:X \to Y$ determines a unique morphism $f^*:X^* \to Y^*$ such that $r © f = f^* © s$ where $r:X \to X^*$ and $s:Y \to Y^*$ are the reflections morphisms respectively.

Proof: Since $s © f:X \to Y^*$, it can be factored uniquely over X^* by r. Let $f^* © r$ be this factorization. Then $s © f = f^* © r$ and since f^* is unique, it is the unique morphism that satisfies the theorem. ∎

Theorem 5.15 shows that the correspondences $X \to X^*$ and $f \to f^*$ determine a functor ρ called the **reflection functor**. The object function of ρ is $\rho_O:A \to R$ defined by $\rho_O(X) = X^*$ for each $X \in A$ and the morphism function ρ_M is defined by $\rho_M(f) = f^*$ for each $f \in A$. To see that ρ is a functor, we need to show that ρ_O and ρ_M satisfy (3) and (4) of the definition of a functor. To demonstrate (3) we note that by Theorem 5.15, the identity 1_X has a corresponding unique morphism $1_X^*:X^* \to X^*$ such that $1_X^* © r = r © 1_X = r$. Now $1_{X^*} © r = r$ so $1_{X^*} = 1_X^*$ since 1_X^* is unique. Therefore, $\rho_M(1_X) = 1_{\rho_O(X)}$.

To demonstrate (4), if $g © f$ is a morphism of A, then by Theorem 5.15, there exists unique morphisms g^*, f^* and $(g © f)^*$ such that $(g © f)^* © r = r © (g © f)$, $f^* © r = r © f$ and $g^* © r = r © g$. But then $r © (g © f) = g^* © r © f =$

$g^* \circledcirc f^* \circledcirc r$. Since $(g \circledcirc f)^*$ is unique, $(g \circledcirc f)^* = g^* \circledcirc f^*$. Therefore, $\rho_M(g) \circledcirc \rho_M(f)$.

The dual concept to the concept of a reflection is the concept of a *coreflection* defined as follows. A full subcategory K of a concrete category A is said to be a **coreflection** of A if for each object $X \in A$, there exists an object $X^* \in K$ and a coreflection morphism $k{:}X^* \to X$ such that each morphism $f{:}Z \to X$ where $Z \in K$ can be factored uniquely over X^* by k (i.e., $f = k \circledcirc g$ for a unique morphism $g{:}Z \to X^*$). The space X^* is called the **coreflection of** X in K or simply the K**-reflection** of X.

As with reflections, there are many interesting coreflections of the category of uniform spaces. Possibly the simplest and most useful is the *fine coreflection* that maps each uniform space (X, μ) onto (X, u) where u is the finest uniformity for X. Here, the coreflection morphisms are the identity mappings. To see that the class F of fine uniform spaces is a coreflection of the category U of uniform spaces, it suffices to show that each morphism $f{:}Z \to X$ where Z is a fine space can be factored uniquely over (X, u). But since $f{:}Z \to X$ is continuous and Z is fine, $f{:}Z \to (X, u)$ is uniformly continuous, so f can be factored over (X, u) by the identity mapping.

By the principle of duality, for any coreflection K of a category A, a K-coreflection X^* of an object X is unique up to isomorphism. As with reflections, the uniqueness of K-coreflections allows us to make the assumption that if $X \in K$, the only k-coreflection of X is X itself and the only coreflective morphism $k{:}X \to X$ is the identity morphism, so we have the dual theorem:

THEOREM 5.16 If K is a coreflection of a concrete category $A, X, Y \in A$ and X^*, Y^* are their K-coreflections respectively, then any morphism $f{:}X \to Y$ determines a unique morphism $f^*{:}X^* \to Y^*$ such that $f \circledcirc k = j \circledcirc f^*$ where $k{:}X^* \to X$ and $j{:}Y^* \to Y$ are the coreflective morphisms respectively.

Theorem 5.16 shows that the correspondences $X \to X^*$ and $f \to f^*$ determine a functor κ called the **coreflection functor**. The object function of κ is $\kappa_0{:}A \to K$ defined by $\kappa_0(X) = X^*$ and the morphism function κ_M is defined by $\kappa_M(f) = f^*$ for each $f \in A$.

EXERCISES

1. Show that the definitions of uniform sums and products are equivalent to the categorical definitions of sums and products in the concrete category of uniform spaces.

DUAL THEOREMS

2. Show that the categorical theorem *P*: A retraction *f* that is a monomorphism is an identity implies the dual theorem *Q*: A retraction *f* that is an epimorphism is an identity.

REFLECTIONS

3. Let *U* be the category of uniform spaces and for each $(X, \mu) \in U$ let $p(\mu)$ denote the uniformity consisting of all coverings refined by finite members of μ (see Exercise 6, Section 5.6). Then $(X, p(\mu))$ is a precompact uniform space. Show that the class *P* of all precompact uniform spaces is a reflection of *U* with reflection morphisms being the identity functions. Let *p* denote the reflection functor (called the **precompact reflection**). Show that for each $(X, \mu) \in U$ that $p_O(X, \mu) = (X, p(\mu))$ and for each morphism $f:(X, \mu) \to (Y, \nu)$ that $p_M(f) = f$.

4. For each $(X, \mu) \in U$ let $e(\mu)$ be the collection of all coverings of X refined by a countable member of μ. Show that $e(\mu)$ is a separable uniformity for X. [A uniform space is **separable** if it has a basis consisting of countable coverings.] Show that the class U_W of all separable uniform spaces is a reflection of uniform with reflection morphisms being the identity functions. Let *e* denote the reflection functor (known as the **separable reflection**), and show that for each $(X, \mu) \in U$, $e_O(X, \mu) = (X, e(\mu))$ and for each morphism $f:(X, \mu) \to (Y, \nu)$ that $e_M(f) = f$.

5. Show that the separable uniformities are precisely the countably bounded uniformities.

6. Show that the reflections *p* and *e* commute, i.e., for each $(X, \mu) \in U$, $p(e(X,\mu)) = e(p(X,\mu))$.

7. Let π denote the completion reflection that maps each $(X, \mu) \in U$ into μX. Show that the reflections *p* and π commute.

8. It was long an unsolved problem (due to Ginsberg and Isbell) whether the reflections *e* and π commute. J. Pelant (1974) showed that in general they do not.

THE SAMUEL COMPACTIFICATION

9. Show that the composition π © *p* of the reflection π and *p* is again a reflection. Since for each $X \in U$, $p(X)$ is precompact, it is clear that $\pi(p(X))$ is compact and $p(X)$ is uniformly homeomorphic with a dense subspace of $\pi(p(X))$. $\pi(p(X))$ is called the **Samuel compactification** of X, or the **uniform compactification** of X.

COREFLECTIONS

10. Let U be the category of uniform spaces and for each $X = (X, \mu) \in U$ let $\lambda(\mu)$ denote the locally fine coreflection of μ. Put $X^* = (X, \lambda(\mu))$ and let L be the category of locally fine uniform spaces. Show that $\lambda: U \rightarrow L$ defined by $\lambda(X) = X^*$ for each $X \in U$ and $\lambda(f) = f$ is a coreflective functor so that the locally fine coreflection is really a coreflection categorically.

Chapter 6

PARACOMPACTIFICATIONS

6.1 Introduction

In the late 1950s and during the 1960s, K. Morita and H. Tamano worked (independently) on a number of problems that involved the completions of uniform spaces. We state some of these problems here even though we have not yet defined some of the terms used in the statements of these problems.

(1) (Tamano) Characterize the uniform spaces with paracompact completions.
(2) (Morita) Characterize the Tychonoff spaces with paracompact topological completions.
(3) (Morita) Characterize the Tychonoff spaces with Lindelöf topological completions.
(4) (Morita) Characterize the Tychonoff spaces with locally compact topological completions.
(5) (Tamano) Is there a paracompactification of a Tychonoff space analogous to the Stone-Cech compactification or the Hewitt realcompactification?

By the **topological completion** of a uniform space, we mean the completion of the space with respect to its finest uniformity. By a **paracompactification** of a topological space, we mean a paracompact Hausdorff space Y such that X is homeomorphic with a dense subspace of Y. If we identify X with this dense subspace (which is customary), then Y is a paracompact Hausdorff space containing X as a dense subspace. Certain paracompactifications are compact. We call these **compactifications.** The theory of compactifications is well developed. The theory of paracompactifications is not. Notable among compactifications is the Stone-Čech compactification. We will devote a significant amount of effort in this chapter to its development.

We are not yet in a position to define what a *realcompact* space is. Like paracompactness, realcompactness is a generalization of compactness. We will show in the next chapter that *for all practical purposes*, realcompactness is also a generalization of paracompactness. We will of course, formalize what we mean by "for all practical purposes," but for now, we only mention that this formalization is a set theoretic matter. It turns out that the statement *All paracompact spaces are realcompact* is known to be consistent with the axioms of ZFC, but it is not known if it is independent of ZFC. Furthermore, it is known that if there exists a Tychonoff space that is paracompact but not realcompact, that its cardinality is larger than any cardinal with which we are familiar.

Once we know what a realcompact space is, we will be able to define a *realcompactification* analogously with the way we defined a paracompactification. Like the theory of compactifications, the theory of realcompactifications is well developed. Among the various realcompactifications a space may have, there is one, known as the *Hewitt realcompactification* that plays a role in the theory of realcompactifications analogous to the role played by the Stone-Čech compactification in the theory of compactifications.

Since "for all practical purposes," paracompactifications lie between realcompactifications and compactifications, it is natural to ask if there exists a paracompactification that plays a role analogous to the role played by the Stone-Čech compactification and the Hewitt realcompactification in the theories of compactifications and realcompactifications respectively. This is precisely Problem 5 above that we will refer to as *Tamano's Paracompactification Problem.* In this chapter we will show that, in general, there does not exist such a paracompactification. We will also characterize those spaces for which such a paracompactification exists and examine some of their properties.

In 1937, two papers appeared that characterized the Tychonoff (uniformizable) spaces as those having compactifications. The two papers were: *Application of the Theory of Boolean Rings to General Topology* that appeared in the Transactions of the American Mathematical Society (Volume 41, pp. 375-481) by M. H. Stone and *On bicompact spaces* by E. Čech that appeared in Annals of Mathematics (Volume 38, pp. 823-844). Their approaches were very different, but both exhibited a compactification that is the largest possible compactification of a Tychonoff space.

Clearly every compactification of a Tychonoff space X is the completion of X with respect to some totally bounded uniformity (since we can relativize a compactification's unique uniformity to X to obtain the desired totally bounded uniformity on X). In 1948, P. Samuel published a study of compactifications obtained as completions of uniform spaces (see *Ultrafilters and compactifications*, Transactions of the American Mathematical Society, Volume 64, pp 100-132). As a result, compactifications obtained as the completion of a uniform space are sometimes called the **Samuel compactification** of the uniform space. In this chapter we will employ this approach. The **Stone-Čech** compactification will be obtained as the completion of a Tychonoff space with respect to its β uniformity (introduced in Chapter 3). If X is uniformizable, the completion (X', β') of (X,β) is usually denoted by βX. The study of βX plays a central role in general topology.

Since β is the finest totally bounded uniformity, the Stone-Čech compactification is realizable as the completion with respect to the finest uniformity for which a compactification exists. A similar situation occurs with respect to realcompactifications. We will find in the next chapter that realcompactifications are precisely the completions with respect to the countably bounded uniformities and the Hewitt realcompactification is the completion with respect to the finest countably bounded uniformity (namely the e uniformity). Consequently, if Tamano's Paracompactification Problem is to be answered affirmatively for a given space X, the solution needs to be the completion of X with respect to the finest uniformity for X that has a paracompact completion.

K. Morita was interested in when the topological completion of a Tychonoff space is paracompact. He showed that if X is an M-space (to be defined later in this chapter), then X has this property. Morita called the topological completion of an M-space X the "paracompactification of X." Consequently, we will call a paracompactification that is a solution to Tamano's Paracompactification Problem the **Tamano-Morita paracompactification.**

In studying Tamano's Paracompactification Problem, several questions come to mind. The first is: Which spaces admit a paracompactification? This question is easily answered (for T_1 spaces). Since compactifications are paracompactifications, the existence of the Stone-Čech compactification implies that all Tychonoff spaces have paracompactifications. Conversely, if a space has a paracompactification it must be Tychonoff since paracompact Hausdorff spaces are normal (hence uniformizable) and as we saw in Chapter 5, a subspace of a uniform space is also a uniform space. Consequently, the class of spaces is not expanded by considering those with paracompactifications as opposed to those with compactifications.

The next question that arises is: Are all paracompactifications obtainable as a completion with respect to some uniformity? This question is also easy to answer. If X is a Tychonoff space and PX some paracompactification of X, then X is a dense subspace of PX. Since PX is paracompact, by Theorem 4.4, it is cofinally complete with respect to its finest uniformity u, and hence complete with respect to u. Now u induces the ν (derived) uniformity on X (see Section 4.5). Since completions are unique (Theorem 4.10), (PX,u) is the completion of (X,ν). Consequently, all paracompactifications are obtainable as completions with respect to some uniformity.

A completion of a uniform space that is cofinally complete will be called a **uniform paracompactification**. We have already seen (Theorem 4.12) a necessary and sufficient condition for a uniform space to have a uniform paracompactification. Clearly, a Tychonoff space can have many paracompactifications in the same manner that it can have multiple distinct Samuel compactifications. We begin our study of paracompactifications by first considering compactifications. We wish to obtain the Stone-Čech compactification of βX of a Tychonoff space X as the completion of X with respect to its β uniformity. Since β has a basis λ consisting of all finite normal coverings it is clearly precompact, so the completion of (X, β) is compact. But what makes the completion of (X, β) the Stone-Čech compactification? E. Čech characterized βX in the following way:

THEOREM 6.1 (E. Čech, 1937) Any Hausdorff compactification BX of X is the image of βX under a unique continuous mapping f that keeps X pointwise fixed and such that f(βX - X) = BX - X.

M. H. Stone characterized βX in another way:

THEOREM 6.2 (M. H. Stone, 1937) If f is any continuous mapping of a Tychonoff space X into a compact Hausdorff space Y, then f has a unique continuous extension f^β : βX → Y.

We propose to show that the completion of (X, β) satisfies Theorems 6.1 and 6.2 in the sense that if we replace βX with the completion of (X, β) in the statement of both these theorems, they both remain true. Then we will show that any compactification satisfying Theorems 6.1 and 6.2 is the completion of (X, β). Thus, among the various compactifications a Tychonoff space X may have, the completion of (X, β) is distinguished as the Stone-Čech compactification of X.

6.2 Compactifications

In what follows, we will adopt the following notation: if (X, μ) is a uniform space, then μX will denote its completion. For us then, βX will denote the completion of (X, β). We first prove Theorem 6.2 using this interpretation of βX.

(Proof of Theorem 6.2) Let β′ be the unique uniformity of βX (Theorem 2.8). Then (βX, β′) is the completion of (X, β). Let B be the unique uniformity on Y. By Lemma 3.8, each Tychonoff space admits a uniformity that has a basis consisting of all finite normal coverings. Let $\mathcal{U}$ ∈ B be one of them. For each U ∈ $\mathcal{U}$ put V_U = U∩f(X) and let $\mathcal{V} = \{V_U \mid U \in \mathcal{U}\}$. Then $\mathcal{V}$ is a finite normal covering of f(X) and $f^{-1}(\mathcal{V}) = \{f^{-1}(V_U) \mid U \in \mathcal{U}\}$ is a finite normal covering of X. By definition of β, $f^{-1}(\mathcal{V})$ ∈ β. But then f is uniformly continuous with respect to β and B. By Proposition 4.20, Y is complete, so by Theorem 4.9, f has a unique uniformly continuous extension f^β:βX → Y. ■

(Proof of Theorem 6.1) X can be identified with dense subspaces of both βX and BX. Define f:X ⊂ βX → X ⊂ BX by f(x) = x for each x ∈ X. Then, as in the proof of Theorem 6.2, B′ (the unique uniformity on BX) has a basis consisting of all finite normal coverings. If $\mathcal{U}' \in$ B′ is one of these finite normal coverings, we have just seen that $f^{-1}(\mathcal{U} \cap f(X))$ is a finite normal covering of X, so f is uniformly continuous. Therefore, we can apply Theorem 6.2 and obtain a unique continuous extension $f^{\beta}:\beta X \to BX$. Since βX is compact, $f(\beta X)$ is closed in BX. But $X \subset f(\beta X) \subset BX$ which implies $f(\beta X) = BX$ since $Cl(X) = BX$. It remains to show that $f^{\beta}(\beta X - X) \subset BX - X$. For this let $x \in \beta X - X$ and let $\{x_{\alpha}\} \subset X$ be the fundamental Cauchy net in X (see Theorem 4.9) that is used to define $x' = f'(x)$. Suppose $x' \in X$. Now $\{x_{\alpha}\}$ converges to x and $\{f(x_{\alpha})\}$ converges to x'. But since $f(x) = x$ for each $x \in X$, $\{f(x_{\alpha})\} = \{x_{\alpha}\}$. But then $\{x_{\alpha}\}$ converges to $x' \in X$ and $\{x_{\alpha}\}$ converges to x which is not in X which is a contradiction. Hence x' does not belong to X. Therefore, $f^{\beta}(\beta X - X) \subset BX - X$. Since $f^{\beta}(X) = X$, we have $f^{\beta}(\beta X - X) = BX - X$. ■

Finally, assume BX is a compactification of X that satisfies Theorem 6.1. Then there exists a unique uniformly continuous function $f:BX \to \beta X$ such that $f(BX) = \beta X$, $f(x) = x$ for each $x \in X$, and $f(BX - X) = \beta X - X$. Let $\mathcal{U} \in \beta$ be a finite normal covering of X. Then there exists a finite normal covering $\mathcal{U}'$ of β' such that $\mathcal{U}' \cap X = \mathcal{U}$. Since f is uniformly continuous, $f^{-1}(\mathcal{U}')$ is a finite normal covering of BX that belongs to B′. But then $f^{-1}(\mathcal{U}') \cap X \in$ B. Since $f(x) = x$ for each $x \in X$ and $f(BX - X) = \beta X - X$, $f^{-1}(\mathcal{U}') \cap X = f^{-1}(\mathcal{U}' \cap X) = f^{-1}(\mathcal{U}) = \mathcal{U}$. Hence B contains all finite normal coverings of X, so $\beta \subset$ B. But by Lemma 3.8 and Theorem 2.8, the unique uniformity B′ on BX has a basis consisting of all finite normal coverings (on BX). Consequently B has a basis consisting of (perhaps not all) finite normal coverings so $B \subset \beta$. But then $B = \beta$, so BX is βX.

In order to demonstrate the utility and importance of the Stone-Cech compactification, this section and the next two will be devoted to an analysis of some of the most important topological properties of Tychonoff spaces in terms of subsets of βX. The approach we will follow is due to H. Tamano, and appeared in his 1962 paper titled *On compactifications* which appeared in the Journal of Mathematics of Kyoto University (Volume 1, Number 2). Tamano's approach is usually thought of as characterizing topological properties of a Tychonoff space X in terms of the behavior of subsets of $\beta X - X$ (which it does). But as we shall see, his approach is deeply involved with certain normal coverings and with extending uniformities of X into βX. In this section we build the tools needed for this development. In the next section we prove *Tamano's Completeness Theorem*, and in Section 6.4, we present *Tamano's Theorem.*

In Chapter 2, we defined an open set $A \subset X$ to be regularly open if $Int_X(Cl_X(A)) = A$. If in turn, X is a subspace of Y, we can generalize this concept by taking the closure and the interior in Y rather than in X in order to get what is called an *extension* of A over Y. In general, if U is open in X and U′ is open in Y such that $U = U' \cap X$, then U′ is said to be an **extension** of U over Y. Put $U^{\varepsilon(Y)} = Y - Cl_Y(X - U)$. Then $U^{\varepsilon(Y)}$ is an extension of U over Y since

$U^{\varepsilon(Y)} \cap X = (Y - Cl_Y(X - U)) \cap X = X - Cl_Y(X - U) \cap X = X - (X - U) = U$. We call $U^{\varepsilon(Y)}$ the **proper extension** of U over Y. It is clear from this definition that $U \subset V$ implies $U^{\varepsilon(Y)} \subset V^{\varepsilon(Y)}$.

LEMMA 6.1 Let X be a dense subspace of Y and let U be open in X. Then $U^{\varepsilon(Y)}$ is the largest extension of U over Y.

Proof: Let U′ be an extension of U over Y. If y does not belong to $U^{\varepsilon(y)}$, then $y \in Cl_Y(X - U)$. Therefore, every open set $V'(y) \subset Y$ containing y meets X - U which implies $V'(y) \cap X$ is not a subset of $U = U' \cap X$. Consequently, $V'(y)$ is not contained in U′ which implies y does not belong to U′ since U′ is open. Therefore, $U' \subset U^{\varepsilon(Y)}$. ∎

LEMMA 6.2 Let X be a dense subspace of Y. Then for any $A \subset X$, we have $Int_X(Cl_X(A)) = Int_Y(Cl_Y(A)) \cap X$.

Proof: Let $x \in Int_X(Cl_X(A))$. Then there exists an open set $U(x)$ containing x such that $U(x) \subset Cl_X(A) \subset Cl_Y(A)$. But then there exists an open set $U'(x) \subset Y$ such that $U'(x) \cap X = U(x)$. Suppose $U'(x)$ is not a subset of $Cl_Y(A)$. Then there exists a $y \in U(x)$ with $y \in Y - Cl_Y(A)$ which is open. Since X is dense in Y, $U'(x) \cap [Y - Cl_Y(A)] \cap X \neq \varnothing$. Let $z \in U'(x) \cap [Y - Cl_Y(A)] \cap X$ which implies $z \in U'(x) \cap X = U(x) \subset Cl_Y(A)$. But on the other hand, $z \in Y - Cl_Y(A)$ which implies z does not belong to $Cl_Y(A)$ which is a contradiction. Therefore, $U'(x) \subset Cl_Y(A)$ which implies $x \in Int_Y(Cl_Y(A)) \cap X$ so $Int_X(Cl_X(A)) \subset Int_Y(Cl_Y(A)) \cap X$.

Conversely, if $x \in Int_Y(Cl_Y(A)) \cap X$ then $x \in X$ and there exists an open set $U'(x) \subset Y$ containing x such that $U'(x) \subset Cl_Y(A)$ which implies $U'(x) \cap X \subset Cl_Y(A) \cap X$. Put $U(x) = U'(x) \cap X$. Then $U(x)$ is an open set in X containing x with $U(x) \subset Cl_Y(A) \cap X$. Suppose $y \in Cl_Y(A) \cap X$. Then every open set $V'(y) \subset Y$ containing y meets A. But then $V'(y) \cap X$ meets A. Since $A \subset X$, every open set $V(y) \subset X$ containing y meets A which implies $y \in Cl_X(A)$. Therefore, $Cl_Y(A) \cap X \subset Cl_X(A)$. Hence $U(x) \subset Cl_Y(A) \cap X \subset Cl_X(A)$. Therefore, $x \in Int_X(Cl_X(A))$, so $Int_Y(Cl_Y(A)) \cap X \subset Int_X(Cl_X(A))$. ∎

LEMMA 6.3 Let X be a dense subspace of Y. If U′ is an extension of the open set $U \subset X$, then $Cl_Y(U') = Cl_Y(U)$.

Proof: Since $U = U' \cap X$ we have $U \subset U'$ which implies $Cl_Y(U) \subset Cl_Y(U')$. Suppose $y \in Cl_Y(U')$. Then every open set $V'(y) \subset Y$ containing y contains a point of U′. Since $V'(y) \cap U' \neq \varnothing$ and since X is dense in Y, there exists an $x \in X$ such that $x \in V'(y) \cap U'$ which implies $x \in U$. Therefore, every open set $V'(y) \subset Y$ containing y meets U which implies $y \in Cl_Y(U)$. Hence $Cl_Y(U') \subset Cl_Y(U)$ so $Cl_Y(U') = Cl_Y(U)$. ∎

LEMMA 6.4 Let X be a dense subspace of Y. Then $U^{\varepsilon(Y)}$ is regularly open in Y if and only if U is regularly open in X. Also, if U is regularly open in X, then $U^{\varepsilon(Y)} = Int_Y(Cl_Y(U))$.

Proof: If $U^{\varepsilon(Y)}$ is regularly open, then $U = U^{\varepsilon(Y)} \cap X = Int_Y(Cl_Y(U^{\varepsilon(Y)})) \cap X$. Since $U^{\varepsilon(Y)}$ is an extension of U, by Lemma 6.3 we have $U = Int_Y(Cl_Y(U)) \cap X$. Then by Lemma 6.2, $U = Int_X(Cl_X(U))$ so U is regularly open. Conversely, suppose U is regularly open. Then $U = Int_X(Cl_X(U)) = Int_Y(Cl_Y(U)) \cap X$. Hence $U' = Int_Y(Cl_Y(U))$ is a regularly open extension of U over Y. Therefore, $U' \subset U^{\varepsilon(Y)}$ by Lemma 6.1. We can complete the proof by showing $U^{\varepsilon(Y)} \subset U'$. For this, we suppose on the contrary that $U^{\varepsilon(Y)}$ is not a subset of U' which implies $U^{\varepsilon(Y)}$ is not a subset of $Cl_Y(U')$. Since X is dense in Y we have $[U^{\varepsilon(Y)} \cap (Y - Cl_Y(U'))] \cap X \neq \emptyset$. Therefore, $U = U^{\varepsilon(Y)} \cap X$ is not a subset of $Cl_Y(U')$ which is a contradiction. Hence $U^{\varepsilon(Y)} \subset U'$ so $U^{\varepsilon(Y)} = Int_Y(Cl_Y(U))$. ∎

PROPOSITION 6.1 Let X be a dense subspace of Y and Y a dense subspace of Z. Let U and V be open sets in X and Y respectively. Then the following are valid:

(1) $U^{\varepsilon(Z)} \cap Y = U^{\varepsilon(Y)}$,

(2) $[U^{\varepsilon(Y)}]^{\varepsilon(Z)} = U^{\varepsilon(Z)}$ and

(3) $V^{\varepsilon(Z)} \subset [V \cap X]^{\varepsilon(Z)}$. If V is regularly open then $V^{\varepsilon(Z)} = [V \cap X]^{\varepsilon(Z)}$.

Proof: To prove (1) we note that $U^{\varepsilon(Z)} \cap Y = [Z - Cl_Z(X - U)] \cap Y = Y - [Cl_Z(X-U) \cap Y] = Y - Cl_Y(X - U) = U^{\varepsilon(Y)}$. To prove (2) observe that $[U^{\varepsilon(Y)}]^{\varepsilon(Z)} = Z - Cl_Z(Y - U^{\varepsilon(Y)}) = Z - Cl_Z[Y - (Y - Cl_Y(X - U))] = Z - Cl_Z(Cl_Y(X - U)) = Z - Cl_Z(X-U) = U^{\varepsilon(Z)}$. To prove (3), note that V is an extension of $V \cap X$ so by Lemma 6.1, $V \subset (V \cap X)^{\varepsilon(Y)}$ and hence $V^{\varepsilon(Z)} \subset [V \cap X]^{\varepsilon(Z)}$ by (2) above. If V is regularly open, then $V \cap X$ is also regularly open since $V \cap X = Int_Y(Cl_Y(V)) \cap X = Int_Y(Cl_Y(V \cap U)) \cap X$ by Lemma 6.3. But then by Lemma 6.2, $V \cap X = Int_X(Cl_X(V \cap X))$. Therefore, $V = (V \cap X)^{\varepsilon(Y)}$ by Lemma 6.4. Again by (2) above, $V^{\varepsilon(Z)} = (V \cap X)^{\varepsilon(Z)}$. ∎

If we take an open covering $\mathcal{U}$ of X and then take the proper extension of each member of $\mathcal{U}$ to βX, we get a new covering $\mathcal{U}^{\varepsilon(\beta X)}$ of X in βX called the **proper extension** of $\mathcal{U}$ into βX and we call the set $\cup \mathcal{U}^{\varepsilon(\beta X)}$, denoted $E_{\mathcal{U}}$, the **extent** of $\mathcal{U}$ in βX. An open covering $\mathcal{V}$ of X is said to be **stable** if there exists a normal sequence $\{\mathcal{V}_n\}$ of open coverings of X such that $\mathcal{V}_1 < \mathcal{V}$ and $E_{\mathcal{V}} = E_{\mathcal{V}_n}$ for each n. The first of Tamano's theorems that we prove shows that an open covering is stable if and only if its extent is paracompact. In general, if X is a dense subspace of Y, we can define the proper extension $\mathcal{V}^{\varepsilon(Y)}$ of $\mathcal{V}$ over Y and the extent $\cup \mathcal{V}^{\varepsilon(Y)}$ of $\mathcal{V}$ in Y analogously.

Let f:X → **R** (reals). We denote by **Z**(f) the **zero set** of f defined by **Z**(f) = $\{x \in X \mid f(x) = 0\}$ and by **0**(f), the complement of **Z**(f) in X. We denote by **Z**(X) the set of all zero sets of X. In Section 1.1, **0**(f) was called the *support* of f. Also recall from Section 1.1 that a *partition of unity* on a space X is a family Φ

$= \{\phi_\lambda\}$ of continuous functions ϕ_λ:X $\to$ [0,1] such that $\Sigma_\lambda \phi_\lambda(x) = 1$ for each $x \in$ X, and such that all but a finite number of members of Φ vanish inside some neighborhood of x. It is clear that $\{\mathbf{0}(\phi_\lambda)\}$ is a locally finite covering of X. If $\{\mathbf{0}(\phi_\lambda)\}$ is star-finite, we say Φ is a **star-finite partition of unity** on X. A partition of unity $\{\phi_\lambda\}$ is said to be **subordinate** to a covering $\{U_\alpha\}$ if each $\mathbf{0}(\phi_\lambda)$ is contained in some U_α. In the paper *Une generalization des espaces compacts* by J. Dieudonné, referenced in Section 1.2, he established the following:

THEOREM 6.3 (J. Dieudonné, 1944) For each locally finite open covering $\{U_\alpha\}$ of a normal space, there exists a partition of unity $\{\phi_\alpha\}$ subordinate to $\{U_\alpha\}$ such that $\mathbf{0}(\phi_\alpha) \subset U_\alpha$ for each α.

Proof: Let X be a normal space and $\{U_\alpha\}$ a locally finite open covering of X. By Lemma 1.3, $\{U_\alpha\}$ is shrinkable to an open covering $\{V_\alpha\}$. By Theorem 0.3, for each α there exists a real valued continuous function f_α:X $\to$ [0,1] with $f_\alpha(Cl(V_\alpha)) = 1$ and $f_\alpha(\mathrm{X} - U_\alpha) = 0$. Since $\{U_\alpha\}$ is locally finite, so is $\{V_\alpha\}$. Hence for each $p \in$ X, $f_\alpha(p) \neq 0$ for at most finitely many α. Thus F:X $\to$ [0,1) defined by $F(x) = \Sigma_\alpha f_\alpha(x)$ for each $x \in$ X is well defined. To show F is continuous, let $p \in$ X. Then there exists an open W(p) containing p such that W(p) meets only finitely many U_α, say $U_{\alpha_1} \ldots U_{\alpha_n}$. Therefore, $F(x) = \Sigma_{i=1}^{n} f_{\alpha_i}(x)$ for each $x \in$ W(p). Since each f_{α_i} is continuous on W(p), so is F. But then F is continuous at each point of X. For each α, define ϕ_α by $\phi_\alpha(x) = f_\alpha(x)/F(x)$. Clearly $\{\phi_\alpha\}$ is a partition of unity subordinate to $\{U_\alpha\}$ such that $\mathbf{0}(\phi_\alpha) \subset U_\alpha$ for each α. ∎

The first theorem of Tamano that we prove not only determines stable coverings as those with paracompact extents, but also shows they are determined by star-finite partitions of unity. In K. Morita's paper *Star-Finite Coverings and the Star-Finite Property* referenced in Section 2.4, he showed that a locally compact space X is paracompact if and only if X has the star-finite property (every open covering has a star-finite refinement). We will need this result to prove the first of Tamano's theorems, so we include Morita's development here.

PROPOSITION 6.2 A regular Lindelöf space has the star-finite property.

Proof: Let X be a regular Lindelöf space and let $\mathcal{U}$ be an open covering of X. For each $p \in$ X, there exists an open set V(p) such that Cl(V(p)) $\subset$ U for some U $\in \mathcal{U}$. Then $\mathcal{V} = \{\mathrm{V}(p) \mid p \in \mathrm{X}\}$ is an open covering of X and as such has a countable subcovering say $\{\mathrm{V}(p_n)\}$. For each positive integer n let $U_n \in \mathcal{U}$ such that $Cl(\mathrm{V}(p_n)) \subset U_n$. Since X is normal by Lemmas 1.2 and 3.6, there exist real valued continuous functions f_n on X such that $f_n(Cl(\mathrm{V}(p_n))) = 0$ and $f_n(\mathrm{X}-U_n) = 1$, so we can apply Theorem 2.16 which yields a star-finite refinement of $\mathcal{U}$. Consequently, a regular Lindelöf space has the star-finite property. ∎

PROPOSITION 6.3 If X is a regular space and there exists an open covering $\mathcal{U}$ of X such that each member of $\mathcal{U}$ is Lindelöf and $\mathcal{U}$ admits a star-finite refinement $\mathcal{V}$, then X can be decomposed into a sum of disjoint open sets, each of which is Lindelöf and X is fully normal and has the star-finite property.

Proof: If $x,y \in$ X and there exists a positive integer n with $y \in Star^n(x,\mathcal{V})$ then we put $x \sim y$. As was seen in Section 2.4, this relation can be used to decompose X into disjoint sets X_α, $\alpha \in$ A, such that x and y belong to the same X_α if and only if $x \sim y$ [see Theorem 2.15]. Then for each $\alpha \in$ A, $Star^n(x,\mathcal{V}) \subset X_\alpha$ for any $x \in X_\alpha$. Since $X_\alpha \cap X_\beta = \varnothing$ if $\alpha \neq \beta$, X_α is both open and closed.

To show a particular X_α is Lindelöf, we first show that for any open covering $\mathcal{W}$ of X and any pair (x,n) such that $x \in X_\alpha$ and n a positive integer, that there exists a countable subset of $\mathcal{W}$ covering $Star^n(x,\mathcal{V})$. We induct on n. First note that $Star(x,\mathcal{V}) \subset$ U for some U $\in \mathcal{U}$ and U is Lindelöf. Next assume $Star^n(x,\mathcal{V})$ has this property. Then there exists a countable collection $\{V_i\} \subset \mathcal{V}$ that covers $Star^n(x,\mathcal{V})$ which implies $Star^{n+1}(x,\mathcal{V}) \subset \cup_{i=1}^{\infty} Star(V_i,\mathcal{V})$. But $Star(V_i,\mathcal{V}) \subset U_i$ for some $U_i \in \mathcal{U}$, so $Star^{n+1}(x,\mathcal{V}) \subset \cup_{i=1}^{\infty} U_i$. Since each U_i is Lindelöf, $Star^{n+1}(x,\mathcal{V})$ can be covered by countably many members of $\mathcal{W}$.

But then, for any open covering $\mathcal{W}$ of X, $X_\alpha = \cup_{n=1}^{\infty} Star^n(x,\mathcal{V})$ can be covered by countably many members of $\mathcal{W}$. Since X_α is closed, X_α is Lindelöf. This proves the first part of the proposition. The second part follows from Lemma 3.6 and Proposition 6.2. ∎

A space X is said to be *locally compact* if each $p \in$ X is contained in a compact neighborhood.

LEMMA 6.5 A locally compact Hausdorff space is completely regular.

Proof: Let X be a locally compact space and let p be a point of X contained in some open set W. Since X is locally compact, there exists an open set V containing p such that Cl(V) is compact. Put U = V∩W. Then Cl(U) is also a compact neighborhood of p. Since Cl(U) is a compact Hausdorff space, it is normal, so there exists a continuous function g:Cl(U) → [0,1] such that $g(p) = 1$ and g(Cl(U) - U) = 0. Let h be the constant function on X - U defined by $h(x) =$ 0 for each $x \in$ X - U. Define f on X by $f(x) = g(x)$ if $x \in$ U or $f(x) = h(x)$ otherwise. Clearly f:X → [0,1] is continuous, $f(p)$ = 1 and f(X - W) = 0. Consequently X is completely regular. ∎

LEMMA 6.6 If $\mathcal{U}$ is an open covering of a Hausdorff space X such that the closure of each member of $\mathcal{U}$ is compact, and if $\mathcal{U}$ admits a star-finite refinement $\mathcal{V}$, then X is fully normal and has the star-finite property.

Proof: Clearly X is locally compact so by Lemma 6.5, X is completely regular. Moreover, any set U of $\mathcal{U}$ has the property that for any open covering $\mathcal{W}$ of X, U can be covered by countably many members of $\mathcal{W}$. This was the only use made of the Lindelöf property in Proposition 6.3. Hence the same method of proof can be used to prove this lemma. ∎

THEOREM 6.4 (K. Morita, 1948) A necessary and sufficient condition for a locally compact Hausdorff space to be paracompact is that it possess the star-finite property.

Proof: Let X be a locally compact Hausdorff space. Assume first that X possesses the star-finite property. Then each open covering has a star-finite refinement. Since a star-finite refinement is locally finite, we conclude that X is paracompact. Conversely assume X is paracompact. By Stone's Theorem (1.1), X is fully normal. Since X is locally compact, there exists an open covering $\mathcal{U}$ such that the closure of each member of $\mathcal{U}$ is compact. Since X is fully normal, $\mathcal{U}$ has a star refinement. Then by Lemma 6.6, X has the star-finite property. ∎

E. Čech, in his paper **On bicompact spaces** referenced in Section 6.1, showed that a T_1 space X is normal if and only if $Cl_{\beta X}(E) \cap Cl_{\beta X}(F) = \varnothing$ for each pair of disjoint closed sets E, F $\subset$ X. This useful result is needed to characterize the locally compact Hausdorff spaces as those spaces X such that BX - X is compact for any compactification BX of X. This result will also be needed in the proof of Tamano's characterization of stable coverings.

THEOREM 6.5 (E. Čech, 1937) A T_1 space X is normal if and only if $Cl_{\beta X}(E) \cap Cl_{\beta X}(F) = \varnothing$ for each pair of disjoint closed sets E, F $\subset$ X.

Proof: Let X be a T_1 space and assume that for disjoint closed sets E, F $\subset$ X, $Cl_{\beta X}(E) \cap Cl_{\beta X}(F) = \varnothing$. Since βX is normal, there exist disjoint open sets U,V $\subset$ βX such that $Cl_{\beta X}(E) \subset U$ and $Cl_{\beta X}(F) \subset V$. Then U$\cap$X and V$\cap$X are disjoint open sets in X containing E and F respectively, so X is normal.

Conversely, assume X is normal and that E and F are disjoint closed sets in X. Then there exists a real valued continuous function f:X $\to$ [0,1] such that f(E) = 0 and f(F) = 1. By Theorem 6.2, f can be extended to a continuous function $f^\beta:\beta X \to [0,1]$. Since E is dense in $Cl_{\beta X}(E)$ and f(E) = 0, $f^\beta(Cl_{\beta X}(E)) = 0$. Similarly $f^\beta(Cl_{\beta X}(F)) = 1$. Then $U = \{x \in \beta X \mid f^\beta(x) < 1/2\}$ and $V = \{x \in \beta X \mid f^\beta(x) > 1/2\}$ are disjoint open sets containing $Cl_{\beta X}(E)$ and $Cl_{\beta X}(F)$ respectively, so $Cl_{\beta X}(E) \cap Cl_{\beta X}(F) = \varnothing$. ∎

PROPOSITION 6.4 A Tychonoff space X is locally compact if and only if BX - X is compact for any compactification BX of X.

Proof: For each $x \in$ X, there exists a neighborhood U(x) of x such that

$Cl_X(U(x))$ is compact, if X is locally compact. Then $Cl_X(U(x))$ is closed in BX so $Cl_X(U(x)) = Cl_{\beta X}(U(x))$. Therefore, no point of BX - X is contained in $Cl_{\beta X}(U(x))$, so X is open in BX. Hence BX - X is closed and therefore compact.

Conversely, if BX - X is compact, then it is closed so for each $x \in X$, there exists a neighborhood $U(x)$ of x in βX such that $Cl(U(x)) \cap [BX - X] = \emptyset$. But then $Cl_{\beta X}(U(x)) = Cl_X(U(x))$. Hence $Cl_X(U(x))$ is compact. We conclude that X is locally compact. ∎

We are finally in a position to prove the first major theorem from Tamano's 1962 paper. This theorem not only establishes the connection among stable coverings, star-finite partitions of unity and coverings with paracompact extents, but it is essential in the proof of Tamano's characterization of the Lindelöf property in terms of compact subsets of BX - X (where BX is a compactification of X) that we will encounter later.

THEOREM 6.6 (H. Tamano, 1962) The following statements are equivalent for a Tychonoff space X:

(1) $\mathcal{V}$ is a stable covering of X,

(2) the extent $E_{\mathcal{V}}$ of $\mathcal{V}$ in βX is paracompact and

(3) there exists a star-finite partition of unity $\{\phi_\lambda\}$ on X such that $\mathcal{W} = \{W(x) \mid x \in X\} < \mathcal{V}$ where $W(x) = \{y \in X \mid \Sigma_\lambda |\phi_\lambda(x) - \phi_\lambda(y)| < 1\}$, and such that $E_{\mathcal{W}} = E_{\mathcal{V}}$.

Proof: To simplify the proof, we write V^ε for $V^{\varepsilon(\beta X)}$ and $\mathcal{V}^\varepsilon$ for $\mathcal{V}^{\varepsilon(\beta X)}$. Suppose $\mathcal{V}$ is a stable covering of X and $\{\mathcal{V}_n\}$ is a normal sequence of open coverings with $\mathcal{V}_1 <^* \mathcal{V}$ and $E_{\mathcal{V}_n} = E_{\mathcal{V}}$ for each n. Then $\mathcal{V}_n^\varepsilon \cap X = \{V_n^\varepsilon \cap X \mid V_n \in \mathcal{V}_n\} = \mathcal{V}_n$ for each n, and by Lemma 6.1, $\{Star(V_n^\varepsilon, \mathcal{V}_n^\varepsilon) \mid V_n \in \mathcal{V}_n\} < \{Star(V_n, \mathcal{V}_n)^\varepsilon \mid V_n \in \mathcal{V}_n\} = \{Star(V_n, \mathcal{V}_n) \mid V_n \in \mathcal{V}_n\}^\varepsilon$. Consequently, $\{Star(V_{n+1}^\varepsilon, \mathcal{V}_{n+1}^\varepsilon) \mid V_{n+1} \in \mathcal{V}_{n+1}\} < \{Star(V_{n+1}, \mathcal{V}_{n+1}) \mid V_{n+1} \in \mathcal{V}_{n+1}\}^\varepsilon < \mathcal{V}_n^\varepsilon$, so $\mathcal{V}_{n+1}^\varepsilon <^* \mathcal{V}_n^\varepsilon$ for each n. Therefore, $\{\mathcal{V}_n^\varepsilon\}$ is a normal sequence of open coverings of $E_{\mathcal{V}}$.

Let d be the pseudo-metric on $E_{\mathcal{V}}$ defined (as in the proof of Theorem 2.5) with respect to the normal sequence $\{\mathcal{V}_n^\varepsilon\}$. Then clearly, the collection of spheres of radius 2^{-n} coincides with the point star coverings associated with $\mathcal{V}_n^\varepsilon$, i.e., $S(x, 2^{-n}) = S(x, \mathcal{V}_n^\varepsilon)$ for each n, and $d: E_{\mathcal{V}} \times E_{\mathcal{V}} \to [0,1]$. Let τ denote the topology of $(E_{\mathcal{V}}, d)$. By Corollary 1.1, $(E_{\mathcal{V}}, \tau)$ is paracompact. Let $x \in E_{\mathcal{V}}$ and put $U(x) = \{y \in E_{\mathcal{V}} \mid d(x,y) < 1/4\}$. Then $Cl_{\beta X}(U(x)) \subset E_{\mathcal{V}}$. To show this will take a little effort.

Let $r \in Cl_{\beta X}(U(x))$ and let $d_x(y) = d(x,y)$ for each $y \in E_{\mathcal{V}}$. Now $d_x: E_{\mathcal{V}} \to [0,1]$ is uniformly continuous with respect to the pseudo-metric uniformity on $E_{\mathcal{V}}$. To see this let $\varepsilon > 0$ and suppose $d(x,y) < \varepsilon$. There are two cases $[d(x,z) < d(x,y)$ and $d(x,z) \geq d(x,y)]$ to argue, but both yield the following inequality:

$$|d_x(y) - d_x(z)| = |d(x,y) - d(x,z)| \leq |d(x,z) + d(z,y) - d(x,z)| = |d(y,z)| < \varepsilon.$$

By Theorem 6.2, the restriction of d_x to X can be uniquely extended to a continuous function $d_x^\beta:\beta X \to [0,1]$. To show that d_x^β agrees with d_x on E_V we pick $y \in E_V - X$, and assume $d_x(y) \neq d_x^\beta(y)$. Now there exists a net $\{y_\alpha\} \subset X$ that converges to y since d_x and d_x^β are both continuous, $\{d_x(y_\alpha)\}$ converges to $d_x(y)$ and $\{d_x^\beta(y_\alpha)\}$ converges to $d_x^\beta(y)$. But since $\{y_\alpha\} \subset X$, $\{d_x(y_\alpha)\} = \{d_x^\beta(y_\alpha)\}$. Consequently, $\{d_x(y_\alpha)\}$ must converge to two distinct points in a Hausdorff space which is impossible. We conclude that d_x can be uniquely extended to βX.

Now $d_x^\beta(r) \leq 1/4$ since $r \in Cl_{\beta X}(U(x))$, so there exists an open W(r) in βX containing r such that $d_x^\beta(y) < 1/2$ for each $y \in W(r) \cap E_V$. Therefore, $W(r) \cap E_V \subset S(x, 2^{-1}) = S(x, \mathcal{V}_1^\varepsilon) \subset V^\varepsilon$ for some $V^\varepsilon \in \mathcal{V}^\varepsilon$. By Lemma 6.1, $W(r) \subset V^\varepsilon$ so $r \in E_V$. Consequently, $Cl_{\beta X}(U(x)) \subset E_V$.

Put $\mathcal{U} = \{U(x) \mid x \in E_V\}$. Since (E_V, τ) is paracompact, there exists a locally finite open refinement $\{U_\lambda\}$ covering E_V. Since the coverings $\mathcal{V}_n^\varepsilon$ consist of open sets in βX, the set $B_V = \{V_n^\varepsilon \mid V_n^\varepsilon \in \mathcal{V}_n^\varepsilon$ for some $n\}$ is a basis for τ. But B_V is a subset of the topology on βX so the topology induced on E_V by βX is finer than τ. Consequently, the covering $\{U_\lambda\}$ is also open in βX. An argument similar to the one above can be used to show that $Cl_{\beta X}(U_\lambda) \subset E_V$ for each λ. To show that E_V is a paracompact subset of βX, let $\{G_\alpha\}$ be an open covering of E_V. Then for each λ, $Cl_{\beta X}(U_\lambda)$ is covered by finitely many of the G_α's say $G_1 \ldots G_m$. Put $H_\lambda^\kappa = U_\lambda \cap G_\kappa$ for $\kappa = 1 \ldots m$. Then the family $\{H_\lambda^\kappa\}$ is a locally finite open refinement of $\{G_\alpha\}$. This completes the proof of (1) → (2).

(2) → (3). Assume that E_V is paracompact. Since E_V is open in βX, βX - E_V is closed and hence compact. Since βX is also a compactification of E_V, by Proposition 6.4, E_V is locally compact. Then by Theorem 6.4, E_V possesses the star-finite property. Since E_V is paracompact, there exists a normal sequence of open coverings $\{\mathcal{V}_n'\}$ in E_V such that $\mathcal{V}_1' < \mathcal{V}^\varepsilon$. Consider the open covering $\mathcal{U}' = \{S(z, \mathcal{V}_2') \mid z \in E_V\}$. Since E_V has the star-finite property, by Theorem 6.3, there exists a star-finite partition of unity $\Phi' = \{\phi_\lambda'\}$ that is subordinate to $\mathcal{U}'$. Since $0(\phi_\lambda') \subset S(z, \mathcal{V}_2')$ for some $z \in E_V$, the union of all $0(\phi_\mu')$ for which $0(\phi_\mu') \cap 0(\phi_\lambda') \neq \varnothing$ is contained in some $V_1'(z) \in \mathcal{V}_1'$. It follows that $\mathcal{W}' = \{W'(z) \mid z \in E_V\}$, where $W'(z) = \{y \in E_V \mid \Sigma_\lambda |\phi_\lambda'(z) - \phi_\lambda'(y)| < 1\}$ refines $\mathcal{V}_1'$. Let ϕ_λ denote the restriction of ϕ_λ' to X for each λ. Then $\Phi = \{\phi_\lambda\}$ is a star-finite partition of unity on X such that $\mathcal{W}^\wedge = \mathcal{W}' \cap X$ refines $\mathcal{V}_1' \cap X < \mathcal{V}^\varepsilon \cap X = \mathcal{V}$. For each $x \in X$ put $W(x) = \{y \in X \mid \Sigma_\lambda |\phi_\lambda(x) - \phi_\lambda(y)| < 1\}$. Then each $W(x) \in \mathcal{W}^\wedge$ so $\mathcal{W} = \{W(x) \mid x \in X\}$ refines $\mathcal{V}$.

It remains to show that $E_{\mathcal{W}} = E_V$. For this first note that $\mathcal{W} < \mathcal{V}$ implies $E_{\mathcal{W}} \subset E_V$. To show that $E_V \subset E_{\mathcal{W}}$, let $z \in E_V$. Pick $x \in W'(z) \cap X$, which is possible since X is dense in βX. Then $\Sigma_\lambda |\phi_\lambda'(z) - \phi_\lambda'(x)| < 1$ implies $z \in W'(x)$. But then $E_V \subset \cup\{W'(x) \mid x \in X\} \subset \cup\{W^\varepsilon(x) \mid x \in X\} = E_{\mathcal{W}}$. This completes the proof of (2) → (3).

(3) → (1). Let $\Phi = \{\phi_\lambda\}$ be a star-finite partition of unity on X such that the open covering $\mathcal{V}$ of X is refined by $\mathcal{W} = \{W(x) \mid x \in X\}$ where $W(x) = \{y \in X \mid \Sigma_\lambda \mid \phi_\lambda(x) - \phi_\lambda(y) \mid < 1\}$ and $E_{\mathcal{W}} = E_{\mathcal{V}}$. For each λ let ϕ_λ^β denote the extension of ϕ_λ to βX. Since $\{\mathbf{0}(\phi_\lambda)\}$ is star-finite, so is $\{\mathbf{0}(\phi_\lambda^\beta\}$. Let X′ be the subspace of βX consisting of all points $p \in \beta X$ such that all but finitely many ϕ_λ^β vanish outside some neighborhood of p. Clearly X′ is open in βX and $\{\mathbf{0}(\phi_\lambda^\beta) \cap X'\}$ is a star-finite covering of X′. For each λ let ϕ_λ' denote the restriction of ϕ_λ^β to X′. Then $\Phi' = \phi_\lambda'$ is a star-finite partition of unity on X′.

For each pair n, x where n is a non-negative integer and $x \in X$, put $V_n(x) = \{y \in X \mid \Sigma_\lambda \mid \phi_\lambda(x) - \phi_\lambda(y) \mid < 2^{-n}\}$ and let $\mathcal{V}_n = \{V_n(x) \mid x \in X\}$. Also, for each pair n, x' where n is a non-negative integer and $x' \in X'$, put $V_n'(x') = \{y' \in X' \mid \Sigma_\lambda \mid \phi_\lambda'(x') - \phi_\lambda'(y') \mid < 2^{-n}\}$ and let $\mathcal{V}_n' = \{V_n'(x') \mid x' \in X'\}$. Clearly $\mathcal{V}_n' \cap X = \mathcal{V}_n$ and we have

$$\mathcal{V}_n' \cap X = \mathcal{V}_n \subset \mathcal{V}_n^\varepsilon \cap X = [\mathcal{V}_n^\varepsilon \cap X'] \cap X$$

by Proposition 6.1.(1). Therefore, $\mathcal{V}_n^\varepsilon \cap X'$ contains $\mathcal{V}_n'$ so $\cup \mathcal{V}_n^\varepsilon$ contains $\cup \mathcal{V}_n'$ which implies $E_{\mathcal{V}_n}$ contains X′ for each non-negative integer n. To show that $E_{\mathcal{V}_n} \subset X'$ for each n, let $p \in E_{\mathcal{V}_n}$. Then $p \in V_n^\varepsilon(x)$ for some $x \in X$. Let $y \in V_n^\varepsilon(x) \cap X$. Then $\Sigma_\lambda \mid \phi_\lambda(x) - \phi_\lambda(y) \mid < 2^{-n}$. Let $\phi_1 \ldots \phi_m$ be the finite set of members of Φ such that $\phi_\kappa(x) \neq 0$ for each $\kappa = 1 \ldots m$. Then $y \in \cup_{\kappa=1}^m \mathbf{0}(\phi_\kappa)$ for each $y \in V_n^\varepsilon(x) \cap X$, for otherwise, $\phi_\kappa(y) = 0$ for each $\kappa = 1 \ldots m$ which implies

$$\Sigma_{\kappa=1}^m \mid \phi_\kappa(x) - \phi_\kappa(y) \mid = \Sigma_{\kappa=1}^m \phi_\kappa(x) = \Sigma_\lambda \phi_\lambda(x) = 1.$$

Now $\Sigma_\lambda \mid \phi_\lambda(x) - \phi_\lambda(y) \mid \geq \Sigma_{\kappa=1}^m \mid \phi_\kappa(x) - \phi_\kappa(y) \mid = 1 > 2^{-n}$ which implies y does not belong to $V_n^\varepsilon(x) \cap X$ which is a contradiction. Therefore, $V_n^\varepsilon(x) \cap X \subset \cup_{\kappa=1}^m \mathbf{0}(\phi_\kappa)$.

If p does not belong to X′ then $V_n^\varepsilon(x)$ meets infinitely many $\mathbf{0}(\phi_\lambda^\beta)$'s. If $V_n^\varepsilon(x) \cap \mathbf{0}(\phi_m^\beta) \neq \varnothing$ then $\phi_\mu(y) > 0$ for some $y \in V_n^\varepsilon(x)$ which implies there exists a $z \in V_n^\varepsilon(x) \cap X$ such that $\phi_\mu^\beta(z) > 0$ since ϕ_μ^β is continuous and X is dense in $V_n^\varepsilon(x)$. Consequently, $V_n^\varepsilon(x) \cap X$ meets infinitely many $\mathbf{0}(\phi_\lambda)$'s which implies $\cup_{\kappa=1}^m \mathbf{0}(\phi_\kappa)$ meets infinitely many $\mathbf{0}(\phi_\lambda)$'s which is a contradiction. Hence $p \in X'$. We conclude $E_{\mathcal{V}_n} = X'$ for each non-negative integer n. Since $\mathcal{V}_0 = \mathcal{W}$ we have $E_{\mathcal{V}} = E_{\mathcal{W}} = E_{\mathcal{V}_0} = X' = E_{\mathcal{V}_n}$ for each positive integer n. Therefore, $\mathcal{V}$ is a stable covering of X which completes the proof of (3) → (1). ■

The next step in Tamano's development is to strengthen the hypothesis and conclusion in Theorem 6.6 in the following way: a locally compact Hausdorff space is said to be **σ-compact** if it is the union of countably many compact spaces. In what follows we show that the locally compact paracompact Hausdorff spaces are precisely the topological sums of σ-compact spaces. Hence the extents of stable coverings are topological sums of σ-compact spaces as can be seen in the proof of (2) → (3) in Theorem 6.6. Let us agree to call a stable covering **strongly stable** if its extent is σ-compact. Tamano's next theorem establishes the connection between strongly stable coverings and *countable* star-finite partitions of unity.

There are a number of alternate definitions of both local compactness and σ-compactness that are very useful. It will facilitate our development to establish these equivalent definitions. The proofs of the next two lemmas are left as exercises (Exercises 1 and 2).

LEMMA 6.7 The following four properties are equivalent:
(1) X is a locally compact Hausdorff space.
(2) For each $x \in X$ and neighborhood U of x, there exists a neighborhood V of x with $Cl(V) \subset U$ and Cl(V) compact.
(3) For each compact K and open U containing K, there exists an open V with $K \subset V \subset Cl(V) \subset U$ and Cl(V) is compact.
(4) X has a basis consisting of open sets with compact closures.

LEMMA 6.8 The following three properties are equivalent:
(1) X is a σ-compact space.
(2) There exists a sequence $\{U_n\}$ of open sets with each $Cl(U_n)$ compact such that $Cl(U_n) \subset U_{n+1}$ for each n, and $X = \cup_{n=1}^{\infty} U_n$.
(3) X is a locally compact Lindelöf space.

At first glance it might appear that any space that is the union of at most countably many compact spaces is σ-compact, but the rational subset *R* of **R** (reals) is a counter example because *R* is not locally compact. Consequently, local compactness is an essential part of the definition of σ-compactness. Since a σ-compact space is Lindelöf by Lemma 6.8 and completely regular by Lemma 6.5, then by Lemma 3.6 it is paracompact. Therefore, we get locally compact paracompact spaces by forming topological sums of σ-compact spaces. The next proposition shows that this is the only way to get such spaces.

LEMMA 6.9 If a covering $\{U_\alpha\}$, $\alpha \in A$, has a point (locally) finite refinement $\{V_\beta\}$, $\beta \in B$, then $\{U_\alpha\}$ has a precise point (locally) finite refinement $\{W_\alpha\}$. If each V_β is open, then each W_α can be chosen to be open.

Proof: Define a function f:B → A by assigning to each $\beta \in B$ some $\alpha \in A$ such that $V_\beta \subset U_\alpha$. For each α let $W_\alpha = \cup\{V_\beta | f(\beta) = \alpha\}$. It is possible that $W_\alpha = \emptyset$ for some of the α's. Clearly $W_\alpha \subset U_\alpha$ for each α, and $\{W_\alpha\}$ is a covering because each V_β is a subset of some W_α. If $\{V_\beta\}$ is point (locally) finite, then each (some neighborhood of) $x \in X$ lies in at most finitely many of the V_β and consequently cannot meet more than finitely many of the W_α's. Therefore, $\{W_\alpha\}$ is a point (locally) finite refinement of $\{U_\alpha\}$. ∎

PROPOSITION 6.5 The locally compact paracompact Hausdorff spaces are precisely the topological sums of σ-compact spaces.

Proof: In view of the remarks preceding Lemma 6.9, it is clear that we only need to prove that a locally compact, paracompact, Hausdorff space X is a topological sum of σ-compact spaces. Let X be covered by open sets U(x) such that $x \in$ U(x) and *Cl*(U(x)) is compact for each $x \in$ X. By Lemma 6.9, there exists a precise locally finite open refinement {V(x) | $x \in$ X}. Since *Cl*(V(x)) is compact for each $x \in$ X, it meets at most finitely many other V(y)'s. To see this, for each $z \in$ *Cl*(V(x)), V(z) meets at most finitely many other V(y)'s and since *Cl*(V(x)) is compact, it can be covered by finitely many V(y)'s, say V(z_1) . . . V(z_n). Then W = $\cup_{i=1}^{n}$ V(z_i) contains *Cl*(V(x)) and meets at most finitely many V(y)'s.

For each pair $x,y \in$ X put $x \sim y$ if there exists a finite family of sets V(z_1) . . V(z_n) such that $x \in$ V(z_1), $y \in$ V(z_n) and V(z_i)∩V(z_{i+1}) ≠ ∅ for $i = 1 \ldots n-1$. Clearly ~ is an equivalence relation in X. Let {X_α | $\alpha \in$ A} denote the family of equivalence classes with respect to ~. Then the X_α cover X and are pairwise disjoint. Each X_α is open since it is the union of V(y)'s. Therefore, each X_α is a locally compact Hausdorff space. It remains to show that each X_α is σ-compact. Let K_1 = *Cl*(V(x)) for some $x \in X_\alpha$. For each positive integer n, inductively define K_n as the union of all *Cl*(V(y))'s with V(y) meeting K_{n-1}. Since K_{n-1} is compact, it meets at most finitely many V(y)'s (as shown above) so K_n is compact. Since for each $y \in X_\alpha$ there are finitely many V(z)'s, say V(z_1) . . . V(z_n) with $x \in$ V(z_1), $y \in$ V(z_n) and V(z_i)∩V(z_{i+1}) ≠ ∅ for each i = 1 . . . n - 1, it is clear that $y \in K_j$ for some positive integer j. Therefore, $X_\alpha = \cup_{n=1}^{\infty} K_n$. By Lemma 6.8, X_α is σ-compact. Therefore, X is a topological sum of σ-compact spaces. ∎

THEOREM 6.7 (H. Tamano, 1962) The following statements are equivalent for a Tychonoff space X:

(1) V is a strongly stable covering of X,

(2) there is a countable star-finite partition of unity {ϕ_n} *on X with W* = {$W(x)$ | $x \in X$} < *V where* $W(x) = \{y \in X \mid \Sigma_n |\phi_n(x) - \phi_n(y)| < 1\}$, *and such that* $E_{\mathcal{W}} = E_{\mathcal{V}}$.

The proof of this theorem can be obtained by modifying the proof of Theorem 6.6 and is left as an exercise (Exercise 3).

EXERCISES

1. Prove Lemma 6.7.

2. Prove Lemma 6.8.

3. Prove Theorem 6.7.

6.3 Tamano's Completeness Theorem

In 1944, in a paper referenced in Section 1.2, J. Dieudonné showed that the product $X \times Y$ of a paracompact space X with a compact space Y is paracompact. Hence $X \times Y$ is normal. Thereafter, many researchers tried to determine if the reverse implication held; i.e., does the normality of $X \times Y$ imply X is paracompact when Y is compact? By 1960, many partial results had been published. Then in 1960, in a paper titled *On paracompactness* published in the Pacific Journal of Mathematics (Volume 10, pp. 1043-1047), Tamano showed that it does if $Y = \beta X$ (more generally if BX is any compactification of X, the normality of $X \times BX$ implies that X is paracompact). This result has come to be known as *Tamano's Theorem.*

In Section 4.5 it was mentioned that H. Tamano and K. Morita had considered the problem: "What is a necessary and sufficient condition for a uniform space to have a paracompact completion?" Tamano published four other papers in 1960. One of them titled *Some properties of the Stone-Čech compactification* published in the Journal of the Mathematical Society of Japan (Volume 12, pp. 104-117) details his results in this area on a number of problems having to do with the completion of uniform spaces. One of the theorems in this paper gives a necessary and sufficient condition for a uniform space to be complete in terms of the *radical* of a uniform space. The concept of the radical is based on the concept of the extent of a covering in βX. If (X, μ) is a uniform space, the **radical** of (X, μ) is defined to be $\cap\{E_{\mathcal{U}} \mid \mathcal{U} \in \mu\}$, i.e., the intersection of all the extents in βX of uniform coverings. This theorem states that if R is the radical of (X, μ) then X is complete if and only if $R = X$. We will refer to this result as *Tamano's Completeness Theorem.* This section is devoted to its proof.

LEMMA 6.10 Let A be a regularly open set in a space X. Let B be an open set which is not contained in A. Then there is an open $C \subset B$ with $C \cap Cl_X(A) = \varnothing$.

Proof: If $B \subset Cl_X(A)$ then $B \subset Int_X(Cl_X(A)) = A$ since A is regularly open. But B is not contained in A so B is not contained in $Cl_X(A)$. Then $C = B \cap [X - Cl_X(A)]$ is the desired subset of B such that $C \cap Cl_X(A) = \varnothing$. ■

LEMMA 6.11 Let X be a dense subspace of a space Y. Then
(1) If A is regularly open in Y then $A \cap X$ is regularly open in X,
(2) B is regularly open in X if and only if $B = Int_Y(Cl_Y(B)) \cap X$,
(3) if A is regularly open in Y and B is open in Y, and if $A \cap X$ contains $B \cap X$ then $B \subset A$ and
(4) Two regularly open sets A,B in Y are identical if and only if $A \cap X = B \cap X$.

Proof: Assume A is regularly open in Y. $A \cap X \subset Int_X(Cl_X(A \cap X)) =$

$Int_Y(Cl_Y(A \cap X)) \cap X$ by Lemma 6.2. Now $Int_Y(Cl_Y(A \cap X)) \cap X \subset Int_Y(Cl_Y(A)) \cap X = A \cap X$ since A is regularly open in Y. Therefore, $A \cap X = Int_X(Cl_X(B)) \cap X$ so $A \cap X$ is regularly open in X. This proves (1).

To show (2) suppose $B = Int_Y(Cl_Y(B)) \cap X$. Since $Int_Y(Cl_Y(B))$ is regularly open in Y, by (1) above, B is regularly open in X. Conversely, suppose B is regularly open in X. Then $B = Int_X(Cl_X(B))$ so by Lemma 6.2, $B = Int_Y(Cl_Y(B)) \cap X$. Therefore, B is regularly open in X if and only if $B = Int_Y(Cl_Y(B)) \cap X$.

To show (3) assume A is regularly open in Y and B is open in Y. Suppose B is not contained in A. By Lemma 6.10 there is an open $C \subset B$ with $C \cap A = \varnothing$. Since X is dense in Y, there exists a $p \in C$ such that $p \in X$. But then $p \in B \cap X$ but p is not in $A \cap X$ which implies $B \cap X$ is not contained in $A \cap X$. Therefore, $B \cap X \subset A \cap X$ which implies $B \subset A$.

To show (4) assume both A and B are regularly open in Y and $A \cap X = B \cap X$. Then by (3) above $A \subset B$ and $B \subset A$ so $A = B$. Conversely, if $A = B$ then $A \cap X = B \cap X$. Hence two regularly open sets A, B in Y are identical if and only if $A \cap X = B \cap X$ ∎

Tamano worked exclusively with entourage uniformities. His original proof was somewhat different than the proof we give here but the proof given here preserves his approach. The following development is a good example of how working with covering uniformities leads to simpler statements of theorems as opposed to working with entourage uniformities. Tamano's proof was subdivided into a number of propositions. Each of the next five propositions, with the exception of Proposition 6.11, is equivalent to one of his propositions.

PROPOSITION 6.6 Let X be a dense subspace of a Tychonoff space Y and let λ be a basis for a uniformity μ consisting of regularly open coverings. For each $\mathcal{U} \in \lambda$, $Star(\mathcal{U}^{\varepsilon(Y)}, \mathcal{U}^{\varepsilon(Y)}) \subset [Star(U, \mathcal{U})]^{\varepsilon(Y)}$ so $\mathcal{V} <^* \mathcal{U}$ implies $\mathcal{V}^{\varepsilon(Y)} <^* \mathcal{U}^{\varepsilon(Y)}$ for each $\mathcal{V} \in \lambda$.

Proof: Since each $\mathcal{U} \in \lambda$ consists of regularly open sets, $\mathcal{U}^{\varepsilon(Y)} = \{Int_Y(Cl_Y(U)) \mid U \in \mathcal{U}\}$ by Lemma 6.4. Suppose $y \in Star(U^{\varepsilon(Y)}, \mathcal{U}^{\varepsilon(Y)})$. Then there exists a $U_1 \in \mathcal{U}$ such that $y \in U_1^{\varepsilon(Y)}$ and $U_1^{\varepsilon(Y)} \cap U^{\varepsilon(Y)} \neq \varnothing$. Therefore, there is a $z \in U_1^{\varepsilon(Y)} \cap U^{\varepsilon(Y)} \cap X$ so $U_1 \subset Star(U, \mathcal{U})$. Now $y \in U_1^{\varepsilon(Y)}$ implies $y \in Int_Y(Cl_Y(U_1)) \subset Int_Y(Cl_Y(Star(U, \mathcal{U}))) \subset [Star(U, \mathcal{U})]^{\varepsilon(Y)}$. Next let $\mathcal{U}$, $\mathcal{V} \in \lambda$ with $\mathcal{V} <^* \mathcal{U}$. Then $Star(V, \mathcal{V}) \subset U$ for some $U \in \mathcal{U}$. By the previous argument, $Star(V^{\varepsilon(Y)}, \mathcal{V}^{\varepsilon(Y)}) \subset [Star(V, \mathcal{V})]^{\varepsilon(Y)}$ and since $Star(V, \mathcal{V}) \subset U$, we have $[Star(V, \mathcal{V})]^{\varepsilon(Y)} \subset U^{\varepsilon(Y)}$. Therefore, $Star(V^{\varepsilon(Y)}, \mathcal{V}^{\varepsilon(Y)}) \subset U^{\varepsilon(Y)}$ so $\mathcal{V}^{\varepsilon(Y)} <^* \mathcal{U}^{\varepsilon(Y)}$. ∎

Let (X, μ) be a uniform space and (X', μ') its completion. Let βX and $\beta X'$ be the Stone-Cech compactifications of X and X′ respectively. Then $\beta X'$ is also a compactification of X so by Theorem 6.1, there exists a continuous function ϕ:

$\beta X \to \beta X'$ such that ϕ is onto, $\phi(X) = X$ and $\phi(\beta X - X) = \beta X' - X$. Let λ be a regularly open basis for μ.

PROPOSITION 6.7 For each $\mathcal{U} \in \mu$, $\mathcal{U}^{\varepsilon(\beta X')} = \mathcal{U}'^{\varepsilon(\beta X')}$.

Proof: Let $\mathcal{U} \in \mu$. It suffices to show that $U^{\varepsilon(\beta X')} = U'^{\varepsilon(\beta X')}$ for each $U \in \mathcal{U}$. Since X is dense in X' which is in turn dense in $\beta X'$, by Lemma 6.1, $U = U' \cap X$ which implies $U' \subset U^{\varepsilon(X')}$. Therefore, $U'^{\varepsilon(\beta X')} \subset [U^{\varepsilon(X')}]^{\varepsilon(\beta X')} = U^{\varepsilon(\beta X')}$ by Proposition 6.1.(2). But $U \subset U'$ which implies $U^{\varepsilon(\beta X')} \subset U'^{\varepsilon(\beta X')}$. Therefore, $U^{\varepsilon(\beta X')} = U'^{\varepsilon(\beta X')}$. ∎

PROPOSITION 6.8 For each $\mathcal{U} \in \lambda$, $\phi^{-1}(\mathcal{U}'^{\varepsilon(\beta X')}) < \mathcal{U}^{\varepsilon(\beta X)}$.

Proof: Let $\mathcal{U} \in \lambda$. It suffices to show that $\phi^{-1}(U'^{\varepsilon(\beta X')}) \subset U^{\varepsilon(\beta X)}$ for each $U \in \mathcal{U}$. By Proposition 6.7, $U'^{\varepsilon(\beta X')} = U^{\varepsilon(\beta X')}$ so $\phi^{-1}(U'^{\varepsilon(\beta X')}) = \phi^{-1}(U^{\varepsilon(\beta X')})$. Now $\phi^{-1}(U^{\varepsilon(\beta X')})$ is open in βX and $U^{\varepsilon(\beta X)}$ is regularly open in βX since $\mathcal{U} \in \lambda$. Therefore, by Lemma 6.11.(3) it suffices to show that $\phi^{-1}(U^{\varepsilon(\beta X')}) \cap X \subset U^{\varepsilon(\beta X)} \cap X = U$. For this let $y \in \phi^{-1}(U^{\varepsilon(\beta X')}) \cap X$ which implies $y \in X$ and $\phi(y) \in U^{\varepsilon(\beta X')}$. Now $y \in X$ implies $\phi(y) = y \in U^{\varepsilon(\beta X')} \cap X = U$. We conclude that $\phi^{-1}(U^{\varepsilon(\beta X')}) \cap X \subset U$. ∎

LEMMA 6.12 $R = \cap_\lambda[\cup\{S(x,\mathcal{U}^{\varepsilon(\beta X)}) \mid x \in X\}] = \cap_\lambda[\cup\{S(x,\mathcal{U})^{\varepsilon(\beta X)} \mid x \in X\}]$.

Proof: From the definition of the radical, $R = \cap_\lambda[\cup\{U^{\varepsilon(\beta X)} \mid U \in \mathcal{U}\}] \subset \cap_\lambda[\cup\{S(x,\mathcal{U}^{\varepsilon(\beta X)}) \mid x \in X\}] \subset \cap_\lambda[\{\cup S(x,\mathcal{U})^{\varepsilon(\beta X)} \mid x \in X\}]$. Conversely, suppose y does not belong to $\cap_\lambda[\cup\{S(x,\mathcal{U})^{\varepsilon(\beta X)} \mid x \in X\}]$. Then there exists a $\mathcal{V} \in \lambda$ such that y does not belong to $S(x,\mathcal{V})^{\varepsilon(\beta X)}$ for each $x \in X$ which implies y does not belong to $U^{\varepsilon(\beta X)}$ for each $U \in \mathcal{U}$ which in turn implies y does not belong to R. Hence, $R = \cap_\lambda[\cup\{S(x,\mathcal{U}^{\varepsilon(\beta X)}) \mid x \in X\}] = \cap_\lambda[\cup\{S(x,\mathcal{U}^{\varepsilon(\beta X)} \mid x \in X\}]$. ∎

PROPOSITION 6.9 A point $p \in \beta X$ *is contained in the radical if and only if* $\phi(p) \in X'$.

Proof: If $p \in R$ then for each $\mathcal{U} \in \lambda$, $p \in S(x, \mathcal{U}^{\varepsilon(\beta X)})$ for some $x \in X$. Put $\psi(\mathcal{U}) = x$. Hence $K_{\mathcal{U}} = S(p, \mathcal{U}^{\varepsilon(\beta X)}) \cap X \neq \varnothing$. Let $V(p)$ be an open neighborhood of p in βX. Then $V(p) \cap S(p,\mathcal{U}^{\varepsilon(\beta X)}) \neq \varnothing$ which implies $V(p) \cap K_{\mathcal{U}} \neq \varnothing$. Consequently, $p \in Cl_{\beta X}[K_{\mathcal{U}}]$ for each $\mathcal{U} \in \lambda$. Since ϕ is continuous, $\phi(p) \in Cl_{\beta X'}[K_{\mathcal{U}}]$. We wish to show that $\cap_\lambda Cl_{\beta X'}[K_{\mathcal{U}}]$ is a single point $q \in X'$ which will mean $\phi(p) \in X'$.

For this consider the net $\psi: \lambda \to X$. For each $\mathcal{U} \in \lambda$, $\psi(\mathcal{U}) \in K_{\mathcal{U}}$. Let $\mathcal{V}$, $\mathcal{W} \in \lambda$ such that $\mathcal{W}^* < \mathcal{V}^* < \mathcal{U}$. Then $p \in S(\psi(\mathcal{W}),\mathcal{W}^{\varepsilon(\beta X)}) \cap S(\psi(\mathcal{V}),\mathcal{V}^{\varepsilon(\beta X)})$ which implies $[S(\psi(\mathcal{W}),\mathcal{W})]^{\varepsilon(\beta X)} \cap [S(\psi(\mathcal{V}),\mathcal{V})]^{\varepsilon(\beta X)} \neq \varnothing$ which in turn implies $S(\psi(\mathcal{W}),\mathcal{W}) \cap S(\psi(\mathcal{V}),\mathcal{V}) \neq \varnothing$. Hence $\psi(\mathcal{W}) \in Star\,(S(\psi(\mathcal{V}),\mathcal{V}),\mathcal{V}) \subset S(\psi(\mathcal{V}),\mathcal{U})$.

So, ψ is Cauchy in X and in X′. Therefore, ψ converges to some $q \in X'$. We want to show that $\{q\} = \cap_\lambda Cl_{\beta X'}[K_{\mathcal{U}}]$.

For this let $F_{\mathcal{U}} = \{\psi(\mathcal{V}) \mid \mathcal{V} \in \lambda \text{ and } \mathcal{V}^* < \mathcal{U}\}$. Then $F_{\mathcal{U}} \subset K_{\mathcal{U}}$ for each $\mathcal{U} \in \lambda$. ψ converges to q implies $q \in Cl_{X'}(F_{\mathcal{U}}) \subset Cl_{\beta X'}(F_{\mathcal{U}}) \subset Cl_{\beta X'}(K_{\mathcal{U}})$ for each $\mathcal{U} \in \lambda$. Therefore, $q \in Cl_{\beta X'}(K_{\mathcal{U}})$. Suppose $y \neq q$ is another point of $\beta X'$. Then there exists an open $A \subset \beta X'$ containing q such that y does not belong to $Cl_{\beta X'}(A)$. Then $A \cap X'$ is an open set in X′ containing q so there exists a $\mathcal{U}' \in \lambda'$ with $S(q,\mathcal{U}') \subset A \cap X'$. Let $\mathcal{V}' \in \lambda'$ with $\mathcal{V}' <^* \mathcal{U}'$.

ψ converges to q implies there exists a $\mathcal{W}_0' \in \lambda'$ with $\mathcal{W}_0' <^* \mathcal{V}$ such that $\mathcal{W}' \in \lambda'$ with $\mathcal{W}' <^* \mathcal{W}_0'$ which implies $\psi(\mathcal{W}) \in S(q,\mathcal{V}')$ for each $\mathcal{W}' \in \lambda'$ with $\mathcal{W}' <^* \mathcal{W}_0'$. Now $\psi(\mathcal{W}) \in S(p,\mathcal{W}^{\varepsilon(\beta X)})$ so $p \in S(\psi(\mathcal{W}),\mathcal{W}^{\varepsilon(\beta X)})$. For each $x \in K_{\mathcal{W}}$, $p \in S(x,\mathcal{W}^{\varepsilon(\beta X)})$ so there exists $W_1, W_2 \in \mathcal{W}$ with $p \in W_1^{\varepsilon(\beta X)} \cap W_2^{\varepsilon(\beta X)}$ and $\psi(\mathcal{W}) \in W_1 \cup W_2$. Since $W_1^{\varepsilon(\beta X)} \cap W_2^{\varepsilon(\beta X)} \neq \varnothing$, $W_1 \cap W_2 \neq \varnothing$ which implies there exists a $V_1 \in \mathcal{V}$ with $W_1 \cup W_2 \subset V_1$. Since $\psi(\mathcal{W}) \in S(q,\mathcal{V}')$, there exists a $V_2' \in \mathcal{V}'$ such that $q, \psi(\mathcal{W}) \in V_2'$. Then $V_1' \cap V_2' \neq \varnothing$. Consequently, there exists a $U' \in \mathcal{U}'$ with $V_1' \cup V_2' \subset U'$. Then $q, x \in U'$ which implies $K_{\mathcal{W}} \subset S(q, \mathcal{U}') \subset A \cap X' \subset A$. Hence y does not belong to $K_{\mathcal{W}}$ which implies y does not belong to $\cap_\lambda Cl_{\beta X'}[K_{\mathcal{U}}]$. Therefore, $\cap_\lambda Cl_{\beta X'}[K_{\mathcal{U}}] = \{q\}$.

Conversely, if $\phi(p) \in X'$ then $\phi(p) \in U'$ for some $U' \in \mathcal{U}'$, for each $\mathcal{U}' \in \lambda'$. But then $\phi(p) \in U'^{\varepsilon(\beta X')} \subset U^{\varepsilon(\beta X)}$ by the proof of Proposition 6.8. Hence $p \in E_{\mathcal{U}}$ for each $\mathcal{U} \in \lambda$ which implies $p \in R$. We conclude that $p \in R$ if and only if $\phi(p) \in X'$. ∎

THEOREM 6.8 (H. Tamano, 1960) A uniform space (X, μ) is complete if and only if R = X.

Proof: Assume X is complete. It is evident that $X \subset R$. Let $p \in R$ and for each $\mathcal{U} \in \lambda$ let $\psi(\mathcal{U})$, $K_{\mathcal{U}}$ and $F_{\mathcal{U}}$ be defined as in the proof of Proposition 6.9. Let $\mathcal{V} \in \lambda$ such that $\mathcal{V}^* < \mathcal{U}$. By Proposition 6.6, $\mathcal{V}^{\varepsilon(\beta X)} <^* \mathcal{U}^{\varepsilon(\beta X)}$. Let $y \in K_{\mathcal{V}} = S(p,\mathcal{V}^{\varepsilon(\beta X)}) \cap X$. Then there exists a $V_1 \in \mathcal{V}$ with $y, p \in V_1^{\varepsilon(\beta X)}$ and a $V_2 \in \mathcal{V}$ with $p,\psi(\mathcal{V}) \in V_2^{\varepsilon(\beta X)}$. Then $V_1^{\varepsilon(\beta X)} \cap V_2^{\varepsilon(\beta X)} \neq \varnothing$ and since $\mathcal{V}^{\varepsilon(\beta X)} <^* \mathcal{U}^{\varepsilon(\beta X)}$, there exists a $U \in \mathcal{U}$ with $V_1^{\varepsilon(\beta X)} \cup V_2^{\varepsilon(\beta X)} \subset U^{\varepsilon(\beta X)}$. Therefore, $y \in U^{\varepsilon(\beta X)} \cap X = U$ so $y \in S(\psi(\mathcal{V}),\mathcal{U})$ for each $\mathcal{V} \in \lambda$ with $\mathcal{V}^* < \mathcal{U}$. Hence, $F_{\mathcal{V}} \subset K_{\mathcal{V}} \subset S(\psi(\mathcal{V}), \mathcal{U})$ for each $\mathcal{V} \in \lambda$ with $\mathcal{V}^* < \mathcal{U}$, so ψ is Cauchy in X. Since X is complete ψ converges to some $q \in X$. By an argument similar to the proof of Proposition 6.9, $\cap\{Cl_{\beta X}[K_{\mathcal{U}}] \mid \mathcal{U} \in \lambda\} = \{q\}$ and $p \in Cl_{\beta X}[K_{\mathcal{U}}]$ for each $\mathcal{U} \in \lambda$. Therefore, $p = q \in X$ so $R = X$.

Conversely, assume $R = X$ and suppose (X, μ) is not complete. Then there exists a point $q \in X'$ such that q does not belong to $\phi(X)$. Let $p \in \beta X$ such that $\phi(p) = q$. Then $p \in R$ by Proposition 6.9. Since p does not belong to X we have $R \neq X$ which is a contradiction. We conclude that (X, μ) is complete. ∎

A Tychonoff space X is said to be **topologically complete** if there exists a uniformity μ for X such that (X, μ) is complete. Clearly a Tychonoff space is topologically complete if and only if it is complete with respect to its finest uniformity.

THEOREM 6.9 (H. Tamano, 1960) X is topologically complete if and only if for each $p \in \beta X - X$ there is a partition of unity $\Phi = \{\phi_\lambda\}$ such that $Cl_{\beta X}(0(\phi_\lambda))$ does not contain p for each $\phi_\lambda \in \Phi$.

Proof: Assume X is complete with respect to a uniformity μ, and that $p \in \beta X - X$. By Tamano's Completeness Theorem, $X = \cap\{E_{\mathcal{U}} \mid \mathcal{U} \in \mu\}$. Then, there is a $\mathcal{U} \in \mu$ such that $p \in \beta X - E_{\mathcal{U}}$. Let $\{\mathcal{U}_n\}$ be a normal sequence with $\mathcal{U}_1 <^* \mathcal{U}$. Let d be the pseudo-metric on X defined (as in the proof of Theorem 2.5) with respect to $\{\mathcal{U}_n\}$. Then the collection of spheres of radius 2^{-n} coincides with the point-star coverings associated with $\mathcal{U}_n$ [i.e., $S(x, 2^{-n}) = S(x, \mathcal{U}_n)$ for each n] and d:X × X → [0,1]. Let τ denote the topology on X induced by d. By Stone's Theorem, (X,τ) is paracompact. Now consider the open covering $\mathcal{V} = \{V(x) \mid x \in X\}$ of (X,τ) where $V(x) = \{y \in X \mid d(x,y) < 1/4\}$ and let $\{V_\lambda\}$ be an open locally finite refinement of $\mathcal{V}$. By Theorem 6.3 there exists a partition of unity $\Phi = \{\phi_\lambda\}$ subordinate to $\{V_\lambda\}$ such that $0(\phi_\lambda) \subset V_\lambda$ for each $\phi_\lambda \in \Phi$. Since τ is coarser than the original topology, Φ is also a partition of unity with respect to the original topology.

We now show that p does not belong to $Cl_{\beta X}(V(x))$ for each $x \in X$ which implies p does not belong to $Cl_{\beta X}(0(\phi_\lambda))$ for each $\phi_\lambda \in \Phi$ which will establish the necessity part of the theorem. We can show this by demonstrating that $Cl_{\beta X}(V(x)) \subset E_{\mathcal{U}}$ for each $x \in X$. For this let $z \in X$ and for each $x \in X$ put $d_z(x) = d(z, x)$. Then d_z:X → [0,1] is uniformly continuous with respect to the pseudo-metric uniformity on X by an argument similar to the one in (1) → (2) of Theorem 6.6. By Theorem 6.2, d_z can be uniquely extended to a continuous function $d_z^\beta:\beta X \to [0,1]$. Let $r \in Cl_{\beta X}(V(x))$. Then $d_z^\beta(r) \le 1/4$ so there exists an open W(r) in βX containing r such that $d_z^\beta(y) < 1/2$ for each $y \in W(r) \cap X$. Therefore, $W(r) \cap X \subset S(z, 1/2) = S(z, \mathcal{U}_1) \subset U$ for some $U \in \mathcal{U}$. But then $W(r) \subset Cl_{\beta X}(U) \subset E_{\mathcal{U}}$. Hence $Cl_{\beta X}(V(z)) \subset E_{\mathcal{U}}$. This completes the necessity part of the proof.

To prove the sufficiency part of the theorem, let $\Phi = \{\phi_\lambda\}$ be a partition of unity on X such that p does not belong to $Cl_{\beta X}(0(\phi_\lambda))$ for each $\phi_\lambda \in \Phi$ where $p \in \beta X - X$. For each $x \in X$ and each positive integer n, put $V_n(x) = \{ y \in X \mid \Sigma_\lambda |\phi_\lambda(x) - \phi_\lambda(y)| < 2^{-n}\}$ and let $\mathcal{V}_n = \{V_n(x) \mid x \in X\}$. Then $\{\mathcal{V}_n\}$ is a normal sequence in X, so $\mathcal{V}_n \in u$ for each positive integer n. Let $p \in \beta X - X$ and let $x \in X$. Then $\phi_\lambda(x) = 0$ for all but finitely many ϕ_λ's, say $\phi_{\lambda_1} \ldots \phi_{\lambda_m}$. Now p does not belong to $Cl_{\beta X}(0(\phi_{\lambda_i}))$ for each $i = 1 \ldots m$ which implies p does not belong to $\cup_{i=1}^m Cl_{\beta X}(0(\phi_{\lambda_i})) = Cl_{\beta X}[\cup_{i=1}^m 0(\phi_{\lambda_i})]$. If $y \in V_n(x)$ then $\Sigma_\lambda |\phi_\lambda(x) - \phi_\lambda(y)| < 2^{-n}$ which implies $\Sigma_{i=1}^m |\phi_{\lambda_i}(x) - \phi_{\lambda_i}(y)| < 2^{-n}$ so $\phi_{\lambda_j}(y) > 0$ for some $j \in \{1 \ldots$

$m\}$ or else $\Sigma_{i=1}^{m} |\phi_{\lambda_i}(x) - \phi_{\lambda_i}(y)| = \Sigma_{i=1}^{m} |\phi_{\lambda_i}(x)| = 1$. Therefore, $y \in \mathbf{0}(\phi_{\lambda_j})$ which implies $V_n(x) \subset \cup_{i=1}^{m} \mathbf{0}(\phi_{\lambda_i})$ which in turn implies $Cl_{\beta X}[V_n(x)] \subset Cl_{\beta X}[\cup_{i=1}^{m} \mathbf{0}(\phi_{\lambda_i})]$, so p does not belong to $Cl_{\beta X}[V_n(x)]$. Therefore, p does not belong to R (the radical) with respect to the u uniformity. Hence $R = X$. By Tamano's Completeness Theorem, (X, u) is complete. Consequently, X is topologically complete. ■

Let X be a Tychonoff space. For each $f \in C^*(X)$ define $|f|$ to be $sup\{|f(x)| x \in X\}$. Then $|\ |$ is called the **supremum norm** on $C^*(X)$. We can define a metric d on $C^*(X)$ by $d(f,g) = |f - g| = sup\{|f(x) - g(x)| \, | x \in X\}$. It can be shown (Exercise 1) that $(C^*(X), d)$ is a complete metric space. The following lemma appeared in Irving Glicksberg's 1959 paper *Stone-Čech Compactifications of Products* (Transactions of the American Mathematical Society, Volume 90, pp. 369-382).

LEMMA 6.13 (I. Glicksberg, 1959) Let X and Y be Tychonoff spaces and let $f \in C^*(X \times Y)$. If the mapping $\phi: Y \to C^*(X)$, defined by $\phi(y) = f_y$ where $f_y(x) = f(x,y)$ for each $x \in X$, is continuous, then f has a continuous extension f^β to $\beta X \times Y$.

Proof: For each $y \in Y$, let f_y^β denote the extension of f_y from X to βX, and put $\phi^\beta(y) = f_y^\beta$. Then $\phi^\beta: Y \to C^*(\beta X)$. Now $C^*(\beta X)$ with the metric d^β generated by the supremum norm is isometric with $C^*(X)$ with the metric d. To see this, note that there is a natural one-to-one correspondence $\theta: C^*(X) \to C^*(\beta X)$ defined by $\theta(f) = f^\beta$. It is easily shown (Exercise 2) that $d^\beta(f^\beta, g^\beta) = d(f,g)$ for any pair $f, g \in C^*(X)$ so θ is an isomorphism. Consequently, the continuity of ϕ implies the continuity of ϕ^β.

Notice that for each $(x,y) \in X \times Y$ that $f(x,y) = f_y(x)$. This motivates our definition of f^β as follows: for each $(x,y) \in \beta X \times Y$ put $f^\beta(x,y) = f_y^\beta(x)$. Clearly, f^β is an extension of f. Since $sup\{f_y^\beta(x) | x \in X\} = sup\{f_y(x) | x \in X\}$ for each $y \in Y$, we have that f^β is bounded. To show f^β is continuous, let $(x_0, y_0) \in \beta X \times Y$ and let $\varepsilon > 0$. Since $f_{y_0}^\beta \in C^*(\beta X)$, there exists a neighborhood U of x_0 in βX with $|f_{y_0}^\beta(x) - f_{y_0}^\beta(x_0)| < \varepsilon/2$ for each $x \in U$. Since ϕ^β is continuous, there is a neighborhood V of y_0 in Y with $d^\beta(\phi^\beta(y) - \phi^\beta(y_0)) < \varepsilon/2$ for each $y \in V$. Therefore, $|f_y^\beta - f_{y_0}^\beta| < \varepsilon/2$ for each $y \in V$, so $|f_y^\beta(x) - f_{y_0}^\beta(x)| < \varepsilon/2$ for each $x \in \beta X$. Thus for each $(x,y) \in U \times V$, $|f^\beta(x,y) - f^\beta(x_0,y_0)| = |f_y^\beta(x) - f_{y_0}^\beta(x)| \leq |f_y^\beta(x) - f_{y_0}^\beta(x)| + |f_{y_0}^\beta(x) - f_{y_0}^\beta(x_0)| < \varepsilon/2 + \varepsilon/2 = \varepsilon$, which establishes the continuity of f^β at (x_0,y_0). Since (x_0,y_0) was chosen arbitrarily, $f^\beta \in C^*(\beta X)$. ■

THEOREM 6.10 (H. Tamano, 1962) The following conditions on a Tychonoff space are equivalent:

(1) X is topologically complete.

(2) For each $p \in \beta X - X$ there is a normal sequence $\{\mathcal{V}_n\}$ of open coverings of X such that for each $V_0 \in \mathcal{V}_0$, $Cl_{\beta X}(V_0)$ does not contain p.

(3) If $p \in \beta X - X$ then $\{p\} \times X$ and ΔX are functionally separated (by a member of $C^*(X \times \beta X)$).

Proof: Assume X is topologically complete and let $p \in \beta X - X$. By Theorem 6.9, there exists a partition of unity $\Phi = \{\phi_\lambda\}$ such that $Cl_{\beta X}(0(\phi_\lambda))$ does not contain p for each $\phi_\lambda \in \Phi$. For each $x \in X$ and each non-negative integer n, let $V_n(x) = \{y \in X \mid \Sigma_\lambda |\phi_\lambda(x) - \phi_\lambda(y)| < 2^{-n}\}$ and put $\mathcal{V}_n = \{V_n(x) \mid x \in X\}$. Then $\{\mathcal{V}_n\}$ is a normal sequence of open coverings of X. To show $Cl_{\beta X}[V_0(x)]$ does not contain p for each x, let $y \in V_0(x)$. Now $\phi_\lambda(x) = 0$ for all but finitely many ϕ_λ's, say $\phi_{\lambda_1} \ldots \phi_{\lambda_m}$. If $\phi_{\lambda_i}(y) = 0$ for each $i = 1 \ldots m$, then $\Sigma_{i=1}^m |\phi_{\lambda_i}(x) - \phi_{\lambda_i}(y)| = 1$ which implies y does not belong to $V_0(x)$ which is a contradiction. Therefore, $\phi_{\lambda_j}(y) > 0$ for some j which implies $y \in 0(\phi_{\lambda_j})$ so $V_0(x) \subset \cup_{i=1}^m 0(\phi_{\lambda_i})$. Hence

$$Cl_{\beta X}[V_0(x)] \subset Cl_{\beta X}[\cup_{i=1}^m 0(\phi_{\lambda_i})] \subset \cup_{i=1}^m Cl_{\beta X}[0(\phi_{\lambda_i})].$$

But then $Cl_{\beta X}[V_0(x)]$ does not contain p for each $x \in X$. Thus (1) $\rightarrow$ (2).

To show that (2) $\rightarrow$ (1), let $\{\mathcal{V}_n\}$ be a normal sequence of open coverings of X such that for each $V_0 \in \mathcal{V}_0$, $Cl_{\beta X}(V_0)$ does not contain p. Using Corollary 2.2, we can construct a pseudo-metric d on X such that $\mathcal{W} < \mathcal{V}_0$ where $\mathcal{W} = \{W(x) \mid x \in X\}$ and $W(x) = \{y \in X \mid d(x,y) < 1\}$ for each $x \in X$. Let τ denote the topology of (X, d). Then (X,τ) is paracompact. Clearly W(x) is open with respect to τ and W(x) does not contain p for each $x \in X$. Since (X,τ) is paracompact, we can use Theorem 6.3 to construct a partition of unity $\Phi = \{\phi_\lambda\}$ of X that is subordinate to $\mathcal{W}$. Since τ is coarser than the original topology, Φ is a partition of unity with respect to the original topology. Therefore, there exists a partition of unity $\Phi = \{\phi_\lambda\}$ on X with $Cl_{\beta X}[0(\phi_\lambda)]$ not containing p for each $\phi_\lambda \in \Phi$. It follows from Theorem 6.9 that X is topologically complete. This completes the proof of (2) $\rightarrow$ (1).

To show (1) $\rightarrow$ (3), Assume X is topologically complete and let $p \in \beta X - X$. Then by Theorem 6.9, there exists a partition of unity $\Phi = \{\phi_\lambda\}$ on X such that $Cl_{\beta X}[0(\phi_\lambda)]$ does not contain p for each $\phi_\lambda \in \Phi$. Define a pseudo-metric d on X by $d(x,y) = \Sigma_\lambda |\phi_\lambda(x) - \phi_\lambda(y)|$. Then d is continuous with respect to the topology associated with d, which is coarser than the original topology, and hence continuous with respect to the original topology. For each y let d_y be defined by $d_y(x) = d(x,y)$. Then d_y is continuous since d is continuous, so $d_y \in C^*(X)$. Since d satisfies the triangle inequality, the mapping $\phi: X \rightarrow C^*(X)$, defined by $\phi(y) = d_y$ for each $y \in X$, is continuous. Therefore, by Lemma 6.13, d has a continuous extension $d^*(z,y)$ over $\beta X \times X$. It is clear that $d^* = 1$ on $\{p\} \times X$ and $d^* = 0$ on ΔX. Thus (1) → (3).

To show (3) → (1), let $p \in \beta X - X$ and let $\{p\} \times X$ be functionally separated from ΔX by $f \in C^*(\beta X \times X)$. Without loss of generality, we may assume $f(\Delta X) = 0$ and $f(\{p\} \times X) = 1$. For each $x \in X$ let f_x be defined by $f_x(z) = f(z,x)$ for each $z \in \beta X$. Then put $d(x,y) = sup_{z \in \beta X} |f_x(z) - f_y(z)|$. Then d is a pseudo-metric on X. Let τ denote the topology of (X, d), and consider the space (X,τ) which is paracompact. For each $x \in X$, let $U(x) = \{y \in X | d(x,y) < 1/2\}$. Then $U(x)$ is open in (X,τ). Put $\mathcal{U} = \{U(x) | x \in X\}$. Then there exists a partition of unity $\Phi = \{\phi_\lambda\}$ that is subordinate to $\mathcal{U}$. Since τ is coarser than the original topology, $\mathcal{U}$ is an open covering with respect to the original topology and Φ is a partition of unity with respect to the original topology.

Now $d(x,y) < 1/2$ implies that $sup_{z \in \beta X} |f_x(z) - f_y(z)| < 1/2$ so $|f_x(y) - f_y(y)| < 1/2$. But $f_y(y) = f(y,y) = 0$ so $|f_x(y)| < 1/2$. Therefore, $f_x(z) \leq 1/2$ for each $z \in Cl_{\beta X}[U(x)]$ since f_x is continuous on βX. On the other hand, $f_x(p) = f(p, x) = 1$ for each $x \in X$. It follows that $Cl_{\beta X}[U(x)]$ does not contain p for each $x \in X$. Since Φ is subordinate to $\mathcal{U}$, $Cl_{\beta X}[0(\phi_\lambda)]$ does not contain p for each $\phi_\lambda \in \Phi$. By Theorem 6.9, this implies X is topologically complete. ■

EXERCISES

1. Let d be the metric constructed from the supremum norm on $C^*(X)$ for a Tychonoff space X. Show that $(C^*(X), d)$ is a complete metric space.

2. Show that $d^\beta(f^\beta, g^\beta) = d(f, g)$ in the proof of Lemma 6.13, for any pair $f, g \in C^*(X)$.

6.4 Points at Infinity and Tamano's Theorem

Although Tamano's Theorem is not about uniform spaces, the proofs of Tamano's Completeness Theorem and Tamano's characterization of topological completeness (Theorem 6.9) hold the key to its proof. Also, it is an important example of using *points at infinity* to prove things about the space itself. If we consider the space **R** of real numbers, we notice that **R** has a compactification **R*** that is constructed by adding two additional points that we denote by $-\infty$ and $+\infty$. **R** is sometimes denoted by $(-\infty, +\infty)$ while **R*** is often referred to as the *extended real numbers* and denoted $[-\infty, +\infty]$. The point $-\infty$ is assumed to precede all members of **R** and $+\infty$ is assumed to follow all members of **R**. In this way, the natural ordering on **R** is extended to **R***. The topology of **R*** is formed by taking as a basis, all open intervals in **R** together with all sets in **R*** of the form $[-\infty, r)$ and $(r, +\infty]$ where $[-\infty, r) = \{y \in \mathbf{R}^* | y < r\}$ and $(r, +\infty] = \{y \in \mathbf{R}^* | y > r\}$.

The points $-\infty$ and $+\infty$ are called points at infinity. In general, if (X, μ) is a uniform space and (X', μ') is its completion, the set $X' - X$ is called the set of **points at infinity** of X. Similarly, if P is a paracompactification of X, $PX - X$ is referred to as the set of points at infinity of X. Points at infinity are always *relative to some uniformity*. Points at infinity with respect to precompact uniformities (e.g., $\beta X - X$) play an important role in general topology. If μ is a uniformity for X and $H \subset \mu X - X$, then H is called a **set at infinity**. Compact sets at infinity will be important to us in what follows. Sets at infinity can be very large with respect to the original uniform space. For example, the space N of positive integers with the usual topology is countable whereas $\beta N - N$ is not.

We have already seen a characterization of local compactness in terms of compact sets at infinity, namely, Proposition 6.4. Our next characterization of paracompactness in terms of compact sets at infinity relies on a well known characterization of paracompactness by E. Michael that appeared in the Proceedings of the American Mathematical Society in 1953 (Volume 4, Number 3, pp. 831-838).

LEMMA 6.14 (E. Michael, 1953) The following properties of a regular topological space are equivalent:

(1) X is paracompact,

(2) every open covering of X has a locally finite (not necessarily open) refinement and

(3) every open covering of X has a locally finite closed refinement.

Proof: (1) $\rightarrow$ (2) is obvious. To show (2) $\rightarrow$ (3) let $\mathcal{U}$ be an open covering of X. Since X is regular, there exists an open covering $\mathcal{V}$ of X such that the closures of elements of $\mathcal{V}$ is a refinement of $\mathcal{U}$. By assumption, there exists a locally finite refinement $\mathcal{W}$ of $\mathcal{V}$. Then the closures of elements of $\mathcal{W}$ is the desired locally finite closed refinement of $\mathcal{U}$.

To show that (3) $\rightarrow$ (1), let $\mathcal{U}$ be an open covering of X. Let $\mathcal{V}$ be a locally finite refinement of $\mathcal{U}$ and let $\mathcal{W}$ be a covering of X consisting of closed sets, each one of which intersects only finitely many members of $\mathcal{V}$. Now let $\mathcal{Z}$ be a locally finite closed refinement of $\mathcal{W}$. For each $V \in \mathcal{V}$, let $V' = X - \cup\{Z \in \mathcal{Z} \mid V \cap Z = \varnothing\}$. Then V' is an open set containing V such that if $Z \in \mathcal{Z}$, then Z intersects V' if and only if Z intersects V. For each $V \in \mathcal{V}$ pick a $U_V \in \mathcal{U}$ such that $V \subset U_V$. Let $\mathcal{U}' = \{V' \cap U_V \mid V \in \mathcal{V}\}$. Then $\mathcal{U}'$ is an open refinement of $\mathcal{U}$ and since each element of a locally finite covering $\mathcal{Z}$ intersects only finitely many elements of $\mathcal{U}'$, $\mathcal{U}$, is locally finite. ∎

Notice the similarity between statement (2) of the following theorem and the statement of Theorem 6.9. By replacing points at infinity with compact sets at infinity in the hypothesis, the strength of the conclusion is raised from topological completeness to paracompactness.

THEOREM 6.11 (H. Tamano, 1960) Let BX denote any compactification of X. Then the following statements are equivalent:

(1) X is paracompact.

(2) For each compact set C at infinity, there exists a partition of unity $\Phi = \{\phi_\lambda\}$ on X such that $Cl_{\beta X}[0(\phi_\lambda)] \cap C = \varnothing$ for each $\phi_\lambda \in \Phi$.

(3) For each compact set C at infinity, there exists a collection $\{G_\alpha\}$ of compact subsets of BX with $X \subset \cup G_\alpha$, $C \cap G_\alpha = \varnothing$ for each α, and for each $x \in X$ there exists an open (in BX) neighborhood $U'(x)$ of x that meets only finitely many of the G_α's.

Proof: Assume X is paracompact and let C be a compact set at infinity with respect to B. Then BX - C is open so for each $x \in X$ there is an open neighborhood $U'(x)$ of x with $Cl_{BX}[U'(x)] \cap C = \varnothing$. For each $x \in X$ put $U(x) = U'(x) \cap X$ and let $\mathcal{U} = \{U(x) \mid x \in X\}$. By Theorem 6.3 we can find a partition of unity $\Phi = \{\phi_\lambda\}$ of X subordinate to $\mathcal{U}$. Since for each λ, there exists an x with $0(\phi_\lambda) \subset U(x)$, $Cl_{BX}[0(\phi_\lambda)] \cap C = \varnothing$ for each $\phi_\lambda \in \Phi$. Thus (1) $\rightarrow$ (2). To show (2) $\rightarrow$ (3), it is only necessary to observe that since $\{0(\phi_\lambda)\}$ is a locally finite open covering of X in BX, that $\{Cl_{BX}[0(\phi_\lambda)]\}$ is also a locally finite covering of X in BX and consequently $\{Cl_{\beta X}[0(\phi_\lambda)]\}$ is the collection $\{G_\alpha\}$ that is desired.

To show (3) $\rightarrow$ (1), we use Lemma 6.13. Let $\{U_\beta\}$ be an open covering of X and for each β let U_β^ξ denote the proper extension of U_β over BX. Put C = BX - $\cup U_\beta^\xi$. Then C is a compact set at infinity. By hypothesis there is a collection $\{G_\alpha\}$ of compact subsets of BX with $X \subset \cup G_\alpha$, $C \cap G_\alpha = \varnothing$ for each α, and for each $x \in X$ there is an open neighborhood $U'(x)$ of x that meets only finitely many of the G_α's. Now each G_α is covered by a finite number of the U_β^ξ's, say $U_1^\xi \dots U_n^\xi$. Put $H_\alpha^k = G_\alpha \cap U_k^\xi \cap X$. Then each H_α^k is contained in some $U_\beta = U_\beta^\xi \cap X$ and $G_\alpha \cap X = \cup_{k=1}^n H_\alpha^k$. Constructing the H_α^k's for each G_α in this manner yields a locally finite refinement $\{H_\alpha^k\}$ of $\{U_\beta\}$. It follows that X is paracompact. ∎

Let C(X) denote the collection of all real valued continuous functions on a space X and let C′(X) be a subset of C(X). If there is an $f \in C'(X)$ such that $f(x) = 1$ for each $x \in F$ and $f(x) = 0$ for each $x \in G$, then F and G are said to be **functionally separated** by a member of C′(X). The following theorem contains Tamano's Theorem, i.e., (1) and (3) are equivalent.

THEOREM 6.12 (H. Tamano, 1962) Let BX be any compactification of X. Then the following statements are equivalent:

(1) X is paracompact.

(2) For each compact $C \subset \beta X - X$, there is a normal sequence of open coverings $\{\mathcal{V}_n\}$ of X such that for each $V_0 \in \mathcal{V}_0$, $Cl_{BX}(V_0) \cap C = \varnothing$.

(3) $X \times BX$ is normal.

(4) If $G = X \times C$ is closed in $X \times BX$ and $G \cap \Delta X = \varnothing$, then G and ΔX are functionally separated (by a member of $C^*(X \times BX)$).

Proof: The proof of the equivalence of (1) and (2) is almost identical to the proof of the equivalence of (1) and (2) in Theorem 6.10 and therefore will not be included here. It consists of exchanging the point $p \in \beta X - X$ with the compact set $C \subset \beta X - X$ and adjusting the proof accordingly.

To show that (1) → (3) we use the original proof of Dieudonné (1944). For this let $\mathcal{U}$ be an open covering of X × BX where X is assumed to be paracompact. For each point $(x,y) \in X \times BX$ there are neighborhoods $V(x,y)$ of x in X and $W(x,y)$ of y in BX with $(x,y) \in V(x,y) \times W(x,y) \subset U$ for some $U \in \mathcal{U}$. For a fixed $x \in X$, the sets $V(x,y) \times W(x,y)$ such that $y \in BX$ form an open covering of the compact set $\{x\} \times BX$. Therefore, there exists a finite number of points $y_1 \ldots y_n \in BX$ such that $\{V(x,y_i) \times W(x,y_i) \mid i = 1 \ldots n\}$ covers $\{x\} \times BX$. Put $V_x = \cap_{i=1}^n V(x,y_i)$. Then $\{V_x \times W(x,y_i) \mid i = 1 \ldots n\}$ is a covering of $\{x\} \times BX$ and V_x is a neighborhood of x in X. Let $\mathcal{V} = \{V_x \mid x \in X\}$. Then by assumption there exists a locally finite open refinement $\mathcal{V}'$ of $\mathcal{V}$. For each $V' \in \mathcal{V}'$ pick $x_{V'} \in V'$. Then $x_{V'} \in V' \subset V_{x_o}$ for some $x_0 \in X$. Let $y_1 \ldots y_m$ be the finite collection of points in BX corresponding to x_0 such that $\{V_{x_o} \times W(x_0,y_i) \mid i = 1 \ldots m\}$ covers $\{x_0\} \times BX$. Then the collection $\{V' \times W(x_{V'},y_i) \mid i = 1 \ldots m\}$ covers $\{x_{V'}\} \times BX$. Hence $\mathcal{U}' = \{V' \times W(x_{V'},y_i) \mid V' \in \mathcal{V}\}$ is an open covering of X × BX that clearly refines $\mathcal{U}$. Furthermore, $\mathcal{U}'$ is locally finite since if $(a,b) \in X \times BX$, there exists an open set A in X containing a such that A meets only finitely many members of $\mathcal{V}'$. But then A × BX meets only finitely many members of $\mathcal{U}'$. This completes the proof that (1) → (3).

To prove (3) → (4) note that ΔBX (the diagonal of BX) is closed in BX × BX so that $\Delta BX \cap [X \times BX]$ is closed in X × BX. But $\Delta BX \cap [X \times BX] = \Delta X$, so ΔX is closed in X × BX. Since X × BX is normal, there exists a continuous function $f \in C^*(X \times BX)$ with $f(\Delta X) = 0$ and $f(G) = 1$, so G and ΔX are functionally separated.

The proof that (4) → (1) is very similar to the proof the (3) → (1) of Theorem 6.10 and therefore will not be included here. It consists of exchanging the point p in $\beta X - X$ with the compact set $C \subset \beta X - X$, and adjusting the proof accordingly. ∎

EXERCISES

1. Show that (1) and (2) of Theorem 6.12 are equivalent.

2. Show that (4) implies (1) in Theorem 6.12.

3. [A. Hohti, 1981] A uniform space (X, μ) is uniformly paracompact if and only if for each compact $K \subset \beta X - X$ there is a $\mathcal{U} \in \mu$ with $Cl_{\beta X}(S(x,\mathcal{U})) \subset \beta X - K$ for each $x \in X$.

4. [A. Hohti, 1981] A uniform space (X, μ) is uniformly paracompact if and only if $(X \times \beta X, \mu \times \beta)$ is *C*-normal. *C*-normality is defined as follows: (X, μ) is ***C*-normal** if for each pair A, B of disjoint closed sets, there exists a $\mathcal{U} \in \mu$ such that given any $x \in X$, there is a $\mathcal{U}_x \in \mu$ with $Star(A,\mathcal{U}_x) \cap S(x,\mathcal{U}) \cap Star(B,\mathcal{U}_x) = \varnothing$.

6.5 Paracompactifications

If there is a paracompactification πX of X that plays a role analogous to the role of the Stone-Čech compactification and the Hewitt realcompactification in the theories of compactification and realcompactification respectively, what properties should π have in order that it be considered to play an analogous role? First, if X is a paracompact space then we should expect that πX = X and second, πX should be realizable as the completion of X with respect to some uniformity π such that π is the finest uniformity to have a paracompact completion. Every paracompactification of X is realizable as the completion of X with respect to some uniformity as shown in Section 6.1.

We will show that in general, such a paracompactification does not exist, and that a necessary and sufficient condition for π to exist is for the finest uniformity *u* to be preparacompact, in which case *u* = π. Consequently, if X is paracompact then πX exists and πX = X. In the case π exists, we call πX the **Tamano-Morita paracompactification** of X. We begin our discussion by proving a lemma about the relationship of any Tychonoff space Y that contains X as a dense subspace and the subspaces of βX that contain X.

Let X and Y be Tychonoff spaces. A function f:X → Y is said to be a **perfect** mapping if f is closed, continuous and the inverse image of each point of Y is compact in X. Let i:X → Y be an imbedding of X into a dense subspace of Y. Let j:Y → βY be the imbedding of Y into its Stone-Čech compactification and let k: βX → βY be the unique continuous extension of j © i such that $k(\beta X - X) = \beta Y - X$. Finally, let $Z = k^{-1}(Y)$.

LEMMA 6.15 (1) The restriction $k_Z = k|Z$ is a closed mapping, (2) Z is the largest subspace of βX to which j © i has a continuous extension to Y, and (3) Z is the only subspace of βX containing X for which k_Z is a perfect mapping.

Proof: Since βX and βY are both compact, the mapping k is closed and the inverse image of each compact set in βY is compact in βX. Then k_Z has the same properties because it is the restriction of k to the *total* inverse image of Y. By definition of Z, Z is the largest subset of βX to which j © i has a continuous extension to Y. Furthermore, we claim that Z is the only subspace of βX for which $k|Z$ is a closed mapping such that the inverse image of each point is compact. To show this suppose W is another subspace of βX containing X having the property that $k|W$ is a closed mapping and the inverse image of each point is compact. Put $S = W \cap Z$. Then $X \subset S \subset Z$ and since $Z \neq W$, S is a proper dense subspace of Y. Moreover, $k^{-1}(p) \cap S$ is compact for each $p \in Y$. Choose $z \in Z - S$ and let $q = k(z)$. Then z has an open neighborhood V such that $Cl_Z(V) \cap [k^{-1}(q) \cap S] = \emptyset$. Thus q does not belong to $k(Cl_Z(V) \cap S)$. But $Cl_Z(Cl_Z(V) \cap S) = Cl_Z(V)$ and $q \in k(Cl_Z(V))$. Hence $q \in k(Cl_Z(Cl_Z(V) \cap S)) \subset Cl_Y(k(Cl_Z(V) \cap S))$. Therefore, $k(Cl_Z(V) \cap S)$ is not closed in Y. Hence $k|S$ takes the closed set $Cl_Z(V) \cap S$ into a set that is not closed in Y. Therefore, $k|S$ is not a closed mapping which is a contradiction. Consequently Z is the only subspace of βX containing X for which the extension k of $j \circledcirc i$ is a closed mapping such that the inverse image of each point is compact. ∎

The following development of the Tamano-Morita paracompactification appeared in the author's paper titled *Paracompactifications, Preparacompactness and Some Problems of K. Morita and H. Tamano* that appeared in Questions and Answers in General Topology, Volume 10, pp. 191-204.

THEOREM 6.13 Each completion μX of a uniform space (X, μ) where μ is finer than β is homeomorphic to a subspace of βX containing X.

Proof: Let $i{:}X \to \mu X$ be the imbedding of X into its completion and $j{:}\mu X \to \beta\mu X$ the imbedding of μX into its Stone-Čech compactification. By the above lemma, there exists a unique continuous extension k of $j \circledcirc i$ such that $k(\beta X - X) = \beta\mu X - X$ and such that if $Z = k^{-1}(\mu X)$ then (1) the restriction $k_Z = k|Z$ is a closed mapping and (2) the inverse image of each point under k_Z is compact.

Let μ' denote the uniformity of μX and let μ^+ be the coarsest uniformity on Z that makes k_Z uniformly continuous. Then a basis for μ^+ is the collection $\{k_Z^{-1}(\mathcal{U}') \mid \mathcal{U}' \in \mu'\}$. Clearly X is dense in Z. To see that (X, μ) is a uniform subspace of (Z, μ^+), let $\mathcal{U}^+ \in \mu^+$ such that $\mathcal{U}^+ = k_Z^{-1}(\mathcal{U}')$ for some $\mathcal{U}' \in \mu'$. Since k_Z keeps X pointwise fixed, $\mathcal{U}^+ \cap X = \mathcal{U}' \cap X = \mathcal{U}$ for some $\mathcal{U} \in \mu$. Conversely, if $\mathcal{U} \in \mu$ then there exists a $\mathcal{U}' \in \mu'$ such that $\mathcal{U} = \mathcal{U}' \cap X$. But then $\mathcal{U} = k_Z^{-1}(\mathcal{U}') \cap X$ and $k_Z^{-1}(\mathcal{U}') \in \mu^+$. Consequently (X, μ) is a dense uniform subspace of (Z, μ^+). Clearly μ^+ generates the same topology on Z as the one induced on Z by βX since μ^+ is finer than β.

Finally we claim that (Z, μ^+) is complete and therefore (Z, μ^+) is the completion of (X, μ). Since completions are unique, we can identify μX with Z so that $X \subset \mu X \subset \beta X$. To see (Z, μ^+) is complete, let $\psi{:}D \to Z$ be Cauchy with respect to μ^+. Since k_Z is uniformly continuous, $k_Z \circledcirc \psi$ is Cauchy in μX and hence converges to some $q \in \mu X$. Suppose ψ does not cluster to any $p \in k_Z^{-1}(q)$. Since $k_Z^{-1}(q)$ is compact, there exists an open set O containing $k_Z^{-1}(q)$ such that ψ is eventually in $Z - O$. Then $k_Z(Z - O)$ is closed in μX and does not contain q. Then $\mu X - k_Z(Z - O)$ is an open set containing q such that $k_Z \circledcirc \psi$ is eventually in its complement. Therefore ψ does not converge to q which is a contradiction. Hence ψ must cluster to some $p \in k_Z^{-1}(q)$. Since ψ is Cauchy, ψ converges to p. Therefore, (Z, μ^+) is complete. But then (Z, μ^+) is the completion of (X, μ). ∎

PROPOSITION 6.10. The Stone-Čech compactification $\beta\mu X$ *of a completion* $\mu X \subset \beta X$ *of X with respect to some uniformity* μ *of X is* βX *(i.e.,* $\beta\mu X = \beta X$*).*

Proof: Let $i{:}\mu X \to \beta X$ be the identity imbedding of μX into βX and let j be the identity mapping on βX. By Lemma 6.14, there exists a unique continuous extension $k{:}\beta\mu X \to \beta X$ such that $k(\beta\mu X - \mu X) = \beta X - \mu X$ and if $Z - k^{-1}(\beta X) = \beta\mu X$ then $k|Z = k$ is a closed mapping such that the inverse image of each point is compact. We claim that by an argument similar to the one in the proof of Theorem 6.13, we can see that βX is homeomorphic to $\beta\mu X$.

For this let β' denote the uniformity of βX and put $\beta^+ = \beta'_k$ (the coarsest uniformity on $\beta\mu X$ that makes k uniformly continuous). Then (X, β) is a dense uniform subspace of $(\beta\mu X, \beta^+)$ by an argument similar to the one in the proof of Theorem 6.13 that showed (X, μ) to be a dense uniform subspace of (Z, μ'). Finally, we claim $(\beta\mu X, \beta^+)$ is complete by an argument similar to the one in the proof of Theorem 6.13 that showed (Z, μ') to be complete. Since completions are unique, $\beta\mu X$ is homeomorphic to βX (i.e., we can identify $\beta\mu X$ with βX so that $\beta\mu X = \beta X$.) ∎

If X is a subspace of Y such that each $f \in C(X)$ can be extended to a continuous function over Y, then X is said to be **C-imbedded** in Y. If each $g \in C^*(X)$ can be extended to a continuous function over Y, then X is **C*-imbedded** in Y. Then every closed subspace of a normal space is C*-imbedded (by Theorem 0.4). The following theorem has many applications. Notice that Proposition 6.10 is a special case of the equivalence of (5) and (6) in this theorem.

THEOREM 6.14 Let X be dense in the Tychonoff space Y. Then the following statements are equivalent:

(1) Every continuous $f{:}X \to K$*, where K is a compact Hausdorff space, has a continuous extension* $f^Y{:}Y \to K$*.*

(2) X is C-imbedded in Y.*

(3) Any two disjoint zero sets in X have disjoint closures in Y.

(4) If $Z, Z' \in \mathbf{Z}(X)$ *then* $Cl_Y(Z \cap Z') = Cl_Y(Z) \cap Cl_Y(Z')$*.*

(5) $\beta X = \beta Y$*.*

(6) $X \subset Y \subset \beta X$*.*

Proof: (1) → (2) If $f \in C^*(X)$ then f is a continuous mapping from X into the compact subset $Cl_{\mathbf{R}}(f(X))$. Hence (2) is a special case of (1), so (1) implies (2).

(2) → (3) Let $Z, Z' \in \mathbf{Z}(X)$ with $Z \cap Z' = \varnothing$. Then there exists $f, f' \in C^*(X)$ such that $Z = \mathbf{Z}(f)$ and $Z' = \mathbf{Z}(f')$. Let $g = max\{|f|,1\}$ and $g' = 1 - max\{|f'|,1\}$. Put $h' = min\{g,g'\}$ and let $h = max\{h',1\}$. Then $0 \le h \le 1$, $h(Z) = 0$ and $h(Z') = 1$. Let h^Y be the continuous extension of h to Y. Let $U = \{y \in Y \mid h(y) < 1/2\}$ and $V = \{y \in Y \mid h(y) > 1/2\}$. Then U and V are disjoint open sets

in Y such that $Z \subset U$ and $Z' \subset V$. Consequently, Z and Z′ have disjoint closures in Y.

(3) → (4) Clearly $Cl_Y(Z \cap Z') \subset Cl_Y(Z) \cap Cl_Y(Z')$. To show the reverse inclusion, let $p \in Cl_Y(Z) \cap Cl_Y(Z')$. Then there exists a zero set neighborhood V of p in Y, $p \in Cl_Y(V \cap Z)$ and $p \in Cl_Y(V \cap Z')$. Thus (3) implies $(V \cap Z) \cap (V \cap Z') \neq \emptyset$ which in turn implies $V \cap (Z \cap Z') \neq \emptyset$. Therefore, $p \in Cl_Y(Z \cap Z')$. Consequently, $Cl_Y(Z) \cap Cl_Y(Z') \subset Cl_Y(Z \cap Z')$.

(4) → (5) We show $\beta X = \beta Y$ by showing that (X, β) is a uniform subspace of (Y, β) and appealing to the uniqueness of the completion (Theorem 4.10). To do this, it suffices to show that every finite normal covering $\mathcal{U}$ of X can be extended to a finite normal covering $\mathcal{U}'$ of Y such that $\mathcal{U} = \mathcal{U}' \cap X$. If $\mathcal{U}$ is a finite normal covering of X, then it can be shown (Exercise 4) that there exists a normal sequence $\{\mathcal{U}_n\}$ of finite open coverings of X such that $\mathcal{U}_1 < \mathcal{U}$, and for each positive integer n and $U \in \mathcal{U}_n$, $X - U$ is a zero set of X. Next, notice that we can extend statement (4) of the theorem to any finite collection $Z_1 \ldots Z_n$ so that $Cl_Y(\cap_{i=1}^n Z_i) = \cap_{i=1}^n Cl_Y(Z_i)$.

Now $\mathcal{U}^{\varepsilon(Y)}$ is an extension of $\mathcal{U}$ into Y for each positive integer n, and $\mathcal{U}_n^{\varepsilon(Y)}$ is an extension of $\mathcal{U}_n$ into Y. Furthermore, by the extended version of (4), we have for each positive integer n, $\cap\{Y - V^{\varepsilon(Y)} \mid V \in \mathcal{U}_n\} = \cap\{Cl_Y(X - V) \mid V \in \mathcal{U}_n\} = Cl_Y(\cap\{X - V \mid V \in \mathcal{U}_n\}) = Cl_Y(\emptyset) = \emptyset$. Consequently, $\mathcal{U}_n^{\varepsilon(Y)}$ is a covering of Y for each positive integer n, so $\mathcal{U}^{\varepsilon(Y)}$ is a covering of Y by Proposition 6.6. Also, by Proposition 6.6, $\mathcal{U}_{n+1} <^* \mathcal{U}_n$ for each positive integer n so $\mathcal{U}_{n+1}^{\varepsilon(Y)} <^* \mathcal{U}_n^{\varepsilon(Y)}$ for each n. Hence $\mathcal{U}^{\varepsilon(Y)}$ is a finite normal covering of Y and $\mathcal{U} = \mathcal{U}^{\varepsilon(Y)} \cap X$.

(5) → (6) is obvious.

(6) → (1) By Theorem 6.2, a continuous $f: X \to K$, where K is a compact Hausdorff space, has a continuous extension $f^{\beta}: \beta X \to K$. But then $f^Y = f^{\beta} \mid Y$ is a continuous extension to Y. ■

THEOREM 6.15 The completion of a Tychonoff space X with respect to its finest uniformity u is $uX = \cap\{\mu X \mid \mu$ *is a uniformity for X and* μX *is a subspace of* $\beta X\}$.

Proof: To see that $\cap \mu X$ is topologically complete, observe that for each point $p \in \beta X - \cap \mu X$, p must belong to $\beta X - \lambda X$ for some uniformity λ of X. Since λX is complete, λX is topologically complete. By Theorem 6.9, for each $q \in \beta\lambda X - \lambda X$, there exists a partition of unity $\Phi = \{\phi_\lambda\}$ on λX such that $Cl_{\beta\lambda X}(0(\phi_\lambda))$ does not contain q for each $\phi_\lambda \in \Phi$. By Proposition 6.10, $\beta\lambda X = \beta X$ so $Cl_{\beta X}(0(\phi_\lambda))$ does not contain p for each $\phi_\lambda \in \Phi$. For each $\phi_\lambda \in \Phi$ put $\psi_\lambda = \phi_\lambda \mid \cap \mu X$. Then $\Psi = \{\psi_\lambda\}$ is a partition of unity on $\cap \mu X$ such that $Cl_{\beta X}(0(\psi_\lambda))$ does not contain p for each $\psi_\lambda \in \Psi$. Since this holds for each $p \in \beta X - \cap \mu X$, we have that $\cap \mu X$ is topologically complete.

Let ν be the finest uniformity on $\cap\mu X$ and let ν' be the uniformity induced on X by ν. Since ν is the finest uniformity on $\cap\mu X$, ν is finer than the uniformity u' induced on $\cap\mu X$ by u. But then ν' is finer than u which implies that $\nu' = u$. Since $\cap\mu X$ is topologically complete, $\cap\mu X$ is the completion of X with respect to ν'. Since completions are unique, $\cap\mu X = uX$. ∎

LEMMA 6.16 A perfect preimage of a paracompact space is paracompact.

Proof: Let $\{U_\alpha\}$ be an open covering of X. For each $y \in Y$, let $\{U_{\alpha_i} \mid i = 1 \ldots n(y)\}$ be a finite subcovering for the compact set $f^{-1}(y)$. Then $F(y) = \cup_{i=1}^{n(y)} U_{\alpha_i}$ is closed in X so $f(F(y))$ is closed in Y. Put $V(y) = Y - f(F(y))$. Then $V(y)$ is an open neighborhood of y and $f^{-1}(V(y)) \subset \cup_{i=1}^{n(y)} U_{\alpha_i}$. Next let $\{W(y)\}$ be a precise locally finite open refinement of $\{V(y) \mid y \in Y\}$. For each $i = 1 \ldots n(y)$, put $W(y,i) = f^{-1}(W(y)) \cap U_{\alpha_i}$. Then the collection $\{W(y,i) \mid y \in Y \text{ and } i = 1 \ldots n(y)\}$ is an open covering of X that refines $\{U_\alpha\}$. Moreover, $\{W(y,i)\}$ is locally finite since if $p \in X$, there exists a neighborhood V of $f(p)$ that meets at most finitely many of the $W(y)$ and then the neighborhood $f^{-1}(V)$ of p meets at most finitely many of the $W(y,i)$. ∎

John Mack sketched an outline of the proof of the following theorem for the author late one night at the 1991 Northeast Topology Conference after a lengthy discussion of paracompact subspaces of βX containing X. It plays a major role in the author's solution to Tamano's Paracompactification Problem.

THEOREM 6.16 (J. Mack, 1991) Let u be the finest uniformity for the space X. Then for each $p \in \beta X - uX$ there is a paracompact $Y \subset \beta X - \{p\}$ that contains uX.

Proof: By Theorem 6.13, $X \subset uX \subset \beta X$ so by Theorem 6.14, $\beta uX = \beta X$. Since uX is topologically complete and $\beta uX = \beta X$, by Theorem 6.10, if $p \in \beta X - uX$, there exists an $F \in C^*(uX \times \beta X)$ such that $F(x, x) = 0$ and $F(x, p) = 1$ for each $x \in uX$. Put $f = max\{F,1\}$ and for each pair $x,y \in uX$ put

$$\delta(x,y) = sup\{|f(x, q) - f(y, q)| \mid q \in \beta X\}.$$

Then $max\{f(x, y), f(y, x)\} \le \delta(x,y) \le 1$. It is easily shown that δ is a pseudometric on uX. For each pair $x,y \in uX$ define $x \sim y$ if and only if $\delta(x,y) = 0$ and set $[x] = \{y \in uX \mid \delta(x,y) = 0\}$. Let $M = \{[x] \mid x \in uX\}$ and put $d([x],[y]) = \delta(x,y)$. Then (M, d) is a metric space and the quotient mapping $\phi' : uX \to M$ defined by $\phi'(x) = [x]$ for each $x \in uX$ is continuous. Let (M^*, d^*) be the completion of (M, d).

Now put $\phi = j \circledcirc i \circledcirc \phi'$ where $i : M \to M^*$ is the imbedding of M into its completion and $j : M^* \to \beta M^*$ is the imbedding of M^* into its Stone-Čech compactification. Then $\phi: uX \to \beta M^*$. Let ϕ^β be the extension of ϕ over $\beta uX = \beta X$. Since βX and βM^* are both compact, ϕ^β is closed and the inverse image of each compact set in βM^* is compact in βX. Put $Y = (\phi^\beta)^{-1}(M^*)$ and let $\phi_Y^\beta = \phi^\beta | Y$. Then $\phi_Y^\beta: Y \to M^*$ and since Y is the *total* inverse image of M*, ϕ_Y^β is also closed and the inverse image of each compact set in M* is compact in Y. Therefore, ϕ_Y^β is a perfect mapping. By Lemma 6.15, Y is paracompact. Clearly, $uX \subset Y$.

By a proof similar to the proof of Lemma 6.12, we can show that $d^*: M^* \times M^* \to [0,1]$ has a continuous extension $d^\beta: M^* \times \beta M^* \to [0,1]$. Set $g = d^\beta \circledcirc (\phi_Y^\beta \times \phi^\beta)$. Then $g \in C^*(Y \times \beta X)$ and $g(Y \times \beta X) \subset [0,1]$. Let $(x,y) \in X \times X$. Then $g(x,y) = d^\beta(\phi_Y^\beta(x), \phi^\beta(y)) = d^\beta(x,y) = d^*(x,y) = \delta(x,y) \geq f(x,y)$. Therefore,

$$f|X \times X \leq g|X \times X \text{ which implies } f|X \times \beta X \leq g|X \times \beta X.$$

Hence $1 = f(x, p) \leq g(x, p)$ for each $x \in X$. Therefore, $1 \leq g(x, p)$ for each $y \in Y$. Since $g(x, x) = \delta(x, x) = 0$ for each $x \in X$, $g(y,y) = 0$ for each $y \in Y$. Consequently, p does not belong to Y. Hence Y is a paracompact subset of $\beta X - \{p\}$ containing uX. ■

THEOREM 6.17. If π is the finest uniformity for a space X such that the completion πX is paracompact, then $\pi = u$ (the universal uniformity for X).

Proof: Put $PX = \cap\{Y \subset \beta X | X \subset Y \text{ and } Y \text{ is paracompact}\}$. Clearly $PX \neq \emptyset$ and since βX is paracompact, π is finer than β. Therefore, $\pi X \subset \beta X$ which implies $PX \subset \pi X$. Now, for each paracompact space Y such that $X \subset Y \subset \beta X$, it is easily shown that Y is the completion of X with respect to some uniformity λ that is finer than β. Then by Theorem 6.15, $uX \subset PX$. Therefore, $PX = uX$.

There are two uniformities on X that bound π, namely, u (the finest) and $sup\{\lambda | \lambda X \subset \beta X \text{ and } \lambda X \text{ is paracompact}\}$, which we denote by $sup\lambda$. Clearly $sup\lambda \subset \pi \subset u$. Now πX is paracompact and $PX \subset \pi X$. We claim that the completion $sup\lambda X$ of X with respect to $sup\lambda$ is PX. Since $X \subset PX \subset \lambda X$ for each uniformity λ such that λX is a paracompact subset of βX, λ' (the uniformity of the completion λX) induces a uniformity λ^+ on PX and the uniformity λ on X. Consequently, $sup\lambda^+ = sup\{\lambda^+\}$ is a uniformity on PX that induces the uniformity $sup\lambda$ on X. Thus $(X, sup\lambda)$ is a dense uniform subspace of $(PX, sup\lambda^+)$. To see that $(PX, sup\lambda^+)$ is complete, let ψ be a Cauchy net in PX with respect to $sup\lambda^+$. Then ψ is Cauchy in PX with respect to each λ^+. But then ψ is Cauchy in λX which implies ψ converges to some $p_\lambda \in \lambda X \subset \beta X$ for each λ. Since βX is Hausdorff, $p_\lambda = p_\mu$ for each pair of uniformities λ, μ such that λX and μX are paracompact subsets of βX. Therefore, there exists a $p \in \beta X$ such that ψ converges to p and $p \in PX$ since $p = p_\lambda \in \lambda X$ for each λ. Hence $(PX, sup\lambda^+)$ is complete, and since completions are unique, $sup\lambda X = PX$.

Finally, we show that $\pi X = uX$. Since $PX \subset \pi X$, π induces a uniformity π^+ on PX that is finer than $sup\lambda^+$ since π is finer than $sup\lambda$. Then (X, π) is a dense uniform subspace of (PX, π^+). Now (PX, π^+) is complete since $(PX, sup\lambda^+)$ is complete and $\pi^+ \subset sup\lambda^+$, and since completions are unique, $\pi X = PX = uX$. But then uX is paracompact, so by hypothesis, $\pi = u$. ∎

Analogously to the definition of topologically complete, a space will be called **topologically preparacompact** if it is preparacompact with respect to its finest uniformity.

THEOREM 6.18 (N. Howes, 1992) A necessary and sufficient condition for the existence of the Tamano-Morita paracompactification is topological preparacompactness.

Proof: Suppose X is topologically preparacompact and let u be the universal uniformity on X. Then uX is paracompact which implies that π exists and equals u. Conversely, if π exists, then by Theorem 6.17, $\pi = u$ which implies that uX is paracompact which in turn implies that u' is cofinally complete. Therefore, u is preparacompact which implies X is topologically preparacompact. ∎

We now show that some of the results in Chapter 4 can be generalized by replacing cofinally Cauchy nets with almost Cauchy nets. Part (1) of the following theorem is due to J. Isbell and appeared in his paper *Supercomplete spaces* referenced in Chapter 5.

THEOREM 6.19 (J. Isbell, 1962, N. Howes, 1992) Let X be a Tychonoff space and let u be the finest uniformity on X. Then

(1) X is paracompact if and only if (X, u) is supercomplete,

(2) X is Lindelöf if and only if (X, e) is supercomplete and

(3) X is compact if and only if (X, β) is supercomplete.

Proof: (1) is essentially the equivalence of (1) and (2) in Theorem 5.8. To prove (2), we need only show that if (X, e) is supercomplete then X is Lindelöf since by Theorem 4.4 and Corollary 5.1, a Lindelöf space is supercomplete with respect to the e uniformity. By Lemma 4.1 it will suffice to show that each ω-directed (countably directed) net is almost Cauchy with respect to e. For this let $\psi: D \to X$ be an ω-directed net. Let $\mathcal{U} \in e$. Then there exists a countable normal covering $\{U_i\}$ that refines $\mathcal{U}$. For each i put $C_i = \{d \in D \mid \psi(d) \in U_i\}$ and let $E = \cup\{C_i \mid C_i \text{ is not cofinal in } D\}$. If $E = \emptyset$ pick any $d \in D$. Then $\{C_i\}$ is a collection of cofinal subsets of D such that each $d' \geq d$ is in some C_i and for each i, $\psi(C_i) \subset U_i \subset U$ for some $U \in \mathcal{U}$.

If $E \neq \emptyset$, then for each i such that C_i is not cofinal in D, there exists a $d_i \in D$ with $R(d_i) \cap C_i = \emptyset$ where $R(d_i) = \{d \in D \mid d_i \leq d\}$. Since ψ is ω-directed, there exists a $d' \in D$ with $d_i \leq d'$ for each i such that C_i is not cofinal in D.

Hence $R(d')\cap C_i = \emptyset$ for each i such that C_i is not cofinal in D. Therefore, $R(d')\cap E = \emptyset$. For each j put $D_j = C_j \cap R(d')$. Then $\{D_j\}$ is a collection of cofinal subsets of D such that each $d^+ \geq d'$ is in some D_j and for each j, $\psi(D_j) \subset U_j \subset U$ for some $U \in \mathcal{U}$. Since $\mathcal{U}$ was chosen arbitrarily, ψ is almost Cauchy.

To prove (3), we need only show that if (X, β) is supercomplete then X is compact since by Theorem 4.4 and Corollary 5.1, a compact space is supercomplete with respect to β. For this, it will suffice to show that each net in X is almost Cauchy with respect to β. Let ψ:D → X be a net in X. Let $\mathcal{U} \in \beta$. Then there exists a finite normal covering $\{U_1 \ldots U_n\}$ that refines $\mathcal{U}$. For each $i = 1 \ldots n$ put $C_i = \{d \in D \mid \psi(d) \in U_i\}$. Then an argument similar to the above can be used to show that there exists a $d \in D$ such that the collection $\{C_i \mid C_i$ is cofinal in D$\}$ is a collection of cofinal subsets of D such that each $d' \geq d$ is in some C_i and for each i, $\psi(C_i) \subset U_i \subset U$ for some $U \in \mathcal{U}$. Therefore, each net in X is almost Cauchy with respect to β. ∎

If instead of Theorem 4.4, we were to use Theorem 6.18 to motivate our definition of preparacompactness, then we would define preparacompactness to be the property that each almost Cauchy net has a Cauchy subnet. To distinguish between these two notions we define a uniform space to be **almost paracompact** if each almost Cauchy net has a Cauchy subnet. Then we can prove:

THEOREM 6.20 (N. Howes, 1992) Let (X, μ) be a uniform space and let ν be the uniformity on X derived from μ. Then

(1) (X, μ) has a paracompact completion if and only if (X,ν) is almost paracompact.

(2) (X, μ) has a Lindelöf completion if and only if (X,ν) is countably bounded and almost paracompact.

Proof: To prove (1) let (X′, μ′) be the completion of (X, μ) and let u' be the finest uniformity for X′. Assume (X,ν) is almost paracompact and that ψ:D → X′ is an almost Cauchy net with respect to u'. Let $E = D \times u'$ and define < on E by $(d,\mathcal{U}') < (e,\mathcal{V}')$ if $d < e$ and $\mathcal{V}' <^* \mathcal{U}'$. For each $(d, \mathcal{U}') \in E$ put $\theta(d, \mathcal{U}') = a$ for some $a \in X$ such that a and $\psi(d)$ both belong to some $U' \in \mathcal{U}'$. Then the correspondence $(d, \mathcal{U}') \to \theta(d, \mathcal{U}')$ defines a net θ:E → X.

Let $\mathcal{U}' \in u'$ and pick $\mathcal{V}' \in u'$ such that $\mathcal{V}' <^* \mathcal{U}'$. Since ψ is almost Cauchy, there exists a $d \in D$ and a collection $\{C_\alpha\}$ of cofinal subsets of D such that each $d' \geq d$ belongs to some C_α and for each index α, $\psi(C_\alpha) \subset V'_\alpha$ for some $V'_\alpha \in \mathcal{V}'$. For each index α put

$$E_\alpha = \{(e,\mathcal{W}') \mid e \in C_\alpha \text{ and } \mathcal{W}' <^* \mathcal{V}'\}.$$

Then E_α is cofinal in E for each index α. Let $(e,\mathcal{W}') \in E_\alpha$. Then $\theta(e,\mathcal{W}') = y \in X$ such that y and $\psi(e)$ both belong to some $W' \in \mathcal{W}'$. Since $(e,\mathcal{W}') \in E_\alpha$, $e \in$

C_α which puts $\psi(e)$ in V'_α. Consequently,

$$y \in Star(V'_\alpha, W') \subset Star(V'_\alpha, V') \subset U'_\alpha$$

for some $U'_\alpha \in \mathcal{U}'$. Therefore, $\theta(E_\alpha) \subset U'_\alpha$ for each index α. To show θ is almost Cauchy it remains to show there exists an $e \in E$ such that each $e' \geq e$ belongs to some E_α. For this choose $W' \in u'$ such that $W' <^* V'$ and put $e = (d, W')$. Now $d \in C_\beta$ for some index β which implies $e \in E_\beta$. If $e' > e$ then $e' = (d', Z')$ for some $d' > d$ and $Z' <^* W'$. Then $d' \in C_\gamma$ for some index γ and $Z' <^* V'$ so $e' \in E_\gamma$. Hence θ is almost Cauchy with respect to u'.

Since θ lies in X, θ is almost Cauchy with respect to ν and therefore has a Cauchy subnet ϕ. But then ϕ is Cauchy in X′ and hence converges to some $p \in$ X′. Therefore, θ clusters to p. It remains to show that ψ clusters to p. This can be done by applying an argument similar to the proof that ψ clusters to p in Theorem 6.13. Hence each almost Cauchy net in X′ clusters which implies that (X′, u') is supercomplete. By Theorem 6.19, X′ is paracompact so (X, μ) has a paracompact completion.

Conversely, assume X′ is paracompact. By Theorem 6.19, (X′, u') is supercomplete. Let ψ be an almost Cauchy net in X with respect to ν. Then ψ is almost Cauchy in (X′, u'). Since (X′, u') is supercomplete, ψ clusters to some $p \in$ X′. But then ψ has a subnet θ that converges to p. Then θ is Cauchy in (X′,u') and hence in (X,ν). Consequently, (X,ν) is almost paracompact.

To prove (2), assume (X,ν) is countably bounded and almost paracompact. To show that (X, μ) has a Lindelöf completion, by Theorem 4.11.(2), it will suffice to show that (X,ν) is preparacompact. By part (1), (X′, μ') is paracompact which implies (X,ν) is preparacompact by Theorem 4.11.(1). Hence (X, μ) has a Lindelöf completion. Conversely, if X′ is Lindelöf, by Theorem 4.10.(2), (X,ν) is countably bounded and preparacompact which implies (X,ν) is countably bounded and almost paracompact (by comparing the definitions of preparacompactness and almost paracompactness). ■

The following corollaries and propositions may be of some interest. They are stated without proof since they are easily proved using the techniques presented herein. A uniform space is said to be *topologically almost paracompact* if it is almost paracompact with respect to its finest uniformity.

COROLLARY 6.1 The completion of an almost paracompact uniform space is supercomplete.

COROLLARY 6.2 A countably bounded supercomplete uniform space is Lindelöf.

PROPOSITION 6.11. A Tychonoff space is paracompact if and only if it is complete and topologically almost paracompact.

PROPOSITION 6.12. A uniform space is countably bounded if and only if each ω-directed net is almost Cauchy

By a **supercompletion** of a uniform space (X, μ), we mean a supercomplete uniform space (Y, λ) such that X is uniformly homeomorphic to a dense subspace of Y. It is natural to inquire about the existence and uniqueness of a supercompletion for a given uniform space. In fact, A. Hohti, posed this question in a paper titled *Uniform hyperspaces* (Proc. 1983 Hungarian Topology Colloquium, pp. 333-334). The following theorem provides an answer.

THEOREM 6.21. (N. Howes, 1992) A necessary and sufficient condition for a uniform space to have a supercompletion is that it be almost paracompact. In this case, the supercompletion is unique and identical with the ordinary completion.

The proof is left as an exercise (Exercise 3).

EXERCISES

1. Prove Proposition 6.11.

2. Prove Proposition 6.12.

3. Prove Theorem 6.21. [Hint: Adapt the proof of Theorem 4.11 using Corollary 6.1]

4. Show that if $\mathcal{U}$ is a finite normal covering of a Tychonoff space X that there exists a normal sequence $\{\mathcal{U}_n\}$ of finite open coverings of X such that $\mathcal{U}_1 < \mathcal{U}$ and for each positive integer n and U ∈ $\mathcal{U}_n$, X - U is a zero set of X.

RELATIVELY FINE SPACES

In 1993, B. Burdick gave an example of a uniform space that is cofinally complete with respect to transfinite sequences but not cofinally complete (uniformly paracompact) thereby showing the existence of uniform properties that cannot be characterized by transfinite sequences in the same way they are characterized by nets. The class of relatively fine uniform spaces is a class of uniforms spaces in which the use of transfinite sequences is equivalent to the use of nets for studying several uniform properties like uniform paracompactness.

Let κ be an infinite cardinal. A uniformity μ is said to be **κ-bounded** if each uniform covering has a subcovering of cardinality less than κ. For a uniformizable space X and for each infinite cardinal κ there exists a finest κ-bounded uniformity. It is the supremum of all κ-bounded uniformities and is denoted by u_κ. The uniformities u_κ where κ is an infinite *regular* cardinal are called the **relatively fine** uniformities. Note that u, e and β are all relatively fine uniformities.

5. [Howes, 1994] For a relatively fine space (X, u_κ), the following are equivalent:

(1) (X, u_κ) is uniformly paracompact.
(2) (X, u_κ) is cofinally complete.
(3) (X, u_k) is cofinally complete with respect to transfinite sequences.
(4) X is paracompact and each transfinite sequence with no subsequence of cardinality less than κ clusters.
(5) X is paracompact and each κ-directed net clusters.
(6) Each directed open covering is uniform.

6. [Howes, 1994] Let (X, μ) be relatively fine where $\mu \neq \beta$. If there do not exist measurable cardinals then:

(1) The following are equivalent:
  (a) (X, μ) has a paracompact completion.
  (b) X is topologically preparacompact.
  (c) X is topologically almost paracompact.

(2) The following are equivalent:
  (a) (X, μ) has a Lindelöf completion.
  (b) X is topologically preparacompact and $u = e$.
  (c) X is topologically almost paracompact and $u = e$.

7. [Burdick, Howes, 1994] For a locally fine space (X, μ) the follwing are equivalent:

(1) (X, μ) is cofinally complete.
(2) (X, μ) is cofinally complete with respect to transfinite sequences.

RESEARCH PROBLEM

8. What other uniform properties can be characterized in locally fine spaces by transfinite sequences and are these characterizations analogous to their net characterizations?

6.6 The Spectrum of βX

We denote the weak completion $u(X)$ of X with respect to u, developed in Section 5.5, by uX since $u(X)$ is homeomorphic to uX. To see this, recall that

(X, u) is uniformly homeomorphic to a dense subspace of $u(X)$. Hence the uniformity of $u(X)$ is the finest uniformity for $u(X)$. Then by Theorem 5.5, $u(X)$ is complete. Since completions are unique, we have that $u(X)$ is uniformly homeomorphic with uX. In this section $\{\Phi_\lambda \mid \lambda \in \Lambda\}$ will denote the collection of all normal sequences of open coverings of X.

Let $\{\Psi_\gamma \mid \gamma \in \Gamma\}$ be the collection of all normal sequences of open coverings of βX such that Ψ_γ begins with a finite covering. If $\gamma \in \Gamma$ then $\Psi_\gamma = \{\mathcal{W}_n^\beta\}$ for some normal sequence of open coverings such that $\mathcal{W}_1^\beta$ is finite. Then for each n, $\mathcal{W}_n = \{W^\beta \cap X \mid W^\beta \in \mathcal{W}_n^\beta\}$ is an open covering of X and $\Phi_\gamma = \{\mathcal{W}_n\}$ is a normal sequence of open coverings of X such that $\mathcal{W}_1$ is finite. Hence we consider Γ to be a subset of Λ and as in Section 5.5, we have a canonical map $\theta_\gamma:\beta X \to \beta X/\Psi_\gamma$ where $\theta_\gamma = \theta_\gamma^* \circledcirc i_\gamma$ where i_γ is the identity map on βX considered as a uniformly continuous mapping with respect to the uniformity on $[\beta X]_\gamma$ (the pseudo-metric space induced on βX by Ψ_γ) and the uniformity of βX, and where $\theta_\gamma^*:[\beta X]_\gamma \to \beta X/\Psi_\gamma$ is the quotient mapping of $[\beta X]_\gamma$ onto $\beta X/\Psi_\gamma$ (as defined in Section 5.5). Also, as in Section 5.5, X_γ is the pseudo-metric space induced by Ψ_γ restricted to X, i.e., by $\Phi_\gamma = \Psi_\gamma \cap X$.

LEMMA 6.17 (1) For two points $x,y \in X$ we have $y \in \cap_{n=1}^\infty S(x,\mathcal{W}_n^\beta)$ if and only if $y \in \cap_{n=1}^\infty S(x,\mathcal{W}_n)$, (2) if H is open in $[\beta X]_\gamma$ then $H \cap X$ is open in X_γ and (3) if G is open in X_γ, then there exists an open set H in $[\beta X]_\gamma$ such that $G = H \cap X$.

Proof: The proofs of (1) and (2) are trivial. To see (3) let $x \in G$. Then there exists a positive integer $n(x)$ such that $S(x,\mathcal{W}_{n(x)}) \subset G$. Put $K = \cup\{S(x,\mathcal{W}_{n(x)}^\beta) \mid x \in G\}$. Clearly $K \cap X = G$. Let H be the interior of K in $[\beta X]_\gamma$. Clearly $G \subset H \subset K$ so $H \cap X = G$. This proves (3). ∎

For any continuous $f:X \to Y$ where X and Y are both Tychonoff spaces, there exists a continuous extension $f^\beta:\beta X \to \beta Y$. By Lemma 6.16, X/Φ_γ is a uniform subspace of $\beta X/\Psi_\gamma$. Let $\eta_\gamma:X/\Phi_\gamma \to \beta X/\Psi_\gamma$ denote the inclusion mapping. Then we have the following diagram:

$$\begin{array}{ccccc} X & \xrightarrow{\phi_\gamma} & X/\Phi_\gamma & \xrightarrow{\eta_\gamma} & \beta X/\Psi_\gamma \\ \downarrow & & \downarrow & & \| \\ \beta X & \xrightarrow[\phi_\gamma^\beta]{} & \beta(X/\Phi_\gamma) & \xrightarrow[\eta_\gamma^\beta]{} & \beta(\beta X/\Psi_\gamma) \end{array}$$

Clearly $\beta X/\Psi_\gamma$ is compact since it is the quotient of a compact space, so $\beta(\beta X/\Psi_\gamma) = \beta X/\Psi_\gamma$. Now θ_γ restricted to X is $\eta_\gamma \circledcirc \phi_\gamma$ so by the above diagram we see that $\theta_\gamma = \eta_\gamma^\beta \circledcirc \phi_\gamma^\beta$.

Let Ψ_λ, Ψ_μ be two normal sequences of open members of β' (the uniformity of βX) such that $\lambda < \mu$. Then, just as in Section 5.5, there exists a bonding map $(\phi_\mu^\lambda)^\beta: \beta X/\Psi_\mu \to \beta X/\Psi_\lambda$. Clearly $(\phi_\mu^\lambda)^\beta$ restricted to X/Φ_μ is just ϕ_μ^λ. Then, just as in Section 5.5, $\{\beta X/\Psi_\lambda, (\phi_\mu^\lambda)^\beta\}$ is an inverse limit system of uniform spaces, and the inverse limit of this system, which we denote by $\beta'(\beta X)$ is the weak completion of βX with respect to β'. Since βX is complete, by Theorem 5.4, $\beta'(\beta X) = \beta X$, so βX is the inverse limit of $\{\beta X/\Psi_\lambda, (\phi_\mu^\lambda)^\beta\}$.

If Ψ_λ is a normal sequence of open members of β' then there exists a Ψ_μ such that $\mu \in \Gamma$ with $\lambda < \mu$. To see this let $\Psi_\lambda = \{\mathcal{W}_n^\beta\}$ and let $\gamma \in \Gamma$. Let $\Psi_\gamma = \{\mathcal{V}_n^\beta\}$. Then $\mathcal{V}_1^\beta$ is finite and $\{\mathcal{W}_n^\beta \cap \mathcal{V}_n^\beta\}$ is a normal sequence. Put $\mathcal{U}_1^\beta = \mathcal{V}_1^\beta$ and for each $n > 1$ put $\mathcal{U}_n^\beta = \mathcal{W}_n^\beta \cap \mathcal{V}_n^\beta$. Then let μ be the element of Γ such that $\Psi_\mu = \{\mathcal{U}_n^\beta\}$. Clearly, $\lambda < \mu$. Therefore, $\{\Psi_\gamma \mid \gamma \in \Gamma\}$ is cofinal in the collection of all normal sequences of open members of β'. Consequently, by Exercise 5 of Section 5.5, βX is the inverse limit space of the inverse limit system $\{\beta X/\Psi_\gamma, (\phi_\mu^\lambda)^\beta\}$ where $\gamma \in \Gamma$.

Now since $\eta_\gamma: X/\Phi_\gamma \to \beta X/\Psi_\gamma$ is the inclusion mapping, η_γ^β must be one-to-one, so $\beta(X/\Phi_\gamma)$ can be identified with the subset $\eta_\gamma^\beta[\beta(X/\Phi_\gamma)]$ of $\beta X/\Psi_\gamma$. But then $\beta X/\Psi_\gamma$ is a compactification of X/Φ_γ containing a copy of $\beta(X/\Phi_\gamma)$ so $\beta X/\Psi_\gamma$ can be identified with $\beta(X/\Phi_\gamma)$. Moreover, for each pair $\lambda, \mu \in \Gamma$ with $\lambda < \mu$, $\phi_\mu^\lambda: X/\Phi_\mu \to X/\Phi_\lambda$ has a uniformly continuous extension $\beta(\phi_\mu^\lambda): \beta(X/\Phi_\mu) \to \beta(X/\Phi_\lambda)$. Since $\beta(X/\Phi_\mu)$ and $\beta(X/\Phi_\lambda)$ can be identified with $\beta X/\Psi_\mu$ and $\beta X/\Psi_\lambda$ respectively, we can consider $\beta(\phi_\mu^\lambda)$ as a mapping from $\beta X/\Psi_\mu$ into $\beta X/\Psi_\lambda$. Then $\beta(\phi_\mu^\lambda)$ restricted to X/Φ_μ is the same as $(\phi_\mu^\lambda)^\beta$ restricted to X/Φ_μ. Therefore, $\beta(\phi_\mu^\lambda) = (\phi_\mu^\lambda)^\beta$. Consequently, we have the following:

THEOREM 6.22 (K. Morita, 1970) *$\{\beta(X/\Phi_\lambda), (\phi_\mu^\lambda)^\beta\}$ is an inverse limit system whose inverse limit can be identified with βX.*

Proof: The proof of this theorem has essentially been proved in the preceding discussion. It remains to observe that since Γ is cofinal in Λ, the inverse system in the hypothesis of the theorem has the same inverse limit as the system where λ ranges over Γ, namely, βX (see Exercise 5, Section 5.5). ∎

Consequently, βX has the spectrum $\{\beta(X/\Phi_\lambda), (\phi_\mu^\lambda)^\beta\}$. Recall from Section 5.5 that $uX = u(X)$ has the spectrum $\{X/\Phi_\lambda, \phi_\mu^\lambda\}$. Also recall the uniform imbedding $\phi: X \to uX$ defined by $\phi(x) = \{\phi_\lambda(x)\} \in uX$. It's extension $\phi^\beta: \beta X \to \beta uX = \beta X$ can be seen to be the mapping defined by $\phi^\beta(x) = \{\phi_\lambda^\beta(x)\} \in \beta X$.

THEOREM 6.23 (K. Morita, 1970) For any uniformizable space X, $uX \subset \beta X$ and $uX = \cap\{(\phi_\lambda^\beta)^{-1}(X/\Phi_\lambda) \mid \lambda \in \Lambda\} = \cap\{(f^\beta)^{-1}(T) \mid f: X \to T$ is continuous with T metrizable$\}$.

Proof: That $uX \subset \beta X$ follows from either Theorem 6.13 or Exercise 1. To prove the first equality, suppose $x \in uX$. Then $x \in \Pi_\lambda X/\Phi_\lambda$ such that for each pair λ, $\mu \in \Lambda$ with $\lambda < \mu$, $\phi^\lambda_\mu(x_\mu) = x_\lambda$. Hence $x \in (\phi^\beta_\lambda)^{-1}(X/\Phi_\lambda)$ for each $\lambda \in \Lambda$. Therefore, $uX \subset \cap\{(\phi^\beta_\lambda)^{-1}(X/\Phi_\lambda) \mid \lambda \in \Lambda\}$. Conversely, suppose $x \in \cap\{(\phi^\beta_\lambda)^{-1}(X/\Phi_\lambda) \mid \lambda \in \Lambda\}$. Then $\phi^\beta_\lambda(x) \in X/\Phi_\lambda$ for each $\lambda \in \Lambda$. Therefore, $\phi^\beta(x) = \{\phi^\beta_\lambda(x)\} \in \Pi_\lambda X/\Phi_\lambda$. Suppose x does not belong to uX. Then there exists a pair λ, $\mu \in \Lambda$ with $\lambda < \mu$ such that $\phi^\lambda_\mu(x_\mu) \neq x_\lambda$ which implies $(\phi^\lambda_\mu)^\beta(x_\mu) \neq x_\lambda$ which in turn implies x does not belong to βX. But this is a contradiction since for each $\lambda \in \Lambda$, $x \in (\phi^\beta_\lambda)^{-1}(X/\Phi_\lambda) \subset \beta X$. Therefore, $\cap\{(\phi^\beta_\lambda)^{-1}(X/\Phi_\lambda) \mid \lambda \in \Lambda\} \subset uX$. Consequently, $uX = \cap\{(\phi^\beta_\lambda)^{-1}(X/\Phi_\lambda) \mid \lambda \in \Lambda\}$.

To prove the second equality, let f be a continuous mapping of X into a metric space T. For each positive integer n, let $\mathcal{U}_n = \{S(t, 2^{-n}) \mid t \in T\}$. Then the sequence $\{\mathcal{U}_n\}$ forms a basis for the metric uniformity of T. For each n put $\mathcal{V}_n = f^{-1}(\mathcal{U}_n)$. Then $\{\mathcal{V}_n\}$ is a normal sequence of open coverings of X, so $\{\mathcal{V}_n\} = \Phi_\lambda$ for some $\lambda \in \Lambda$. Then $\phi_\lambda: X \to X/\Phi_\lambda$. Let $x^* \in X/\Phi_\lambda$ and let $x_1, x_2 \in x^*$ so that $\phi_\lambda(x_1) = x^* = \phi_\lambda(x_2)$. Then for each positive integer n, $x_2 \in S(x_1, \mathcal{V}_n)$ which implies $f(x_2) \in S(f(x_1), \mathcal{U}_n)$. Therefore, $f(x_1) = f(x_2)$. Define $g: X/\Phi_\lambda \to$ T by $g(x^*) = f(x_1)$ where $x_1 \in x^*$. Then $f = g \circledcirc \phi_\lambda$ and g is continuous.

To see that g is continuous let $t \in T$ and let U be an open neighborhood of t in T. Then there exists a positive integer n such that $S(t, 2^{-n}) \subset U$ in T which implies $S(t, \mathcal{U}_n) \subset U$ which in turn implies $Star(f^{-1}(t), \mathcal{V}_n) \subset F^{-1}(U)$. Pick $x \in f^{-1}(t)$. Then $S(x, \mathcal{V}_n)$ is open in X_λ. Put $V = S(x, \mathcal{V}_n)$. Then $f(V) \subset U$, so $(g \circledcirc \phi^*_\lambda \circledcirc i_\lambda)(V) \subset U$. Since $i_\lambda(V) = V$, $f(V) = (g \circledcirc \phi^*_\lambda)(V)$. For any open $W \subset X_\lambda$ we have:

$$W = \{x \in X_\lambda \mid S(x, \mathcal{V}_n) \subset W \text{ for some positive integer } n\}.$$

From this it is easily seen that $(\phi^*_\lambda)^{-1}(\phi_\lambda(W)) = W$ so that ϕ^*_λ is also an open mapping. Hence $Z = \phi^*_\lambda(V)$ is open in X/Φ_λ and $g(Z) \subset (g \circledcirc \phi^*_\lambda)(V) = f(V) \subset U$. Moreover, $\phi_\lambda(x) = x^*$ so $x^* \in Z$ and $g(x^*) = f(x) = t$. Therefore, g is continuous.

Consequently, $(\phi^\beta_\lambda)^{-1}(X/\Phi_\lambda) \subset (\phi^\beta_\lambda)^{-1}[(g^\beta)^{-1}(T)] = (f^\beta)^{-1}(T)$. Therefore, $\cap\{(\phi^\beta_\lambda)^{-1}(X/\Phi_\lambda) \mid \lambda \in \Lambda\} \subset \cap\{(f^\beta)^{-1}(T) \mid f: X \to T$ with f continuous and T metrizable$\}$. Since the opposite inclusion is obvious, we have the second equality. ■

Let $f: X \to Y$ be a continuous mapping and let $g: Y \to T$ be continuous where T is a metrizable space. Then by Theorem 6.23, $uX \subset ((g \circledcirc f)^\beta)^{-1}(T) = (f^\beta)^{-1}[(g^\beta)^{-1}(T)]$ which implies $f^\beta(uX) \subset (g^\beta)^{-1}(T)$. Since this holds for all such g, again by Theorem 6.23, $f^\beta(uX) \subset uY$. Therefore, f^β maps uX into uY. Consequently, we can define an extension $f^u: uX \to uY$ of f by $f^u = f^\beta \mid uX$. Then if $g: Y \to Z$ is a continuous function, $(g \circledcirc f)^u = f^u \circledcirc g^u$ and $(1_X)^u = 1_{uX}$.

LEMMA 6.18 If $X \subset Y \subset uX$ then $uY = uX$.

Proof: If g:Y $\to$ T is continuous where T is a metric space, then $f = g|X$: $\to$ T is continuous. Then f^u:uX $\to$ uT, g^u:uY $\to$ uT and $(i_X)^u$:uX $\to$ uY, where i_X:X $\to$ Y is the inclusion (identity) mapping. Since $(i_X)^u = i_{uX}$, uX $\subset$ uY. Since $g^u|u$X:uX $\to$ uT, $f^u = g^u|u$X. But $f^u = f^\beta|u$X and $g^u = g^\beta|u$Y so $f^\beta|u$X $= (g^\beta|u$Y$)|u$X $= g^\beta|u$X which implies $f^\beta = g^\beta$.

Conversely, for any continuous f:X $\to$ T where T is a metric space, f^u:uX $\to$ uT. Since T is metric, T is weakly complete with respect to the metric uniformity which implies T is weakly complete with respect to u so uT = T by Theorem 5.4. Therefore, $g = f^u|$Y is a continuous mapping from Y into T such that $g|$X $= f$.

Then by Theorem 6.22, uY $= \cap\{(g^\beta)^{-1}$(T)$|\, g$: Y $\to$ T where g is continuous and T is metrizable$\}$ $\subset \cap\{(f^\beta)^{-1}(T)|\, f$:X $\to$ T where f is continuous and T is metrizable$\}$ = uX. Therefore, uY $\subset$ uX. Similarly, uX $\subset$ uY so uY = uX. ∎

THEOREM 6.24 (K. Morita, 1970) uX is characterized as a space Y with the following properties:

(1) Y is topologically complete and contains X as a dense subspace.

(2) Any continuous f:X → T where T is a metric space can be extended to a continuous mapping from Y into T.

Proof: uX is the inverse limit of the inverse system of metric spaces $\{X/\Phi_\lambda, \phi^\lambda_\mu\}$ and hence by Theorem 5.5 is topologically complete. Thus uX satisfies (1). To show uX satisfies (2), let f:X $\to$ T where T is a metric space. Then f^u:uX $\to$ uT. But since T is a metric space, T is weakly complete, so by Theorem 5.4, uT = T. Therefore, f^u is a continuous mapping from uX into T.

Conversely, let Y be a space satisfying (1) and (2). Let $g \in C^*$(X). Then g(X) is a bounded subset of **R** and hence metrizable, so g:X $\to$ g(X) is a mapping of X into a metric space and by (2) g can be extended to Y. But then X is C^*-imbedded in Y so by Theorem 6.14, βX = βY and X $\subset$ Y $\subset$ βX. Again by (2), for any continuous mapping f from X into a metric space T, there is a continuous extension g:Y $\to$ T. Since βX = βY we have $f^\beta = g^\beta$ where f^β and g^β are the unique continuous extensions from βX and βY into βT given by Theorem 6.2, and $g^\beta|$Y $= g$. Now g^β(Y) $\subset$ T so f^β(Y) $\subset$ T which in turn implies Y $\subset (f^\beta)^{-1}$(T). Therefore, by Theorem 6.23, Y $\subset$ uX. Then by Lemma 6.17, uY = uX. But since Y is topologically complete, Y = uX. ∎

EXERCISES

1. If $\{Y, F\}$ is an inverse limit system where $Y = \{Y_\alpha | \alpha \in A\}$ and $F = \{f^\beta_\alpha : Y_\alpha \to Y_\beta | \alpha,\beta \in A$ with $\beta < \alpha\}$ and if for each $\alpha \in A$, $X_\alpha \subset Y_\alpha$, then the inverse limit of $\{X,G\}$ is a subset of the inverse limit of $\{Y, F\}$ where $X = \{X_\alpha | \alpha \in A\}$ and $G = \{f^\beta_\alpha | X_\alpha | \alpha,\beta \in A$ with $\beta < \alpha\}$.

2. Show that uX is characterized as a topologically complete space Y which is the smallest with respect to properties (1) and (2) below:

(1) Y contains X as a dense subspace.
(2) Every real-valued bounded continuous function on X can be extended to a continuous function over Y.

3. Show that if X is the topological sum of $\{X_\lambda \mid \lambda \in \Lambda\}$, then uX is the topological sum of $\{uX_\lambda \mid \lambda \in \Lambda\}$.

6.7 The Tamano-Morita Paracompactification

As an example of the Tamano-Morita paracompactification of a space X, we consider the case where X is an M-space. This example was given by K. Morita in his paper *Topological completions and M-spaces* referenced in Section 5.5, as an example of a space whose topological completion is paracompact. M-spaces were introduced in 1964 by Morita in a paper titled *Products of Normal Spaces with Metric Spaces* (Math. Annelen, Volume 154, pp. 365-382). Paracompact M-spaces, unlike paracompact spaces, are countably productive. M-spaces are defined as follows: a space X is said to be an **M-space** if there exists a normal sequence $\{\mathcal{U}_n\}$ of open coverings of X satisfying the condition (M) below:

(M) If $\{K_n\}$ is a decreasing sequence of non-empty closed sets in X such that $K_n \subset S(x, \mathcal{U}_n)$ for each positive integer n, and for some $x \in X$, then $\cap K_n \neq \varnothing$.

Let X be an M-space and let $\{\Phi_\lambda \mid \lambda \in \Lambda'\}$ be the collection of all normal sequences of open coverings of X satisfying condition (M). It is left as an exercise to show that Λ' is cofinal in Λ. Hence by Exercise 1 of Section 6.6, uX is the inverse limit of the inverse limit subsystem $\{X/\Phi_\lambda, \phi^\lambda_\mu \mid \lambda \in \Lambda'\}$ of $\{X/\Phi_\lambda, \phi^\lambda_\mu \mid \lambda \in \Lambda\}$. Recall that a continuous function f:X → Y is said to be a perfect mapping if f is closed and for each $p \in Y$, $f^{-1}(p)$ is a compact subset of X. f is said to be **quasi-perfect** if f is closed and $f^{-1}(p)$ is countably compact for each $p \in Y$.

LEMMA 6.19 For each $\lambda \in \Lambda'$, ϕ_λ:X → X/Φ_λ is a quasi-perfect mapping.

Proof: Let $\lambda \in \Lambda'$ and let $\Phi_\lambda = \{\mathcal{U}_n\}$. Then ϕ_λ:X → X/Φ_λ is the mapping defined in Section 5.5 by $\phi_\lambda = \phi^*_\lambda$ © i_λ where i_λ:X → X_λ is the identity mapping on X viewed as a mapping from X onto X_λ and ϕ^*_λ is the quotient mapping from X_λ onto X/Φ_λ. X_λ is the pseudo-metric space (X, d_λ) where d_λ is the pseudo-metric generated by Φ_λ. From the proof of Theorem 6.23, ϕ^*_λ is an open mapping. To show that ϕ_λ is closed, let F be a closed subset of X and let $y^* \in Cl(\phi_\lambda(F))$. Let $x \in \phi_\lambda^{-1}(y^*)$. Since $Star(S(x, \mathcal{U}_{n+1}), \mathcal{U}_{n+1}) \subset S(x, \mathcal{U}_n)$, we have:

$$S(x, \mathcal{U}_{n+1}) \subset Int_{X_\lambda}(S(x, \mathcal{U}_n)) \subset S(x, \mathcal{U}_n).$$

Hence $V = \phi_\lambda^*(Int_{X_\lambda}(S(x, \mathcal{U}_n)))$ is open in X/Φ_λ containing y^* and $V \cap \phi_\lambda(F) \neq \varnothing$ which implies $(\phi_\lambda^*)^{-1}(V) \cap F \neq \varnothing$. But then $S(x, \mathcal{U}_n) \cap F \neq \varnothing$ for each positive integer n. Therefore, the family $\{S(x, \mathcal{U}_n) \cap F\}$ has the finite intersection property. For each positive integer n put $K_n = \cap_{i=1}^n Cl(S(x, \mathcal{U}_n) \cap F)$. Then $\{K_n\}$ is a decreasing sequence of non-empty closed sets in X such that $K_n \subset S(x, \mathcal{U}_n)$ for each positive integer n, so by (M), $\cap K_n \neq \varnothing$. Let $z \in \cap K_n$. Then $z \in F$ and $z \in S(x, \mathcal{U}_n)$ for each n. Consequently, $y^* = \phi_\lambda(x) = \phi_\lambda(z) \in \phi_\lambda(F)$. Hence ϕ_λ is a closed mapping.

Next let $\{U_n\}$ be a countable open covering of $\phi_\lambda^{-1}(w^*)$ for some $w^* \in X/\Phi_\lambda$. For each n put $K_n = \phi_\lambda^{-1}(w^*) - \cup_{i=1}^n U_i$. Suppose $\{U_n\}$ has no finite subcovering. Then $\{K_n\}$ is a decreasing sequence of non-void closed sets. Let $x \in \phi_\lambda^{-1}(w^*)$. Then $K_n \subset S(x, \mathcal{U}_n)$ for each positive integer n. Then by (M) we have $\cap K_n \neq \varnothing$ which implies $\{U_n\}$ does not cover $\phi_\lambda^{-1}(w^*)$ which is a contradiction. Hence $\phi_\lambda^{-1}(w^*)$ is countably compact, so ϕ_λ is quasi-perfect. ∎

LEMMA 6.20 If $f:X \to Y$ and $g:Y \to Z$ such that both f and $g \circledcirc f$ are quasi-perfect mappings, then g is a quasi-perfect mapping.

Proof: Let F be closed in Y. Then $f^{-1}(F)$ is closed in X so $(g \circledcirc f)(f^{-1}(F)) = g(f(f^{-1}(F))) = g(F)$ is closed in Z. Therefore, g is a closed mapping. Next let $z \in Z$. Then $(g \circledcirc f)^{-1}(z)$ is countably compact in X which implies $f((g \circledcirc f)^{-1}(z))$ is countably compact in Y. But $f((g \circledcirc f)^{-1}(z)) = f(f^{-1}(g^{-1}(z))) = g^{-1}(z)$ so g is a quasi-perfect mapping. ∎

As a consequence of Lemmas 6.18 and 6.19, if $\lambda, \mu \in \Lambda'$ with $\lambda < \mu$, we have $\phi_\lambda = \phi_\mu^\lambda \circledcirc \phi_\mu$; and since ϕ_λ and ϕ_μ are quasi-perfect, we must have that ϕ_μ^λ is quasi-perfect. Let $(\phi_\mu^\lambda)^\beta$ denote the Stone extension of ϕ_μ^λ. Then $(\phi_\mu^\lambda)^\beta(X/\Phi_\mu) = X/\Phi_\lambda$ and $(\phi_\mu^\lambda)^\beta(\beta(X/\Phi_\mu) - X/\Phi_\mu) = \beta(X/\Phi_\lambda) - X/\Phi_\lambda$. Hence $X/\Phi_\mu = ((\phi_\mu^\lambda)^\beta)^{-1}(X/\Phi_\lambda)$ for each $\lambda \in \Lambda'$. Then $(\phi_\mu^\beta)^{-1}(X/\Phi_\mu) = (\phi_\mu^\beta)^{-1}(((\phi_\mu^\lambda)^\beta)^{-1}(X/\Phi_\lambda)) = (((\phi_\mu^\lambda)^\beta \circledcirc \phi_\mu^\beta)^{-1}(X/\Phi_\lambda) = (\phi_\lambda^\beta)^{-1}(X/\Phi_\lambda)$. Thus $(\phi_\lambda^\beta)^{-1}(X/\Phi_\lambda) = (\phi_\mu^\beta)^{-1}(X/\Phi_\mu)$ for each $\lambda, \mu \in \Lambda'$ with $\lambda < \mu$. Combining this result with Theorem 6.23 it is easily shown that:

$$uX = (\phi_\lambda^\beta)^{-1}(X/\Phi_\lambda) \text{ for each } \lambda \in \Lambda'.$$

This fact allows us to establish the following theorem:

THEOREM 6.25 (K. Morita, 1970) If X is a Tychonoff M-space then $uX = (f^\beta)^{-1}(T)$ for any quasi-perfect mapping f from X onto a metric space T.

Proof: The proof consists in showing that a quasi-perfect mapping f from X onto a metric space T coincides with ϕ_λ:X $\to$ X/Φ_λ for some $\lambda \in \Lambda'$. From the proof of Theorem 6.23, we already know that $f = g \circledcirc \phi_\lambda$ for some $\lambda \in \Lambda$ and for some continuous g:X/$\Phi_\lambda \to$ T defined by $g(x^*) = f(z)$ where $z \in x^*$. Let $\{\mathcal{U}_n\}$ and $\{\mathcal{V}_n\}$ be as defined in the proof of Theorem 6.23. Clearly g is onto since for each $t \in$ T we can pick an $x \in f^{-1}(t)$ which implies $g(x^*) = f(x) = t$. To show g is one-to-one, let x^* and y^* be distinct elements of X/Φ_λ. Then there exists a positive integer n such that $S(x,\mathcal{V}_n) \cap S(y,\mathcal{V}_n) = \varnothing$ for some $x \in x^*$ and $y \in y^*$. Therefore, $S(f(x), \mathcal{U}_n) \cap S(f(y), \mathcal{U}_n) = \varnothing$. Since $g(x^*) = f(x)$ and $g(y^*) = f(y)$, $g(x^*) \neq g(y^*)$ so g is one-to-one.

Finally, to show g is open, let U be open in X/Φ_λ and let $x^* \in$ U. Then there exists a positive integer n such that $S(x^*, 2^{-n}) \subset$ U so $g(S(x^*, 2^{-n})) \subset g$(U) and contains $g(x^*)$. Therefore, if $g(S(x^*, 2^{-n}))$ is open in T then x^* is an interior point of g(U). Since x^* was chosen arbitrarily, g(U) is open. Let $x \in x^*$. Then

$$g(S(x^*, 2^{-n})) = \{t \in \mathrm{T} \mid y^* \in S(x^*, 2^{-n}), y \in y^*, \text{and } f(y) = t\} =$$

$$\{t \in \mathrm{T} \mid d_\lambda(x,y) < 2^{-n} \text{ and } f(y) = t\} = f(S(x,\mathcal{V}_n)) = S(f(x), \mathcal{U}_n).$$

Now X - $S(x,\mathcal{V}_n)$ is closed so $f(\mathrm{X} - S(x,\mathcal{V}_n)) = \mathrm{T} - S(f(x), \mathcal{U}_n)$ is closed in T since f is a closed mapping. Therefore, $S(f(x), \mathcal{U}_n)$ is open in T so g is an open mapping. Hence g is a homeomorphism so we can identify T with X/Φ_λ which implies f coincides with ϕ_λ.

It remains to show that $\lambda \in \Lambda'$. We do this by showing that $\{\mathcal{V}_n\}$ satisfies (M). Let $K = \{K_n\}$ be a decreasing collection of closed sets in X such that for some $p \in$ X, $K_n \subset S(p,\mathcal{V}_n)$ for each positive integer n. Put $q = f(p)$. Then $f(K_n) \subset S(q, \mathcal{U}_{n-1})$ for each n since $K_n \subset S(p,\mathcal{V}_n)$. Now the $f(K_n)$ are closed subsets of T and $\{f(K_n)\}$ is a decreasing sequence so $q \in f(K_n)$ for each positive integer n. Therefore, $f^{-1}(q) \cap K_n \neq \varnothing$ for each n. Hence $\{f^{-1}(q) \cap K_n\}$ is a decreasing sequence of closed sets in X. Since $f^{-1}(q)$ is countably compact, $\cap_n[f^{-1}(q) \cap K_n] \neq \varnothing$ so $\cap K_n \neq \varnothing$. Therefore, $\{\mathcal{V}_n\}$ satisfies (M). ∎

LEMMA 6.21 Let f:X $\to$ T where f is continuous and T is metric. If X is an M-space then f^u:uX $\to$ T and the following assertions hold:

(1) f is onto if and only if f^u is onto,
(2) f is closed if and only if f^u is closed and
(3) f is quasi-perfect if and only if f^u is perfect.

Proof: From the proof of Theorem 6.23, we know there exists a continuous function g:X/$\Phi_\lambda \to$ T such that $f = g \circledcirc \phi_\lambda$ for some $\lambda \in \Lambda$. The proof can be modified to show that λ can be chosen in such a way that $\lambda \in \Lambda'$. For this let $\{\mathcal{U}_n\}$ and $\{\mathcal{V}_n\}$ be as in the proof of Theorem 6.23. Then $\Phi_\lambda = \{\mathcal{V}_n\}$. Pick $\mu \in \Lambda'$ such that $\lambda < \mu$ and let $\Phi_\mu = \{\mathcal{W}_n\}$. Then ϕ_μ:X $\to$ X/Φ_μ and $\phi_\lambda = \phi_\mu^\lambda \circledcirc \phi_\mu$. Let $y^* \in$ X/Φ_μ and let $y_1, y_2 \in y^*$ so that $\phi_\mu(y_1) = y^* = \phi_\mu(y_2)$. Then for each positive integer n, $y_2 \in S(y_1, \mathcal{W}_n)$. Since $\lambda < \mu$, $y_2 \in S(y_1, \mathcal{V}_n)$ for each positive integer n which implies $f(y_2) \in S(f(y_1), \mathcal{U}_n)$ for each n. Therefore, $f(y_1) = f(y_2)$. Define $h\colon X/\Phi_\mu \to T$ by $h(y^*) = f(y_1)$ where $y_1 \in y^*$. Then $f = h \text{ © } \phi_\mu$ where $\mu \in \Lambda'$. A proof similar to the one in Theorem 6.22 can now be used to show that h is continuous. Since $\phi^u_\mu\colon uX \to u(X/\Phi_\mu) = X/\Phi_\mu$ we have the following diagram:

$$\begin{array}{ccc} X & \xrightarrow{\ f\ } & T \\ \downarrow & \searrow^{\phi_\mu} & \uparrow \\ uX & \xrightarrow[\ \phi^u_\mu\ ]{} & X/\Phi_\mu \end{array}$$

Clearly if f is onto then f^u is onto. If f^u is onto then h is onto since $f^u = g^u \text{ © } \phi^u_\mu$ and $g^u = g$ because $u(X/\Phi_\mu) = X/\Phi_\mu$. Then $f = h \text{ © } \phi_\mu$ is onto since ϕ_μ is onto. This proves (1).

To prove (2), first assume f is closed and let F be closed in uX. Put H = F$\cap$X. Then f(H) is closed in T which implies $h^{-1}(f(H))$ is closed in X/Φ_μ which in turn implies $u((\phi_\mu)^{-1}(f(H)))$ is closed in uX and contains F. Therefore, $f^u(F) \subset f(F)$. But clearly $f(F) \subset f^u(F)$. Therefore, $f^u(F) = f(F)$ which implies f^u is closed. Conversely, if f^u is closed then h must be closed since $f^u = h \text{ © } \phi^u_\mu$. But then $f = h \text{ © } \phi_\mu$ is closed since ϕ_μ is quasi-perfect by Lemma 6.18. This proves (2).

To prove (3), assume f is quasi-perfect. Then by (2) above, f^u is closed. If $p \in u$T then $p \in \beta$T which implies $(f^\beta)^{-1}(p)$ is closed and hence compact in βX. Since X is an M-space, by Theorem 6.25, $uX = (f^\beta)^{-1}(T)$ so $(f^u)^{-1}(p) = (f^\beta)^{-1}(p)$. Hence $(f^u)^{-1}(p)$ is compact so f^u is perfect. Conversely, assume f^u is perfect. By (2) above, f is closed. Hence, as shown in the proof of Theorem 6.25, h is a homeomorphism of X/Φ_μ onto $f(X) \subset T$. Identifying X/Φ_μ with f(X), we see that f agrees with ϕ_μ on X so by Lemma 6.18, f is quasi-perfect. This proves (3). ∎

THEOREM 6.26 If X is a Tychonoff M-space, then X is topologically preparacompact and the Tamano-Morita paracompactification of X is $(f^\beta)^{-1}(T)$ for any quasi-perfect mapping from X onto a metric space T.

Proof: Let $\lambda \in \Lambda'$. Then $\phi_\lambda\colon X \to X/\Phi_\lambda$ is a quasi-perfect mapping by Lemma 6.17. By Lemma 6.20, $\phi^u_\lambda\colon uX \to X/\Phi_\lambda$ is perfect. By Lemma 6.15, uX is paracompact so X is topologically preparacompact by Corollary 4.2. Then uX is the Tamano-Morita paracompactification of X and by Theorem 6.25, $uX = (f^\beta)^{-1}(T)$ for any quasi-perfect mapping from X onto a metric space T. ∎

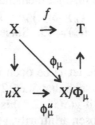

EXERCISES

1. Let $\{\Phi_\lambda \mid \lambda \in \Lambda'\}$ be the collection of all normal sequences of open coverings of the M-space X that satisfy condition (M). Show that Λ' is cofinal in Λ.

2. A space X is said to be **pseudo-compact** if every real-valued continuous function on X is bounded. Show that:

   (a) the continuous image of a pseudo-compact space is pseudo-compact.
   (b) for normal spaces, pseudo-compactness is equivalent to countable compactness.

3. Show that for a Tychonoff space X, uX is compact if and only if X is pseudo-compact. For this reason, a space X such that uX is paracompact is sometimes called **pseudo-paracompact.**

4. Let f be a quasi-perfect mapping from an M-space X onto an M-space Y. Show that $f^u: u\mathrm{X} \to u\mathrm{Y}$ is a perfect mapping.

5. Let f:X → Y be a perfect mapping. Show that if Y is Lindelöf (compact), then X is Lindelöf (compact).

6. Let f:X → Y be a perfect mapping. Show that if Y is countably compact then so is X.

Chapter 7

REALCOMPACTIFICATIONS

7.1 Introduction

From Stone's characterization of βX (Theorem 6.2), it can be seen that if f is a real valued bounded continuous function on X, then f can be uniquely extended over βX. Consequently, any continuous function f:X → [0,1] has a unique continuous extension over βX. In fact, βX can be characterized by this property. To see this we need only establish that if each continuous function f:X → [0,1] has a unique continuous extension over βX, then any continuous function f:X → Y where Y is a compact Hausdorff space has a continuous extension over βX.

For this let g:X → Y be a continuous function from X into a compact Hausdorff space Y. Let F be the collection of all continuous functions from Y into [0,1] and for each $f \in F$ put $I_f = [0,1]$. Let $P = \Pi_{f \in F} I_f$ and define the function ϕ:Y → P by $\phi(y) = \Pi_{f \in F} f(y) \in P$. Then ϕ is a continuous function so ϕ © g:X → P is a continuous function. Now for each $f \in F$, f © g:X → [0,1] can be uniquely extended to a continuous function h_f:βX → [0,1]. Define h:βX → P by

$$h(p) = (\ldots, h_f(p), \ldots) = \Pi_{f \in F} h_f(p).$$

Then h is also continuous. Let $\{x_\alpha\}$ be a net in X that converges to p. Then $\{h(x_\alpha)\}$ is a net in P that converges to $h(p)$. Also, $\{g(x_\alpha)\}$ is a net in Y that converges to some $y_p \in$ Y. Hence $\{\phi$ © $g(x_\alpha)\}$ is a net in P that converges to $\phi(y_p)$. Since $x_\alpha \in$ X for each α, and since h_f is an extension of f © g, f © $g(x_\alpha) = h_f(x_\alpha)$ for each α, so $\{h(x_\alpha)\} = \{\phi$ © $g(x_\alpha)\}$. Consequently, $\{\phi$ © $g(x_\alpha)\}$ converges to $h(p)$ which means $\phi(y_p) = h(p)$. Define γ:βX → Y by $\gamma(p) = y_p$ for each $p \in$ βX.

To show that γ is a continuous extension of g, suppose $x \in$ X. Then clearly $y_x = g(x)$ so $\gamma(x) = g(x)$. Therefore, γ is an extension of g over βX. To show that γ is continuous, first notice that ϕ:Y → ϕ(Y) ⊂ P is a homeomorphism. This is because the family F separates points and can distinguish points from closed sets [see Lemma 1.6]. Then $\gamma(p) = y_p = \phi^{-1}(h(p))$ for each $p \in$ βX so $\gamma = \phi^{-1}$ © h. Therefore, γ is continuous. Finally, to show that γ is unique, suppose δ:βX → Y is another continuous extension of g over βX. By Corollary 2.3, both γ and δ are uniformly continuous with respect to β' (the uniformity of βX). By an argument similar to the proof of Theorem 6.2, $g{:}X \to Y$ is also uniformly continuous with respect to β. Since βX is the completion of (X, β), by Theorem 4.9, g has a unique uniformly continuous extension over βX. Therefore, $\gamma = \delta$. We record this result as:

THEOREM 7.1 βX is characterized among compactifications of X by the property that each continuous function $f{:}X \to [0,1]$ has a unique continuous extension $f^{\beta}{:}\beta X \to [0,1]$.

Since βX is the only compactification of X for which each real valued bounded continuous function can be continuously extended, it is natural to inquire about the nature of spaces Y such that X is dense in Y and every real valued (not necessarily bounded) continuous function on X has a continuous extension over Y. This investigation leads to the concept of a *realcompact* space.

In a 1948 paper titled *Rings of real valued continuous functions I* (Transactions of the American Mathematical Society, Volume 64, pp. 45-99), E. Hewitt introduced the concept of a *Q-space*. Q-spaces are defined as follows: let **Z**(X) denote the zero sets of real valued continuous functions on the Tychonoff space X [cf. Section 6.2]. A non-empty subfamily Z of **Z**(X) is said to be **CZ-maximal** if Z contains no countable subset with empty intersection and Z is maximal with respect to this property. A **Q-space** is a Tychonoff space such that the intersection of any CZ-maximal family is non-void. This definition of a Q-space is due to Shirota. It is equivalent to Hewitt's original definition [see Definition 13 and Theorem 50, p. 85 of Hewitt's paper]. Shirota, in his 1952 paper *A Class of Topological Spaces* (referenced in Section 3.2), showed the equivalence between Q-spaces and Tychonoff spaces that are complete with respect to their e uniformity which he called **e-complete spaces.**

7.2 Realcompact Spaces

Today, Q-spaces and e-complete spaces are known as **realcompact** spaces, a term introduced by L. Gillman and M. Jerison in their book *Rings of Continuous Functions* published by Van Nostrand in 1960. Hewitt showed that realcompact spaces are precisely the closed subspaces of Cartesian products of real lines. He also showed that for each Tychonoff space X, there exists a realcompact space υX, called the **Hewitt realcompactification** of X such that $X \subset \upsilon X \subset \beta X$ and υX has the properties: (1) υX is the largest subspace of βX containing X such that each real valued continuous function can be extended to υX and (2) υX is the smallest realcompact space between X and βX (i.e., X is realcompact if and only if $X = \upsilon X$). Further, Hewitt showed that υX is completely determined (up to homeomorphism) by the following properties: (1) υX is realcompact, (2) υX contains X as a dense subspace, and (3) every real valued continuous function on X can be continuously extended to υX. Hewitt proved parts (2) and (3) of the next theorem. The proof given below is due to Shirota.

THEOREM 7.2 (E. Hewitt, 1948 and T. Shirota, 1952) For a Tychonoff space X, the following conditions are equivalent:

(1) X is e-complete,

(2) X is realcompact,

(3) X is homeomorphic to a closed subset of a Cartesian product of real lines with the usual topology.

Proof: (1) $\rightarrow$ (2). Let (X, e) be complete and let $\mathcal{Z}$ be a CZ-maximal family of X. Let $\mathcal{U} = \{U_n\}$ be a countable normal covering of X. We first show that there exists a $Z \in \mathcal{Z}$ and a $U \in \mathcal{U}$ such that $Z \subset U$. As shown in the proof of Lemma 3.5, we can construct a precise closed refinement $\{F_n\}$ of $\{U_n\}$, and then use Corollary 2.2 and Proposition 1.4 to construct real valued continuous functions f_n:X $\rightarrow$ [0,1] such that $f_n(F_n) = 1$ and $f_n(X - U_n) = 0$. For each n, the set $Z_n = \{x \in X \mid f_n(x) \geq 1/2\}$ can be shown to be a zero set and $F_n \subset Z_n \subset U_n$. If for each positive integer n, Z_n does not belong to $\mathcal{Z}$ then for each n there exists a countable collection $\{Z^n_m\} \subset \mathcal{Z}$ such that $Z_n \cap [\cap^{\infty}_{m,n=1} Z^n_m] = \varnothing$. Then $\cap^{\infty}_{m,n=1} Z^n_m = [\cap^{\infty}_{m,n=1} Z^n_m] \cap X = [\cap^{\infty}_{m,n=1} Z^n_m] \cap [\cup^{\infty}_{n=1} Z_n] = \cup^{\infty}_{n=1}(Z_n \cap [\cap^{\infty}_{m,n=1} Z^n_m]) = \varnothing$. But this contradicts the assumption that $\mathcal{Z}$ is a CZ-maximal family. Hence for some positive integer n, $Z_n \in \mathcal{Z}$. Then Z_n is the desired $Z \subset U \in \mathcal{U}$.

If $Z_1, Z_2 \in \mathcal{Z}$ then $Z_1 \cap Z_2 \neq \varnothing$ and if Z_1 is the zero-set of f and Z_2 is the zero-set of g, then $Z_1 \cap Z_2$ is the zero-set of fg. Clearly $Z_1 \cap Z_2 \in \mathcal{Z}$. Since $\mathcal{Z} \cup \{Z_1 \cap Z_2\}$ contains $\mathcal{Z}$ and satisfies properties (1) through (3) of the definition of a CZ-maximal family. Let $E = \{(Z, x) \mid Z \in \mathcal{Z}$ and $x \in Z\}$ and define $\leq$ on E by $(Z_1, x) \leq (Z_2, y)$ if $Z_2 \subset Z_1$. Then (E, $\leq$) is a directed set. For each $(Z, x) \in E$ put $\psi(Z, x) = x$. Then ψ:E $\rightarrow$ X is a net. If $\mathcal{U} \in e$, there is a $Z \in \mathcal{Z}$ and a $U \in \mathcal{U}$ with $Z \subset U$. Then $\psi(Z, x) \in U$ for each $x \in Z$ and if $(Z', y) \in E$ with $(Z, x) \leq (Z', y)$, then $Z' \subset Z$ so $\psi(Z', y) \in U$. Therefore, ψ is Cauchy with respect to e and hence converges to some $p \in X$. Let $Z \in \mathcal{Z}$ and let U(p) be an open set containing p. Since ψ is eventually in U(p), there is a $(Z', x) \in E$ with $Z' \subset Z$ and $\psi(Z', x) = x \in U(p)$. Then $U(p) \cap Z \neq \varnothing$ so $p \in Cl_X(Z) = Z$. Hence $p \in \cap\{Z \mid Z \in \mathcal{Z}\}$. Therefore, X satisfies condition (2). This completes the proof of (1) $\rightarrow$ (2).

To show that (2) $\rightarrow$ (3), let C denote the family of all real valued continuous functions on X and for each $f \in C$ let $R_f = \mathbf{R}$. Set $P = \Pi_{f \in C} R_f$ and let h:X $\rightarrow$ P be the product mapping defined in Section 1.3 by $[h(x)]_f = f(x)$. Since X is Tychonoff, the family C can distinguish points and can distinguish points from closed sets. Consequently, by Lemma 1.6, h is an imbedding, so h(X) is homeomorphic with X. We may therefore assume $X \subset P$. Let π be the product uniformity on P and let (X, π') denote the uniform subspace of (P, π). It remains to show that X is closed in P. By Exercise 3 of Section 4.4, it will suffice to show that (X, π') is complete.

For this let $\{x_\alpha\}$ be a Cauchy net in (X, π'). For each α put $H_\alpha = Cl\{x_\beta \mid \alpha < \beta\}$. It is easily shown that for each $\mathcal{U} \in \pi$, there is an α such that $H_\alpha \subset U$ for some $U \in \mathcal{U}$. Moreover, since there exists a continuous function $f: X \to [0,1]$ with $f(H_\alpha) = 0$ and $f(X - U) = 1$, we have for the zero-set $\mathbf{Z}(f)$ that $H_\alpha \subset \mathbf{Z}(f) \subset U$. Let $\mathcal{Z}'$ be the family of zero-sets of X that contain some H_α. Then if $Z_1, Z_2 \in \mathcal{Z}'$, $Z_1 \cap Z_2 \in \mathcal{Z}'$. By Zorn's Lemma (Theorem 0.1(4)), it is possible to find a maximal subfamily $\mathcal{Z}$ of $\mathbf{Z}(X)$ with respect to the finite intersection property that contains $\mathcal{Z}'$. We now show that $\mathcal{Z}$ is also CZ-maximal.

Suppose, on the contrary, there is a countable subfamily $\{Z_i\}$ of $\mathcal{Z}$ with $\cap_{i=1}^\infty Z_i = \varnothing$. Then for each positive integer i, there is an $f_i \in C$ with $Z_i = \mathbf{Z}(f_i)$. In fact, we can assume that for each i, $f_i: X \to [0,1]$. Now for each positive integer n, define $g_n: X \to [0,1]$ by $g_n(x) = max\{\,|f_i(x)|\,|\, i = 1 \ldots n\}$ for each $x \in X$. Put $g = \Sigma_{n=1}^\infty 2^{-n} g_n$. Then since $\cap_{i=1}^\infty Z_i = \varnothing$, we have $g(x) > 0$ for each $x \in X$. Consequently, g^{-1} defined by $g^{-1}(x) = 1/g(x)$ is continuous. Therefore, $g^{-1} \in C$. Since π is the product uniformity on P, π is the coarsest uniformity on P that makes all the canonical projections uniformly continuous. But for each $f \in C$, $p_f = f$, so g^{-1} is uniformly continuous. Since for each $\mathcal{U} \in \pi$, there is a $Z \in \mathcal{Z}$ with $Z \subset U$ for some $U \in \mathcal{U}$, and since g^{-1} is uniformly continuous, then for each $\varepsilon > 0$, there is a $Z \in \mathcal{Z}$ such that $g^{-1}(Z)$ is contained in a set of diameter ε.

Now if $x \in \mathbf{Z}(g_n)$, then $g(x) < 2^{-n+1}$. Hence there exists a sufficiently large n with $g^{-1}(x) > max\{g^{-1}(y) \mid y \in Z\}$ for any $x \in \mathbf{Z}(g_n)$. This implies that $\mathbf{Z}(g_n) \cap Z = \varnothing$. But $\mathbf{Z}(g_n) = \cap_{i=1}^n \mathbf{Z}(f_i)$ which implies that $\mathbf{Z}(g_n) \in \mathcal{Z}$. Then $\mathcal{Z}$ does not satisfy the finite intersection property which is a contradiction. Therefore, $\mathcal{Z}$ is a CZ-maximal family. Consequently, by property (2), $\mathcal{Z}$ has a non-empty intersection. Let $p \in \cap\{Z \mid Z \in \mathcal{Z}\}$. Let $\mathcal{U} \in \pi'$ and choose $\mathcal{V} \in \pi'$ such that $\mathcal{V}^* < \mathcal{U}$. Suppose no $Z \in \mathcal{Z}$ meets $S(p,\mathcal{V})$. Since there exists a $Z(\mathcal{V}) \in \mathcal{Z}$ and a $V \in \mathcal{V}$ with $Z(\mathcal{V}) \subset V$, we see that p does not belong to $Z(\mathcal{V})$ which is a contradiction. Therefore, $Z(\mathcal{V}) \cap S(p,\mathcal{V}) \neq \varnothing$ which implies there is an open neighborhood V' of p such that $V' \cap V \neq \varnothing$ so there is a $U \in \mathcal{U}$ with $V' \cup V \subset U$. But then $Z(\mathcal{V}) \subset S(p, \mathcal{U})$ and $Z(\mathcal{V})$ contains H_α for some index α. Hence, $\{x_\alpha\}$ converges to p so (X, π') is complete. This completes the proof of (2) $\to$ (3).

To show (3) $\to$ (1), let X be a closed subset of a product $P = \Pi_\alpha R_\alpha$ where $R_\alpha = \mathbf{R}$ for each index α. Since R_α is complete for each α, so is P. Let π denote the product uniformity on P and let π' denote π restricted to X. Since X is closed in P, (X, π') is complete. It will suffice to show that e is a basis for π'. For this let $\mathcal{U}$ be a basic member of π. Then

$$\mathcal{U} = p_{\alpha_1}^{-1}(\mathcal{U}_{\alpha_1}) \cap \ldots \cap p_{\alpha_n}^{-1}(\mathcal{U}_{\alpha_n})$$

for some finite collection of uniform coverings $\mathcal{U}_{\alpha_i}$ of R_{α_i} where p_{α_i} denotes the canonical projection of P onto R_{α_i} for each $i = 1 \ldots n$. Now each $\mathcal{U}_{\alpha_i}$ has a countable subcovering $\mathcal{V}_{\alpha_i}$ since R_{α_i} is Lindelöf. Furthermore, each $\mathcal{V}_{\alpha_i}$ is normal since R_{α_i} is paracompact. Therefore, $p_{\alpha_1}^{-1}(\mathcal{V}_{\alpha_1}) \cap \ldots \cap p_{\alpha_n}^{-1}(\mathcal{V}_{\alpha_n})$ is a countable normal covering of P that refines $\mathcal{U}$. Hence $\mathcal{U}$ is a member of the e uniformity for P. But then e is a basis for the π' uniformity on X. Therefore, $e = \pi'$, so X is e-complete. ∎

Now that we know realcompact spaces are precisely the e-complete spaces, we can use this fact to derive some interesting properties about realcompact spaces.

THEOREM 7.3 Every Lindelöf space is realcompact.

This is because Lindelöf spaces are cofinally complete with respect to the e uniformity by Theorem 4.4 and hence complete with respect to e by Corollary 4.1.

THEOREM 7.4 A closed subspace of a realcompact space is realcompact.

This is because if Y is a closed subspace of a realcompact space X, then Y is complete with respect to the e uniformity of X relativized to Y by Proposition 4.18. But the e uniformity of X relativized to Y is precisely the e uniformity of Y.

THEOREM 7.5 A product of realcompact spaces is realcompact.

Proof: Let $X = \Pi_\alpha X_\alpha$ where each X_α is realcompact and for each α, let e_α be the e uniformity of X_α. Then for each α, (X_α, e_α) is complete so X is complete with respect to the product uniformity by Exercise 4 of Section 5.3. Let μ denote the product uniformity of the e_α's. Then a basis element of μ is of the form

$$\mathcal{U} = p_{\alpha_1}^{-1}(\mathcal{U}_{\alpha_1}) \cap \ldots \cap p_{\alpha_n}^{-1}(\mathcal{U}_{\alpha_n})$$

for some finite collection of uniform coverings $\mathcal{U}_{\alpha_i}$ of e_{α_i} where each $\mathcal{U}_{\alpha_i}$ is countable and where p_{α_i} denotes the canonical projection of X onto X_{α_i} for each $i = 1 \ldots n$. Then $\mathcal{U}$ is also countable and therefore belongs to e (the e uniformity of X). Hence $\mu \subset e$. Since (X, μ) is complete and e is finer than μ, we have that (X, e) is complete so X is realcompact. ∎

We have already seen in Sections 2.2 and 5.5 how a normal sequence of open coverings can be used to construct a pseudo-metric. If we consider the family $\Phi = \{\Phi_\lambda \mid \lambda \in \Lambda\}$ of all normal sequences of open uniform coverings of a uniform space (X, μ) we obtain by this construction process, a family $D = \{d_\lambda \mid \lambda \in \Lambda\}$ of pseudo-metrics that satisfies the following properties:

(1) if $d, e \in D$ then $d \vee e \in D$,
(2) if d is a pseudo-metric, and if for each $\varepsilon > 0$ there exists a $d' \in D$

and a $\delta > 0$ such that $d'(x,y) < \delta$ implies $d(x,y) < \varepsilon$ for all $x,y \in X$, then $d \in D$,

(3) whenever $x \neq y$, there exists a $d \in D$ with $d(x,y) \neq 0$.

where $d \vee e$ is defined by $d \vee e(x,y) = min\{d(x,y), e(x,y)\}$ and is called the **join** of d and e. Property (3) only holds for Hausdorff uniform spaces and property (2) implies that if $d \in D$, then $rd \in D$ for each $r > 0$; and if $d \in D$ and $e \leq d$ then $e \in D$.

To show property (1) holds let $d_\lambda, d_\theta \in D$ and let Φ_λ and Φ_θ be the normal sequences from which they were constructed. Let $\Phi_\lambda = \{\mathcal{U}_n^\lambda\}$ and $\Phi_\theta = \{\mathcal{U}_n^\theta\}$. Then $\Phi_\nu = \Phi_\lambda \cap \Phi_\theta = \{\mathcal{U}_n^\lambda \cap \mathcal{U}_n^\theta\}$ is a normal sequence of open uniform coverings that generates a pseudo-metric d_ν. Moreover, $d_\nu \leq d_\lambda$ and $d_\nu \leq d_\theta$ as can be seen by the method of construction of these pseudo-metrics [see proof of Theorem 2.5]. But then $d_\nu \leq d_\lambda \vee d_\theta$. Let Φ_ξ be the normal sequence of open coverings $\{\mathcal{V}_n\}$ where $\mathcal{V}_n = \{S(x, 2^{-n}) \mid x \in X\}$ where $S(x, 2^{-n})$ is the sphere about x of radius 2^{-n} with respect to the pseudo-metric $d_\lambda \vee d_\theta$. For each n, $\mathcal{U}_n^\lambda \cap \mathcal{U}_n^\theta < \mathcal{V}_n$ since $d_\nu \leq d_\lambda \vee d_\theta$ and since $S(x, \mathcal{U}_n^\lambda \cap \mathcal{U}_n^\theta) = S_\nu(x, 2^{-n})$ and $S(x,\mathcal{V}_n) = S(x, 2^{-n})$ where $S_\nu(x, 2^{-n})$ is the sphere about x of radius 2^{-n} with respect to the pseudo-metric d_ν.

Hence $\Phi_\xi \in \Phi$. But from the construction of d_ξ, it can be seen that $d_\xi = d_\lambda \vee d_\theta$ since the definition of d_ξ depends on the definition of the function $\delta(x,y)$ which is defined to be 2^{-n} where n is the largest index such that $y \in S(x,\mathcal{V}_n)$ or to be 0 if there is no largest index. But since $S(x,\mathcal{V}_n) = S(x, 2^{-n})$, and $S(x, 2^{-n})$ is the sphere about x of radius 2^{-n} with respect to $d_\lambda \vee d_\theta$, we have $S_\nu(x, 2^{-n}) = S(x, 2^{-n})$for each positive integer n. [A complete proof that Φ_ξ generates $d_\lambda \vee d_\theta$ involves showing that if two pseudo-metrics d and d' have the same spheres of radius 2^{-n} for each positive integer n, then $d = d'$ which we leave as an exercise.]

To show that property (2) holds, let ρ be a pseudo-metric and suppose there exists a $d_\lambda \in D$ such that for each $\varepsilon > 0$ there exists a $\delta > 0$ with $d_\lambda(x,y) \leq \delta$ implies $\rho(x,y) \leq \varepsilon$. As we saw from the above discussion, d is generated by the normal sequence of uniform coverings $\{\mathcal{U}_n\}$ where $\mathcal{U}_n = \{S_\lambda(x, 2^{-n}) \mid x \in X\}$ and ρ is generated by the normal sequence of open coverings $\{\mathcal{V}_n\}$ where $\mathcal{V}_n = \{S(x, 2^{-n}) \mid x \in X\}$ where $S(x, 2^{-n})$ is the sphere about x of radius 2^{-n} with respect to the pseudo-metric ρ. Then for each positive integer n, there exists a positive integer m such that $\mathcal{U}_m < \mathcal{V}_n$. Hence $\{\mathcal{V}_n\} \in \Phi$ which implies $\rho \in D$.

To show that property (3) holds when (X, μ) is a Hausdorff uniform space, notice that if $x \neq y$ there exists a $\mathcal{U} \in \mu$ with $S(x, \mathcal{U}) \cap S(y, \mathcal{U}) = \varnothing$. Let $\Phi_\lambda = \{\mathcal{U}_n^\lambda\}$ be a normal sequence of open coverings such that $\mathcal{U}_1^\lambda < \mathcal{U}$. Let d_λ be the pseudo-metric constructed from Φ_λ. Then $d_\lambda \in D$ and $d_\lambda(x,y) \neq 0$.

From the above discussion it should come as no surprise that uniformities can be studied as families of pseudo-metrics that satisfy the above properties (1)

- (3). In such a case, we call a family D of pseudo-metrics satisfying these properties a **pseudo-metric uniformity**. It is left as an exercise (Exercise 2) to show that if D is a pseudo-metric uniformity, then the family of coverings $\mu = \{\mathcal{U}_\varepsilon^d \mid d \in D \text{ and } \varepsilon > 0\}$ where $\mathcal{U}_\varepsilon^d = \{S_d(x, \varepsilon) \mid x \in X\}$, is a basis for a covering uniformity on X.

It is easily shown (Exercise 3) that the intersection of any collection of pseudo-metric uniformities is again a pseudo-metric uniformity. Since **0** belongs to every pseudo-metric uniformity, the intersection is never empty. Consequently, if S is a non-empty family of pseudo-metrics on X, there exists a smallest pseudo-metric uniformity D containing S. We say S is a **subbase** for D and that D is **generated** by S. If B is a subbase for D such that for each $d \in D$ and $\varepsilon > 0$, there exists a $d' \in B$ and a $\delta > 0$ with $d'(x,y) \leq \delta$ implies $d(x,y) \leq \varepsilon$ for each pair $x,y \in$ X, then B is called a **base** for D. It is left as an exercise (Exercise 4) to show that if S is a subbase for D, then the family B of all finite joins $d_1 \vee \ldots \vee d_n$ such that $d_i \in S$ for each $i = 1 \ldots n$ is a base for D.

The family C(X) of all real valued continuous functions on X and the family C*(X) of all real valued bounded continuous functions on X can be used to generate two pseudo-metric uniformities C and C^* as follows: for each $f \in$ C(X) let d_f be defined as

$$d_f(x,y) = |f(x) - f(y)|$$

for each pair $x,y \in$ X. It is an easy exercise (Exercise 5) to show that d_f is a pseudo-metric on X. Let $C = \{d_f \mid f \in \mathrm{C(X)}\}$ and $C^* = \{d_f \mid f \in C^*(\mathrm{X})\}$. Let c and c^* denote the covering uniformities associated with C and C^* respectively. Then c and c^* have bases of the form

$$b = \{\mathcal{U}_\varepsilon^f \mid d_f \in C \text{ and } \varepsilon > 0\}$$

$$b^* = \{\mathcal{U}_\varepsilon^f \mid d_f \in C^* \text{ and } \varepsilon > 0\}$$

respectively, where $\mathcal{U}_\varepsilon^f = \{S_f(x, \varepsilon) \mid x \in X\}$ and where $S_f(x, \varepsilon)$ is the sphere about x of radius ε with respect to the pseudo-metric d_f.

We now show that $c \subset e$ and $c^* \subset \beta$. For this, let $\mathcal{U} \in c^*$. Then $\mathcal{U} = \mathcal{U}_\varepsilon^f$ for some $f \in$ C*(X) and $\varepsilon > 0$. Since $f \in$ C*(X) there is some *inf* a and *sup* b such that $f(\mathrm{X}) \subset [a,b]$. f may or may not assume the end points a and b. But since f is continuous, it assumes each point p such that $a < p < b$. Let $\delta = \varepsilon/2$ and pick $p_1 \in$ X such that $f(p_1) = a$ if $f(x) = a$ for some $x \in$ X or else pick $p_1 \in$ X such that $f(p_1) = a + \varepsilon$. If $a + \varepsilon > b$ pick any $x \in (a,b)$ and put $p_1 = x$. For each positive integer k pick $p_{k+1} \in$ X such that $f(p_{k+1}) = f(p_k) + \varepsilon$ if $f(p_k) + \varepsilon \leq b$. Otherwise put $p_{k+1} = b$ unless f does not assume the value b in which case put $p_{k+1} = p_k$. Let n be the least positive integer such that $f(p_{n+1}) = f(p_n)$.

Clearly $\{S_f(p_i, \varepsilon) \mid i = 1 \ldots n\}$ is a finite subcovering of $\mathcal{U}^f_\varepsilon$. It is also clear that $\mathcal{U}^f_\delta < \mathcal{U}^f_\varepsilon$ and $\mathcal{U}^f_\delta \in c$. Therefore, every $\mathcal{U}^f_\varepsilon$ where $f \in C^*(X)$ has a finite normal refinement and consequently belongs to β. This shows that $c^* \subset \beta$. This argument can be modified to show that $c \subset e$ (see Exercise 6). Consequently, if X is complete with respect to c, then X is realcompact and if X is complete with respect to c^* then X is compact. In what follows we will show that the completion with respect to c is the Hewitt realcompactification, and the completion with respect to c^* is the Stone-Čech compactification, but for now we merely record the following:

PROPOSITION 7.1 If a Tychonoff space is complete with respect to the c uniformity, then it is realcompact. If it is complete with respect to c^*, it is compact.

How to translate the concept of uniform continuity into the terminology of pseudo-metric uniformities follows from the next proposition, whose proof we leave as an exercise (Exercise 7).

PROPOSITION 7.2 Let (X, M) and (Y, N) be pseudo-metric uniform spaces and let (X, μ) and (Y,ν) be their associated covering uniform spaces. Then $f{:}X \to Y$ is uniformly continuous with respect to μ and ν if and only if for each $e \in N$ and $\varepsilon > 0$, there exists a $d \subset M$ and a $\delta > 0$ such that if $d(x,y) < \delta$ then $e(f(x), f(y)) < \varepsilon$.

Proposition 7.2 has the following useful implication: $f{:}X \to \mathbf{R}$ is uniformly continuous with respect to some pseudo-metric uniformity D if and only if $d_f \in D$. Consequently, f is uniformly continuous with respect to c or c^* if and only if $d_f \in C$ or C^* respectively. For C^* this means that if $f \in C^*(X)$ then $d_f \in C^*$ so f is uniformly continuous with respect to c^* and hence can be continuously extended to the completion of X with respect to c^*. By Proposition 7.1, c^*X is compact. But then by Theorem 7.1, c^*X is the Stone-Čech compactification of X. Since βX has a unique uniformity, we have

THEOREM 7.6 $c^* = \beta$.

EXERCISES

1. Let d and d' be two pseudo-metrics on X having the same spheres of radius 2^{-n} for each positive integer n. Show that $d = d'$.

PSEUDO-METRIC UNIFORMITIES

2. Let D be a pseudo-metric uniformity on X and let $\nu = \{\mathcal{U}^d_\varepsilon \mid d \in D \text{ and } \varepsilon > 0\}$ where $\mathcal{U}^d_\varepsilon = \{S_d(x, \varepsilon) \mid x \in X\}$. Show that ν is the basis for a covering

uniformity that generates the same topology on X as the pseudo-metric uniformity.

3. Show that the intersection of any collection of pseudo-metric uniformities is again a pseudo-metric uniformity.

4. Let S be a subbase for a pseudo-metric uniformity D and let B be the family of all finite joins $d_1 \vee \ldots \vee d_n$ such that $d_i \in S$ for each $i = 1 \ldots n$. Show that B is a base for D.

5. Let $f \in C(X)$ and define d_f by $d_f(x,y) = |f(x) - f(y)|$ for each pair $x,y \in X$. Show that f is a pseudo-metric on X.

6. Show that the e uniformity is finer than the c uniformity.

7.3 Realcompactifications

By a **realcompactification** of a Tychonoff space X, we mean a realcompact space Y in which X can be imbedded as a dense subspace. From Theorem 7.2 it is easily seen that a realcompactification Y of X can be realized as the completion of X with respect to some countably bounded uniformity. To see this, note that if Y is realcompact, it is e-complete, and the e uniformity of Y relativized to X is countably bounded. The completion of X with respect to this uniformity is Y. Furthermore, e is the finest countably bounded uniformity on X.

For this, first notice that the e uniformity is countably bounded since it has a basis consisting of countable normal coverings. Next, if μ is a countably bounded uniformity of X and $\mathcal{U}$ is an open member of μ, then there is a normal sequence $\{\mathcal{U}_n\}$ of open members of μ such that $\mathcal{U}_1 < \mathcal{U}$. Let $\mathcal{V}$ be a countable subcovering of $\mathcal{U}$ and for each positive integer n put $\mathcal{V}_n = \mathcal{U}_n \cap \mathcal{V}$. Then $\{\mathcal{V}_n\}$ is a normal sequence of open coverings of X such that $\mathcal{V}_1 < \mathcal{V} \subset \mathcal{U}$. Therefore, $\mathcal{U} \in e$ so e is finer than μ.

Also, this shows that the countable normal coverings $\mathcal{V}$ such that $\mathcal{V} < \mathcal{U}$ for some $\mathcal{U} \in \mu$ form a basis for μ, so μ has a basis consisting of (perhaps not all) countable normal coverings. Therefore, the countably bounded uniformities are precisely the ones that have bases consisting of countable normal coverings (i.e., they are the separable uniformities).

Shirota showed that if X is a Tychonoff space, that eX is precisely the Hewitt realcompactification υX. He did this by proving the following:

THEOREM 7.7 (T. Shirota, 1951) Let X be a Tychonoff space. Then eX has the following properties:

(1) X is dense in eX,

(2) eX is realcompact,

(3) each $f \in C(X)$ can be continuously extended to eX.

Also, any space satisfying these three properties is homeomorphic with eX.

Clearly eX satisfies property (1) of Theorem 7.7. Shirota states in his proof that it is obvious that eX satisfies property (2) also. This would be the case if we knew, for instance, that e' (the e uniformity of eX) is the e uniformity of eX. Although it is not surprising that this should be so, the proof is not what we usually think of as being obvious. Consequently, we first prove this as a lemma before proving Theorem 7.7.

LEMMA 7.1 Let e' be the uniformity of eX. Then e' is the e uniformity of eX.

Proof: Each $\mathcal{U}' \in e'$ has a countable normal refinement. To see this, let $\mathcal{V}' = \{V'_\beta\}$ be a closed uniform refinement of $\mathcal{U}'$. Then $\mathcal{V} = \{V'_\beta \cap X\} = \{V_\beta\} \in e$ and therefore has a countable uniform refinement say $\{V_i\}$. For each V_i there is a $V_{\beta_i} \in \mathcal{V}$ with $V_i \subset V_{\beta_i}$ so

$$Cl_{eX}(V_i) \subset Cl_{eX}(V_{\beta_i}) \subset Cl_{eX}(V'_{\beta_i}) = V'_{\beta_i}$$

since $\mathcal{V}'$ is a closed covering. Thus $\{Cl_{eX}(V_i)\}$ refines $\mathcal{V}'$ so $\mathcal{W}' = \{Int_{eX}(Cl_{eX}(V_i))\}$ refines $\mathcal{U}'$. Now $\mathcal{W}' \in e'$. To see this, note that $\{V_i\} = \{V'_i \cap X\}$ for some uniform covering $\{V'_i\}$ of eX. Pick $y \in V'_i$. Then

$$y \in Cl_{eX}(V'_i) = Cl_{eX}(V'_i) \cap Cl_{eX}(X) = Cl_{eX}(V'_i \cap X) = Cl_{eX}(V_i).$$

Therefore, $V'_i \subset Cl_{eX}(V_i)$ for each i so $\{Cl_{eX}(V_i)\}$ is a uniform covering of eX. But then $\{Cl_{eX}(V_i)\}$ has an open refinement in e' which implies $\mathcal{W}' \in e'$. Hence $\mathcal{U}'$ has a countable normal refinement.

It remains to show that all countable normal coverings of eX belong to e'. For this let $\{W'_i\}$ be a countable normal covering of eX. Then there exists a normal sequence $\{\mathcal{U}'_n\}$ of open coverings such that $Cl_{eX}(\mathcal{U}'_1) < \{W'_i\}$. But then $\{\mathcal{U}'_n \cap X\}$ is a normal sequence of open coverings of X such that $Cl_{eX}(\mathcal{U}'_1) \cap X$ refines $\{W'_i \cap X\}$. Hence, $Cl_{eX}(\mathcal{U}'_1) \cap X \in e$. But for each $\mathcal{V} \in e$, $Cl_{eX}(\mathcal{V}) = \{Cl_{eX}(V) \mid V \in \mathcal{V}\} \in e'$. Therefore

$$Cl_{eX}[Cl_{eX}(\mathcal{U}'_1) \cap X] = Cl_{eX}(\mathcal{U}'_1) \cap eX = Cl_{eX}(\mathcal{U}'_1) \in e'.$$

Consequently, $\{W'_i\} \in e'$. Therefore, e' is the e uniformity of eX. ∎

Proof of Theorem 7.7: It remains to show that eX satisfies property (3) and that if Y is another Tychonoff space satisfying properties (1) - (3) then Y and eX are homeomorphic. That eX satisfies property (3) is easily seen from the remarks preceding Proposition 7.1 and those following Proposition 7.2. If $f \in C(X)$ then $d_f \in C$ so that f is uniformly continuous with respect to c. Since $c \subset e$, f is uniformly continuous with respect to e and consequently can be uniquely extended to a uniformly continuous function $f^{\upsilon}: eX \to \mathbf{R}$.

To show that if Y is another Tychonoff space satisfying (1) - (3) then Y and eX are homeomorphic is a little more difficult. For this we will first show that any Tychonoff space Y satisfying properties (1) - (3) also satisfies the following property:

(3′) If $Z_n \in \mathbf{Z}(X)$ for each positive integer n, then $\cap Cl_Y(Z_n) = Cl_Y(\cap Z_n)$.

For this suppose $\cap Cl_Y(Z_n) \neq Cl_Y(\cap Z_n)$. Then there exists a $y \in \cap Cl_Y(Z_n) - Cl_Y(\cap Z_n)$. For each positive integer n, let $f_n \in C(X)$ such that $Z_n = \mathbf{Z}(f_n)$ and such that $|f_n| \le 1$. Since y does not belong to $Cl_Y(\cap Z_n)$, there exists a zero set $Z \subset X$ such that $y \in Cl_Y(Z)$ and $Z \cap (\cap Z_n) = \varnothing$. Let $f \in C(X)$ such that $Z = \mathbf{Z}(f)$ and put $g = \Sigma g_n$ where $g_n = 2^{-n}(|f_n| + |f|)$. Then g is strictly positive, i.e., $g(x) > 0$ for each $x \in X$.

Let g' be the extension of g over Y and for each positive integer n let g_n' be the extension of g_n over Y. Now $g'|X = g = \Sigma g_n = \Sigma g_n'|X$. By property (1), X is dense in Y so $g' = \Sigma g_n'$. Also, for each positive integer n, $g_n' \le |f_n'| + |f'|$ where f_n' is the extension of f_n to Y and f' is the extension of f to Y. Consequently, $g_n'(y) = 0$ for each positive integer n, so $g(y) = 0$. Hence there exists an $h \in C(X)$ such that $h(x)g(x) = 1$ for each $x \in X$. Let h' be the extension of h over Y. Since $(hg)'|X = hg = (h'|X)(g'|X)$ and since X is dense in Y, we have $(hg)' = h'g' = 1$. Hence $g'(y) = g(y) \neq 0$ which is a contradiction. Therefore, $\cap Cl_Y(Z_n) = Cl_Y(\cap Z_n)$.

We now use property (3′) to show that Y and eX are homeomorphic. For this we show that (Y, e) is the completion of (X, e). We need only show that every countable normal covering $\mathcal{U}$ of X can be extended to a countable normal covering $\mathcal{U}'$ of Y such that $\mathcal{U} = \mathcal{U}' \cap X$. This will show that (X, e) is a uniform subspace of the complete uniform space (Y, e), and since completions are unique, Y is homeomorphic with eX. Now there exists a normal sequence $\{\mathcal{U}_n\}$ of countable open coverings of X such that $\mathcal{U}_1 < \mathcal{U}$. $\mathcal{U}^{\varepsilon(Y)}$ is an extension of $\mathcal{U}$ into Y and for each positive integer n, $\mathcal{U}_n^{\varepsilon(Y)}$ is an extension of $\mathcal{U}_n$ into Y. By property (3′), for each positive integer n,

$$\cap\{Y - U^{\varepsilon(Y)} \mid U \in \mathcal{U}_n\} = \cap\{Cl_Y(X - U) \mid U \in \mathcal{U}_n\} = Cl_Y(\cap\{X - U \mid U \in \mathcal{U}_n\}) = \varnothing.$$

Consequently, $\mathcal{U}_n^{\varepsilon(Y)}$ is a covering of Y for each positive integer n, so $\mathcal{U}^{\varepsilon(Y)}$ is a covering of Y. Furthermore, since $\mathcal{U}_{n+1} <^* \mathcal{U}_n$ for each positive integer n, we have that $\mathcal{U}_{n+1}^{\varepsilon(Y)} <^* \mathcal{U}_n^{\varepsilon(Y)}$ for each n. Hence $\mathcal{U}^{\varepsilon(Y)}$ is a countable normal covering of Y, and clearly $\mathcal{U} = \mathcal{U}^{\varepsilon(Y)} \cap X$. ∎

COROLLARY 7.1 Let X be a Tychonoff space. Then eX has the following properties:

(1) X is dense in eX,

(2) eX is realcompact,

(3) if $Z_n \in \mathbf{Z}(X)$ for each positive integer n then $\cap Cl_{eX}(Z_n) = Cl_{eX}(\cap Z_n)$.

Also, any space satisfying these three properties is homeomorphic with eX.

COROLLARY 7.2 Let X be a Tychonoff space. Then $eX = \upsilon X$.

That $eX = \upsilon X$, the Hewitt realcompactification of X, follows from the fact that in Hewitt's development of υX, he characterized υX by the three properties of Theorem 7.7. It should be noted that Hewitt's development of υX was very different from ours. In fact, it was done in terms of the maximal ideals in the ring C(X) of real valued continuous functions on X, in a manner similar to Stone's construction of βX.

We now prove a theorem that is the analogue of Theorem 6.11 for realcompact spaces. This theorem is very useful for proving things about realcompact spaces and realcompactifications.

THEOREM 7.8 Let X be a dense subspace of the Tychonoff space Y. Then the following statements are equivalent:

(1) Every continuous $f: X \to A$, where A is a realcompact Tychonoff space, has a unique continuous extension to Y.

(2) X is C-imbedded in Y.

(3) If $Z_n \in \mathbf{Z}(X)$ for each positive integer n and $\cap Z_n = \varnothing$ then $\cap(Cl_Y(Z_n)) = \varnothing$.

(4) If $Z_n \in \mathbf{Z}(X)$ for each positive integer n then $Cl_Y(\cap Z_n) = \cap(Cl_Y(Z_n))$.

(5) $eX = eY$.

(6) $X \subset Y \subset eX$.

Proof: (1) → (2) By Theorem 7.3, **R** is realcompact, so (2) is just a special case of (1). Hence (1) implies (2).

(2) → (4) This is the same as the proof that (3) implies (3′) in Theorem 7.7.

(4) → (5) We can show $eX = eY$ by showing that (X, e) is a uniform subspace of (Y,e) and appealing to the uniqueness of the completion. But this is similar to the proof that (X, β) is a uniform subspace of (Y, β) in (4) → (5) of Theorem 6.10.

(5) → (6) is obvious.

(6) → (1) This proof is similar to the proof in Section 7.1 that if every continuous f:X → [0,1] has a continuous extension over βX, then each function g:X → K where K is a compact Hausdorff space, has a continuous extension over βX. In this case we have Y ⊂ eX so each continuous f:X → **R** has a continuous extension f^{υ}:eX → **R**, so $f^Y = f^{\upsilon}|Y$ is a continuous extension of f over Y. If we let F denote the collection of all real valued continuous functions on X and for each $f \in F$ we put R_f = **R**, then we can define $P = \Pi_{f \in F} R_f$ and ϕ:Y → P by $\phi(y) = \Pi_{f \in F} f(y) \in P$. The proof in 7.1 then goes through with minor modifications, to show that a function g:X → A, where A is a realcompact Hausdorff space, has a unique continuous extension to Y.

(4) → (3) is obvious.

(3) → (4) Suppose $Cl_Y(\cap Z_n) \neq \cap(Cl_Y(Z_n))$. Then there exists a point $p \in \cap(Cl_Y(Z_n)) - Cl_Y(\cap Z_n)$. Let Z be a zero set neighborhood of p in Y such that $Z \cap Cl_Y(\cap Z_n) = \emptyset$. Then $Cl_Y(\cap(Z \cap Z_n)) = \emptyset$, so by (3), $\cap Cl_Y(Z \cap Z_n) = \emptyset$ which implies $Z \cap Cl_Y(\cap Z_n) = \emptyset$ which is a contradiction since $p \in Z \cap Cl_Y(\cap Z_n)$. ∎

COROLLARY 7.3 eX is the largest subspace of βX in which X is C-imbedded.

Proof: By Theorem 6.13, eX ⊂ βX. Let Y ⊂ βX such that X is C-imbedded in Y. Then by (6) of Theorem 7.8, Y ⊂ eX. ∎

COROLLARY 7.4 eX is the smallest realcompact space between X and βX. In particular, X is realcompact if and only if X = eX.

Proof: Clearly X is realcompact if and only if X = eX since realcompactness is e-completeness. To show e is the smallest realcompact space between X and βX, let X ⊂ Y ⊂ βX and suppose Y is realcompact. By Theorem 6.11, βY = βX and by Corollary 7.3, eY is the largest subspace of βY = βX in which Y is C-imbedded. Since Y is realcompact, Y = eY. Hence Y is the largest subspace of βX in which Y is C-imbedded. But X ⊂ Y implies every real valued continuous function on Y can be extended to Y∪eX since if $f \in C(Y)$ then $f|X$ can be extended to eX and since X is dense in Y∪eX, f can be extended to Y∪eX. But then Y is C-imbedded in Y∪eX which implies Y∪eX ⊂ Y which in turn implies eX ⊂ Y. ∎

COROLLARY 7.5 Every continuous function f:X → Y, where Y is realcompact, has a unique continuous extension f^{υ}:υX → Y. Moreover, if f^{β} is the Stone extension of f:X → Y ⊂ βY, then $f^{\upsilon} = f^{\beta}|\upsilon X$.

Proof: Since υX = eX, by Theorem 7.8.(1) and (5), f has a unique continuous extension f^{υ}:υX → Y. By Theorem 6.13, X ⊂ υX ⊂ βX. Since X is dense in both υX and βX, and both f^{υ} and f^{β} are determined by their values on X, and since both f^{υ} and f^{β} are unique extensions of f into βY, it is clear that $f^{\upsilon} = f^{\beta} | \upsilon X$. ■

PROPOSITION 7.3 An arbitrary intersection of realcompact subspaces X_α of a given space X is realcompact.

Proof: Put $Y = \cap X_\alpha$. For each α, the identity mapping $i{:}Y \to X$ has (by Theorem 7.8) a continuous extension $i_\alpha{:}eY \to X_\alpha \subset X$. Since i can have only one continuous extension from eY into X, all the i_α must coincide. Hence this common extension maps eY into $\cap X_\alpha = Y$. But then $eY = Y$ so Y is realcompact. ■

PROPOSITION 7.4 If Y is C-imbedded in X, then $Cl_{eX}(Y) = eY$.

Proof: Since Y is C-imbedded in X and X is C-imbedded in eX, by the transitivity of C-imbedding, Y is C-imbedded in eX. Then $Cl_{eX}(Y)$ is realcompact by Theorem 7.4, and Y is C-imbedded in $Cl_{eX}(Y)$. By Theorem 7.8, $Cl_{eX}(Y) \subset eY$. By Corollary 7.4, $eY \subset Cl_{eX}(Y)$. Therefore, $Cl_{eX}(Y) = eY$. ■

LEMMA 7.2 If f is a continuous function from a space Y into a space Z, whose restriction to a dense subset X is a homeomorphism, then $f(Y - X) = Z - f(X)$.

Proof: Suppose $f(x) = f(y)$ where $x \in X$ and $y \neq x$. Let U be an open neighborhood of X in Y such that $Cl_Y(U) \subset \{p\} = \emptyset$. The homeomorphism $f|X$ maps $U \cap X$ onto an open neighborhood of $f(x)$ in $f(X) \subset Z$. Therefore, $f(U \cap X) = V \cap f(X)$ for some open neighborhood V of $f(x)$ in Z. Since X is dense in Y, every neighborhood of y contains a point of X - U. The homeomorphism $f|X$ maps such a point into Z - V. Therefore, no open neighborhood of y is mapped by f into V. Hence f is not continuous at y which is a contradiction. Consequently, each point $p \in Y - X$ cannot be mapped onto $f(x)$ for any $x \in X$. Therefore, $f(Y - X) = Z - f(X)$. ■

PROPOSITION 7.5 Let $f{:}X \to Y$ be a continuous function from the realcompact space X into the space Y. If K is a realcompact subset of Y then $f^{-1}(K)$ is realcompact in X.

Proof: Let $H = f^{-1}(K)$ and let i be the identity mapping on H. By Theorem 7.8.(1) and (5), $i{:}H \to H \subset X$ has a unique extension $i^{\upsilon}{:}\upsilon H \to X$. Also, $f_H = f|H$ has a unique extension $f_H^{\upsilon}{:}\upsilon H \to K$. Since H is dense in υH and both of these unique extensions are determined by their values on H, we have $f_H = f \circledcirc i$ which implies $f_H^{\upsilon} = (f \circledcirc i)^{\upsilon} = f \circledcirc i^{\upsilon}$. Now by Lemma 7.2, $i^{\upsilon}(\upsilon H - H) = X - H$ so that $(f \circledcirc i)^{\upsilon}(\upsilon H - H) \subset Y - K$. But $f_H^{\upsilon}(\upsilon H - H) \subset K$. Since $f_H^{\upsilon} = (f \circledcirc i)^{\upsilon}$ this implies $\upsilon H - H = \emptyset$ so by Corollary 7.4, $H = f^{-1}(K)$ is realcompact in X. ■

A space is said to have a given property **hereditarily** if every subspace also has the property. For instance, a space is **hereditarily realcompact** if every subspace is realcompact.

COROLLARY 7.6 A realcompact space X, every point of which is a G_δ, is hereditarily realcompact.

Proof: For each $p \in$ X, there exists an $f \in$ C(X) such that **Z**(f) = $\{p\}$ since $\{p\}$ is a G_δ in X. Therefore, X - $\{p\}$ = X - **Z**(f) = f^{-1}(**R** - $\{0\}$). Since **R** - $\{0\}$ is Lindelöf, it is realcompact, so by Proposition 7.5, X - $\{p\}$ is realcompact. Now any proper subset of X is an intersection of subsets of the form X - $\{p\}$ such that $p \in$ X. Therefore, by Proposition 7.3, every subspace of X is realcompact. ∎

PROPOSITION 7.6 In any space, the union of a compact subset with a realcompact subset is realcompact.

Proof: Let X = R∪K where R is realcompact and K is compact. Suppose X is not realcompact. Then eX ≠ X. Pick $p \in e$X - X. Since K is compact, we must have $p \in Cl_{eX}$(R). The next step will be to show that R is C-imbedded in R∪$\{p\}$. For this, let $f \in$ C(R). It is possible to construct a function $g \in$ C(eX) that is zero on a neighborhood U of K and 1 on a neighborhood V of p. Then (g|R)(f) can be extended to a continuous function h on X by defining $h(y) = 0$ for each $y \in$ K. h can be further extended to X∪$\{p\}$ since $p \in e$X.

Since f agrees with h on V∩R (which is dense in Cl(V∩R) and which contains p), f can also be extended to p. Consequently, R is C-imbedded in R∪$\{p\}$. By Theorem 7.8.(5), eR = e[R∪$\{p\}$]. Since R is realcompact, eR = R which implies $p \in$ R which is a contradiction since $p \in e$X - X. Therefore, X must be realcompact. ∎

THEOREM 7.9 The following statements about a space Y are equivalent:

(1) For each space X, if there exists a continuous f:X → Y with $f^{-1}(y)$ compact for each $y \in$ Y then X is realcompact.

(2) Every space X of which Y is a one-to-one continuous image is realcompact.

(3) Y is hereditarily realcompact.

(4) For each $y \in$ Y, Y - $\{y\}$ is realcompact.

Proof: (1) → (2) is obvious.

(2) → (3) Let X be a subspace of Y and let τ be the topology of Y. Define another topology τ' on Y by adding the sets X and Y - X to τ. Then τ' is also completely regular and both τ and τ' induce the same relative topology on X. Since the identity map i:(Y, τ') → (Y, τ) is one-to-one, by (2), (Y, τ') is realcompact. Then X, which is closed in (Y, τ'), is realcompact by Theorem

7.4. But since τ and τ' induce the same relative topology on X, X is a realcompact subspace of Y.

(3) $\to$ (4) is obvious.

(4) $\to$ (1) Let f:X $\to$ Y be a continuous mapping such that $f^{-1}(y)$ is compact for each $y \in$ Y. By Proposition 7.6, (4) implies Y is realcompact. By Corollary 7.5, f has a unique continuous extension f^{υ}:υX $\to$ Y. Now let $y \in$ Y. By (4), Y - $\{y\}$ is realcompact, so by Proposition 7.5, Z = $(f^{\upsilon})^{-1}$(Y - $\{y\}$) is realcompact. Therefore, by Proposition 7.6, Z$\cup f^{-1}(y)$ is realcompact. Since Z$\cup f^{-1}(y)$ lies between X and υX, by Corollary 7.4 we have Z$\cup f^{-1}(y)$ = υX. Therefore, $f^{-1}(y) = (f^{\upsilon})^{-1}(y)$ for each $y \in$ Y so f^{υ} maps no point of υX - X onto y. Since this holds for each $y \in$ Y, we have υX - X = $\varnothing$ so X is realcompact. ■

COROLLARY 7.7 If f:X → Y is one-to-one and continuous, and if Y is hereditarily realcompact, then X is hereditarily realcompact.

EXERCISES

1. Show that a space X is realcompact if and only if whenever X is imbedded in T such that X is dense in T but X $\neq$ T, then X is not C-imbedded in T.

2. Show that eX = βX if and only if X is pseudocompact.

3. Show that $f^{\upsilon}(e\text{X}) = f(\text{X})$ for each $f \in$ C(X).

4. Let f be a continuous function from X into the realcompact space Y. Let f^{β} be the Stone extension into βY. Show that $(f^{\beta})^{-1}$(Y) is realcompact.

5. Show that if U is open in eX, then X$\cap$U is C-imbedded in U.

6. Show that if K is compact in X, then X - K is C-imbedded in eX - K.

7.4 Realcompact Spaces and Lindelöf Spaces

There is a great deal of similarity between realcompact spaces and Lindelöf spaces. We have already seen that Lindelöf spaces are realcompact, that the realcompact spaces are the ones that are complete with respect to their e uniformity and that Lindelöf spaces are the ones that are cofinally complete with respect to their e uniformity. In this section we continue to investigate this similarity. Of particular interest in this regard are two theorems of H. Tamano that appeared in his 1962 paper *On compactifications* referenced in Chapter 6.

Let $f \in$ C(X) and let $\mathbf{R}^* = \mathbf{R}\cup\{\infty\}$ be the one-point compactification of $\mathbf{R}$. Let f^* be the Stone extension of f:X $\to$ $\mathbf{R}^*$ to βX. Let X_f be the maximal subspace of βX to which $f{:}X \to \mathbf{R}$ can be continuously extended. Then for each $p \in \beta X - X_f$ we must have $f^*(p) = \infty$ or else we could extend $f{:}X \to \mathbf{R}$ to $X_f \cup \{p\}$. Consequently $X_f = (f^*)^{-1}(\mathbf{R})$ so X_f is open in βX. Moreover, by Corollary 7.3, $\upsilon X = \cap\{X_f \mid f \in C(X)\}$. Therefore, by Corollary 7.4, X is realcompact if and only if $X = \cap X_f$. We record this as:

PROPOSITION 7.7 X is realcompact if and only if $X = \cap X_f$.

Let d_f be the pseudo-metric on X generated by f (i.e., $d_f(x,y) = |f(x) - f(y)|$) and let $\mathcal{V}_f = \{S_f(x,1) \mid x \in X\}$ where $S_f(x,1)$ is the sphere about x of radius 1 with respect to d_f. Let E_f denote the extent of $\mathcal{V}_f$ in βX (cf. Section 6.2).

PROPOSITION 7.8 $X_f = E_f$.

Proof: Let f^0 be the extension of f over X_f and let d_{f^0} be the pseudo-metric on X_f generated by f^0. Let $S_{f^0}(x,1)$ be the sphere about x of radius 1 with respect to d_{f^0}. Since each member of $\mathcal{V}^{\varepsilon(\beta X)}$, the proper extension of $\mathcal{V}_f$ into βX, is regularly open in βX, by Lemma 6.10.(1), each member of $\mathcal{V}_f^{\varepsilon(\beta X)} \cap X_f$ is regularly open in X_f. It is evident that for each $p \in X$,

$$S_{f^0}(p,1) \cap X = S_f(p,1) \subset S_f(p,1)^{\varepsilon(\beta X)} \cap X = [S_f(p,1)^{\varepsilon(\beta X)} \cap X_f] \cap X$$

so by Lemma 6.10.(3) $S_{f^0}(p,1) \subset S_f(p,1)^{\varepsilon(\beta X)} \cap X_f$. Now the collection of all spheres $S_{f^0}(p,1)$ where $p \in X$ covers X_f, so $X_f \subset E_f$.

Conversely, if $p \in \beta X$ is contained in E_f, then $p \in S(x,1)^{\varepsilon(\beta X)}$ for some $x \in X$. Then $|f(x) - f(y)| < 1$ for each $y \in S_f(x,1) \cap X$ which implies f is bounded on $S_f(x,1)^{\varepsilon(\beta X)} \cap X$. It follows that $p \in X_f$, for if not then $f^*(p) = \infty$ which implies that for each positive integer n, there exists a $z \in S_f(x,1)^{\varepsilon(\beta X)} \cap X$ with $f(z) > n$ which is a contradiction. Therefore, $E_f \subset X_f$. ■

COROLLARY 7.8 $\upsilon X = \cap E_f$.

THEOREM 7.10 (H. Tamano, 1962) The following statements about a Tychonoff space are equivalent:

(1) X is realcompact.

(2) For each $p \in \beta X - X$, *there is a closed* G_δ *set C in* $\beta X - X$ *containing p.*

(3) For each $p \in \beta X - X$, *there is a countable star-finite partition of unity* $\Phi = \{\phi_n\}$ *such that* $Cl_{\beta X}(\mathbf{0}(\phi_n))$ *does not contain p for each n.*

Proof: (1) → (2) Let X be realcompact and let $f \in C(X)$. Put $C_f = \beta X - X_f$. Then C_f is closed in βX since X_f is open. Evidently, C_f is a G_δ in βX since $C_f = \{p \in \beta X \mid |f(p) - f(x)| \geq 1 \text{ for each } x \in X\}$ because $C_f = X - E_f$ and E_f is the extent of $\mathcal{V}_f$ in βX. By Proposition 7.5, $C_f \subset \beta X - X$ for each $f \in C(X)$. If $p \in$

$\beta X - X$, since X is realcompact, there exists an $f \in C(X)$ that cannot be extended over p. Hence x does not belong to X_f which implies $x \in C_f$. Then C_f is a closed G_δ in βX containing p.

(2) $\to$ (1) It is easily shown that each closed G_δ in $\beta X - X$ is a C_f for some $f \in C(X)$. Hence $\cup C_f = \beta X - X$ which implies $X = \cap X_f$. By Proposition 7.5, this implies X is realcompact.

(2) $\to$ (3) If C is a closed G_δ in $\beta X - X$, we can construct a continuous f:X $\to [0,1]$ such that $\mathbf{Z}(f^*) = C$ and such that f^* assumes all values in [0,1]. For each positive integer $n \geq 2$ put $f_n = min\{f, 1/(n-1)\}$ and $g_n = max\{f_n, 1/(n+1)\}$. Define h_n:X $\to [0,1]$ by

$$h_n(x) = 0 \text{ if } g_n(x) < 1/(n-1) \text{ and } h_n(x) = 1/(n-1) - 1/(n+1) \text{ otherwise.}$$

Then put $k_n = g_n - 1/(n+1)$, $l_n = k_n - h_n$ and $\phi_n = l_n/\Sigma l_n$. It is left as an exercise (Exercise 1) to show that $\Phi = \{\phi_n\}$ is a countable star-finite partition of unity, and that $Cl_{\beta X}(\mathbf{0}(\phi_n)) \cap C = \emptyset$ for each n.

(3) $\to$ (2) If $\Phi = \{\phi_n\}$ is a countable star-finite partition of unity such that $Cl_{\beta X}(\mathbf{0}(\phi_n))$ does not contain p for each n, then $f = \Sigma\phi_n 2^{-n}$ is a continuous function on X such that $p \in \mathbf{Z}(f^*) \subset \beta X - X$. ■

We have already seen that a Tychonoff space is paracompact if and only if it is cofinally complete with respect to the u uniformity and that it is Lindelöf if and only if it is cofinally complete with respect to the e uniformity. Furthermore, we have seen that it has a Tamano-Morita paracompactification if and only if it is preparacompact with respect to the u uniformity. It should therefore come as no surprise that the Hewitt realcompactification is Lindelöf if and only if the space is preparacompact with respect to the e uniformity.

THEOREM 7.11 (N. Howes, 1970) The Hewitt realcompactification of X is Lindelöf if and only if X is preparacompact with respect to the e uniformity.

Proof: Let (X, e) be preparacompact and let (eX, e') be the completion of (X, e). Then eX is the Hewitt realcompactification of X. By Corollary 4.3, (eX, e') is cofinally complete, so by Corollary 4.4, eX is Lindelöf.

Conversely, if eX is Lindelöf and e' is the e uniformity for eX, then by Theorem 4.4, (eX, e') is cofinally complete which implies (X, e) is preparacompact. Consequently, the proof reduces to showing that e' is the e uniformity for eX. But this has already been proved (Lemma 7.1). ■

THEOREM 7.12 (H. Tamano, 1962) Let BX be any compactification of the Tychonoff space X. Then the following statements are equivalent:

(1) X is Lindelöf.

(2) For each compact $C \subset BX - X$, there is a countable star-finite partition of unity $\phi = \{\phi_n\}$ on X such that $Cl_{BX}(0(\phi_n)) \cap C = \varnothing$ for each n.

(3) For each compact $C \subset BX - X$, there is a closed G_δ set G of BX such that $C \subset G \subset BX - X$.

(4) For each compact $C \subset BX - X$, there is a countable family $\{G_n\}$ of compact subsets of BX such that $G_n \cap C = \varnothing$ for each n and $X \subset \cup G_n$.

Proof: (1) → (2) Assume X is Lindelöf and suppose C is a compact subset of BX - X. For each $x \in X$ let U(x) be an open neighborhood of x such that $Cl_{BX}(U(x)) \cap C = \varnothing$ and consider the covering $\mathcal{U} = \{U(x) \mid x \in X\}$. By Proposition 6.2, $\mathcal{U}$ has a countable star-finite refinement $\mathcal{V} = \{V_n\}$. By Lemma 3.6, X is paracompact and hence normal. By Lemma 1.3, $\mathcal{V}$ is shrinkable to a countable star-finite refinement $\mathcal{W} = \{W_n\}$. Therefore, by Theorem 6.3, there exists a countable star-finite partition of unity $\Phi = \{\phi_n\}$ such that $Cl_X(0(\phi_n)) \subset W_n$ for each n. Then $Cl_{BX}(0(\phi_n)) \cap C = \varnothing$ for each $\phi_n \in \Phi$.

(2) → (3) Assume statement (2) is valid. For each positive integer n let f_n be a continuous function on BX such that $0 \le f_n \le 1$, $f_n = 0$ on C and $f_n = 1$ on $Cl_{BX}(0(\phi_n))$. Put $f = \Sigma f_n 2^{-n}$. Then $\mathbf{Z}(f)$ is a closed G_δ in BX such that $C \subset \mathbf{Z}(f) \subset BX - X$.

(3) → (4) is obvious.

(4) → (1) Assume statement (4) is valid. Let $\mathcal{U} = \{U_\alpha\}$ be an open covering of X. For each α let U_α^ε denote the proper extension of U_α to BX. Put $C = BX - \cup U_\alpha^\varepsilon$. Then C is a compact subset of BX - X. By (2), there exists a countable family $\{G_n\}$ of compact subsets of BX such that $G_n \cap C = \varnothing$ for each n and $X \subset \cup G_n$. Evidently, each G_n is covered by a finite subfamily of $\{U_\alpha^\varepsilon\}$ and therefore $\cup G_n$ is covered by a countable subfamily of $\{U_\alpha^\varepsilon\}$. Since $X \subset \cup G_n$, we see that X is covered by a countable subfamily of $\mathcal{U}$. Therefore, X is Lindelöf. ∎

EXERCISE

1. Show that the collection of functions $\{\phi_n\}$ in the proof of (2) → (3) of Theorem 7.9 is a countable star-finite partition of unity such that $Cl_{\beta X}(0(\phi_n)) \cap \mathbf{Z}(f^*) = \varnothing$.

7.5 Shirota's Theorem

Clearly the discrete space ω of positive integers is realcompact since ω is Lindelöf. In fact, if we denote the cardinality $|\mathbf{R}|$ of the real numbers by c, it is easily shown that any discrete space X with $|X| \leq c$ is realcompact. To see this note that if $|X| \leq c$ there exists a one-to-one function $f:X \rightarrow \mathbf{R}$. Since X is discrete, f is continuous and since $\mathbf{R}$ is separable, $f(X)$ is Lindelöf and hence realcompact. Then by Proposition 7.5, $f^{-1}(f(X)) = X$ is realcompact.

It is natural to ask if all discrete spaces are realcompact. In the light of Shirota's Theorem, this straightforward question takes on unexpected significance. Shirota's Theorem essentially states that for any Tychonoff space X, if all discrete spaces of cardinality $\leq 2^{|X|}$ are realcompact, then X is realcompact if and only if it is topologically complete. Consequently, if all discrete spaces are realcompact, topological completeness is the same as realcompactness and $eX = uX$ for any Tychonoff space X.

The problem as to whether all discrete spaces are realcompact reduces to a (still unsolved) set theoretic problem as to whether there exists a certain type of cardinal number called a *measurable cardinal*. However, it is known that if a measurable cardinal exists, it must be larger than any known cardinal (i.e., any cardinal that can be constructed from the rules of cardinal arithmetic). All known examples of uniform spaces can be constructed from ω and the rules of cardinal arithmetic, so for all practical purposes the existence of measurable cardinals does not concern us in our study of uniform spaces. The problem of the existence of measurable cardinals has only philosophical interest at the present time. It is known for example that the statement "There do not exist measurable cardinals." is consistent with the axioms of ZFC.

Shirota's Theorem is one of the most striking theorems in the theory of uniform spaces. J. Isbell, in his book *Uniform Spaces* published by the American Mathematical Society in 1964, calls Shirota's Theorem the "first deep theorem of uniform spaces." It is interesting to note that Shirota's proof extends the (now famous) construction in Stone's Theorem of the closed covering $\{F_n\}$ where each F_n is a discrete collection $\{F_{n_\alpha} \mid \alpha \in A_{n_\alpha}\}$ of closed sets.

We begin our investigation of this problem by defining what we mean by a measurable cardinal. The concept of a *measure* will be introduced and investigated in the next chapter, but for now it suffices to introduce a special type of measure called a **zero-one measure** or $\{0,1\}$ measure. A measure μ on a set X is always defined to be a set function on X whose range is a subset of $[0,\infty]$ or of the complex numbers, i.e., μ is defined on some collection of subsets of X (called the measurable subsets of X) and the value $\mu(E)$ of a measurable set $E \subset X$ is called the measure of E. For a discrete space X, the collection of subsets on which a $\{0,1\}$ measure will be defined is the set 2^X of all subsets of X, and the range will be the subset $\{0,1\}$ of $[0,\infty]$.

There is more to a measure than just being a set function however. Measures are distinguished from set functions by the property of being *additive*. A set function $\mu:X \to [0,\infty]$ is said to be **σ-additive** if for each pairwise disjoint countable collection $\{E_n\}$ of measurable subsets of X,

$$\mu(\cup E_n) = \Sigma_n \mu(E_n).$$

Consequently, we see that the countable union of measurable subsets must also be measurable. In the case of {0,1} measures, $\mu(\cup E_n)$ can only be 0 or 1, so at most one of the E_n can have measure 1, i.e., $\mu(E_n) = 1$ for at most one n. Hence the series $\Sigma_n \mu(E_n)$ always converges.

There are two trivial {0,1} measures on a discrete space X that we can give as examples of a {0,1} measure. The first is the *zero measure* μ_0 defined by $\mu_0(E) = 0$ for each $E \subset X$. The other is the *point measure* μ_p defined for each $p \in X$ by $\mu_p(E) = 1$ if $p \in E$ and $\mu_p(E) = 0$ otherwise. For a point measure μ_p, it is clear that $\mu_p(\{p\}) = 1$, i.e., the measure of a single point is 1. The question is: Does there exist a {0,1} measure μ on the discrete space X that is not one of these two trivial types, i.e., such that $\mu(\{x\}) = 0$ for each $x \in X$, but $\mu(E) = 1$ for some $E \subset X$? If there does, X is said to be **measurable.** If α is a cardinal number, we can give α the discrete topology. If α, with the discrete topology is measurable, then α is called a **measurable cardinal.**

There is an interesting relationship between the non-zero {0,1} measures on a discrete space X and the CZ-maximal families on X. If Z is a CZ-maximal family on X, let χ_Z be the **characteristic function** of Z defined on all subsets of X by $\chi_Z(E) = 1$ if $E \in Z$ and $\chi_Z(E) = 0$ otherwise. We want to show that χ_Z is a non-zero (i.e., $\chi_Z(A) = 1$ for some $A \subset X$) {0,1} measure on X. We first show that if $E \subset X$ then either $E \in Z$ or $X - E \in Z$. If E does not belong to Z then $E \cap (\cap E_n) = \varnothing$ for some sequence $\{E_n\} \subset Z$, for otherwise we could add E to the collection Z to get a new collection Z′ that has the countable intersection property and that contains Z which is a contradiction since Z is maximal. Now $\cap E_n \in Z$ for the same reason. Since $(\cap E_n) \cap E = \varnothing$, $\cap E_n \subset X - E$. But then $X - E \in Z$ since $(X - E) \cap (\cap Z_n) \neq \varnothing$ for any collection $\{Z_n\} \subset Z$ because $(\cap E_n) \cap (\cap Z_n) = \varnothing$.

Then clearly $X \in Z$ so $\chi_Z(X) = 1$ which implies χ_Z is a non-zero set function. Suppose $\{E_n\}$ is a pairwise disjoint collection of subsets of X. If m, k is a pair of positive integers such that $\chi_Z(E_m) = 1 = \chi_Z(E_k)$ then $E_m, E_k \in Z$ but $E_m \cap E_k = \varnothing$ which is a contradiction. Therefore, at most one of the E_n can belong to Z. If exactly one of the E_n, say $E_k \in Z$ then $\Sigma_n \chi_Z(E_n) = 1$ and $E_k \subset \cup E_n$ which implies $\cup E_n \in Z$ so $\chi_Z(\cup E_n) = 1$. Thus

$$\chi_Z(\cup E_n) = 1 = \Sigma_n \chi_Z(E_n).$$

If $\chi_Z(E_n) = 0$ for each positive integer n then E_n does not belong to Z for each n

which implies $X - E_n \in Z$ for each n. But then $\cap_n(X - E_n) \in Z$ and $(\cup E_n)\cap[\cap_n(X - E_n)] = \varnothing$ so $\cup E_n \subset X - \cap_n(X - E_n)$ which does not belong to Z. Hence $\cup E_n$ cannot belong to Z so $\chi_Z(\cup E_n) = 0$. Also, since $\chi_Z(E_n) = 0$ for each n, $\Sigma_n\chi_Z(E_n) = 0$. Thus

$$\chi_Z(\cup E_n) = 0 = \Sigma_n\chi_Z(E_n).$$

Consequently, χ_Z is a non-zero $\{0,1\}$ measure on X.

Conversely, if μ is a non-zero $\{0,1\}$ measure on a discrete space X we can put $Z_\mu = \{E \subset X \mid \mu(E) = 1\}$. We will show that Z_μ is a CZ-maximal family on X. Let $\{E_n\} \subset Z_\mu$ and suppose $\cap E_n = \varnothing$. For each positive integer n let $A_n = E_n - \cup_{k=n+1}^{\infty}E_k$. Then $\{A_n\}$ is a sequence of pairwise disjoint subsets of X and since $\cap E_n = \varnothing$, $\cup A_n = \cup E_n$. Since μ is σ-additive it is monotone, i.e., if $A \subset B$ then $\mu(A) \subset \mu(B)$. To see this note that A and B - A are disjoint subsets of X so that:

$$\mu(A) + \mu(B - A) = \mu(A\cup[B - A]) = \mu(B).$$

Since $\mu(B - A) = 0$ or 1 it is clear that $\mu(A) \le \mu(B)$. Since $\mu(E_n) = 1$ for each n, and $E_n \subset \cup E_n$ for each n, we have $\mu(\cup E_n) = 1$ so $\mu(\cup A_n) = 1$. But $\mu(\cup A_n) = \Sigma_n\mu(A_n)$ which implies exactly one of the A_n, say A_m has $\mu(A_m) = 1$. Now $A_m\cup E_{m+1} \subset E_n$ and $\mu(\cup E_n) = 1$ which implies $\mu(A_m\cup E_{m+1}) = 1$ since μ is monotone. Since $A_m\cap E_{m+1} = \varnothing$ and μ is σ-additive, we have $\mu(A_m\cup E_{m+1}) = \mu(A_m) + \mu(E_n)$ which is a contradiction since $\mu(A_m) = 1 = \mu(E_{m+1})$. Therefore, $\cap E_n \neq \varnothing$.

Consequently, Z_μ has the countable intersection property. Next we show that Z_μ is maximal with respect to this property, i.e., that Z_μ is a CZ-maximal family. For this, suppose Z_μ is a proper subset of Z and that Z has the countable intersection property. Let $Z \in Z - Z_\mu$. Then $\mu(Z) = 0$ which implies $\mu(X - Z) = 1$ which in turn implies $X - Z \in Z_\mu$. But $Z\cap(X - Z) = \varnothing$ so Z does not even have the finite intersection property which is a contradiction. Therefore, Z_μ is maximal with respect to the countable intersection property. We record this as:

PROPOSITION 7.9 There is a one-to-one correspondence ϕ between the collection of all CZ-maximal families on a discrete space X and the collection of all non-zero $\{0,1\}$ measures on X. ϕ is defined by $\phi(Z) = \chi_Z$ where Z is a CZ-maximal family and χ_Z its characteristic function.

THEOREM 7.13 A discrete space X is realcompact if and only if $|X|$ is a non-measurable cardinal.

Proof: Let X be realcompact and suppose μ is a non-zero $\{0,1\}$ measure on X. Then $Z = \{E \subset X \mid \mu(E) = 1\}$ is a CZ-maximal family on X, as shown in the proof of Proposition 7.9. Since X is realcompact, $\cap Z \neq \varnothing$. Let $x \in \cap Z$. Then

$\{x\} \in Z$ which implies $\mu(\{x\}) = 1$. Therefore, each non-zero $\{0,1\}$ measure on X is the point measure μ_x for some $x \in$ X so $|X|$ is non-measurable.

Conversely, let $|X|$ be non-measurable and suppose Z is a CZ-maximal family on X. Then χ_Z is a non-zero $\{0,1\}$ measure on X as shown in the proof of Proposition 7.9. Since $|X|$ is non-measurable, $\chi_Z(\{p\}) = 1$ for some $p \in$ X which implies $\{p\} \in Z$ which in turn implies $p \in \cap Z$. Therefore, X is realcompact. ∎

We say that a $\{0,1\}$ measure μ is **M-additive** if $\mu(\cup E_\alpha) = 0$ whenever $\{E_\alpha \mid \alpha \in A\}$ is a family of pairwise disjoint sets of measure zero with $|A| = M$. Clearly an M-additive $\{0,1\}$ measure is N-additive for each cardinal $N \leq M$. It is also easily seen that μ is M-additive if and only if the intersection of any collection of M sets, each having measure 1 is a set of measure 1.

PROPOSITION 7.10 Each $\{0,1\}$ *measure* μ *on a non-measurable cardinal M (with the discrete topology) is M-additive.*

Proof: If μ is not M-additive, there exists a family of pairwise disjoint sets $\{E_\alpha \subset M \mid \alpha \in M\}$ such that $\mu(E_\alpha) = 0$ for each α and $\mu(\cup E_\alpha) = 1$. Define the measure λ on M by $\lambda(E) = \mu(\cup\{E_\alpha \mid \alpha \in E\})$. Then $\lambda(M) = 1$ and $\lambda(\{\alpha\}) = \mu(E_\alpha) = 0$ for each α. Therefore, M is measurable which is a contradiction. Consequently, each $\{0,1\}$ measure μ on a non-measurable cardinal M is M-additive. ∎

A $\{0,1\}$ measure μ is said to be **free** if $\mu(\{x\}) = 0$ for each $x \in$ X. The intuitive notion behind this definition is that the measure of the whole space is not *bound* to some $p \in$ X in the sense that $\mu(\{p\}) = 1$. Clearly, only the free measures are useful in proving that a discrete space is measurable.

PROPOSITION 7.11 If μ *is a free measure on a discrete space X and* $|E|$ *is non-measurable for some* $E \subset X$*, then* $\mu(E) = 0$.

Proof: Define the $\{0,1\}$ measure λ on E by $\lambda(A) = \mu(A)$ for each $A \subset E$. Clearly λ is also free. Since $|E|$ is non-measurable, we cannot have $\lambda(E) = 1$. Hence $\mu(E) = \lambda(E) = 0$. ∎

We say that a class of cardinal numbers is **closed** if it is closed under all the rules (of cardinal arithmetic) for forming new cardinals from given ones: addition, multiplication, exponentiation, taking suprema of a collection of cardinals, and the passage from a given cardinal to its successor or any of its predecessors. Clearly the class of all cardinals (this class is not a set) is closed since any cardinal formed from given cardinals using these rules is again a member of the class. The class of all finite cardinals is also a closed class (it is understood that the cardinal number of any index set used in applying these rules is a member of the class in question). Of primary interest to us are the closed classes that contain $|\omega|$.

LEMMA 7.3 A non-empty class of cardinals C is closed if and only if, whenever $M \in C$, then:

(1) $N \in C$ for each cardinal $N < M$,

(2) the sum of any M members of C is in C,

(3) $2^M \in C$.

Proof: The necessity of conditions (1) - (3) is obvious. Conversely, assume C satisfies conditions (1) - (3). We need to show that C contains products, exponentials, suprema and successors. For this let $M \in C$. Then $M+1 \leq 2^M$ so by (3) and (1), $M+1 \in C$. Therefore, successors are contained in C. To show that suprema belong to C, let $\{N_\alpha\}$, where $\alpha \in M$, be a collection of cardinals and put $N = \Sigma N_\alpha$. By (2), $N \in C$. Since $supN_\alpha \leq N$, we have $supN_\alpha \in C$ by (1). Also,

$$\Pi N_\alpha \leq \Pi 2^{N_\alpha} = 2^N \in C$$

so C contains products. Finally, if $N \in C$, then $N^M = \Pi N_\alpha$ where $N_\alpha = N$ for each $\alpha \in M$. Since products are contained in C, this means N^M is contained in C. ∎

THEOREM 7.14 The class of all non-measurable cardinals is a closed class containing $|\omega|$.

Proof: Since ω with the discrete topology is realcompact, $|\omega|$ is non-measurable. We need to show that if C is the class of all non-measurable cardinals, then C satisfies the three properties of Lemma 7.3. To show (1), it will suffice to show that every subspace of a discrete realcompact space is realcompact. But every subspace of a discrete space is closed, so by Theorem 7.4, we have property (1).

To show property (2), let $|X|$ be the sum of M non-measurable cardinals where $M \in C$. Then X is the union of M pairwise disjoint subsets say $\{E_\alpha\}$ such that $\alpha \in M$. Let μ be a free measure on X. By Proposition 7.11. $\mu(E_\alpha) = 0$ for each α. Since $M \in C$, M is M-additive by Proposition 7.10. Hence $\mu(X) = \mu(\cup E_\alpha) = 0$. Therefore, μ is the zero measure so $|X|$ is non-measurable. Hence C contains sums, so property (2) holds.

To show property (3), let $|X| = M \in C$ and let P be the set of all subsets of X. Then $|P| = 2^M$. Suppose μ is a non-zero measure on P. By Proposition 7.10, μ is M-additive. For each $x \in X$ let $H_x = \{A \subset X | x \in A\}$ and let $K_x = \{B \subset X | x$ does not belong to B$\}$. Put $Z = \{x \in X | \mu(H_x) = 1\}$. Then the set

$$E = [\cap_{x \in Z} H_x] \cap [\cap \{K_x | x \text{ does not belong to } Z\}]$$

is the intersection of no more than M sets of measure 1. Therefore, $\mu(E) = 1$ since μ is M-additive. But E contains only one element. To see this note that

$\cap_{x \in Z} H_x = \{A \subset X | Z \subset A\}$ and $\cap\{K_x | x$ does not belong to $Z\} = \{B \subset X | B \subset Z\}$. Hence

$$E = \{A \subset X | Z \subset A\} \cap \{B \subset X | B \subset Z\} = \{Z\}.$$

Therefore, μ is not free which is a contradiction. Thus $|P| = 2^M$ is non-measurable, so $2^M \in C$. This shows that property (3) holds. ∎

The smallest closed class containing $|\omega|$ contains all the cardinals that we know how to construct using the rules of cardinal arithmetic. This class of cardinals is truly immense. A cardinal number is said to be **strongly inaccessible** if the set of all smaller cardinals is a closed class containing $|\omega|$. Therefore, by Theorem 7.14, the smallest measurable cardinal (if such a cardinal exists) must be strongly inaccessible. It is impossible to prove from the axioms of ZFC that strongly inaccessible cardinals exist. This is because if we have a model for set theory that includes an inaccessible cardinal, then the set of all cardinals less than the first inaccessible cardinal is a model for set theory that does not include strongly inaccessible cardinals and, in which, all the axioms of ZFC hold. Consequently, there is a model for set theory that does not include inaccessible cardinals, so their non-existence is consistent with the axioms of ZFC.

It is conceivable that one can prove that no such cardinals exist from the axioms of ZFC, but this is still an open question. A similar situation exists if one tries to prove the existence of infinite cardinals with the axioms of ZFC minus the axiom of infinity. Consequently, most mathematicians believe another axiom must be added to ZFC to prove their non-existence. Clearly their existence requires another axiom. But even if strongly inaccessible cardinals exist, it may still be the case that measurable cardinals do not.

Consequently, for all practical purposes, i.e., for working with any space X we can construct from the axioms of ZFC, $|X|$ is non-measurable. Therefore, the question appears to be of philosophical rather than practical interest.

LEMMA 7.4 Let Z be a CZ-maximal family on a Tychonoff space X and let $f \in C(X)$ such that $f \neq 0$ and $\mathbf{Z}(f) \in Z$. If $Y = \{x | f(x) \leq a\}$ for some $a > 0$, then $Z' = \{\mathbf{Z}(g) \in \mathbf{Z}(Y) | \emptyset \neq \mathbf{Z}(f) \cap Z \subset \mathbf{Z}(g)$ for some $Z \in Z\}$ is a CZ-maximal family on Y.

Proof: It is easily shown that Z' is a non-empty subfamily of $\mathbf{Z}(Y)$ with the countable intersection property. Therefore, it suffices to show that Z' is maximal with respect to the finite intersection property in $\mathbf{Z}(Y)$. For this suppose Z' is not maximal, i.e., suppose W is a non-empty subfamily of $\mathbf{Z}(Y)$ with the finite intersection property that contains Z' as a proper subset. Let $\mathbf{Z}(g) \in W - Z'$ for some $g \in C(Y)$ and put $F_0 = \mathbf{Z}(g) \cap \mathbf{Z}(f)$. Then $F_0 \in W$ so $F_0 \neq \emptyset$. Now put $F_1 = Y$.

For each rational $r \in [0, 1]$ let $U_r = \{x \mid f(x) < r\}$, $Z_r = \{x \mid f(x) \leq r\}$, $U'_r = \{x \mid g(x) < r\}$ and $Z'_r = \{x \mid g(x) \leq r\}$. Put $G_r = U_r \cap U'_r$ and $F_r = Z_r \cap Z'_r$. Then G_r is open in X and F_r is closed. For any pair of rational numbers $r, s \in [0, 1]$ with $r < s$ we have $F_r \subset G_s \subset F_s$. For each $x \in X$ put $g'(x) = sup\{r \mid x$ does not belong to $F_r\}$. Then $g' \in C(X)$ and $\mathbf{Z}(g') = \cap F_r = \cap_r(Z_r \cap Z'_r) = \mathbf{Z}(f) \cap \mathbf{Z}(g) = F_0$. For each r, $F_r \in \mathbf{Z}(X)$ and $\mathbf{Z}(f) \subset F_r$. Therefore, each $F_r \in Z$ which implies $\cap_r F_r \in Z$. Hence $F_0 \in Z$. Then by the definition of Z', $\mathbf{Z}(g) \in Z'$ which is a contradiction. Therefore, Z' is measurable after all. ▪

THEOREM 7.15 (T. Shirota, 1951) Let X be a Tychonoff space such that $|X|$ is non-measurable. Then X is realcompact if and only if X is complete with respect to u.

Proof: Assume X is complete with respect to u. It suffices to show that if Z is a CZ-maximal family of X and $\mathcal{U} \in u$, there is a $B \in Z$ and a $U \in \mathcal{U}$ with $B \subset U$. The reason for this is that if this is the case, then a proof similar to (1) → (2) of Theorem 7.2 will imply that there is a $p \in X$ with $p \in \cap Z$. Thus X will be realcompact. For this, observe that $\mathcal{U} \in u$ implies there is a normal sequence $\{\mathcal{U}_n\}$ with $\mathcal{U}_1 <^* \mathcal{U}$. Let $\mathcal{U} = \{U_\alpha \mid \alpha < \gamma\}$ for some cardinal γ. In the proof of Stone's Theorem (Theorem 1.1), it was shown that there is a closed covering $H = \{H_{n\alpha} \mid \alpha < \gamma$ and n is a positive integer$\}$ of X satisfying the properties:

(1) If $\alpha \neq \beta$, no member of $\mathcal{U}_n$ meets both $H_{n\alpha}$ and $H_{n\beta}$.
(2) $Star(H_{n\alpha}, \mathcal{U}_n) \subset U_\alpha$ for each $\alpha < \gamma$ and positive integer n.

Also, in Stone's Theorem, the closed covering $\{F_n\}$ was defined as follows: for each $\alpha < \gamma$ and positive integer n put $E_{n\alpha} = Star(H_{n\alpha}, \mathcal{U}_{n+3})$ and $G_{n\alpha} = Star(H_{n\alpha}, \mathcal{U}_{n+2})$. Then $H_{n\alpha} \subset E_{n\alpha} \subset Cl(E_{n\alpha}) \subset G_{n\alpha}$ for each pair α, n and if $\alpha \neq \beta$, no member of $\mathcal{U}_{n+2}$ meets both $G_{n\alpha}$ and $G_{n\beta}$. Put $F_n = \cup\{Cl(E_{n\alpha}) \mid \alpha < \gamma\}$ for each positive integer n. If for each pair α, n we put $F_{n\alpha} = Cl(E_{n\alpha})$, we have a closed covering $F = \{F_{n\alpha}\}$ of X having the following properties:

(3) If $\alpha \neq \beta$, no member of $\mathcal{U}_{n+2}$ meets both $F_{n\alpha}$ and $F_{n\beta}$.
(4) $Star(F_{n\alpha}, \mathcal{U}_{n+1}) \subset U_\alpha$ for each $\alpha < \gamma$ and positive integer n.

Then for each positive integer n, $F_n = \cup\{F_{n\alpha} \mid \alpha < \gamma\}$. Suppose that for each n $F_n \cap Z_n = \varnothing$ for some $Z_n \in Z$. Then $\cap Z_n = \cap Z_n \cap X = (\cap Z_n) \cap (\cup F_n) = \cup[F_n \cap (\cap Z_n)] \subset \cup(F_n \cap Z_n) = \varnothing$. But this contradicts the assumption that Z is CZ-maximal. Therefore, there exists a positive integer m such that $F_m \cap B \neq \varnothing$ for each $B \in Z$.

Let $f \in C^*(X)$ with $f(F_m) = 0$, $f(X - Star(F_m, \mathcal{U}_4)) = 2$ and $0 \leq f \leq 2$. Put $Z_0 = \{x \mid f(x) = 0\}$ and $Z_1 = \{x \mid f(x) \leq 1\}$. Since $F_m \subset Z_0$ and Z is CZ-maximal, $Z_0 \in Z$. By Lemma 7.4, $Z' = \{\mathbf{Z}(g) \in \mathbf{Z}(Z_1) \mid \varnothing \neq B \cap Z_0 \subset \mathbf{Z}(g)$ for some $B \in$

$Z\}$ is a CZ-maximal family on Z_1. Then, just as in the proof (1) → (2) of Theorem 7.2, for each countable normal covering $\mathcal{V} = \{V_n\}$ of Z, there is a $B' \in Z'$ and a positive integer k with $B' \subset V_k$.

For each $\alpha < \gamma$ put $Z_\alpha = Z_1 \cap Star(F_{m_\alpha}, \mathcal{U}_{m+4})$. Clearly $Z_1 \subset Star(F_m, \mathcal{U}_{m+4})$ and by (3) above, $Star(F_{m_\alpha}, \mathcal{U}_{m+4}) \cap Star(F_{m_\beta}, \mathcal{U}_{m+4}) = \varnothing$ so we can construct a function $f_\alpha \in C^*(X)$ from f (by putting $f_\alpha(x) = f(x)$ if $x \in Star(F_{m_\alpha}, \mathcal{U}_{m+4})$ and $f(x) = 2$ otherwise) such that $Z_\alpha = \mathbf{Z}(f_\alpha)$. Then $Z_1 = \cup Z_\alpha$ and by (3) above,

$$Star(Z_\alpha, \mathcal{U}_{m+4}) \cap Star(Z_\beta, \mathcal{U}_{m+4}) \subset Star(F_{m_\alpha}, \mathcal{U}_{m+3}) \cap Star(F_{m_\beta}, \mathcal{U}_{m+3}) = \varnothing.$$

Let Y be the discrete space $\{\alpha \mid Z_\alpha \neq \varnothing\}$ and let $h: Z_1 \to Y$ be defined by $h(x) = \alpha$ where $x \in Z_\alpha$. It is easily shown that h is a uniformly continuous mapping from (Z_1, e) into (Y, e). Since $|Y| \leq |A| \leq 2^{|X|}$, by Lemma 7.3, $|Y|$ is non-measurable so Y is realcompact by Theorem 7.13. Now for each $\mathcal{W} \in e$ in Y, $h^{-1}(\mathcal{W}) \in e$ in Z_1. Therefore, there is a $B' \in Z'$ with $B' \subset h^{-1}(W)$ for some $W \in \mathcal{W}$. But then $h(B') \subset W$. Hence for each $\mathcal{W} \in e$ in Y, there exists $h(B') \in h(Z')$ with $h(B') \subset W$ for some $W \in \mathcal{W}$. Then just as in the proof of (1) → (2) of Theorem 7.2, there exists an $\alpha \in Y$ with $\alpha \in h(B')$ for each $B' \in Z'$. But this implies $B' \cap Z_\alpha \neq \varnothing$ for each $B' \in Z'$ so $Z_\alpha \in Z'$.

For any $B \in Z$, $B \cap Z_1 \in Z'$ by the definition of Z', so $Z_\alpha \cap (B \cap Z_1) \neq \varnothing$ which implies $Z_\alpha \cap B \neq \varnothing$ which in turn implies $Z_\alpha \in Z$. But $Z_\alpha \subset Star(F_{m_\alpha}, \mathcal{U}_{m+4}) \subset U_\alpha$ by (4) above, so for each $\mathcal{U} \in u$, there exists a $B \in Z$ with $B \subset U$ for some $U \in \mathcal{U}$. ∎

Part II: Analysis

Chapter 8

MEASURE AND INTEGRATION

8.1 Introduction

In this chapter and the next, the theory of integration in uniform spaces will be developed. This chapter will only be concerned with those aspects of integration theory that do not depend on the uniform structure of the space. In elementary analysis one encounters the concept of the Riemann integral. Intuitively, the process of Riemann integration in one, two and three dimensional Euclidean space corresponds to calculating lengths, areas and volumes respectively. The formalization of the Riemann integral occurred during the nineteenth century. Briefly, the main idea for one dimensional Euclidean space is that the Riemann integral of a function f over an interval $[a,b]$ can be approximated by sums of the form

$$\Sigma_{i=1}^{n} f(x_i)\Delta(I_i)$$

where $I_1 \ldots I_n$ are disjoint intervals whose union is $[a, b]$, $\Delta(I_i)$ denotes the length of I_i and $x_i \in I_i$ for each $i = 1 \ldots n$.

By the end of the nineteenth century and early into the twentieth century, many mathematicians were working on replacements for the Riemann integral that would be more flexible and better suited for dealing with the problems of modern analysis. Some of the more notable researchers were Borel, Jordan, Lebesgue and Young. They realized that a more satisfactory theory of integration could be obtained by replacing the intervals $I_1 \ldots I_n$ in the above approximation with subsets of the real line that belonged to a more general class, e.g., the so-called *measurable sets*. When this is done, the function f which was usually assumed to be continuous (or nearly so) could be significantly generalized, e.g., the so-called *measurable functions*. Ultimately, the construction that proved most successful was the one of Lebesgue, so it is customary to call this new integral the Lebesgue integral even though many participated in developing the modern theory of integration.

In the theory of Lebesgue integration, associated with the concept of measurable sets is the concept of a *measure*. In this chapter, a measure is a non-negative real valued set function (i.e., it maps sets onto non-negative numbers or ∞). Intuitively, the measure tells how "big" (in some sense) the sets are. In this chapter, after a development of the essential properties of measures, we present a development of the Lebesgue integral for topological spaces. In Chapter 9, the theory of Lebesgue integration will be specialized in uniform spaces to *invariant* integrals.

8.2 Measure Rings and Algebras

Let X be a set and *R* a collection of subsets of X with the property that if A, B $\in$ *R* then A$\cup$B and A - B $\in$ *R*. Then *R* is said to be a **ring** in X. It should be noted that this property implies A$\cap$B $\in$ *R* since

$$A \cap B = [A \cup B] - [(A - B) \cup (B - A)].$$

If *R* has the additional property that $A^c \in R$ whenever A $\in$ *R* then *R* is called an **algebra**. Clearly the empty set belongs to every ring *R* since for any A $\in$ *R*, $\varnothing$ = A - A.

PROPOSITION 8.1 A ring R in X is an algebra if and only if X $\in$ R.

Proof: Assume X $\in$ *R*. If A $\in$ *R* then A^c = X - A $\in$ *R*. Therefore *R* is also an algebra. Conversely, assume *R* is an algebra and let A, B $\in$ *R*. Then $\varnothing$ = (A - B) $\cap$ (B - A) $\in$ *R*. Since *R* is an algebra, X = $\varnothing^c \in$ *R*. ∎

If *R* is a ring in X such that whenever $\{A_i\}$ is a sequence in *R* then $\cup A_i \in R$ then *R* is said to be a **σ-ring**. If *A* is both an algebra and a σ-ring then *A* is called a **σ-algebra**.

PROPOSITION 8.2 If α is a collection of subsets of X then there is a smallest ring R containing α.

Proof: The class of all subsets of X is clearly a ring containing α. Let *R* be the intersection of the set of all rings containing α. Clearly *R* must also be a ring. If *M* is another ring containing α then $R \subset M$ since *R* is the intersection of all rings containing α. Therefore *R* is the smallest ring containing α. ∎

COROLLARY 8.1 If α is a collection of subsets of X then there is a smallest algebra containing α.

PROPOSITION 8.3 If α is a collection of subsets of X then there is a smallest σ-ring R containing α.

Proof: As in Proposition 8.2 we can let *R* be the intersection of the family of all σ-rings containing α since there exists at least one such σ-ring. As in the previous proof, *R* is a ring. To show *R* is a σ-ring let $\{A_i\}$ be a sequence in *R*. This implies $\{A_i\}$ is a sequence in each σ-ring containing α which in turn implies $\cup A_i$ is contained in each σ-ring containing α so $\cup A_i \in R$. Therefore R is a σ-ring. If M is any other σ-ring containing α then $R \subset M$ so R is the smallest σ-ring containing α. ■

COROLLARY 8.2 If α is a collection of subsets of X such that X ∈ α then there is a smallest σ-algebra A containing α.

Let (X,τ) be a topological space. Corollary 8.2 implies the existence of a smallest σ-algebra B containing τ. The members of B are called the **Borel sets** of X. In particular, all open and all closed sets belong to B. Furthermore, all F_σ's (countable unions of closed sets) and all G_δ's (countable intersections of open sets) belong to B. Also, Proposition 8.3 implies the existence of a smallest σ-ring B_0 containing all compact subsets of X. The members of B_0 are called the **Baire sets** of X.

Let R be a ring and let $\mu: R \to [0, \infty]$. Let A, B ∈ R. If μ has the following properties:

M1. $\mu(\varnothing) = 0$,
M2. if $A \subset B$ then $\mu(A) \le \mu(B)$,
M3. if $A \cap B = \varnothing$ then $\mu(A \cup B) = \mu(A) + \mu(B)$,

then μ is said to be a **measure** on R and the triple (X, R, μ) is a **measure ring**. The members of R are the **measurable subsets** of X. If R is also an algebra then (X, R, μ) is a **measure algebra**. If R is a σ-ring and μ has the additional property

M4. $\mu(\cup A_i) = \Sigma\mu(A_i)$ whenever $\{A_i\}$ is a sequence in R with $A_i \cap A_j = \varnothing$ whenever $i \ne j$,

then μ is said to be **σ-additive**. If R is a σ-algebra and μ is σ-additive then (X, R, μ) is called a **measure space**. Properties M1 and M2 are easily derived from property M3 (Exercise 1) but it is customary to list them when defining a measure, probably because properties M1 and M2 together with an alteration of M4 are used in defining "outer measures." Outer measures will be introduced in Section 8.4.

As will be seen later in this chapter, sets of measure zero are negligible when integrating. Therefore, it is reasonable to expect subsets of these "negligible sets" to be negligible. However, it may happen that for some A ∈ R, μ(A) = 0 but there exists a B ⊂ A with B not belonging to R and consequently B is not measurable. For such subsets we could define μ(B) = 0, but would this extension of μ to subsets of measure zero still be a measure on some ring in X? If μ is a measure on a ring R in X, μ is said to be **complete** if whenever A ∈ R with μ(A) = 0 and B ⊂ A then B ∈ R.

THEOREM 8.1 If μ is a measure on a ring R in X and if R is defined to be the family of subsets H of X for which there exists sets A, B $\in$ R with A $\subset$ H $\subset$ B and $\mu(B - A) = 0$ then R* is also a ring in X. Define $\mu^*(H) = \mu(A)$ in this case and $\mu^*(B) = \mu(B)$ for B $\in$ R. Then μ^* is a complete measure on R*.*

This extended measure μ^* is called the **completion** of μ. With this new measure, all subsets of measure zero with respect to the original measure μ are now measurable. Consequently, we can essentially assume all measures are complete for all practical purposes.

Proof of Theorem 8.1: If H, K $\in R^*$ then there are sets $A_1, A_2, B_1, B_2 \in R$ such that $A_1 \subset H \subset B_1$, $A_2 \subset K \subset B_2$, $\mu(B_1 - A_1) = 0$ and $\mu(B_2 - A_2) = 0$. To show that H - K $\in R^*$ first observe that $A_1 - B_2 \subset H - K \subset B_1 - A_2$. It needs to be demonstrated that $\mu((B_1 - A_2) - (A_1 - B_2)) = 0$. For this it will first be shown that

$$(B_1 - A_2) - (A_1 - B_2) \subset [(B_1 - A_1) - B_2] \cup [(B_2 - A_2) \cap B_1].$$

Now $y \in (B_1 - A_2) - (A_1 - B_2)$ implies $y \in B_1$ and y does not belong to either A_2 or $(A_1 - B_2)$. This yield two cases:

Case 1: (y does not belong to A_1) Suppose y does not belong to $(B_2 - A_2) \cap B_1$. This implies y does not belong to $(B_2 - A_2)$. But then y does not belong to B_2 since we already know y does not belong to A_2. A_1 does not contain y implies $y \in B_1 - A_1$. Since y does not belong to B_2 we have $y \in [(B_1 - A_1) - B_2]$.

Case 2: ($y \in A_1 \cap B_2$). Then $y \in [(B_2 - A_2) \cap B_1]$. In either case we have $y \in [(B_1 - A_1) - B_2] \cup [(B_2 - A_2) \cap B_1]$. Since μ is a measure we have

$$\mu((B_1 - A_2) - (A_1 - B_2)) \leq \mu([(B_1 - A_1) - B_2] \cup [(B_2 - A_2) \cap B_1]) \leq$$

$$\mu((B_1 - A_1) - B_2) + \mu((B_2 - A_2) \cap B_1) \leq \mu(B_1 - A_1) + \mu(B_2 - A_2) = 0.$$

Consequently $\mu((B_1 - A_2) - (A_1 - B_2)) = 0$ so H - K $\in R^*$. It remains to show that H$\cup$K $\in R^*$. But $A_1 \cup A_2 \subset H \cup K \subset B_1 \cup B_2$ and $(B_1 \cup B_2) - (A_1 \cup A_2) \subset (B_1 - A_1) \cup (B_2 - A_2)$ since if y does not belong to B_1 then $y \in B_2$ and y does not belong to A_2 implies $y \in B_2 - A_2$. Then

$$\mu((B_1 \cup B_2) - (A_1 \cup A_2)) \leq \mu((B_1 - A_1) \cup (B_2 - A_2)) = 0.$$

Thus $\mu((B_1 \cup B_2) - (A_1 \cup A_2)) = 0$ so H$\cup$K $\in R^*$ which implies R^* is a ring.

Next we show that μ^* is well defined on R^*. If $A_1 \subset H \subset B_1$ and $A_2 \subset H \subset B_2$ with $\mu(B_1 - A_1) = 0 = \mu(B_2 - A_2)$ then $A_1 - A_2 \subset B_2 - A_2$ so $\mu(A_1 - A_2) = 0$. Similarly $\mu(A_2 - A_1) = 0$. Therefore $\mu(A_1) = \mu(A_1 \cap A_2) = \mu(A_2)$. Clearly μ^*

satisfies properties M1 through M3. Consequently μ^* is a measure on R^*. By the definition of μ^* and R^*, μ^* is complete. ■

COROLLARY 8.3 If μ is a measure on an algebra A in X and if A is defined to be the family of subsets* H *of X for which there exists sets* A, B $\in$ *A with* A $\subset$ H $\subset$ B *and* μ(B - A) = 0 *then A* is also an algebra in X. Define* μ^*(H) = μ(A) *in this case and* μ^*(B) = μ(B) *for* B $\in$ *A. Then μ^* is a complete measure on A*.*

It is not surprising that Theorem 8.1 and Corollary 8 .3 can be extended to σ-rings and σ-algebras respectively where μ is a σ-additive measure. This follows from the fact that if μ is a σ-additive measure on the σ-ring R in X and if $H_i \in R^*$ for each $i = 1, 2, 3 \ldots$ then it is possible to find $A_i, B_i \in R$ for each i with $\mu(B_i - A_i) = 0$, such that $A_i \subset H_i \subset B_i$. Put $A = \cup A_i$, $B = \cup B_i$ and $H = \cup H_i$. Then A $\subset$ H $\subset$ B and since R is a σ-ring, A, B $\in R$. Since

$$B - A \subset \cup\{B_i - A_i \mid i = 1, 2, 3 \ldots\}$$

we have $\mu(B - A) = 0$ if $\mu(B_i - A_i) = 0$ for each i since μ is σ-additive. This being the case, we conclude H $\in R^*$ so that R^* is a σ-ring. We record this as:

THEOREM 8.2 If μ is a σ-additive measure on a σ-ring R in X and if R is defined to be the family of subsets* H *of X for which there exists sets* A, B $\in$ *R with* A $\subset$ H $\subset$ B *and* μ(B - A) = 0 *then R* is also a σ-ring in X. Define* μ^*(H) = μ(A) *in this case and* μ^*(B) = μ(B) *for* B $\in$ *R. Then μ^* is a complete measure on R*.*

COROLLARY 8.4 If μ is a σ-additive measure on a σ-algebra A in X and if A is defined to be the family of subsets* H *of X for which there exists sets* A, B $\in$ *A with* A $\subset$ H $\subset$ B *and* μ(B - A) = 0 *then A* is also a σ-algebra in X. Define* μ^*(H) = μ(A) *in this case and* μ^*(B) = μ(B) *for* B $\in$ *A. Then μ^* is a complete measure on A*.*

Let (X,τ) be a topological space and let B denote the Borel sets of X with respect to τ. If μ is a measure on B then μ is said to be a **Borel measure** for (X,τ) or simply a Borel measure if τ is understood. By Corollary 8.3, B can be extended to a σ-algebra B^* consisting of the family of subsets H of X for which there exist sets A, B $\in B$ with A $\subset$ H $\subset$ B and $\mu(B - A) = 0$. Furthermore, μ^* defined on B^* by $\mu^*(H) = \mu(A)$ is a complete measure on B^*. μ^* is said to be a **Lebesgue measure** on B^*. More generally, a Lebesgue measure is a complete measure on a σ-algebra containing the Borel sets of X.

Similarly, a measure m_0 on the σ-ring B_0 of Baire sets is said to be a **Baire measure** for (X, τ) or simply a Baire measure if τ is understood. Analogously, B_0 can be extended to a σ-ring B_0^* consisting of subsets K of X for which there are A, B $\in B_0$ with A $\subset$ K $\subset$ B and μ_0(B - A) = 0. Then μ_0^* defined on B_0^* by μ_0^*(K) = μ_0(A) is a complete measure on B_0^*. By a *complete* Borel or Baire measure is meant a complete measure on B^* or B_0^* respectively.

Let L be the collection of members of B that can be covered by countably many compact sets. It is left as an exercise (Exercise 2) to show that L is a σ-ring in X, sometimes called the **Lebesgue ring** in X. This can sometimes lead to confusion since both members of L and B^* are said to be Lebesgue measurable sets. In what follows, members of L will be referred to as Lebesgue measurable (L) whereas members of B^* will be referred to as Lebesgue measurable (B^*). If neither L or B^* is specified, the term **Lebesgue measurable set** will refer to a member of B^*.

If μ is a σ-additive measure on L, it is easily shown (Exercise 3) that μ can be extended to a σ-additive measure on B by setting $\mu^{\sim}$(A) = μ(A) if A $\in L$ and $\mu^{\sim}$(A) = ∞ otherwise. Then $\mu^{\sim}$ is a σ-additive measure on the σ-algebra B, and as such can be extended to a complete σ-additive measure on B^*. Therefore, in some sense it can be said that when dealing with Lebesgue measures it suffices only to consider the σ-additive measures on L. However, this is not precisely true since there exist **finite** Lebesgue measures (i.e., μ(A) < ∞ for each A $\in L$) but the extension process discussed above always leads to an **infinite** measure (i.e., there exists some A $\in B^*$ for which μ(A) = ∞) if $L \neq B$.

EXERCISES

1. Show that the defining properties M1 and M2 of a measure can be derived from the property M3.

2. Let X be an uncountable set and let M be the collection of all sets E $\subset$ X such that E is countable or X - E is countable. Define μ on M by μ(E) = 0 if E is countable and μ(E) = 1 otherwise. Show that M is a σ-algebra and that μ is a measure on M.

3. It is easily shown that if μ is a measure and A_1 and A_2 are measurable sets that $\mu(A_1 \cup A_2) = \mu(A_1) + \mu(A_2) - \mu(A_1 \cap A_2)$ and if A_3 is also measurable then $\mu(A_1 \cup A_2 \cup A_3) = \mu(A_1) + \mu(A_2) + \mu(A_3) - \mu(A_1 \cap A_2) - \mu(A_1 \cap A_3) - \mu(A_2 \cap A_3) + \mu(A_1 \cap A_2 \cap A_3)$. Show that in general, if $\{A_n\}$ where $n = 1 \ldots N$ is a finite sequence of measurable sets that

$$\mu(\cup_{n=1}^{N} A_n) = \Sigma_{n=1}^{N} (-1)^{n+1} (\Sigma_{n\,1<\ldots<nj} \mu[\cap_{n\,1<\ldots<nj} A_{jn}]).$$

In other words, $\mu(\cup_{n=1}^{N} A_n)$ is an alternating finite series of terms such that the nth term is the sum of all the measures of distinct intersections of n members of $\{A_n\}$.

4. Let L be the Lebesgue ring in a topological space (X,τ). Show that L is a σ-ring in X.

5. If μ is a σ-additive measure on L show that μ can be extended to a σ-additive measure $\mu^{\sim}$ on B by setting $\mu^{\sim}(A) = \mu(A)$ if $A \in L$ and $\mu^{\sim}(A) = \infty$ otherwise.

8.3 Properties of Measures

A set function μ on a collection of sets S is said to be **monotone** if whenever A, B $\in S$ and $A \subset B$ then $\mu(A) \le \mu(B)$. The defining property M2 of a measure can be restated as: μ is monotone on R. The defining property M3 can be restated as: μ is **additive** on R. A set function μ on a collection S is called **subtractive** if whenever A, B $\in S$ with $A \subset B$, $B - A \in S$, and $|\mu(A)| < \infty$ then $\mu(B - A) = \mu(B) - \mu(A)$.

LEMMA 8.1 If μ is a measure on a ring R then μ is subtractive.

Proof: If A, B $\in R$ with $A \subset B$, $B - A \in R$ and $|\mu(A)| < \infty$, then by M3, $\mu(B) = \mu(A) + \mu(B - A)$. Since $|\mu(A)| < \infty$, it may be subtracted from both sides of this equation which gives $\mu(B - A) = \mu(B) - \mu(A)$. ∎

LEMMA 8.2 If μ is a σ-additive measure on a σ-ring R and if $A \in R$ with $\{A_n\}$ being a sequence in R covering A then $\mu(A) \le \Sigma_n \mu(A_n)$.

Proof: For each positive integer n put $B_n = A_n - \cup_{i<n} A_i$. Then $B_i \cap B_j = \varnothing$ if $i \ne j$. Consequently $\{A \cap B_n\}$ is a sequence of disjoint sets whose union is A. By M4, we have $\mu(A) = \Sigma\mu(A \cap B_n) \le \Sigma\mu(B_n) \le \Sigma\mu(A_n)$. ∎

LEMMA 8.3 If μ is a σ-additive measure on a σ-ring R and if $\{A_n\}$ is a disjoint sequence of members of R with $\cup A_n \subset A \in R$ then $\Sigma\mu(A_n) \le \mu(A)$.

Proof: By M2, $\mu(\cup A_n) \le \mu(A)$ and by M4 $\Sigma\mu(A_n) \le \mu(\cup A_n)$. Combining these inequalities gives the desired result. ∎

PROPOSITION 8.4 If μ is a measure on a ring R and $\{E_1 \ldots E_N\} \subset R$ covers each point of $E \in R$ at least M times, then $\mu(E) \le (\Sigma_{n=1}^N \mu(E_n))/M$.

Proof: We use induction on N. If $N = 1$ then we must have $M = 1$ and the proof is trivial. Assume it is true for each positive integer $n < N$. If $M = 1$, the result follows from Lemma 8.2, so suppose $M > 1$. Let $\{E_1 \ldots E_N\} \subset R$ cover each point of $E \in R$ at least M times. Put $\mathcal{U} = \{E_1 \ldots E_{N-1}\}$ and let V be the set of points of E that are covered at least M times by $\mathcal{U}$. Let $p \in$ V and let $\mathcal{V}$ be a collection of M members of $\mathcal{U}$ that contain p. Then $p \in \cap\mathcal{V} \in R$. Moreover $\cap\mathcal{V} \subset$ V. Clearly, V is the union of the collection of non-void intersections of M

members of $\mathcal{U}$. Since this collection is finite and each of these intersections is measurable, so are V and E - V. By the induction hypothesis we have $\mu(V) \leq \Sigma_{n=1}^{N-1}\mu(E_n \cap V)/M$ and $\mu(E - V) \leq \Sigma_{n=1}^{N-1}\mu(E_n - V)/(M - 1)$. Since $E - V \subset E_N$ we have $\mu(E - V) \leq \mu(E_N)$. Then $(M - 1)\mu(E - V) \leq \Sigma_{n=1}^{N-1}\mu(E_n - V)$ which implies $M\mu(E - V) \leq \Sigma_{n=1}^{N-1}\mu(E_n - V) + \mu(E_N)$. Therefore,

$$\mu(E) = \mu(V) + \mu(E - V) \leq \Sigma_{n=1}^{N-1}\mu(E_n \cap V)/M + \Sigma_{n=1}^{N-1}\mu(E_n - V)/M + \mu(E_N)/M =$$

$$\Sigma_{n=1}^{N-1}\mu(E_n)/M + \mu(E_N)/M = \Sigma_{n=1}^{N}\mu(E_n)/M. \blacksquare$$

PROPOSITION 8.5 If μ is a measure on a ring R and $\{E_1 \ldots E_N\} \subset R$ is a collection of subsets of $E \in R$ such that no point of E is contained in more than M of the E_n, then $\mu(E) \geq (\Sigma_{n=1}^{N}\mu(E_n))/M$.

Proof: Again, we induct on N. If $N = 1$ the proof is trivial so assume the proposition holds for each positive integer $n < N$. If $M = 1$ the proof is trivial so assume $M > 1$. Then by the induction hypothesis, $\mu(E) \geq \Sigma_{n=1}^{N-1}\mu(E_n)/(M - 1)$ so $(M - 1)\mu(E) \geq \Sigma_{n=1}^{N-1}\mu(E_n)$. Therefore, $M\mu(E) \geq \Sigma_{n=1}^{N-1}\mu(E_n) + \mu(E) \geq \Sigma_{n=1}^{N}\mu(E_n)$ which implies $\mu(E) \geq \Sigma_{n=1}^{N}\mu(E_n)/M$. ■

PROPOSITION 8.6 If μ is a σ-additive measure on a σ-ring R and $\{A_n\}$ is an ascending sequence of sets (i.e., $A_n \subset A_{n+1}$ for each n) then $\mu(\cup A_n) = \lim_{n\to\infty}\mu(A_n)$.

Proof: Put $A_0 = \emptyset$. Then $\mu(\cup A_n) = \mu(\cup_{n=1}^{\infty}(A_n - A_{n-1})) = \Sigma_{n=1}^{\infty}\mu(A_n - A_{n-1}) = \lim_{N\to\infty}\Sigma_{n=1}^{N}\mu(A_n - A_{n-1}) = \lim_{N\to\infty}\mu(\cup_{n=1}^{N}(A_n - A_{n-1})) = \lim_{N\to\infty}\mu(A_N)$. ■

PROPOSITION 8.7 If μ is a σ-additive measure on a σ-ring R and $\{A_n\}$ is a decreasing sequence (i.e., $A_{n+1} \subset A_n$ for each n) of members of R of which at least one has finite measure then $\mu(\cap A_n) = \lim_{n\to\infty}\mu(A_n)$.

Proof: If for some positive integer m, $\mu(A_m) < \infty$ then $\mu(A_n) \leq \mu(A_m) < \infty$ for each $n > m$ and hence $\mu(\cap A_n) < \infty$. By Lemma 8.1 μ is subtractive so $\mu(A_m) - \mu(\cap A_n) = \mu(A_m - \cap A_n) = \mu(\cup(A_m - A_n))$. Now the sequence $\{A_m - A_n\}$ of members of R is increasing, so by Proposition 8.6 we have: $\mu(\cup(A_m - A_n)) =$

$$\lim_{n\to\infty}\mu(A_m - A_n) = \lim_{n\to\infty}(\mu(A_m) - \mu(A_n)) = \mu(A_m) - \lim_{n\to\infty}\mu(A_n).$$

Combining these two equations yields:

$$\mu(A_m) - \mu(\cap A_n) = \mu(A_m) - \lim_{n\to\infty}\mu(A_n)$$

from which the desired result can be obtained. ■

A set function $\mu: C \to [0, \infty]$ where C is a collection of sets is said to be **continuous from below** at a set A if for every increasing sequence $\{A_n\} \subset C$ for which $\cup A_n = A$ we have $\lim_{n\to\infty}\mu(A_n) = \mu(A)$. Similarly, μ is **continuous from above** at A if for each descending sequence $\{A_n\} \subset C$ for which $\cap A_n = A$ and $|\mu(A_m)| < \infty$ for some positive integer m, we have $\lim_{n\to\infty}\mu(A_n) = \mu(A)$. Propositions 8.7 and 8.8 show that σ-additive measures on σ-rings are both continuous from above and below. The following proposition shows that the converse also holds under appropriate conditions.

PROPOSITION 8.8 If μ is a finite, non-negative, additive set function on a σ-ring R and μ is either continuous from below at each $A \in R$ or continuous from above at $\varnothing$, then μ is a σ-additive measure on R.

Proof: Since μ is additive, property M3 of the definition of a measure holds. By Exercise 1 of Section 8.2, properties M1 and M2 hold so it only remains to show that μ is σ-additive. For this, first observe that the additivity of μ and the fact that R is a ring implies (by induction) that μ is finitely additive; i.e., if $A_1 \ldots A_n \in R$ with $A_i \cap A_j = \varnothing$ for $i \neq j$ then $\mu(\cup_{i=1}^n A_i) = \Sigma_{i=1}^n \mu(A_i)$. Let $\{A_n\}$ be a disjoint sequence of members of R and put $A = \cup A_n$. For each positive integer n put $F_n = \cup_{i=1}^n A_i$ and $G_n = A - F_n$. If μ is continuous from below, then since $\{F_n\}$ is increasing and $A = \cup F_n$, by Proposition 8.7 we have:

$$\mu(A) = \lim_{n\to\infty}\mu(F_n) = \lim_{n\to\infty}\Sigma_{i=1}^n \mu(A_i) = \Sigma_{i=1}^\infty \mu(A_i).$$

If μ is continuous from above at $\varnothing$, then since $\{G_n\}$ is decreasing and $\cap G_n = \varnothing$ we have for each positive integer, $\mu(A) = \Sigma_{i=1}^n \mu(A_i) + \mu(G_n)$. Taking the limit of both sides of this equation gives:

$$\lim_{n\to\infty}\mu(A) = \lim_{n\to\infty}\Sigma_{i=1}^n \mu(A_i) + \lim_{n\to\infty}\mu(G_n) = \Sigma_{i=1}^\infty \mu(A_i) + 0$$

by Proposition 8.8. Consequently, in either case, μ is σ-additive which is the desired result. ∎

EXERCISES

1. If μ is a σ-additive measure on a σ-ring R and if $\{A_n\}$ is a sequence of sets in R, and if we define

$$\mathit{lim\ inf}_{n\to\infty} A_n = \cup_{n=1}^\infty \cap_{i=1}^n A_i \quad \text{and} \quad \mathit{lim\ sup}_{n\to\infty} A_n = \cap_{n=1}^\infty \cup_{i=1}^n A_i,$$

show that $\mu(\mathit{lim\ inf}_{n\to\infty} A_n) \leq \mathit{lim\ inf}_{n\to\infty}\mu(A_n)$ and if $\mu(\cup_{i=n}^\infty A_i)$ is finite for at least one value of n, then $\mu(\mathit{lim\ sup}_{n\to\infty} A_n) \geq \mathit{lim\ sup}_{n\to\infty}\mu(A_n)$.

2. Let **R** denote the real line with the interval topology and let B denote the

family of all bounded half open intervals of the form $[a, b)$ in **R**. Let R be the collection of all finite disjoint unions of members of B. Define μ on B by $\mu([a,b)) = b - a$ and observe that $\mu(\emptyset) = \mu([a, a)) = a - a = 0$.

(a) Show that if $\{A_1 \ldots A_n\}$ are disjoint members of B and each A_i is contained in some given set $A \in B$ then $\Sigma_{i=1}^{n}\mu(A_i) \leq \mu(A)$.
(b) Show that if a closed interval $[a_0, b_0]$ is contained in the union of a finite number of bounded open intervals, say (a_i, b_i) for $i = 1 \ldots n$ then $b_0 - a_0 < \Sigma_{i=1}^{n}(b_i - a_i)$.
(c) Show that if $\{A_n\}$ is a sequence of sets in B and $A \in B$ with $A \subset \cup_{i=1}^{\infty} A_i$ then $\mu(A) \leq \Sigma_{i=1}^{\infty}\mu(A_i)$.
(d) Show that R is a σ-ring in **R**.

8.4 Outer Measures

A non-empty collection of sets H in a set X is said to be **hereditary** if whenever $A \in H$ and $B \subset A$ then $B \in H$. Hereditary collections share the property with rings, σ-rings, algebras and σ-algebras that the intersection of any set of hereditary collections is again a hereditary collection and consequently, for any collection S of sets there is a smallest hereditary collection H_S containing S. In what follows, we will be interested in those hereditary collections that are also σ-rings and σ-algebras. Hereditary σ-rings are the class of sets upon which we will define outer measures.

Outer measures, sometimes referred to a Carathéodory measures (after C. Carathéodory who introduced them in a paper titled *Ueber das lineare Mass von Punktmengen - eine Verallgemeinerung des Langenbegriffs* (Nachr. Ges. Wiss. Gottingen) in 1914, are an important generalization of measures. Their importance arises from a standard technique employed when attempting to prove a set function is a measure. This technique is to first prove the function is an outer measure on some hereditary σ-ring and then use a theorem from the theory of outer measures (to be demonstrated below) to conclude that the function is a measure on a suitably restricted subcollection that is a σ-ring.

If E is any collection of sets, then $H(E)$ will denote the smallest hereditary Σ-ring containing E and $H(E)$ will be referred to as the hereditary σ-ring **generated** by E. $H(E)$ is the collection of all sets that can be covered by countably many members of E. That $H(E)$ is indeed a hereditary σ-ring is left as an exercise (Exercise 1). Let R be a hereditary σ-ring and let $\mu^*: R \to [0,\infty]$. If μ^* has the property

M4′. $\mu^*(\cup A_i) \leq \Sigma\mu^*(A_i)$ whenever $\{A_i\}$ is a sequence in R

then μ^* is said to be **countably subadditive**. If μ^* has the property that whenever $A, B \in R$ then $\mu^*(A \cup B) \leq \mu^*(A) + \mu^*(B)$ then μ^* is said to be

subadditive. If μ^* has properties M1 and M2 that define a measure plus property M4′ then μ^* is called an **outer measure.** Measures on rings can be extended to outer measures on the hereditary σ-rings that these rings generate.

THEOREM 8.3 If μ is a σ-additive measure on a σ-ring R and if for each $A \in H(R)$ we define $\mu^*(A) = inf\{\Sigma\mu(A_n) \mid \{A_n\}$ is a sequence in R covering A} then μ^* is an outer measure on H(R) such that $\mu^*(A) = \mu(A)$ for each $A \in R$.

Proof: Let $A \in R$ and put $A_1 = A$ and $A_n = \varnothing$ for each positive integer $n > 1$. Then $\{A_n\} \subset R$ and $A \subset \cup A_n$. By the definition of μ^*, $\mu^*(A) \leq \Sigma\mu(A_n) = \mu(A) + \Sigma_{n>1}\mu(A_n) = \mu(A) + 0$. Conversely, if $\{A_n\}$ is a sequence in R covering A, then by Proposition 8.5 $\mu(A) \leq \Sigma\mu(A_n)$ so by the definition of μ^*, $\mu(A) \leq \mu^*(A)$. Consequently $\mu^*(A) = \mu(A)$ for each $A \in R$.

Since $\varnothing \in R$ this shows that $\mu^*(\varnothing) = 0$ so μ^* satisfies property M1 of a measure. To show μ^* satisfies property M2, let $A, B \in H(R)$ with $A \subset B$. Let $\{B_n\}$ be a sequence in R covering B. Then $\{B_n\}$ also covers A so $\mu^*(A) \leq \Sigma\mu(B_n)$. But since this is the case for all sequences $\{B_n\}$ in R covering B, from the definition of μ^* we have $\mu^*(A) \leq \mu^*(B)$. So μ^* satisfies M2.

To show μ^* satisfies M4′ let $A \in H(R)$ and let $\{A_n\}$ be a sequence of members of $H(R)$ covering A. Let $\varepsilon > 0$. By the definition of μ^*, it is possible to choose a sequence $\{A^n_m\}$ in R covering A_n such that:

$$\Sigma_{m=1}^{\infty}\mu(A^n_m) \leq \mu^*(A_n) + \varepsilon/2^n.$$

Since $\{A^n_m \mid n = 1, 2, 3 \ldots \text{ and } m = 1, 2, 3 \ldots\}$ is a countable collection of members of R covering A it follows that:

$$\mu^*(A) \leq \Sigma_{n=1}^{\infty}\Sigma_{m=1}^{\infty}\mu(A^n_m) \leq \Sigma_{n=1}^{\infty}(\mu^*(A_n) + \varepsilon/2^n) = \Sigma_{n=1}^{\infty}\mu^*(A_n) + \varepsilon.$$

Since ε was chosen arbitrarily, this shows μ^* satisfies M4′. Thus μ^* is an outer measure on $H(R)$. ∎

The outer measure μ^* defined in Theorem 8.3 is said to be an **extension** of μ to $H(R)$. Not only is the question of extending measures to outer measures of interest, but of restricting outer measures to smaller classes of sets on which they become measures. Let μ^* be an outer measure on a hereditary σ-ring H. A set $A \in H$ will be called **μ^*-measurable** if for every pair of sets $P, Q \in H$ with $P \subset A$ and $Q \subset X - A$ we have:

$$\mu^*(P \cup Q) = \mu^*(P) + \mu^*(Q).$$

The concept of μ^*-measurability is of fundamental importance in the theory of outer measures. This is because the μ^*-measurable sets constitute a σ-ring in H and μ^* restricted to these sets is a σ-additive measure on them. It is often difficult to gain an intuitive understanding of the nature of these sets. Outer measures can fail to be measures because they do not satisfy property M3. The definition of μ^*-measurable sets focuses on those sets that in some sense force additivity on all other pairs of sets in *H* that they "separate." An equivalent way of defining μ^*-measurable sets is to define them to be the sets $A \in H$ such that if $B \in H$ then $\mu^*(B) = \mu^*(B \cap A) + \mu^*(B - A)$. This definition may shed some additional light on the nature of these sets. It characterizes them as those members of *H* that, no matter how they "split" another member B of *H*, μ^* is additive on the two subsets into which B was split. In view of property M4′, this last definition of μ^*-measurable sets can be replaced by

$$\mu^*(B) \geq \mu^*(B \cap A) + \mu^*(B - A)$$

in the defining relation. We will denote the class of all μ^*-measurable sets by *M*.

PROPOSITION 8.9 If μ^* is an outer measure on a hereditary σ-ring H and if M is the class of all μ^*-measurable sets then M is a ring.

Proof: If $A, B \in M$ and $C \in H$ then we have the following three equations:

(8.1) $\mu^*(C) = \mu^*(C \cap A) + \mu^*(C - A)$.
(8.2) $\mu^*(C \cap A) = \mu^*(C \cap A \cap B) + \mu^*((C \cap A) - B)$.
(8.3) $\mu(C - A) = \mu^*((C - A) \cap B) + \mu^*((C - A) - B)$.

Substituting (8.2) and (8.3) into (8.1) gives:

(8.4) $$\mu^*(C) = \mu^*(C \cap A \cap B) + \mu^*((C \cap A) - B) + \mu^*((C - A) \cap B) + \mu^*((C - A) - B).$$

If we replace C by $C \cap (A \cup B)$ in equation (8.4), the first three terms on the right remain unchanged while the fourth term simply drops out. Therefore, we have

(8.5) $$\mu^*(C \cap (A \cup B)) = \mu^*(C \cap A \cap B) + \mu^*((C \cap A) - B) + \mu^*((C - A) \cap B).$$

Now substituting (8.5) into (8.4) and using the fact that $(C - A) - B = C - (A \cup B)$ we get:

$$\mu^*(C) = \mu^*(C \cap (A \cup B)) + \mu^*(C - (A \cup B))$$

which shows that $A \cup B \in M$. To show $A - B \in M$, we replace C by $C \cap (X - [A - B])$ in equation (d) and notice that the second term on the right goes to zero while the other three terms remain unchanged. This gives

$$(8.6)\quad \mu^*(C\cap(X - [A - B])) = \mu^*(C\cap A\cap B) + \mu^*((C - A)\cap B) + \mu^*((C - A) - B).$$

Substituting (8.6) into (8.4) and using the fact that $C\cap(X - [A - B]) = C - (A - B)$ we get:

$$\mu^*(C) = \mu^*(C\cap(A - B)) + \mu^*(C - (A - B))$$

which shows $A - B \in M$. Consequently M is a ring in H. ∎

PROPOSITION 8.10 If μ^* is an outer measure on a hereditary σ-ring H and if M is the class of all μ^*-measurable sets in H, then if $B \in H$ and $\{A_n\}$ is a disjoint sequence of sets in M with $A = \cup A_n$, then $\mu^*(B\cap A) = \Sigma_n \mu^*(B\cap A_n)$.

Proof: Replacing C by B in (8.5) in the proof of Proposition 8.10 and replacing A and B by A_1 and A_2 we get:

$$\mu^*(B\cap(A_1\cup A_2)) = \mu^*(B\cap A_1\cap A_2) + \mu^*((B\cap A_1) - A_2) + \mu^*((B - A_1)\cap A_2).$$

Since $A_1\cap A_2 = \varnothing$, this equation reduces to $\mu^*(B\cap(A_1\cup A_2)) = \mu^*(B\cap A_1) + \mu^*(B\cap A_2)$. By induction, we can extend this result to

$$(8.7)\qquad \mu^*(B\cap(\cup_{i=1}^n A_i)) = \Sigma_{i=1}^n \mu^*(B\cap A_i)$$

for each positive integer n. Next put $H_n = \cup_{i=1}^n A_i$ for each n. By Proposition 8.10, M is a ring so $H_n \in M$ for each n. Consequently $\mu^*(B) = \mu^*(B\cap H_n) + \mu^*(B - H_n)$. Substituting (8.7) into this equation gives $\mu^*(B) = \Sigma_{i=1}^n \mu^*(B\cap A_i) + \mu^*(B - H_n)$. Since $H_n \subset A$ for each n, $\mu^*(B - A) \le \mu^*(B - H_n)$ for each n. Thus $\mu^*(B) \ge \Sigma_{i=1}^n \mu^*(B\cap A_i) + \mu(B - A)$ for each n. Hence $\mu^*(B) \ge \Sigma\mu(B\cap A_n) + \mu(B\text{-}A)$. Since μ^* is an outer measure, $\mu^*(B\cap A) = \mu^*(B\cap[\cup A_n]) \le \Sigma\mu^*(B\cap A_n)$. Thus

$$(8.8)\qquad \mu^*(B) \ge \Sigma\mu^*(B\cap A_n) + \mu^*(B - A) \ge \mu^*(B\cap A) + \mu^*(B - A).$$

But then $A \in M$ by the alternative definition of μ^*-measurability, and therefore $\mu^*(B) = \mu^*(B\cap A) + \mu^*(B - A)$. But then by (8.8) we have

$$(8.9)\qquad \Sigma\mu^*(B\cap A_n) + \mu^*(B - A) = \mu^*(B\cap A) + \mu^*(B - A).$$

Since we do not know if $\mu^*(B - A)$ is finite we cannot merely subtract $\mu^*(B - A)$ from each side of (8.9) to get the desired result. Instead we notice that (8.9) holds for any $B \in H$ so we can substitute $B\cap A$ for B in (8.9). This causes the second term on each side of the equation to vanish, giving the desired result. ∎

THEOREM 8.4 If μ^* is an outer measure on a hereditary σ-ring H and if M is the set of all μ^*-measurable sets in H, then M is a σ-ring, every set of outer measure zero belongs to M and the set function μ defined on M by $\mu(A) = \mu^*(A)$ is a complete σ-additive measure on M.

Proof: Let $\{M_n\}$ be a sequence in M and let $B \in H$. Let $A_1 = M_1$ and let $B_n = \cup_{i=1}^{n} M_i$. Then put $A_n = M_n - B_{n-1}$. Since M is a ring (Proposition 8.9), each $A_n \in M$. Clearly $\{A_n\}$ is a disjoint sequence in M. For each positive integer n we want to show

$$\mu^*(B) = \Sigma_{i=1}^{n} \mu(B \cap A_i) + \mu(B - B_n). \tag{8.10}$$

Let K be the set of positive integers for which (8.10) holds. Clearly $1 \in K$. Assume $n \in K$. Since $A_{n+1} \in M$ we have $\mu^*(B) = \mu^*(B \cap A_{n+1}) + \mu^*(B - A_{n+1})$. Since $A_{n+1} \cap B_n = \emptyset$ and $B - A_{n+1} \in H$ we have:

$$\mu^*(B) = \mu^*(B \cap A_{n+1}) + \mu^*((B - A_{n+1}) \cap B_n) + \mu^*((B - A_{n+1}) - B_n) =$$

$$\mu^*(B \cap A_{n+1}) + \mu^*(B \cap B_n) + \mu^*(B - B_{n+1}).$$

Now by Proposition 8.10, $\mu^*(B \cap B_n) = \mu^*(B \cap [\cup_{i=1}^{n} A_i]) = \Sigma_{i=1}^{n} \mu^*(B \cap A_i)$. Therefore

$$\mu^*(B) = \Sigma_{i=1}^{n+1} \mu^*(B \cap A_i) + \mu^*(B - B_{n+1}).$$

Hence $n+1 \in K$ so K is the set of all positive integers. Next observe that for each positive integer n, $\mu^*(B - A) \le \mu^*(B - B_n)$ where $A = \cup A_n = \cup M_n$. Thus

$$\mu^*(B) \ge \Sigma_{i=1}^{n} \mu^*(B \cap A_i) + \mu^*(B - A) \tag{8.11}$$

holds for each positive integer n. Taking the limit of both sides of (8.11) as $n \to \infty$ gives $\mu^*(B) \ge \Sigma_{n=1}^{\infty} \mu^*(B \cap A_n) + \mu^*(B - A)$. Again, by Proposition 8.10, $\mu^*(B \cap A) = \Sigma_{n=1}^{\infty} \mu^*(B \cap A_n)$ so $\mu^*(B) \ge \mu^*(B \cap A) + \mu^*(B - A)$. But then $\mu^*(B) = \mu^*(B \cap A) + \mu^*(B - A)$ which shows $A \in M$. Consequently, M is a σ-ring.

To show μ is σ-additive let $\{A_n\}$ be a disjoint sequence in M and let $A = \cup A_n$. Then substituting A for B in equation (8.9) we get:

$$\Sigma \mu^*(A \cap A_n) + \mu^*(\emptyset) = \mu^*(A) + \mu^*(\emptyset)$$

so $\Sigma \mu(A_n) = \mu(A)$ or in other words $\Sigma \mu(A_n) = \mu(A)$ so μ is σ-additive. Next let $M \in H$ such that $\mu^*(M) = 0$. Then for each $B \in H$ we have

$$\mu^*(B) = \mu^*(M) + \mu^*(B) \ge \mu^*(B \cap M) + \mu^*(B - M)$$

so M ∈ *M*. Hence every set of outer measure zero belongs to *M*. Finally, to show that μ is complete let A ∈ *M* with μ(A) = 0 and let B ⊂ A. Then μ*(A) = μ(A) = 0 and hence μ*(B) = 0. But then B ∈ *M* so μ(B) = 0. Hence μ is complete. ∎

EXERCISE

1. Let *E* be a collection of subsets of a set X. Show that there exists a smallest hereditary σ-ring *H*(*E*) containing *E* and that *H*(*E*) consists of the collection of all sets that can be covered by countably many members of *E*.

8.5 Measurable Functions

Let X be a set and *M* a σ-algebra on the set X. Then X is said to be a **measurable space** and the members of *M* are called **measurable sets.** If X is a measurable space and Y is a topological space, a function *f*:X → Y is said to be a **measurable function** if f^{-1}(U) ∈ *M* whenever U is open in Y. The proof of the following proposition is left as an exercise.

PROPOSITION 8.11 If f:X → Y is a measurable function from a measurable space X into a topological space Y and g:Y → Z is continuous where Z is a topological space, then the composition function g © f:X → Z is measurable.

THEOREM 8.5 If u and v are real measurable functions on a measurable space X and g is a continuous mapping of the plane into a topological space Y then h:X → Y defined by h(x) = g(u(x), v(x)) for each x ∈ X is measurable.

Proof: For each $x \in$ X put $f(x) = (u(x), v(x))$. Then f maps X into the plane. Since g is continuous and $h = g$ © f, by Proposition 8.11, it suffices to demonstrate the measurability of f. For this let I × J be an open rectangle in the plane where I and J are open intervals in **R**. Then $f^{-1}(\mathrm{I} \times \mathrm{J}) = u^{-1}(\mathrm{I}) \cap v^{-1}(\mathrm{J})$ is measurable since both u and v are measurable functions. Every open set U in the plane is a union of these basic open sets I × J of the product topology. Since the plane is separable, U can be constructed from countably many of these basic rectangles B_n. Since $f^{-1}(\mathrm{U}) = f^{-1}(\cup_{n=1}^{\infty} B_n) = \cup_{n=1}^{\infty} f^{-1}(B_n)$ we see that f^{-1}(U) is measurable. Hence f is a measurable function. ∎

*PROPOSITION 8.12 If X is a measurable space and u and v are real valued measurable functions on X then f:X → **C** defined by f(x) = u(x) + iv(x) for each x ∈ X is a complex valued measurable function on X.*

Proof: The proof follows from Theorem 8.5 with $g(z) = z$ for each complex number z. ▪

PROPOSITION 8.13 If $f:X \to \mathbf{C}$ is a complex measurable function where $f(x) = u(x) + iv(x)$ for each $x \in X$ then u,v and $|f|$ are all real valued measurable functions on X.

Proof: The proof follows from Proposition 8.11 with $g(z) = \text{Re}(z)$, $\text{Im}(z)$ and $|z|$ respectively for the functions u,v and $|f|$. ▪

PROPOSITION 8.14 If f and g are complex valued measurable functions on a measurable space X, then so are $f + g$ and fg.

Proof: If u and v are real valued measurable functions, and $g(u,v) = u + v$, then by Theorem 8.5, $h(x) = g(u(x), v(x)) = u(x) + v(x) = (u + v)(x)$ is measurable. Since f and g are complex measurable functions, $f = u_1 + iv_1$ and $g = u_2 + iv_2$ for some real valued measurable functions u_1,v_1 and u_2,v_2. Then $f + g = (u_1+u_2) + i(v_1 + v_2)$. As shown above, $u_1 + u_2$ and $v_1 + v_2$ are real valued measurable functions, so by Proposition 8.12, $f + g$ is a complex valued measurable function. The proof that fg is measurable follows from the same argument with $g(u(x), v(x)) = u(x)v(x)$. ▪

PROPOSITION 8.15 If E is a measurable set in X, then the characteristic function χ_E of E, defined by $\chi_E(x) = 1$ if $x \in E$ and $\chi_E(x) = 0$ otherwise, is a measurable function.

Proof: If U is an open set in **R** containing 1 but not 0 then $f^{-1}(U) = E$. If U contains 0 but not 1 then $f^{-1}(U) = X - E$. If U contains both 1 and 0 then $f^{-1}(U) = X$ and if U contains neither then $f^{-1}(U) = \varnothing$. In all of these cases, $f^{-1}(U)$ is a measurable set so χ_E is a measurable function. ▪

PROPOSITION 8.16 If f is a complex measurable function on X, there is a complex measurable function a on X such that $|a| = 1$ and $f = a|f|$.

Proof: Let $Z = \{x \mid f(x) = 0\}$. Then Z is measurable since f is measurable. Put $Y = \mathbf{C} - \{0\}$. Define $g:Y \to \mathbf{C}$ by $g(y) = y/|y|$ for each $y \in Y$. Then for each $x \in X$ put $a(x) = g(f(x) + \chi_Z(x))$. If $x \in Z$, $a(x) = 1$. If x does not belong to Z then $a(x) = f(x)/|f(x)|$. In either case, $|a| = 1$. Since g is continuous on Y and Z is measurable, a is measurable on X. Finally, if $x \in Z$ then $f(x) = 0 = 1\times 0 = a|f(x)|$ and if x does not belong to Z then $a|f(x)| = [f(x)/|f(x)|]|f(x)| = f(x)$. ▪

Let X be a space and B the Borel σ-algebra on X. If $f:X \to Y$ is a continuous function where Y is a topological space, then clearly $f^{-1}(U)$ is a Borel set whenever U is open in Y. Therefore, every continuous mapping of X into any topological space is Borel measurable. If Y is **R** or **C**, the Borel measurable functions will simply be called **Borel functions.**

THEOREM 8.6 Let M be a σ-algebra in X and let $f:X \to Y$ where Y is a topological space.

(1) If $A = \{E \subset Y \mid f^{-1}(E) \in M\}$ then A is a σ-algebra in Y.

(2) If f is measurable and E is a Borel set in Y then $f^{-1}(E) \in M$.

(3) If $Y = [-\infty, \infty]$ and $f^{-1}((a, \infty]) \in M$ for each real a, then f is measurable.

Proof: If A, B ∈ *A* then $f^{-1}(B - A) = f^{-1}(B) - f^{-1}(A)$. Since $f^{-1}(A)$ and $f^{-1}(B) \in M$ we have $f^{-1}(B) - f^{-1}(A) \in M$ so $f^{-1}(B - A) \in M$ which implies B - A ∈ *A*. If $\{A_n\}$ is a disjoint collection of members of *A*, then $f^{-1}(\cup A_n) = \cup f^{-1}(A_n) \in M$ since $f^{-1}(A_n) \in M$ for each *n* and *M* is σ-additive. Therefore, $\cup A_n \in A$ so *A* is a σ-ring. But $X = f^{-1}(Y)$ and X ∈ *M* since *M* is an algebra. Therefore, Y ∈ *A* so *A* is a σ-algebra. This proves (1).

To prove (2), let *A* be the σ-ring defined in (1). If *f* is measurable, then *A* contains all open sets in Y since by the definition of measurability, $f^{-1}(U) \in M$ for each open U ⊂ Y and therefore U ∈ *A*. But then *A* contains all the Borel sets so $f^{-1}(E) \in M$ by the definition of *A*.

To prove (3), let *A* be the σ-algebra defined in (1). Since *A* is a σ-algebra in $[-\infty, \infty]$ and $(a, \infty] \in A$ for each $a \in \mathbf{R}$, then $[-\infty, a) \in A$ since

$$[-\infty, a) = \cup[-\infty, a - 1/n] = \cup(a - 1/n, \infty]^c$$

where $(a - 1/n, \infty]^c$ is the complement of $(a - 1/n, \infty]$ in $[-\infty, \infty]$. Consequently, $(a, b) = [-\infty, b) \cap (a, \infty] \in A$. Since every open set in $[-\infty, \infty]$ is a countable union of open intervals (a, b), every open set in $[-\infty, \infty]$ is in *A*. Therefore, *f* is measurable. ∎

Let $\{x_n\}$ be a sequence in $[-\infty, \infty]$ and put $s_k = sup\{x_k, x_{k+1}, x_{k+2} \ldots\}$ for each positive integer *k*. Then put $\sigma = inf\{s_k\}$. We say σ is the **limit superior** of $\{x_n\}$ and denote σ by $lim\ sup\ x_n$. The **limit inferior** is defined similarly: for each positive integer *k* let $i_k = inf\{x_k, x_{k+1}, x_{k+2} \ldots\}$ and put $lim\ inf\ x_n = sup\{i_k\}$. Clearly,

$$lim\ inf\ x_n = -lim\ sup\ (-x_n).$$

If $\{x_n\}$ converges to *x* it is easily shown that

$$lim\ sup\ x_n = lim\ inf\ x_n = lim_{n\to\infty} x_n = x.$$

If for each positive integer *n*, $f_n:X \to [-\infty, \infty]$, then the functions $sup\ f_n$ and $lim\ sup\ f_n$ can be defined on X by:

$$(sup\, f_n)(x) = sup\,(f_n(x)) \text{ and } [lim\, sup\, f_n](x) = lim\, sup\,(f_n(x)).$$

If the limit $f(x) = lim_{n\to\infty} f_n(x)$ exists (i.e., if $\{f_n(x)\}$ converges) for each $x \in X$, then we say that f is the **point-wise limit** of $\{f_n\}$.

PROPOSITION 8.17 If $\{f_n\}$ is a sequence of measurable functions from (X, M) into $[-\infty, \infty]$ where M is a σ-algebra, then the functions $g = sup\, f_n$ and $h = lim\, sup\, f_n$ are both measurable.

Proof: By Theorem 8.6.(3), it suffices to show that if $a \in \mathbf{R}$, then $g^{-1}((a, \infty])$ and $h^{-1}((a, \infty])$ are measurable. For this, we first show that $g^{-1}((a, \infty]) = \cup f_n^{-1}((a, \infty])$. If $x \in g^{-1}((a, \infty])$, then $g(x) \in (a, \infty]$, so $g(x) > a$. Therefore, there exists a positive integer k such that $a < f_k(x) \leq g(x)$. Hence $f_k(x) \in (a, \infty]$ which implies $x \in f_k^{-1}((a, \infty])$. Therefore, $g^{-1}((a, \infty]) \subset \cup f_n^{-1}((a, \infty])$.

Conversely, if $x \in \cup f_n^{-1}((a, \infty])$ then $x \in f_k^{-1}((a, \infty])$ for some positive integer k which implies $f_k(x) \in (a, \infty]$ which in turn implies $f_k(x) > a$. Since $g(x) = sup\{f_n(x)\}$, we have $a < f_k(x) \leq g(x)$ which implies $g(x) \in (a, \infty]$. But then $x \in g^{-1}((a, \infty])$ so $\cup f_n^{-1}((a, \infty]) \subset g^{-1}((a, \infty])$. Hence $g^{-1}((a, \infty]) = \cup f_n^{-1}((a,\infty])$. Since each f_n is measurable, $f_n^{-1}((a, \infty])$ is measurable. Since M is a σ-algebra, $\cup f_n^{-1}((a, \infty]) \in M$ so $g^{-1}((a, \infty])$ is measurable. Therefore, g is a measurable function.

A similar argument holds with *inf* replacing *sup*. Now since $h = lim\, sup\, f_n$, for each $x \in X$, $h(x) = inf_{k\geq 1}\{sup_{n\geq k}\{f_n(x)\}\}$. For each positive integer k put $\phi_k = sup_{n\geq k}\{f_n(x)\}\}$. Then for each k, ϕ_k is measurable as shown above. But $h(x) = inf_{k\geq 1}\{\phi_k(x)\}$ and so h is measurable. ■

COROLLARY 8.5 The limit of every point-wise convergent sequence of complex measurable functions is measurable.

Proof: The proof follows immediately from Proposition 8.17 for the real valued case. Proposition 8.12 can then be used to establish the complex case. ■

COROLLARY 8.6 If f and g are measurable functions from a space X into $[-\infty, \infty]$, then so are $max\{f, g\}$ and $min\{f, g\}$.

A special case of Corollary 8.6 is the function $f^+ = max\{f, 0\}$. Another is $f^- = -min\{f, 0\}$. f^+ is called the **positive part** of f while f^- is called the **negative part** of f. Clearly $|f| = f^+ + f^-$ and $f = f^+ - f^-$.

A function s on a measurable space (X, M) whose range is a finite subset of $[0, \infty)$ is called a **simple function.** If $\{a_1 \ldots a_n\}$ is the range of the simple function s and if $E_i = \{x \in X \mid s(x) = a_i\}$, then:

$$s = \Sigma_{i=1}^{n} a_i \chi_{E_i}$$

where χ_{E_i} is the characteristic function of E_i for each $i = 1 \ldots n$. Clearly, s is measurable if and only if E_i is measurable for each i. In this case, s is called a **simple measurable function.**

THEOREM 8.7 If $f:X \to [0, \infty]$ is a measurable function, there exists a sequence $\{s_n\}$ of simple measurable functions on X such that
(1) $0 \le s_1 \le s_2 \le \ldots f$,
(2) $\{s_n(x)\}$ converges to $f(x)$ for each $x \in X$.

Proof: For each positive integer n and for each $i \in [1, n2^n]$, put $E_{n_i} = f^{-1}([a_{n_i}, b_{n_i}))$ where $a_{n_i} = (i - 1)/2^n$ and $b_{n_i} = i/2^n$ and let $F_n = f^{-1}([n, \infty])$. Then, for each positive integer n, define s_n by

$$s_n = \Sigma_{i=1}^{n2^n} a_{n_i}\chi_{E_{n_i}} + n\chi_{F_n}.$$

By Theorem 8.6.(2), the E_{n_i} and F_n are measurable sets in X. Therefore, the s_n are simple measurable functions. Let m, n be positive integers such that $m < n$ and let $x \in X$. To prove (1), there are three cases to consider:

Case 1: ($f(x) \in F_n$) Since $F_n \subset F_m$, $f(x) \in F_m$. Therefore, $s_m(x) = m < n = s_n(x)$.

Case 2: ($f(x) \in F_m$ but $f(x)$ does not belong to F_n) Then $s_m(x) = m$. Let k be the least positive integer such that $x \in E_{n_k}$. Then $k > m2^n$ which implies $a_{n_k} > (m2^n - 1)/2^n = m - 1/2^n$ which in turn implies $a_{n_k} \ge m$. Therefore, $s_n(x) \ge s_m(x)$.

Case 3: ($f(x)$ does not belong to F_m) Let k be the least positive integer such that $x \in E_{n_k}$. Then $x \in E_{m_j}$ where $j = k \bmod (2^{n-m})$. Hence

$$s_n(x) = a_{n_k} = (k - 1)/2^n = (k - 1)/2^{n-m}2^m \ge [k \bmod (2^{n-m})/2^m] - 1/2^n$$

$$s_m(x) = a_{m_j} = [k \bmod (2^{n-m}) - 1]/2^m = [k \bmod (2^{n-m})/2^m] - 1/2^m.$$

Thus $s_m(x) \le s_n(x)$. In all of these cases, $s_m(x) \le s_n(x)$. To show that for each n, $s_n \le f$, first note that if $f(x) = \infty$ for some $x \in X$ that $s_n(x) \le f(x)$ for each n. Therefore, suppose $x \in X$ and $f(x) < \infty$. Let k be the positive integer such that $f(x) \in [k, k + 1)$. Then $s_k(x) = k \le f(x)$. For each positive integer $n > k$, there are 2^n subintervals

$$[(k + i - 1)/2^n, (k + i)/2^n]$$

where $i = 1 \ldots 2^n$ that partition $[k, k + 1)$. Let i be the positive integer such that

$f(x) \in [(k+i-1)/2^n, (k+i)/2^n]$. Then $s_n(x) = (k+i-1)/2^n \le f(x)$. Consequently, $s_n \le f$ for each positive integer n. This proves (1).

To prove (2), first note that if $f(x) = \infty$ then $s_n(x) = n$ for each positive integer n so $\{s_n(x)\}$ converges to $f(x)$. Therefore, assume that $f(x) < \infty$. As shown above, if k is the positive integer such that $f(x) \in [k, k+1)$, then for each positive integer $n > k$, there are 2^n subintervals $[(k+i-1)/2^n, (k+i)/2^n]$ that partition $[k, k+1)$ and $f(x)$ belongs to one of them, say $[(k+j-1)/2^n, (k+j)/2^n]$ and $s_n(x) = (k+j-1)/2^n$. Hence $|f(x) - s_n(x)| < 2^{-n}$. Consequently, the sequence $\{f(x) - s_n(x)\}$ is Cauchy and therefore converges to 0. Therefore, $\{s_n(x)\}$ converges to $f(x)$. This proves (2). ∎

COROLLARY 8.7 Sums and products of measurable functions into $[0,\infty]$ are measurable.

Proof: Let f and g be measurable functions from X into $[0, \infty]$. By Theorem 8.7, there exists sequences $\{f_n\}$ and $\{g_n\}$ of simple measurable functions such that $0 \le f_1 \le f_2 \le \ldots \le f$, $0 \le g_1 \le g_2 \le \ldots \le g$, $\{f_n(x)\}$ converges to $f(x)$ and $\{g_n(x)\}$ converges to $g(x)$ for each $x \in$ X. It is an easy exercise (Exercise 1) to show that sums and products of simple measurable functions are simple measurable functions. Consequently, $\{f_n + g_n\}$ and $\{f_n g_n\}$ are sequences of simple measurable functions. We can also show (Exercise 2) that $\{[f_n + g_n](x)\}$ converges to $[f+g](x)$ and $\{[f_n g_n](x)\}$ converges to $[fg](x)$ for each $x \in$ X. Hence $f + g$ and fg are measurable. ∎

EXERCISES

1. Show that sums and products of simple measurable functions into $[0, \infty]$ are simple measurable functions.

2. Show that if $\{a_n\}$ and $\{b_n\}$ are sequences in $[0, \infty]$ such that $0 \le a_1 \le a_2 \le \ldots$, $0 \le b_1 \le b_2 \le \ldots$, $\{a_n\}$ converges to a and $\{b_n\}$ converges to b, then $\{a_n + b_n\}$ converges to $a + b$ and $\{a_n b_n\}$ converges to ab.

3. Show that the set of points at which a sequence of measurable real valued functions converges is a measurable set.

4. Show that if f is a real valued function on X such that $\{x \in \text{X} \mid f(x) \ge r\}$ is measurable for each rational number r, then f is measurable.

5. Let $\{A_n\}$ be a sequence of measurable sets in a σ-algebra M and let μ be a measure on M. Define the **limit inferior** denoted $\lim\inf A_n$ of the sequence

$\{A_n\}$ by $\lim\inf A_n = \cup_{n=1}^{\infty}(\cap_{k=n}^{\infty}A_n)$ and define the **limit superior** denoted *lim supA*$_n$ by $\cap_{n=1}^{\infty}(\cup_{k=n}^{\infty}A_n)$. Show that

(a) $\mu(\lim\inf A_n) \leq \lim\inf\mu(A_n)$ and
(b) if $\mu(\cup_{n=1}^{\infty}A_n) < \infty$ then $\lim\sup\mu(A_n) \leq \mu(\lim\sup A_n)$.

8.6 The Lebesgue Integral

In the theory of integration we often encounter the concept of *infinity* and the symbols ∞ and $-\infty$. We have already defined a measure μ to be a set function on a set X into $[0, \infty]$. In order not to have to make special provisions for dealing with these concepts and symbols in some of the following theorems, we define addition (+) and multiplication ($\times$) on $[-\infty, \infty]$ as follows:

$$\begin{aligned} a+\infty &= \infty + a = \infty \text{ for each } a \text{ such that } -\infty < a,\\ a-\infty &= -\infty + a = -\infty \text{ for each } a \text{ such that } a < \infty,\\ a\times\infty &= \infty\times a \text{ for each } a \text{ such that } 0 < a,\\ 0\times\infty &= \infty\times 0 = 0,\\ a\times\infty &= \infty\times a = -\infty \text{ for each } a \text{ such that } a < 0. \end{aligned}$$

With these definitions, it can be shown that the *commutative, associative* and *distributive* laws hold for $[0, \infty]$. Since $-\infty + \infty$ and $(-\infty) \times (\infty)$ are not defined, we cannot extend these laws to $[-\infty, \infty]$, but fortunately, we will not need to. The *cancellation* laws also hold in $[0, \infty]$ with the following modifications:

$$\begin{aligned} a+b &= a+c \text{ implies } b = c \text{ if } a < \infty,\\ ab &= ac \text{ implies } b = c \text{ whenever } 0 < a < \infty. \end{aligned}$$

If s is a simple measurable function on X where $\{a_1 \ldots a_n\} \subset [0, \infty)$ is the range of s and for each $i = 1 \ldots n$, $E_i = \{x \in X | s(x) = a_i\}$, and if M is a σ-algebra on X and μ is a measure on M, then for $E \in M$, we define the **Lebesgue integral** of s with respect to μ as:

$$\int_E s\,d\mu = \Sigma_{i=1}^{n} a_i\mu(E\cap E_i).$$

If f:X $\rightarrow [0, \infty]$ is a measurable function, we define the **Lebesgue integral of** f **with respect to** μ to be the supremum of all simple measurable functions s such that $0 \leq s \leq f$, i.e.,

$$\int_E f\,d\mu = \sup\{\int_E s\,d\mu \,|\, 0 \leq s \leq f \text{ and } s \text{ is a simple measurable function}\}.$$

Clearly, $\int_E f\,d\mu \in [0, \infty]$ and the two definitions of Lebesgue integral given above for the case where f is a simple measurable function are equivalent. The Lebesgue integral behaves in the same manner as the Riemann integral, as the following theorem shows.

THEOREM 8.8 Let (X, M) be a measurable space and let $E, F \in M$. Let μ be a measure on M and f and g be measurable functions from X into $[0,\infty]$. Then

(1) If $0 \leq f \leq g$ then $\int_E f d\mu \leq \int_E g d\mu$.
(2) If $E \subset F$ then $\int_E f d\mu \leq \int_F f d\mu$.
(3) If $c \in [0, \infty)$ then $\int_E cf d\mu = c\int_E f d\mu$.
(4) If $f(x) = 0$ for each $x \in E$ then $\int_E f d\mu = 0$.
(5) If $\mu(E) = 0$ then $\int_E f d\mu = 0$.
(6) $\int_E f d\mu = \int_X \chi_E f d\mu$.

The proof of Theorem 8.8 is straightforward and is left as an exercise (Exercise 1). The next proposition reveals an interesting property about certain integrals, namely, that they are also measures.

PROPOSITION 8.18 Let M be a σ-algebra on X, μ a measure on M and s a simple measurable function on X. For each $E \in M$ put $\lambda(E) = \int_E s d\mu$. Then λ is a measure on M.

Proof: Let the range of s be $\{a_1 \ldots a_k\}$ and for each $i = 1 \ldots k$ let $A_i = \{x \in X \mid s(x) = a_i\}$. Suppose $\{E_n\}$ is a sequence of disjoint members of M such that $E = \cup E_n$. Then since μ is σ-additive we have:

$$\lambda(E) = \Sigma_{i=1}^k a_i\mu(E \cap A_i) = \Sigma_{i=1}^k a_i\mu(\cup E_n \cap A_i) = \Sigma_{i=1}^k a_i \Sigma_n \mu(E_n \cap A_i) =$$

$$\Sigma_{i=1}^k \Sigma_n a_i \mu(E_n \cap A_i) = \Sigma_n \Sigma_{i=1}^k a_i \mu(E_n \cap A_i) = \Sigma_n \lambda(E_n).$$

Consequently, λ is also σ-additive so property M3 of a measure is satisfied. Clearly $\lambda(\varnothing) = 0$ so M1 is satisfied. By Theorem 8.8, M2 is satisfied. Therefore, λ is a measure on M. ■

PROPOSITION 8.19 If M is a σ-algebra on X, μ is a measure on M and s and t are simple measurable functions on X, then:

$$\int_X (s + t) d\mu = \int_X s d\mu = \int_X t d\mu.$$

Proof: Let the range of s be $\{a_i \ldots a_m\}$ and the range of t be $\{b_1 \ldots b_k\}$. For each $i = 1 \ldots m$ and each $j = 1 \ldots k$ put $A_i = \{x \in X \mid s(x) = a_i\}$ and $B_j = \{x \in X \mid t(x) = b_j\}$. Now for each pair i, j put $E_{i_j} = A_i \cap B_j$. This yields

$$\int_{E_{i_j}} (s + t) d\mu = (a_i + b_j)\mu(E_{i_j}) = a_i\mu(E_{i_j}) + b_j(E_{i_j}) = \int_{E_{i_j}} s d\mu + \int_{E_{i_j}} t d\mu.$$

Thus the conclusion of this proposition holds for each E_{i_j} in place of X. Then by Proposition 8.18, we have:

$$\int_X (s+t)d\mu = \Sigma_{i,j=1}^{m,k} \int_{E_{ij}} (s+t)d\mu = \Sigma_{i,j=1}^{m,k} [\int_{E_{ij}} sd\mu + \int_{E_{ij}} td\mu] =$$

$$\Sigma_{i,j=1}^{m,k} \int_{E_{ij}} sd\mu + \Sigma_{i,j=1}^{m,k} \int_{E_{ij}} td\mu = \int_X sd\mu + \int_X td\mu. \blacksquare$$

The great success of Lebesgue's definition of the integral is largely due to the ease of passing to the limit of certain sequences of measurable functions. One example of this is the following celebrated theorem.

THEOREM 8.9 (H. Lebesgue, 1904) If M is a σ-algebra on X, μ a measure on M and $\{f_n\}$ a sequence of measurable functions on X such that $0 \le f_i(x) \le f_j(x) \le \infty$ for each pair i, j with $i < j$ and for each $x \in X$, then if $\{f_n(x)\}$ converges to $f(x)$ for each $x \in X$, f is measurable and

$$lim_{n\to\infty} \int_X f_n d\mu = \int_X f d\mu.$$

Proof: By Theorem 8.8(1), $\int_X f_n d\mu \le \int_X f_{n+1} d\mu$ for each positive integer n, so there exists an $a \in [0, \infty]$ such that $\{\int_X f_n d\mu\}$ converges to a. By Proposition 8.17, f is measurable. Since for each $x \in X$, $\{f_n(x)\}$ is a non-decreasing sequence, we have $f_n(x) \le f(x)$ for each positive integer n. Therefore, by Theorem 8.8(1), $\int_X f_n d\mu \le \int_X f d\mu$ for each positive integer n. Consequently, $a \le \int_X f d\mu$ because the convergence of $\{\int_X f_n d\mu\}$ to a implies a is the supremum of $\{\int_X f_n d\mu\}$ since $\{\int_X f_n d\mu\}$ is non-decreasing.

Let s be a simple measurable function such that $0 \le s \le f$ and let $c \in (0,1)$. For each positive integer n put $E_n = \{x \in X \mid f_n(x) \ge cs(x)\}$. Clearly each E_n is measurable and $E_n \subset E_{n+1}$ for each n. Also, $X = \cup E_n$, for if $f(x) = 0$ then $x \in E_1$ whereas if $f(x) > 0$, $cs(x) < f(x)$ since $c < 1$, and since $\{f_n(x)\}$ converges to $f(x)$, there exists a positive integer k such that $cs(x) < f_k(x)$. Therefore, $x \in E_k$. Then

$$c\int_{E_n} sd\mu = \int_{E_n} csd\mu \le \int_{E_n} f_n d\mu \le \int_X f_n d\mu$$

for each positive integer n. By Proposition 8.18, $\lambda(E) = \int_E csd\mu$ is a measure on M. Since $\{E_n\}$ is an ascending sequence with $\cup E_n = X$, by Proposition 8.7, $\{\lambda(E_n)\}$ converges to $\lambda(X)$. Then, since $\{\int_X f_n d\mu\}$ converges to a, $\{\int_{E_n} csd\mu\}$ converges to $\int_X csd\mu$, and for each n $\int_{E_n} csd\mu \le \int_X f_n d\mu$, we have $c\int_X sd\mu = \int_X csd\mu \le a$. Since this is true for each $c < 1$, it is clear that $\int_X sd\mu \le a$. Hence, for each simple measurable function s with $0 \le s \le f$,

$$\int_X sd\mu \le a \le \int_X fd\mu.$$

By definition, $\int_X fd\mu$ is the supremum of all such simple measurable functions, we have $a = \int_X fd\mu$. Therefore, $\{\int_X f_n d\mu\}$ converges to $\int_X fd\mu$. ■

In what follows, some of the sequences of numbers to which we will have occasion to refer will have rather complex representations such as the sequence of integrals $\{\int_X f_n d\mu\}$ in the proof of the preceding theorem. Rather than stating that this sequence converges to the number $\int_X f d\mu$ (we include ∞ as a number) we will adopt a slightly simpler notation by writing

$$\int_X f_n d\mu \to \int_X f d\mu.$$

THEOREM 8.10 If $\{f_n\}$ is a sequence of measurable functions from X into $[0, \infty]$ and $f(x) = \Sigma_n f_n(x)$ for each $x \in X$, then $\int_X f d\mu = \Sigma_n \int_X f_n d\mu$.

Proof: By Theorem 8.7, there exist sequences $\{s_n\}$ and $\{t_n\}$ of simple measurable functions such that $0 \leq s_1 \leq s_2 \leq \ldots \leq f_1$, $0 \leq t_1 \leq t_2 \leq \ldots \leq f_2$, $s_n(x) \to f_1(x)$ and $t_n(x) \to f_2(x)$. For each positive integer n put $u_n(x) = s_n(x) + t_n(x)$ for each $x \in X$. It is easily shown that $\{u_n\}$ is a sequence of simple measurable functions that converges to $(f_1 + f_2)$. By Proposition 8.19,

$$\int_X (s_n + t_n) d\mu = \int_X s_n d\mu + \int_X t_n d\mu$$

for each positive integer n, and by Theorem 8.9,

$$\int_X s_n d\mu \to \int_X f_1 d\mu \text{ and } \int_X t_n d\mu \to \int_X f_2 d\mu.$$

Consequently, by Execise 2 of Section 8.5,

$$\int_X u_n d\mu = \int_X (s_n + t_n) d\mu = \int_X s_n d\mu + \int_X t_n d\mu \to \int_X f_1 d\mu + \int_X f_2 d\mu.$$

But it is also easily shown that $0 \leq u_1 \leq u_2 \leq \ldots \leq (f_1 + f_2)$, so another application of Theorem 8.9 shows that $(f_1 + f_2)$ is a measurable function and

$$\int_X u_n d\mu \to \int_X (f_1 + f_2) d\mu.$$

Since limits are unique, this implies

$$\int_X (f_1 + f_2) d\mu = \int_X f_1 d\mu + \int_X f_2 d\mu.$$

This result can be extended by an induction argument to show that for each positive integer k we have $\Sigma_{n=1}^k f_n$ is a measurable function and

$$\int_X (\Sigma_{n=1}^k f_n) d\mu = \Sigma_{n=1}^k (\int_X f_n d\mu).$$

For each positive integer k put $g_k = \Sigma_{n=1}^k f_n$. Then $\{g_n\}$ is a sequence of measurable functions such that $0 \leq g_1 \leq g_2 \leq \ldots \leq \Sigma_n f_n$ and $g_n(x) \to \Sigma_n f_n(x)$ for each $x \in X$ by the definition of a series $\Sigma_n f_n(x)$. By yet another application of Theorem 8.9, $\Sigma_n f_n$ is a measurable function and

$$\int_X g_n d\mu \to \int_X (\Sigma_n f_n) d\mu.$$

But then the partial sums $\Sigma_{i=1}^n (\int_X f_i d\mu) \to \int_X (\Sigma_n f_n) d\mu$. By the definition of a series, the partial sums $\Sigma_{i=1}^n (\int_X f_i d\mu) \to \Sigma_n (\int_X f_n d\mu)$. Consequently, $\Sigma_n (\int_X f_n d\mu) = \int_X (\Sigma_n f_n) d\mu$. ∎

THEOREM 8.11 If $f_n: X \to [0, \infty]$ is measurable for each positive integer n, then

$$\int_X (\mathit{lim\ inf}\ f_n) d\mu \leq \mathit{lim\ inf} \int_X f_n d\mu.$$

Proof: For each positive integer k put $g_k(x) = \mathit{inf}\{f_n(x) \mid k \leq n\}$ for each $x \in X$. Then for each positive integer n, g_n is a measurable function by Proposition 8.17 and $g_n \leq f_n$, so by Theorem 8.8(1), $\int_X g_n d\mu \leq \int_X f_n d\mu$. Then by the definition of limit inferior we have:

$$\mathit{lim\ inf} \int_X g_n d\mu \leq \mathit{lim\ inf} \int_X f_n d\mu$$

and for each $x \in X$, $\mathit{lim\ inf}\ f_n(x) = \mathit{sup}\{g_n(x)\}$. Therefore, $0 \leq g_1 \leq g_2 \leq \ldots \leq \mathit{lim\ inf}\ f_n$, so by Theorem 8.9, $\int_X g_n d\mu \to \int_X (\mathit{lim\ inf}\ f_n) d\mu$. Since the sequence $\{\int_X g_n d\mu\}$ converges, it converges to its limit inferior, i.e., $\int_X g_n d\mu \to \mathit{lim\ inf} \int_X g_n d\mu$. Therefore, $\int_X (\mathit{lim\ inf}\ f_n) d\mu = \mathit{lim\ inf} \int_X g_n d\mu \leq \int_X f_n d\mu$. ∎

Theorem 8.11 is known as *Fatou's Lemma*. It is possible that the inequality in Theorem 8.11 is strict, i.e., that

$$\int_X (\mathit{lim\ inf}\ f_n) d\mu < \mathit{lim\ inf} \int_X f_n d\mu.$$

To see this let $X = [0, 1]$, let $g(x) = 0$ if $x \leq 1/2$ and 1 otherwise, let $h(x) = 0$ if $x > 1/2$ and 1 otherwise, and put $f_n = g$ if n is even and $f_n = h$ if n is odd. Clearly $\mathit{lim\ inf}\ f_n = 0$ so $\int_X (\mathit{lim\ inf}\ f_n) d\mu = 0$. If $\mu(A) > 0$ for each $A \subset X$ with non-empty interior, then both $\int_X g d\mu > 0$ and $\int_X h d\mu > 0$ which implies $\mathit{lim\ inf} \int_X f_n d\mu > 0$. Hence strict inequality holds. If we define $\mu(A) = \int_0^1 \chi_A dx$ where χ_A is the characteristic function of A and $\int_0^1 \chi_A dx$ is the ordinary Riemann integral, then μ is such a measure.

THEOREM 8.12 If M is a σ-algebra on X and μ is a measure on M, and if $f: X \to [0, \infty]$ is measurable, then λ defined by $\lambda(E) = \int_E f d\mu$ for each $E \in M$ is a measure on M and $\int_X g d\lambda = \int_X g f d\mu$ for each function $g: X \to [0, \infty]$.

Proof: Let $\{E_n\}$ be a sequence of disjoint measurable sets in X with $E = \cup E_n$. Then for each $x \in X$, $[\chi_E f](x) = \Sigma_n [\chi_{E_n} f](x)$ since x can belong to at most one E_n. Therefore, $\chi_E f = \Sigma_n \chi_{E_n} f$. Consequently,

$$\lambda(E) = \int_X \chi_E f d\mu \text{ and } \lambda(E_n) = \int_X \chi_{E_n} f d\mu.$$

By Theorem 8.10 we have

$$\lambda(E) = \int_X \chi_E f d\mu = \Sigma_n \int_X \chi_{E_n} f d\mu = \Sigma_n \lambda(E_n).$$

Since $\mu(\varnothing) = 0$ implies $\lambda(\varnothing) = 0$, we see that λ is a measure on M. Let $E \in M$ and put $g = \chi_E$. Then since χ_E is a simple measurable function,

$$\int_X g d\lambda = \int_X \chi_E d\lambda = \lambda(E) = \int_X \chi_E f d\mu = \int_X g f d\mu.$$

Consequently, the equation in the conclusion of the theorem holds whenever g is the characteristic function of a measurable set. Next let g be the simple measurable function $\Sigma_{n=1}^k a_n \chi_{A_n}$ where $A_n = \{x \in X \mid g(x) = a_n\}$. Then $\int_X g d\lambda =$

$$\Sigma_{n=1}^k a_n \lambda(A_n) = \Sigma_{n=1}^k a_n \int_X \chi_{A_n} d\lambda = \Sigma_{n=1}^k a_n \int_X \chi_{A_n} f d\mu = \Sigma_{n=1}^k \int_X a_n \chi_{A_n} f d\mu.$$

The last equality being attained by Theorem 8.8(3). By Theorem 8.10

$$\Sigma_{n=1}^k \int_X a_n \chi_{A_n} f d\mu = \int_X \Sigma_{n=1}^k a_n \chi_{A_n} f d\mu = \int_X g f d\mu.$$

Hence the conclusion of the theorem also holds whenever g is a simple measurable function. If g is a measurable function, then by Theorem 8.7, there exists a sequence $\{s_n\}$ of simple measurable functions on X such that $0 \leq s_1 \leq s_2 \leq \ldots \leq g$ and such that $s_n(x) \to g(x)$ for each $x \in X$. Then, from what we have already shown, for each positive integer n, $\int_X s_n d\lambda = \int_X s_n f d\mu$. By Theorem 8.9, $\int_X s_n d\lambda \to \int_X g d\lambda$ so $\int_X s_n f d\mu \to \int_X g d\lambda$. But since $0 \leq s_1 \leq s_2 \leq \ldots \leq g$ we have $0 \leq s_1 f \leq s_2 f \leq \ldots \leq gf$ and since $s_n(x) \to g(x)$ for each $x \in X$, we have $[s_n f](x) = s_n(x)f(x) \to g(x)f(x) = [gf](x)$ for each $x \in X$. Again by Theorem 8.9, $\int_X s_n f d\mu \to \int_X g f d\mu$. Therefore, $\int_X g d\lambda = \int_X g f d\mu$. ∎

The equation in the statement of Theorem 8.12 is often written $d\lambda = f d\mu$, or even as $f = d\lambda/d\mu$. The latter notation is suggestive of the role f plays but has no meaning as a ratio.

THEOREM 8.13 (H. Lebesgue) If $\{f_n\}$ is a sequence of complex, measurable functions on X with $f(x) = \lim f_n(x)$ for each $x \in X$ and if there exists a measurable function g on X with $\int_X |g| d\mu < \infty$ such that $|f_n(x)| \leq g(x)$ for each $x \in X$, then $\int_X |f| d\mu < \infty$ and

$$\lim_{n\to\infty} \int_X |f_n - f| d\mu = 0 \text{ and } \lim_{n\to\infty} \int_X f_n d\mu = \int_X f d\mu.$$

Proof: Clearly $|f| \leq g$ and since f is measurable by Corollary 8.5, $\int_X |f| d\mu < \infty$. Since $|f - f_n| \leq 2g$, we can apply Theorem 8.11 to the functions $2g - |f - f_n|$ to get

$$\int_X (\liminf_{n\to\infty} [2g - |f - f_n|]) d\mu \leq \liminf_{n\to\infty} \int_X (2g - |f - f_n|) d\mu.$$

Now $\lim\inf_{n\to\infty}[2g - |f - f_n|] = 2g$ and $\lim\inf_{n\to\infty}\int_X(2g - |f - f_n|)d\mu = \int_X 2g d\mu - \lim\sup_{n\to\infty}\int_X|f - f_n|d\mu$. Hence $\lim\sup_{n\to\infty}\int_X|f - f_n|d\mu \leq 0$. Therefore, $\lim_{n\to\infty}\int_X|f - f_n|d\mu = 0$. By Exercise 2, $|\int_X(f - f_n)d\mu| \leq \int_X|f - f_n|d\mu$ which implies $|\int_X f d\mu - \int_X f_n d\mu| \leq \int_X|f - f_n|d\mu$ for each positive integer n. But since $\lim_{n\to\infty}\int_X|f - f_n|d\mu = 0$ we have $\lim_{n\to\infty}|\int_X f d\mu - \int_X f_n d\mu| = 0$ or in other words, $\lim_{n\to\infty}\int_X f_n d\mu = \int_X f d\mu$. ∎

EXERCISES

1. Define $L^1(\mu)$ to be the collection of complex measurable functions on X for which $\int_X|f|d\mu < \infty$. The members of $L^1(\mu)$ are called the **Lebesgue integrable** functions on X with respect to μ. If $f \in L^1(\mu)$ and $f = u + iv$ where u and v are real measurable functions, define the integral of f (with respect to μ) over E as

$$\int_E f d\mu = \int_E u^+ d\mu - \int_E u^- d\mu + i\int_E v^+ d\mu - i\int_E v^- d\mu$$

for each measurable $E \subset X$. Each of the four integrals on the right are finite, so the integral on the left is a complex number. Show that if $f, g \in L^1(\mu)$ and a and b are complex numbers, then $af + bg \in L^1(\mu)$ and

$$\int_X(af + bg)d\mu = a\int_X f d\mu + b\int_X g d\mu.$$

2. Show that if $f \in L^1(\mu)$ then $|\int_X f d\mu| \leq \int_X|f|d\mu$.

3. Let $\{f_n\}$ be a sequence of complex measurable functions on X such that $f_n(x) \to f(x)$ for each $x \in X$. Show that if there exists a $g \in L^1(\mu)$ such that $|f_n(x)| \leq g(x)$ for each $x \in X$ and each positive integer n, then:

   (a) $f \in L^1(\mu)$.
   (b) $\lim_{n\to\infty}\int_X|f_n - f|d\mu = 0$.
   (c) $\lim_{n\to\infty}\int_X f_n d\mu = \int_X f d\mu$.

4. Suppose $f_n: X \to [0,\infty]$ is measurable for each positive integer n and $f_{n+1} \geq f_n$ for each n. Also assume $f_n(x) \to f(x)$ for each $x \in X$ and $f_1 \in L^1(\mu)$. Show that $\int_X f_n d\mu \to \int_X f d\mu$.

5. Show that the condition $f_1 \in L^1(\mu)$ is essential in Exercise 4.

6. Let $f \in L^1(\mu)$. Show that for each $\varepsilon > 0$ there exists a $\delta > 0$ such that $\int_E|f|d\mu < \varepsilon$ whenever $\mu(E) < \delta$.

8.7 Negligible Sets

If μ is a measure on a σ-algebra M and if $E \in M$ such that $\mu(E) = 0$, then E is said to be a **negligible set** with respect to μ. Negligible sets are negligible in the theory of integration. If P is a property that a point x may have (e.g., some function f may be continuous at x or differentiable at x) we say x has property P and write $P(x)$. If A is a set and E is a negligible subset of A such that for each $x \in A - E$ we have $P(x)$, we say that P holds **almost everywhere** on A. This is frequently abbreviated P **holds a.e. on** A. It is common to make statements like "f is continuous almost everywhere on A," in which case it is meant that the measure of the set of points of discontinuity of f on A has zero measure.

If f and g are measurable functions such that $\mu(E) = 0$ where $E = \{x \mid f(x) \neq g(x)\}$ then X - E and E are disjoint sets whose union is X. Consequently,

$$\int_X f d\mu = \int_{X-E} f d\mu + \int_E f d\mu = \int_{X-E} g d\mu + 0 = \int_{X-E} g d\mu + \int_E g d\mu = \int_X g d\mu.$$

Hence, functions that are the same almost everywhere have equal integrals. This is what is meant by saying that negligible sets are negligible in the theory of integration. Since functions that are equal almost everywhere behave the same with respect to integration, we can generalize our definition of measurable function in the following way. Let $E \in M$ and let $f: E \to \mathbf{R}$. If $\mu(X - E) = 0$ and $f^{-1}(U) \cap E \in M$ for each open U in $\mathbf{R}$ then f is said to be *measurable* on E. Clearly, if we put $f(x) = 0$ for each $x \in X - E$, we extend f to a function on X that is measurable with respect to the old definition. Intuitively, it should not matter what values we assign to f on X - E. We would like to be able to assign values to f on X - E in an arbitrary manner and still get a measurable function with respect to the old definition.

But here, a problem arises! It may be the case that certain subsets of X - E are not measurable. Fortunately, Corollary 8.4 states that every measure can be completed, i.e., every subset of a set of measure zero is itself a subset of measure zero. It will be to our advantage to just deal with complete measures. More measurable sets just mean more measurable functions. Then we can extend the function $f: E \to \mathbf{R}$ above in any arbitrary manner to X and be assured that the extended function is measurable with respect to the old definition of measurability.

This new definition of measurability has many consequences. For example, Theorem 8.10 can be modified to allow the sequence $\{f_n\}$ to be a sequence of measurable functions on X that converges almost everywhere on X. With our new definition of measurability it is easily shown that the limit of $\{f_n\}$ is still a measurable function f and $\int_X f d\mu = \Sigma_n \int_X f_n d\mu$, without having to *restrict* ourselves to the set on which the convergence actually takes place.

EXERCISES

1. Show that if $f:X \to [0, \infty]$ is measurable, $E \in M$ and $\int_E f d\mu = 0$ then $f = 0$ almost everywhere on E.

2. Show that if $f \in L^1(\mu)$ and $\int_E f d\mu = 0$ for each $E \in M$, then $f = 0$ almost everywhere on X.

3. Show that if $f \in L^1(\mu)$ and $|\int_X f d\mu| = \int_X |f| d\mu$, then there exists an $\alpha \in \mathbf{R}$ such that $\alpha f = |f|$ almost everywhere on X.

4. Show that if $\{f_n\}$ is a sequence of complex measurable functions defined almost everywhere on X such that $\Sigma_n \int_X |f_n| d\mu < \infty$, then the function $f(x) = \Sigma_n f_n(x)$ converges almost everywhere on X, $f \in L^1(\mu)$ and $\int_X f d\mu = \Sigma_n \int_X f_n d\mu$.

5. Show that if $\mu(X) < \infty$, $f \in L^1(\mu)$, $F \subset \mathbf{C}$ is closed and the *averages* $A_E = 1/\mu(E)\int_E f d\mu$ lie in F for each $E \in M$ with $\mu(E) > 0$, then $f(x) \in F$ for almost all $x \in X$.

6. Let $\{E_n\} \subset M$ such that $\Sigma_n \mu(E_n) < \infty$. Show that almost all $x \in X$ lie in at most finitely many of the E_n.

8.8 Linear Functionals and Integrals

Recall from linear algebra that a vector space V over the scalar field F is a set V whose elements are called *vectors* and whose two operations are called *addition* and *scalar multiplication*. A *linear transformation* of V into another vector space W is a mapping λ of V into W such that $\lambda(\alpha x + \beta y) = \alpha\lambda(x) + \beta\lambda(y)$ for all $x,y \in V$ and $\alpha,\beta \in F$. In the special case W = F (the field of scalars), λ is called a **linear functional.** Exercise 1 of Section 8.6 shows that $L^1(\mu)$ is a vector space whose scalar field is **C**. It is easily seen that the mapping $I_\mu:L^1(\mu) \to \mathbf{C}$ defined by $I_\mu(f) = \int_X f d\mu$ is a linear functional on $L^1(\mu)$.

In the special case where V is the set of all continuous complex valued functions on the unit interval [0, 1], and F = **R** then the linear functional $J:V \to F$ defined by $J(f) = \int_0^1 f(x)dx$ (the ordinary Riemann integral) is clearly a positive linear functional. Since integrals are linear functionals, it is natural to ask: when are linear functionals integrals? In 1909, F. Riesz provided the following remarkable answer for the vector space *C* of all continuous complex valued functions defined on [0, 1]: for each positive linear functional λ on *C*, there exists a finite positive Borel measure μ on [0, 1] such that $\lambda(f) = \int_0^1 f d\mu$. In fact, we now develop the celebrated Riesz Representation Theorem in a setting more general than the vector space *C*.

LEMMA 8.4 If X is a locally compact Hausdorff space, U is open in X and K ⊂ U is compact, then there exists an open V with compact closure such that K ⊂ V ⊂ Cl(V) ⊂ U.

Proof: Each point of K has an open neighborhood with compact closure, and K is covered by finitely many of them. Therefore, K lies in an open set W with compact closure. If U = X, put V = W. Otherwise, for each $p \in$ X - U let W_p be an open set containing K whose closure does not contain p and put F_p = W$\cap W_p\cap$[X - U]. Then $\{F_p\}$ is a collection of compact sets such that $\cap F_p = \emptyset$.

Then $\{X - F_p\}$ is an open covering of X. Pick $F_q \in \{F_p\}$. Then some finite collection $\{(X - F_{p_1}) \ldots (X - F_{p_n})\}$ covers F_q so $F_q \cap F_{p_1} \cap \ldots \cap F_{p_n} = \emptyset$. Then V = U$\cap W_q \cap W_{p_1} \cap \ldots \cap W_{p_n}$ is open and contains K. Furthermore, Cl(V) $\subset Cl(W_q) \cap Cl(W_{p_1}) \cap \ldots \cap Cl(W_{p_n})$ since $F_q \cap F_{p_1} \cap \ldots \cap F_{p_n} = \emptyset$. ∎

The collection of all complex valued functions f on a space X whose support (denoted **0**(f)) has compact closure is denoted by C_K(X). It is easily shown that C_K(X) is a vector space under the operations of functional addition $[(f + g)(x) = f(x) + g(x)]$ and scalar multiplication $((\alpha f)(X) = \alpha[f(X)])$. The notation K ‹ f will be used to signify that K is a compact subset of X, $f \in C_K$(X), f(X) ⊂ [0, 1] and f(K) = 1. The notation f ‹ V will mean that V is an open subset of X, $f \in C_K$(X), f(X) ⊂ [0, 1] and $Cl(\mathbf{0}(f)) \subset$ V. A linear functional λ on C_K(X) is **positive** if $\lambda(f) \geq 0$ whenever $f \geq 0$ (e.g., $\lambda(f)$ is real valued when f is real valued).

A function f will be called **lower semi-continuous** if $f^{-1}(\alpha, \infty)$ is open for each $\alpha \in$ **R** and **upper semi-continuous** if $f^{-1}(-\infty, \alpha)$ is open for each $\alpha \in$ **R**. The following facts about upper and lower semi-continuous functions follow from the definitions. A real valued continuous function is continuous if and only if it is both upper semi-continuous and lower semi-continuous. Characteristic functions of closed sets are upper semi-continuous and characteristic functions of open sets are lower semi-continuous. The *inf* of any collection of upper semi-continuous functions is upper semi-continuous and the *sup* of any collection of lower semi-continuous functions is lower semi-continuous.

LEMMA 8.5 If X is a locally compact Hausdorff space, V is open in X and K ⊂ V is compact, then there exists an $f \in C_K(X)$ such that K ‹ f ‹ V, $0 \leq f \leq 1$, and f(x) = 1 for each x ∈ K.

Proof: Let $r_1 = 0$ and $r_2 = 1$. Let $\{r_n\}$, $n \geq 3$, be a well ordering of the rationals between 0 and 1. By Lemma 8.4, there exists open sets U_0 and U_1 such that $Cl(U_0)$ is compact and K $\subset U_1 \subset Cl(U_1) \subset U_0 \subset Cl(U_0) \subset$ V. We contend that it is possible to find a sequence $\{U_{r_n}\}$ such that for each n, U_{r_n} is an open set with compact closure such that if $r_i < r_j$ then $Cl(U_{r_i}) \subset U_{r_j}$. For this, let H be the set of all positive integers such that $n \in H$ if it is possible to find $n + 1$ open sets $\{U^n_{r_i} \mid i = 0 \ldots n\}$ with compact closures such that if $r_i < r_j$ $(i, j \leq n)$ then $Cl(U^n_{r_j}) \subset U^n_{r_i}$ and for each $m < n, U^m_{r_i} = U^n_{r_i}$ for $i = 1 \ldots m$.

Clearly, $1 \in H$. Suppose $n \in H$. Then one of the members $r_1 \ldots r_n$, say r_i, is the largest one smaller than r_{n+1}, and one, say r_j, is the smallest one larger than r_{n+1}. By Lemma 8.4, we can find an open $U^{n+1}_{r_{n+1}}$ with compact closure such that

$$Cl(U^n_{r_j}) \subset U^{n+1}_{r_{n+1}} \subset Cl(U^{n+1}_{r_{n+1}}) \subset U^n_{r_i}.$$

Also, for each $m \leq n$ put $U^{n+1}_{r_m} = U^n_{r_m}$. Then $\{U^{n+1}_{r_i} \mid i = 0 \ldots n+1\}$ is the desired collection for $n + 1$, so $n + 1 \in H$. This completes the induction argument. For each positive integer n put $U_{r_n} = U^n_{r_n}$. Then $\{U_{r_n}\}$ is the desired sequence.

For each rational $r \in [0,1]$ define f_r:X $\to [0, 1]$ by $f(x) = r$ if $x \in U_r$ and 0 otherwise. Define g_r:X $\to [0, 1]$ by $g_r(x) = 1$ if $x \in Cl(U_r)$ and r otherwise. Then put $f(x) = sup\{f_r(x)\}$ and $g(x) = inf\{g_r(x)\}$, for each $x \in$ X. Then f is lower semi-continuous and g is upper semi-continuous. Moreover, f(X) $\subset [0, 1]$ and f(K) = 1. Finally, since $0(f_r) \subset Cl(U_r) \subset U_0$ for each rational $r \in [0, 1]$ we conclude $0(f) \subset Cl(U_0)$. Consequently, the proof will be complete if we show $f = g$.

Suppose there exists rationals $r, s \in [0, 1]$ such that $f_r(x) > g_s(x)$ for some $x \in$ X. From the definition of f_r and g_s we see this implies $r > s$, $x \in U_r$ and x does not belong to U_s. But $r > s$ implies $Cl(U_r) \subset U_s$ which is a contradiction. Hence $f_r \leq g_s$ for each pair of rationals, $r, s \in [0, 1]$. Consequently, $f = supf_r \leq infg_r = g$ so $f \leq g$. Conversely, suppose $f(x) < g(x)$ for some $x \in$ X. Then there exists rationals r, s such that $f(x) < r < s < g(x)$. Now $f(x) < r$ implies x does not belong to U_r and $s < g(s)$ implies $x \in Cl(U_s)$. But $r < s$ implies $Cl(U_s) \subset U_r$ which is a contradiction. Therefore, $f(x) = g(x)$ for each $x \in$ X. ∎

LEMMA 8.6 If $U_1 \ldots U_n$ is a finite collection of open sets in a locally compact Hausdorff space and K is a compact subset of $\cup^n_{i=1} U_i$, then there exists a partition of unity on K subordinate to $\{U_1 \ldots U_n\}$.

Proof: For each $x \in$ K, there exists (by Lemma 8.4) a neighborhood V_x with compact closure such that $Cl(V_x) \subset U_i$ for some $i = 1 \ldots n$. Since K is compact, there are finitely many points $p_1 \ldots p_m$ with K $\subset \cap^m_{j=1} V_{p_j}$. For each positive integer $i = 1 \ldots n$, let $F_i = \cup\{Cl(V_{p_j}) \mid p_j \in U_i\}$. By Lemma 8.5, there exists an $f_i \in C_K$(X) with $F_i \prec f_i \prec U_i$, $0 \leq f_i \leq 1$ and $f_i(x) = 1$ for each $x \in F_i$. Put $g_1 = f_1$ and for each $i = 2 \ldots n$ put $g_n = (1 - f_1)(1 - f_2) \ldots (1 - f_{n-1})f_n$. Clearly $g_i \prec U_i$ for each $i = 1 \ldots n$. A simple induction argument can be used to show that for each $x \in$ K, $\Sigma^n_{i=1} g_i(x) = 1 - (1 - f_1)(1 - f_2) \ldots (1 - f_n)$. Since K $\subset \cup^n_{i=1} F_i$, at least one f_i has $f_i(x) = 1$ for each $x \in$ K. Hence $\Sigma^n_{i=1} g_i(x) = 1$ for each $x \in$ K. Therefore, $\{g_1 \ldots g_n\}$ is a partition of unity subordinate to $\{U_1 \ldots U_n\}$. ∎

A Borel (Baire) measure is said to be **outer regular** on a Borel (Baire) set E if $\mu(E) = inf\{\mu(U) \mid E \subset U \text{ and } U \text{ is open}\}$. In case μ is only a Baire measure, we assume U is an open Baire set in the definition of outer regularity. μ is said to be **inner regular** on E if $\mu(E) = sup\{\mu(K) \mid K \subset E \text{ and } K \text{ is compact}\}$. μ is said to be outer regular if it is outer regular on all Borel (Baire) sets and inner regular if it is inner regular on all Borel (Baire) sets. μ is **regular** if it is both outer regular and inner regular. μ is said to be **almost regular** if it is outer regular and inner regular on all Borel (Baire) sets E such that either E is open or $\mu(E) < \infty$. The regularity condition asserts that the values of the measure can be calculated from its values on the topologically important open and compact sets.

The original version of the following theorem was given in 1909 by F. Riesz for the case where the underlying space X is the interval [0, 1]. The more general version of the theorem discussed below is known as the *Riesz Representation Theorem* and was first proven for locally compact Hausdorff spaces by S. Kakutani in 1941 in a paper titled *Concrete representation of abstract (M)-spaces* (Annals of Mathematics, Volume 42, pp. 934-1024). The version of the theorem stated below is due to P. Halmos. Rather than prove this lengthy theorem here, it is developed as a series of exercises at the end of the section.

THEOREM 8.14 (Riesz, 1909, Kakutani, 1941, Halmos, 1950) If X is a locally compact Hausdorff space and λ is a positive linear functional on $C_K(X)$, then there exists a σ-algebra M on X containing all Borel sets, and a unique, complete, almost regular measure μ on M such that $\mu(K) < \infty$ for each compact $K \subset X$, that represents λ in the sense that $\lambda(f) = \int_X f d\mu$ for each $f \in C_K(X)$.

THEOREM 8.15 Let X be a locally compact, σ-compact Hausdorff space and let λ be a positive linear functional on $C_K(X)$. Let M and μ be the σ-algebra and measure respectively from Theorem 8.14. Then M and μ have the following properties:

(1) If $E \in M$ and $\varepsilon > 0$, there exists a closed set F and an open set U such that $F \subset E \subset U$ and $\mu(U - F) < \varepsilon$.

(2) μ is a regular Borel measure on X.

(3) If $E \in M$, there are sets $G, H \subset X$ such that H is an F_σ, G is a G_δ, $H \subset E \subset G$ and $\mu(G - H) = 0$.

Proof: The proof of this theorem is closely related to the proof of the Riesz Representation Theorem. For that reason it is also left as an exercise at the end of the section.

THEOREM 8.16 If X is a locally compact Hausdorff space in which every open set is σ-compact and μ is a Borel measure on X with $\mu(K) < \infty$ for each compact K, then μ is regular.

Proof: For each $f \in C_K(X)$, put $\lambda(f) = \int_X f d\mu$. Clearly, λ is a linear functional. Suppose $f \in C_K(X)$ such that $f \geq 0$. Then f is real valued and hence bounded by some real number r on X. Moreover, $f(X - K) = 0$ for some compact $K \subset X$. Since X is open, there exists an ascending sequence $\{K_n\}$ of compact sets such that $X = \cup K_n$. Since f is real valued, $m(E) = \int_E f d\mu$ for each $E \in M$ is a measure on M by Theorem 8.12. By Proposition 8.6, $m(X) = lim_{n\to\infty} m(K_n)$. Hence $\int_X f d\mu = lim_{n\to\infty} \int_{K_n} f d\mu$. For each n, $\int_{K_n} f d\mu = \int_{K_n \cap K} f d\mu \leq \int_K f d\mu \leq r\mu(K)$. Since $\mu(K) < \infty$, $\{\int_{K_n} f d\mu\}$ is an ascending sequence of non-negative real numbers bounded above by $r\mu(K)$. Hence $\int_X f d\mu$ is a non-negative real number. Therefore, λ is a positive lineal functional. By Theorem 8.14, there exists a measure ν such that for each $f \in C_K(X)$, $\int_X f d\mu = \int_X f d\nu$.

Let U be an open set. Then $U = \cup_{i=1}^{\infty} F_i$ where $\{F_i\}$ is a sequence of compact sets. Pick $f_1 \in C_K(X)$ such that $F_1 \prec f_1 \prec U$. Let K_1 be the closure of the support of f_1. Assume that for each $m \leq n$ that $f_m \in C_K(X)$ has been chosen such that $[\cup_{i=1}^{m-1} F_i] \cup [\cup_{i=1}^{m-1} K_i] \prec f_m \prec U$. Pick f_{n+1} with $[\cup_{i=1}^{n} F_i] \cup [\cup_{i=1}^{n} K_i] \prec f_n \prec U$. Then the sequence $\{f_n\}$ converges to χ_U at each point of X. Hence

$$\nu(U) = \int_X \chi_U d\nu = lim_{n\to\infty} \int_X f_n d\nu = lim_{n\to\infty} \int_X f_n d\mu = \int_X \chi_U d\mu = \mu(U).$$

Let E be a Borel set in X and let $\varepsilon > 0$. By Theorem 8.14, there exists a closed set F and an open set U with $F \subset E \subset U$ and $\nu(U - F) < \varepsilon$. Since $U - F$ is open, this means $\mu(U - F) < \varepsilon$. Hence $inf\{\mu(U - F) \mid F \subset E \subset U$ where U is open and F is closed$\} = 0$ which implies $\mu(E) = inf\{\mu(E) \mid E \subset U$ and U is open$\}$ so μ is regular. ■

EXERCISES

THE RIESZ REPRESENTATION THEOREM

1. Let X be a locally compact Hausdorff space and let λ be a positive linear functional on $C_K(X)$. For each open $U \subset X$ put $\mu(U) = sup\{\lambda(f) \mid f \prec U\}$. Then for each $E \subset X$ let $\mu(E) = inf\{\mu(U) \mid E \subset U$ and U is open$\}$. Let N be the class of all $E \subset X$ such that $\mu(E) < \infty$ and $\mu(E) = sup\{\mu(K) \mid K \subset E$ and K is compact$\}$. Let M be the class of all $E \subset X$ such that $E \cap K \in N$ for each compact K. Show that:

   (a) μ is well defined (on open sets $U \subset X$).
   (b) $\mu(A) \leq \mu(B)$ if $A \subset B$.
   (c) $\mu(E) = 0$ implies $E \in N$ and $E \in M$.
   (e) $f \leq g$ implies $\lambda(f) \leq \lambda(g)$.

2. Show that if $\{E_n\}$ is a sequence of sets in X, then $\mu(\cup E_n) \leq \Sigma_n \mu(E_n)$.

3. Show that N contains all compact sets.

4. Show that N contains every open set U with $\mu(U) < \infty$.

5. Show that if $E = \cup E_n$, where $\{E_n\}$ is a sequence of disjoint members of N, then $\mu(E) = \Sigma_n \mu(E_n)$.

6. Show that if $E = \cup E_n$, where $\{E_n\}$ is a sequence of disjoint members of N, and if $\mu(E) < \infty$ then $E \in N$.

7. Show that if $E \in N$ and $\varepsilon > 0$, there exists a compact K and an open U such that $K \subset E \subset U$ and $\mu(U - K) < \varepsilon$.

8. Show that if $A, B \in N$, then $A - B$, $A \cup B$ and $A \cap B \in N$.

9. Show that M is a σ-algebra on X that contains all Borel sets.

10. Show that N consists of those $E \in M$ such that $\mu(E) < \infty$.

11. Show that μ is a measure on M.

12. Show that for each $f \in C_K(X)$ that $\lambda(f) = \int_X f d\mu$. [Hint: It suffices to prove this for f real, so it is enough to show $\lambda(f) \leq \int_X f d\mu$ for each real $f \in C_K(X)$. Let E be the support of some real $f \in C_K(X)$, and let $[a, b]$ contain the range of f. Let $\varepsilon > 0$ and choose a finite set $y_1 \ldots y_n$ with $y_i - y_{i-1} < \varepsilon$ and $y_0 < a < y_1 < \ldots < y_n = b$. For each $i = 1 \ldots n$ put $E_i = \{x \in X \mid y_{i-1} < f(x) \leq y_i\} \cap E$. Show there exists open sets $V_1 \ldots V_n$ with $E_i \subset V_i$ for each i such that $\mu(V_i) < \mu(E_i) + \varepsilon/n$ for each i and such that $f(x) < y_i + \varepsilon$ for each $x \in V_i$. Then show there are functions $h_i \prec V_i$ with $\Sigma h_i = 1$ on E so $f = \Sigma h_i f$. Use this to finish the proof.]

13. Complete the proof of Theorem 8.14.

14. Prove Theorem 8.15

LEBESGUE MEASURE

Euclidean n-space $\mathbf{R}^n$ is a real vector space with respect to coordinate-wise addition and scalar multiplication. If $x, y \in \mathbf{R}^n$ are given by $(x_1 \ldots x_n)$ and $(y_1 \ldots y_n)$ respectively, there is an inner product defined on $\mathbf{R}^n$ by $(x, y) = \Sigma_{i=1}^n x_i y_i$ and $|x|$ is defined as $(x, x)^{1/2}$. The norm $|\ |$ satisfies the *triangle inequality*; i.e., $|x - y| \leq |x - z| + |z - y|$ for any $z \in \mathbf{R}^n$. This should be familiar to the reader from linear algebra. The function $d(x, y) = |x - y|$ is a metric on $\mathbf{R}^n$. If $E \subset \mathbf{R}^n$ and $x \in \mathbf{R}^n$ we define the **translate** of E by x as the set $E + x = \{y + x \mid y \in E\}$. A set of vectors in $\mathbf{R}^n$ of the form $C = \{x \mid a_i < x_i < b_i\}$ or a set obtained by replacing any or all of the $<$ with $\leq$ in the n inequalities defining C will be called an n-**cell**. The **volume** or *measure* of an n-cell is defined to be $v(C) = \Pi_{i=1}^n (b_i - a_i)$. If $a \in \mathbf{R}^n$ and $\varepsilon > 0$ we call the set $B(a, \varepsilon) = \{x \mid a_i \leq x_i < a_i + \varepsilon\}$

the **box** at a with **side** ε. We will also refer to such a box as an ε-box. For each positive integer m let A_m be the set of all vectors in $\mathbf{R}^n$ whose coordinates are integral multiples of 2^{-n} and let B_m be the collection of all 2^{-n} boxes at vectors $x \in A_n$.

15. Show the following:

(1) For each m, each $x \in \mathbf{R}^n$ lies in one and only one member of B_m.
(2) If $U_1 \in B_m$ and $U_2 \in B_k$ where $m < k$ then either $U_1 \subset U_2$ or $U_1 \cap U_2 = \varnothing$.
(3) If $U \in B_m$ then $v(U) = 2^{-mn}$.
(4) If $m < k$ and $U \in B_m$ then A_n has precisely $2^{(k-m)n}$ vectors in U.
(5) Each non-void open set in $\mathbf{R}^n$ is a countable union of disjoint boxes, each belonging to some B_m.

16. Show that there exists a positive, complete measure m defined on a σ-algebra M in $\mathbf{R}^n$ having the following properties:

(1) $m(U) = v(U)$ for each ε-box U.
(2) M contains all the Borel sets in $\mathbf{R}^n$.
(3) $E \in M$ if and only if there exists an F_σ set F and a G_δ set G with $F \subset E \subset G$ and $m(G - F) = 0$.
(4) m is regular.
(5) For each $x \in \mathbf{R}^n$ and $E \in M$, $m(E + x) = m(E)$.
(6) The property (5) is called **translation invariance**. If μ is a positive, translation invariant Borel measure on $\mathbf{R}^n$ with $\mu(K) < \infty$ for each ompact set K, then there exists a real number r with $\mu(E) = rm(E)$ for each Borel set E in $\mathbf{R}^n$.

Chapter 9

HAAR MEASURE IN UNIFORM SPACES

9.1 Introduction

In 1933, in a paper titled *Die Massbegriff der Theorie der Kontinuierlichen Gruppen* published in the Annals of Mathematics (Volume 34, Number 2), A. Haar established the existence of a *translation invariant* measure in compact, separable, topological groups. Translation invariance of a measure μ in a topological group G means that if E is a measurable set then $\mu(E + x) = \mu(E)$ for each $x \in G$. Here, $E + x = \{y \in G \mid y = a + x \text{ for some } a \in G\}$. $E + x$ is called the *x–translate* of E. The transformation T_x defined on G by $T_x(y) = y + x$ is called the *x–translation* or simply a translation. Topological groups will be defined later in the chapter and these concepts will be developed formally.

In 1934, in a paper titled *Zum Haarschen Mass in topologischen Gruppen* (Comp. Math., Volume 1), J. von Neumann showed the uniqueness of Haar's measure, and in 1940, A. Weil published *L'integration dans les groupes topologiques et ses applications* (Hermann Cie, Paris) where Haar's results were extended to locally compact topological groups. In 1949, I. Segal extended Haar's results to certain uniformly locally compact uniform spaces that we will call *isogeneous* uniform spaces (Journal of the Indian Mathematical Society, Volume 13). In 1958, Y. Mibu, evidently unaware of Segal's work, independently established similar results for this same class of spaces (Mathematical Society of Japan, Volume 10). The Haar measures of both Segal and Mibu were *Baire* measures rather than *Borel* measures. Recall from Chapter 8 that Baire measures are defined on a smaller class of sets than Borel measures, namely, on the smallest σ-ring containing all the compact sets.

In 1972, G. Itzkowitz extended Haar's results to the Borel sets of a class of locally compact uniform spaces (Pacific Journal of Mathematics, Volume 41) that he called *equi-homogeneous* uniform spaces. That equi-homogeneous uniform spaces are equivalent to isogeneous uniform spaces is the subject of Exercise 2 at the end of this section. Itzkowitz showed the existence of a Haar integral (translation invariant, linear functional on the set of real valued continuous functions with compact support) for locally compact equi-homogeneous uniform spaces. His approach was to show that a locally compact equi-homogeneous uniform space (X, μ) is homeomorphic to a quotient G/H of topological groups, where H is a stability subgroup of G, and then apply Weil's theory of invariant measures on these quotients, as recorded in Chapter 3 of L. Nachbin's book *The Haar Integral*, (Van Nostrand, New York, 1965) to obtain a unique Haar measure. Itzkowitz's approach involved showing the modular function on H is constant and then appealing to theorems in Weil's theory to show this implies the existence and uniqueness of a Haar measure.

In Section 4 of his paper, he also presented an alternate proof of the existence part of his development of a Haar measure on the Borel subsets of a locally compact equi-homogeneous uniform space. His proof that locally compact equi-homogeneous uniform spaces are quotients of topological groups contains an error as pointed out by the author in a paper titled *On Haar Measure in Uniform Spaces* (Mathematica Japonica, 1995) with a counter-example to his proof of Lemma 2.1 which is used in an essential manner to prove his Theorem 2.2. However, his alternate existence proof of a Haar measure on the Borel subsets of a locally compact equi-homogeneous uniform space is valid.

This leaves his extended theory of Haar measure (on the Borel subsets of a locally compact isogeneous uniform space) incomplete in the sense that the uniqueness part of the proof has not been established. In 1992, the author, unaware of Itzkowitz's work, showed the existence and uniqueness of a Haar measure on the Borel subsets of locally compact isogeneous uniform spaces and presented that development in a series of lectures at the 1992 Topology Workshop at the University of Salerno. The existence part of the author's development is essentially the same as Itzkowitz's alternate existence proof. But the author's uniqueness proof is a uniform space argument rather than an appeal to Weil's theory of invariant measures on quotients of topological groups. The author's development, as presented at the Salerno Workshop is given in this chapter.

It turns out that Itzkowitz's theorem that locally compact isogeneous uniform spaces are quotients of topological groups is true. We will prove this in Section 3 using a modification of Itzkowitz's approach that allows us to avoid the use of his erroneous Lemma 2.1. We will not show that the rest of Itzkowitz's approach can be corrected because the topology we get with our new proof is finer than Itzkowitz's topology on the group G and this necessitates additional work to straighten out his approach which is beyond the scope of this chapter. What we will show is that the converse of this result is true, i.e., that quotients of topological groups are isogeneous uniform spaces. This characterizes the locally compact isogeneous uniform spaces as locally compact quotients of topological groups and leads to necessary and sufficient conditions for locally compact uniform spaces to have a topological group structure that generates the uniformity or to have an abelian topological group structure that generates the uniformity.

The Segal-Mibu approach uses a generalization of K. Kodaira's construction given in a paper titled *Uber die Beziehung zwischen den Massen und den*

Topologien in einer Gruppe, (Proc. Phys. Math. Soc. Japan, Volume 23, No. 3, 1941, pp. 67-119) whereas the Itzkowitz-Howes approach uses a generalization of A. Weil's technique published in his paper referenced earlier in this section. At the present moment it may appear that the Weil technique is more powerful in isogeneous uniform spaces in that it can be used to obtain a measure on a larger class of sets. However, it is probable that the two methods are equivalent. If this is the case, we would have a way of constructing the measure directly on the Borel sets using a simple combinatoric method.

A uniform space (X, μ) is said to be **isogeneous** if there exists a basis ν for μ and a collection H of uniform homeomorphisms of X onto itself such that:

(1) For each $i \in H$, and each pair of points $x, y \in X$, $y \in S(x, \mathcal{U})$ if and only if $i(y) \in S(i(x), \mathcal{U})$ for each $\mathcal{U} \in \nu$.
(2) For each pair $x, y \in X$, there exists an $i_{xy} \in H$ that carries x onto y.

The members of H are called **isomorphisms with respect to** ν or simply **isomorphisms.** ν is called an **isomorphic basis** for μ. Clearly, if i is an isomorphism and $\mathcal{U} \in \nu$, then $i(S(x, \mathcal{U})) = S(i(x), \mathcal{U})$ for each $x \in X$. Also, it is easily seen that compositions and inverses of isomorphisms are again isomorphisms. A topological space is said to be **homogeneous** if for each pair of points $p, q \in$ X there exists a homeomorphism of X onto itself that carries p onto q. Clearly isogeneous uniform spaces are homogeneous topological spaces.

There are various types of isomorphisms. An isomorphism $t{:}X \to X$ is called a **translation** if t has no fixed points. If the isomorphism $r{:}X \to X$ has a proper subset F $\neq \varnothing$ of fixed points and F does not separate X - F, then r is called a **rotation.** If F separates X - F, r is called a **reflection.** In what follows, we will show that locally compact isogeneous uniform spaces have a unique integral that is not only translation invariant, but also invariant under rotations and reflections. All topological groups are isogeneous uniform spaces with respect to the classical group uniformities and the classical group translation T_x defined by $T_x(y) = y + x$ for each $x \neq 0$ in a topological group satisfies the above definition of translation.

Let $C(X)$ denote the ring of real valued continuous functions on X and $C_K(X)$, the members of $C(X)$ whose support have compact closures. For any $f \in C_K(X)$ and isomorphism $i{:}X \to X$, denote f © i by $f_i \in C_K(X)$. By a **Haar integral** for X, we mean a positive linear functional I on $C_K(X)$ such that $I(f_i) = I(f)$ for each isomorphism $i{:}X \to X$. A **Haar measure** for X is an almost regular, Borel measure m satisfying $m(i(E)) = m(E)$ for each Borel set E and each isomorphism $i{:}X \to X$. The following lemma is left as an exercise.

LEMMA 9.1. If (X, μ) is a locally compact uniform space, then each $f \in C_K(X)$ is uniformly continuous.

Isogeneous uniform spaces were introduced in a series of lectures by the author at the 1992 Topology Workshop, held at the University of Salerno, Italy. The remaining material in this chapter is from the Workshop lecture series.

EXERCISES

1. Prove Lemma 9.1.

EQUI-HOMOGENEOUS UNIFORM SPACES

Let (X, $\mathcal{U}$) be an entourage uniform space. A function f:X → X is said to be **nonexpansive** with respect to a base B for $\mathcal{U}$ if for each $U \in B$ and $(x, y) \in U$, the relation $(f(x), f(y)) \in U$ also holds. By a *B-nonexpansive* homeomorphism f of a uniform space (X, $\mathcal{U}$) onto itself, we mean a homeomorphism f of X onto itself such that f is nonexpansive with respect to a base B for the uniformity $\mathcal{U}$. A uniform space (X, $\mathcal{U}$) will be called an **equi-homogeneous** space if there is a group G of homeomorphisms acting on X such that (i) G is transitive (i.e., given $p, q \in$ X, there is a $g \in G$ such that $g(p) = q$, and (ii) there is a base B for $\mathcal{U}$ such that G is a group of B-nonexpansive homeomorphisms of the uniform space.

2. Show that the equi-homogeneous uniform spaces are precisely the isogeneous uniform spaces.

3. A collection G of functions from a uniform space (X, $\mathcal{U}$) to a uniform space (Y,$\mathcal{V}$) is said to be **equi-continuous** if for each $V \in \mathcal{V}$, there is a $U \in \mathcal{U}$ such that for each $g \in G$, $[g \times g](U) \subset V$.

Show that if G is a group of homeomorphisms acting on a uniform space (X,$\mathcal{U}$), then the following are equivalent: (i) there is a base B for the uniformity such that G is a group of B-nonexpansive homeomorphisms of (X, $\mathcal{U}$), and (ii) G is an equi-continuous group of uniform homeomorphisms on the uniform space (X, $\mathcal{U}$).

4. Show that if (X, $\mathcal{U}$) is a locally compact isogeneous uniform space, then (X,$\mathcal{U}$) is uniformly locally compact.

9.2 Haar Integrals and Measures

In this section, (X, μ) is assumed to be a locally compact isogeneous uniform space. Let ν be an isomorphic basis for μ and H a collection of isomorphisms with respect to ν that satisfy condition (2) in the definition of an isogeneous uniform space. Let $g \geq 0$ be in $C_K(X)$ such that $g(x) \geq b$ for each x in some $\mathcal{U}$-sphere $S(y, \mathcal{U})$ where $\mathcal{U} \in \nu$. If $f \geq 0$ is in $C_K(X)$ with $f(X - K) = 0$ for some compact $K \subset X$ then there is a finite subset $\{x_1 \ldots x_n\}$ of K such that $\{S(x_1, \mathcal{U}) \ldots S(x_n, \mathcal{U})\}$ covers K. If $|f|_j = sup\{f(x) | x \in S(x_j, \mathcal{U})\}$ and $a_j \geq |f|_j/b$ for each $j = 1 \ldots n$, then $f(x) \leq \Sigma_{j=1}^{n} a_j g_{x_j}(x)$ for each $x \in X$ where $g_{x_j} = g © i_{x_jy}$ for some isomorphism $i_{x_jy}:X \to X$ that carries x_j onto y. The finite collection $\{a_1 \ldots a_n\}$ is said to *dominate* f with respect to g. Put $[f|g] = inf\{\Sigma_j a_j | \{a_j\}$ dominates f with respect to $g\}$.

THEOREM 9.1. The number $[f|g]$ is non-negative, finite and satisfies:
(1) $[f_i|g] = [f|g]$ for each isomorphism $i:X \to X$.
(2) $[f_1 + f_2|g] \leq [f_1|g] + [f_2|g]$ for each pair $f_1, f_2 \in C_K(X)$.
(3) $[\alpha f|g] = \alpha[f|g]$ for each $\alpha > 0$.
(4) $f_1 \leq f_2$ implies $[f_1|g] \leq [f_2|g]$ for each $f_1 \leq f_2 \in C_K(X)$.
(5) $[f|h] \leq [f|g][g|h]$ for each $h \in C_K(X)$ with $h \neq 0$.
(6) $[h|f]^{-1} \leq [f|g]/[h|g] \leq [f|h]$ for each $h \in C_K(X)$ with $h \neq 0$.

Proof: We prove only (1). (2) through (6) are left as an exercise. Now $f_i(x) = [f©i](x) \leq \Sigma_{j=1}^{n} a_j[g_{x_j} © i](x)$ for each $\{a_j\}$ that dominates f with respect to g. Also, $g_{x_j} © i = g © [i_{x_jy} © i]$ where $[i_{x_jy} © i]$ is an isomorphism so $\{a_j\}$ dominates f_i with respect to g. Hence $[f_i|g] \leq [f|g]$. Now $f_{i_i^{-1}} = f © i © i^{-1} = f$ and from what has just been proved, we have $[f|g] = [f_{i_i^{-1}}|g] \leq [f_i|g]$. Therefore, $[f_i|g] = [f|g]$ for each isomorphism i. ∎

Let $C_K^+(X)$ denote the non-negative members of $C_K(X)$ and choose some $k \in C_K^+(X)$. For each $g \in C_K^+(X)$, define $I_g:C_K^+(X) \to [0, \infty)$ by $I_g(f) = [f|g]/[k|g]$ for each $f \in C_K^+(X)$.

COROLLARY 9.1. For each $g \in C_K^+(X)$,
(1) $I_g \geq 0$.
(2) $[k|f]^{-1} \leq I_g(f) \leq [f|k]$.
(3) $I_g(f_i) = I_g(f)$ for each isomorphism $i:X \to X$.
(4) $I_g(\alpha f) = \alpha I_g(f)$ for each $\alpha > 0$.
(5) $I_g(f_1 + f_2) \leq I_g(f_1) + I_g(f_2)$.

Since X is locally compact, for each $\mathcal{U} \in \mu$ there exists a $g \in C_K^+(X)$ with $g(X - K) = 0$ for some $\mathcal{U}$-small compact set K.

LEMMA 9.2. For each $f_1, f_2 \in C_K^+(X)$ with $f_1 \neq 0 \neq f_2$ and for each $\varepsilon > 0$, there exists a $\mathcal{U} \in \nu$ such that $I_g(f_1) + I_g(f_2) \leq I_g(f_1 + f_2) + \varepsilon$ for each $g \in C_K^+(X)$ with $g \neq 0$ and such that the support of g is $\mathcal{U}$-small.

Proof: Let $\varepsilon > 0$. Assume $[f_1 + f_2](X - K) = 0$ for some compact K and let $f \in C_K^+(X)$ such that $f(K) = 1$. For any $\delta > 0$ put $\phi = f_1 + f_2 + \delta f$ and for $i = 1, 2$ put $h_i = f_i/\phi$ for $\phi(x) \neq 0$ and 0 otherwise. Clearly ϕ, h_i, $h_2 \in C_K^+$ and $h_1 + h_2 \leq 1$. By Lemma 9.1, for each $\varepsilon' > 0$ there exists a $\mathcal{U} \in \nu$ with $|h_i(x) - h_i(y)| < \varepsilon'$ for each $x, y \in X$ that are $\mathcal{U}$-close. Consider any $g \in C_K^+$ with $g \neq 0$ such that the support of g is $\mathcal{U}$-small. Suppose $\phi \leq \Sigma_j a_j g_{x_j}$. Since $g_{x_j}(x) \neq 0$ implies x and x_j are $\mathcal{U}$-close, we have $|h_i(x_j) - h_i(x)| < \varepsilon'$ for $i = 1, 2$. Therefore,

$$f_i(x) = [\phi h_i](x) \leq [\Sigma_j a_j g_{x_j}(x)]h_i(x) \leq \Sigma_j a_j g_{x_j}(x)[h_i(x_j) + \varepsilon']$$

so $[f_i|g] \leq \Sigma_j a_j[h_i(x_j) + \varepsilon']$ which implies $[f_i|g] + [f_2|g] \leq \Sigma_j a_j(1 + 2\varepsilon')$. Consequently, $[f_1|g] + [f_2|g] \leq [\phi|g](1 + 2\varepsilon')$. Division by $[k|g]$ yields $I_g(f_1) + I_g(f_2) \leq I_g(f_1 + f_2) + 2\varepsilon' I_g(f_1 + f_2) + \delta(1 + 2\varepsilon')I_g(f)$. Since both $I_g(f)$ and $I_g(f_1 + f_2)$ are finite, the last two terms on the right hand side of this inequality can be made less than ε for sufficiently small δ and ε'. ∎

THEOREM 9.2. (G. Itzkowitz, 1972) Each locally compact, isogeneous uniform space has a non-zero Haar integral.

Proof: Let D be the set of all $f \in C_K^+(X)$ with $f \neq 0$. For each $f \in$ D put $H_f = [[k|f]^{-1}, [f|k]] \subset \mathbf{R}$. Then $Y = \Pi H_f$ is a compact Hausdorff space. For each $\mathcal{U} \in \nu$ put $G_{\mathcal{U}} = \{I_g | g \in$ D and the support of g is $\mathcal{U}$-small$\}$. Let $F_{\mathcal{U}} = Cl(G_{\mathcal{U}})$. Then $\{F_{\mathcal{U}}\}$ has the finite intersection property so $\cap F_{\mathcal{U}} \neq \varnothing$. Let $I \in \cap F_{\mathcal{U}}$. Then I can be interpreted as a functional on D where $I(f) = p_f(I)$ and p_f is the projection of Y onto the fth factor space H_f. We now show I satisfies the following:

(1) $I(f) > 0$ for each $f \in$ D.
(2) $I(f_i) = I(f)$ for each isomorphism i and each $f \in$ D.
(3) $I(\alpha f) = \alpha I(f)$ for each $\alpha > 0$ and $f \in$ D.
(4) $I(f_1 + f_2) = I(f_1) + I(f_2)$ for each $f_1, f_2 \in$ D.

First note that for each neighborhood $U(I) = \{\psi \in Y | |\psi(f_j) - I(f_j)| < \varepsilon\}$ for some finite collection $\{f_1 \ldots f_n\} \subset$ D, and each $\mathcal{U} \in \nu$, there exists some $I_g \in U(I) \cap G_{\mathcal{U}}$.

(1) follows from the fact that for each $f \in$ D, $I(f) = p_f(I) \in H_f = [[k|f]^{-1}, [f|k]]$ and hence $I(f) \geq [k|f]^{-1} > 0$ since $k \neq 0$. (2) follows from observing that for each $\varepsilon > 0$, there exists a $g_\varepsilon \in$ D with $I_{g_\varepsilon} \in \{\psi \in Y | |\psi(f_j) - I(f_j)| < \varepsilon\}$ for $j = 1, 2$ where $f_1 = f$ and $f_2 = f_i$. Then $|I(f_i) - I(f)| \leq |I(f_i) - I_{g_\varepsilon}(f_i)| + |I_{g_\varepsilon}(f_i) - I(f)|$ and since $I_{g_\varepsilon}(f_i) = I_{g_\varepsilon}(f)$ we have $|I(f_i) - I(f)| < 2\varepsilon$. (3) follows by using the inequality $|I(\alpha f) - \alpha I(f)| \leq |I(\alpha f) - I_{g_\varepsilon}(\alpha f)| + |I_{g_\varepsilon}(\alpha f) - \alpha I(f)|$ where $I_{g_\varepsilon} \in \{\psi \in Y | |\psi(f_j) - I(f_j)| < \varepsilon\}$ for $j = 1, 2$ where $f_1 = f$ and $f_2 = \alpha f$ and obtaining $|I(\alpha f) - \alpha I(f)| < \varepsilon + \alpha\varepsilon$.

To prove (4), let $\varepsilon > 0$ and put $f_3 = f_1 + f_2$. By Lemma 9.2, there exists a $\mathcal{U} \in \nu$ such that $I_g(f_1) + I_g(f_2) \leq I_g(f_1 + f_2) + \varepsilon$ for each $g \in$ D whose support is $\mathcal{U}$-small. Let $I_{g_\varepsilon} \in \{\psi \in Y | |\psi(f_j) - I(f_j)| < \varepsilon\} \cap G_{\mathcal{U}}$ for $j = 1, 2, 3$. Then

$$|I(f_1) + I(f_2) - I(f_1 + f_2)| \leq |I(f_1) + I(f_2) - I_{g_\varepsilon}(f_1 + f_2)| +$$

$$|I_{g_\varepsilon}(f_1+f_2)-I(f_1+f_2)|$$

and since $I_{g_\varepsilon}(f_1+f_2)=I_{g_\varepsilon}(f_1)+I_{g_\varepsilon}(f_2)$ we have $|I(f_1)+I(f_2)-I(f_1+f_2)| \leq |I(f_1)-I_{g_\varepsilon}(f_1)| + |I(f_2)-I_{g_\varepsilon}(f_2)| + \varepsilon < 3\varepsilon$.

We can extend I to a linear functional on $C_K^+(X)$ by defining $I(0)=0$. To extend I to a linear functional J on $C_K(X)$ note that each $f \in C_K(X)$ can be written as $f^+ - f^-$ for some $f^+, f^- \in C_K^+(X)$. We define $J(f)$ to be $I(f^+) - I(f^-)$. For each $\beta = -\alpha$ (where $\alpha > 0$) we have $J(\beta f) = J(\alpha(-f)) = I(\alpha(-f)^+) - I(\alpha(-f)^-) = \alpha[I((-f)^+) - I((-f)^-)] = \alpha[I(f^-) - I(f^+)] = -\alpha[I(f^+) - I(f^-)] = \beta J(f)$. Clearly $J(f_1) + J(f_2) = J(f_1+f_2)$ for each $f_1, f_2 \in C_K(X)$. Finally, $f_i^+ = f^+ \circledcirc i$ and $f_i^- = f^- \circledcirc i$ so $J(f_i) = I(f_i^+) - I(f_i^-) = I(f^+) - I(f^-) = J(f)$ for each isomorphism $i{:}X \to X$. Therefore, J is a uniform integral for X. ∎

By the Riesz Representation Theorem (Chapter 8), there exists a σ-algebra M in X containing all Borel sets, and a unique, positive, regular measure m on M which represents J in the sense that $J(f) = \int_X f\,dm$ for each $f \in C_K(X)$ and such that $m(K) < \infty$ for each compact $K \subset X$.

THEOREM 9.3 m is a Haar measure for X.

Proof: We first show that if K is compact then $m(i(K)) = m(K)$ for each isomorphism $i{:}X \to X$. Let U be an open set containing K. Then there is an $f_U \in C_K^+(X)$ with $f_U(K) = 1$ and $f_U(X-U) = 0$. Therefore, $\chi_K \leq f_U \leq \chi_U$ so

$$m(K) = \int_X \chi_K dm \leq \int_X f_U dm \leq \int_X \chi_U dm = m(U).$$

Since m is outer regular, this shows that $m(K) = \inf\{\int_X f_U dm \mid K \subset U \text{ and } U \text{ is open}\}$. But then $m(K) = \inf\{\int_X [f_U \circledcirc i^{-1}]dm \mid K \subset U \text{ and } U \text{ is open}\}$. Thus,

$$m(i(K)) \leq \int_X [f_U \circledcirc i^{-1}]dm \leq \int_X \chi_{i(U)} dm = m(i(U)).$$

Now $m(i(K)) = \{m(V) \mid i(K) \subset V \text{ and } V \text{ is open}\}$. For each open V containing $i(K)$, it is possible to find a $V \in \nu$ such that $Star(i(K),V) \subset V$. Since $Star(i(K),V) = i(Star(K,V))$, $m(i(K)) = \inf\{m(i(U)) \mid K \subset U \text{ and } U \text{ is open}\}$, so $m(i(K)) = m(K)$.

Next suppose E is a Borel set with $m(E) < \infty$ and i is an isomorphism on X. Then there is an ascending sequence $\{K_j\}$ of compact subsets of E with $m(E) = \lim_{j\to\infty} m(K_j)$. For each positive integer j let χ_{K_j} be the *characteristic* function of K_j. Then for each j, $\chi_{K_j} \leq \chi_{K_{j+1}}$. Since, the sequence $\{\chi_{K_j}\}$ converges almost everywhere to χ_E, $\{\chi_{K_j} \circledcirc i^{-1}\}$ converges almost everywhere to $\chi_E \circledcirc i^{-1}$. By Theorem 8.9, $\{\int_X \chi_{K_j} dm\}$ converges to $\int_X \chi_E dm$ and $\{\int_X [\chi_{K_j} \circledcirc i^{-1}]dm\}$ converges to $\int_X [\chi_E \circledcirc i^{-1}]dm$. Therefore,

$$m(i(E)) = \int_X \chi_i(E) dm = \int_X [\chi_E \circledcirc i^{-1}]dm = \lim_{j\to\infty} \int_X [\chi_{K_j} \circledcirc i^{-1}]dm =$$

$$lim_{j\to\infty}\int_X \chi_{K_j} dm = \int_X \chi_E dm = m(E).$$

Consequently, $m(i(E)) = m(E)$ for each Borel set E with $m(E) < \infty$, and each isomorphism i. Finally, suppose $m(E) = \infty$. If $m(i(E)) < \infty$, then by what has just been shown, $m(E) = m(i^{-1}(i(E))) = m(i(E))$, so $m(E) < \infty$ which is a contradiction. Therefore, for each Borel set E, $m(E) = m(i(E))$ for each isomorphism i. ∎

We will defer the uniqueness proof for m until Theorem 10.5.

EXERCISES

1. Prove Theorem 9.1.

2. Let (X, μ) be an isogeneous uniform space and let i be an isomorphism of X onto itself. Let m be a regular, Borel measure on X. For each Borel set E put $m_i(E) = m(i(E))$. Show that m_i is an almost regular, Borel measure on X.

3. Show that the results of this section still hold if $C_K(X)$ is replaced by $C_c(X) = \{f{:}X \to \mathbf{C} \mid f \text{ is continuous and } Cl(0(f)) \text{ is compact}\}$.

9.3 Topological Groups and Uniqueness of Haar Measures

In this section we will introduce the concept of a topological group and see that all topological groups are uniformizable and that locally compact topological groups are cofinally complete (uniformly paracompact) and hence complete. We will also see how the classical existence and uniqueness proofs of the Haar integral in locally compact topological groups follow from the results in the previous section and what form they take in topological groups. This section also contains a solution to the problem of characterizing which uniform spaces have a compatible topological group structure that generates the uniformity for the case when the uniform space is locally compact.

The concept of a topological group is simply the combination of the concepts of a group and of a topology on the same underlying set in such a way that the group operation is in some sense compatible with the topology. Precisely, by a **topological group**, we mean a topological space (X, τ) together with a binary operation $+{:}X^2 \to X$ satisfying the following conditions:

G1. $(X, +)$ is a group.
G2. $+$ is continuous with respect to the product topology on X^2.
G3. The *inversion* function $i{:}X \to X$ defined by $i(x) = -x$ for each $x \in X$ is continuous.

It is easy to find interesting examples of topological groups. Probably the most obvious is the additive group of real numbers (**R**, +) with the usual top- ology for **R**. Similarly, coordinatewise addition in $\mathbf{R}^n$ makes $\mathbf{R}^n$ a topological group with respect to the Euclidean n-space topology. Another important example is the *circle group* $\mathbf{T} \subset \mathbf{C}$. The *unit circle* **T** is defined as the subset of complex numbers consisting of members z such that $|z| = 1$. The group operation is usual multiplication of complex numbers and the topology of **T** is the subspace topology of **T** with respect to the usual topology of $\mathbf{R}^2$.

PROPOSITION 9.1 For a group (X, +) with topology τ, the conditions G2 and G3 hold if and only if the function $f: X^2 \to X$ *defined by* $f(x, y) = x - y$ *is continuous.*

Proof: If G2 and G3 hold then f is continuous since $f(x, y) = g(x) + i(y)$ for each pair $x, y \in X$ where g is the identity function on X because $g \times i$ is continuous on X^2 by G3 and addition is continuous by G2 and f is the composition of $g \times i$ and the addition function.

Conversely, if f is continuous then inversion is continuous since $i(x) = f(0,x)$ for each $x \in X$, where 0 is the group identity. Also, since $x + y = f(x, i(y))$ for each pair $(x, y) \in X^2$ it follows that addition is continuous. ∎

PROPOSITION 9.2 For a topological group (X, τ, +), the following functions are homeomorphisms of X onto itself:

(1) inversion i,

(2) right translation r_a *by some* $a \in X$ *defined by* $r_a(x) = x + a$ *for each* $x \in X$,

(3) left translation l_a *by some* $a \in X$ *defined by* $l_a(x) = a + x$ *for each* $x \in X$.

Proof: The functions i, r_a and l_a are clearly continuous for each $a \in X$. Since the compositions $i © i$, $r_a © r_{-a}$, and $l_a © l_{-a}$ are all the identity mapping for each $a \in X$, it is also clear that i, r_a and l_a are all homeomorphisms for each $a \in X$. ∎

For a set $S \subset X$, the notation S^{-1} is used to denote the set consisting of all inverse elements of members of S. S is said to be **symmetric** in X if $S = S^{-1}$. If $T \subset X$, the notation $S + T$ is used to denote the set consisting of all elements of the form $s + t$ such that $s \in S$ and $t \in T$.

PROPOSITION 9.3 If A is open and B is closed in the topological group (X, +), and if $C, D \subset X$ *then:*

(1) $Cl(D^{-1}) = [Cl(D)]^{-1}$.

(2) $Cl(x + D + y) = x + [Cl(D)] + y$ *for each pair* $x, y \in X$.

(3) $D + A$ *and* $A + D$ *are open.*

(4) $B + y$ *and* $y + B$ *are closed for each* $y \in X$.

(5) $Cl(C) + Cl(D) \subset Cl(C + D)$.

Proof: Since i is a homeomorphism, $Cl(D^{-1}) = Cl(i(D)) = i(Cl(D)) = [Cl(D)]^{-1}$. Similarly, since for each pair $x, y \in X$, l_x and r_x are homeomorphisms,

$$Cl[x + D + y] = Cl(l_x(D + y)) = Cl(l_x(r_y(D))) =$$

$$l_x(r_y(Cl(D))) = x + Cl(D) + y.$$

This establishes (1) and (2). (3) follows from Proposition 9.2 since for each $x \in X$, $x + A = l_x(A)$ and $D + A = \cup_{x \in D}(x + A)$. Similarly $A + D$ is open. (4) also clearly follows from Proposition 9.2. (5) follows from the continuity of + on $X \times X$, i.e., $Cl(C) + Cl(D) = +(Cl(C) \times Cl(D)) \subset Cl(+(C \times D)) = Cl(C + D)$. ▪

PROPOSITION 9.4 If $h:X \to Y$ is a group homomorphism between topological groups X and Y then:

(1) $h(A + B) = h(A) + h(B) \subset Cl(h(A)) + Cl(h(B)) \subset h(Cl(A + B))$ for each $A, B \subset X$.

(2) $h^{-1}(C) + h^{-1}(D) \subset h^{-1}(C + D)$ for each $C, D \subset Y$.

(3) $Cl(h^{-1}(C)) + Cl(h^{-1}(D)) \subset Cl(h^{-1}(C + D))$ for each $C, D \subset Y$.

(4) If A is symmetric in X then $h(A)$ and $Cl(h(A))$ are symmetric in Y.

(5) If C is symmetric in Y then $h^{-1}(C)$ and $Cl(h^{-1}(C))$ are symmetric in X.

The proof of Proposition 9.4 is left as an exercise. A subgroup H of a group G is said to be **normal** or *invariant* if for each $x \in H$, $a + x + a^{-1} \in H$ for each $a \in G$. G is said to be **Abelian** if $a + b = b + a$ for each pair $a, b \in G$.

PROPOSITION 9.5 Let H be a subgroup of a topological group G. Then:

(1) H and Cl(H) are topological groups.

(2) If H is normal then Cl(H) is normal.

(3) If G is Hausdorff and H is Abelian then Cl(H) is Abelian.

(4) If H is open then H = Cl(H).

Proof: H is clearly a topological group since the continuous function $f:G^2 \to G$ defined in Proposition 9.1 by $f(x, y) = x - y$ has a continuous restriction to $H^2 \subset G^2$. That $Cl(H)$ is also a topological group follows from the fact that since H is a subgroup of G, $H + H = H = H^{-1}$ and so $Cl(H) + [Cl(H)]^{-1} = Cl(H) + Cl(H^{-1}) \subset Cl(H + H^{-1}) = Cl(H)$ by Proposition 9.3. Hence $Cl(H)$ is also a subgroup of G.

To show (2), assume H is normal. Then $x + H - x = H$ for each $x \in G$, so $x + Cl(H) - x = Cl(x + H - x) = Cl(H)$ for each $x \in G$ by Proposition 9.3 which implies $Cl(H)$ is normal. To show (3), assume G is Hausdorff and H is Abelian. Define $g:G^2 \to G$ by $g(x, y) = x + y - x - y$ for each pair $x, y \in G$. Then g is continuous since + and i (inversion) are continuous. Since G is Hausdorff, $\{0\}$

is closed where 0 is the group identity element. Since g is continuous, $g^{-1}(0)$ is also closed in G. Now $H^2 \subset h^{-1}(0)$ since H is Abelian, so $Cl(H) \times Cl(H) = Cl(H \times H) \subset h^{-1}(0)$ is closed. Hence $Cl(H)$ is also Abelian.

To show (4), assume H is open in G. Then we have $G - H = (G - H) + H = \cup_{x \in G-H}(x + H)$ which is open since $x + H$ is open for each $x \in G - H$ by Proposition 9.3. Therefore, H is closed in G so $H = Cl(H)$. ∎

Let X and Y be topological groups having neighborhood systems N and B at 0 and e respectively, where 0 is the identity element of X and e is the identity element of Y. A homomorphism $h:X \to Y$ is **open at 0** if for each $U \in N$, there is a $V \in B$ with $V \subset h(U)$.

PROPOSITION 9.6 A homomorphism $h:X \to Y$ is continuous (open) if and only if it is continuous at 0 (respectively open at 0).

Proof: Clearly if h is continuous or open, then it is continuous or open respectively at 0. Assume first that h is continuous at 0 and let N be the neighborhood system at 0. For each $x \in X$ and open V containing $y = h(x)$, e (the identity of Y) is in $V - y$ and $V - y$ is open by Proposition 9.3. Since h is continuous at 0, there is a $U \in N$ with $h(U) \subset V - y$. Since $x \in U + x$ and $h(U+x) = h(U) + h(x) \subset (V - y) + y = V$, we have that h is continuous.

Next assume h is open at 0 and let B be the neighborhood system at e. For each $x \in X$ and open U containing x, $U - x$ is an open set containing 0, so there exists a $V \in B$ with $V \subset h(U - x) = h(U) - h(x)$ which implies $V + y \subset h(U)$ where $y = h(x)$. Since $e \in V$, $y \in V + y$ which is open. Therefore, $h(x)$ is an interior point of $h(U)$. Consequently, h is an open mapping. ∎

A key feature of topological groups is that knowing the behavior of the neighborhood system of the identity element is equivalent to knowing the behavior of all neighborhood systems at each point of the group. This derives from the fact that if g is any element of the group and U is a neighborhood of the group identity, then both $g + U$ and $U + g$ are neighborhoods of g and if V is a neighborhood of g, then $g^{-1} + V$ and $V + g^{-1}$ are both neighborhoods of the identity. Consequently, the following results are especially useful in topological groups.

PROPOSITION 9.7 If $(X, +)$ is a topological group and N the neighborhood system of the identity 0, then:

(1) For each $U \in N$ there is a $V \in N$ with $V + V \subset U$.

(2) For each $U \in N$, $U^{-1} \in N$.

(3) For each $U \in N$ and $x \in X$, there is a $V \in N$ with $[x + V - x] \subset U$.

(4) Each filter N of subsets of X containing 0 and satisfying (1) - (3) determines a unique topology that makes $(X, +)$ a topological group and has N as its neighborhood system at 0.

Proof: (1) follows from the continuity of + and (2) from the continuity of inversion. Since for each $x \in X$, l_{-x} and r_x are homeomorphisms, $l_{-x} \circledcirc r_x(U) = [(-x) + U + x] \in N$ for each $U \in N$. Hence $V \subset [(-x) + U + x]$ for some $V \in N$ which implies $[x + V + (-x)] \subset U$. Now (1) implies the existence of such a V, so (3) follows.

To show (4), let $\tau = \{U \subset X \mid \text{for each } p \in U, \text{ there exists } V \in N \text{ with } V + p \subset U\}$. τ is easily shown to be a topology on X. Clearly X, $\emptyset \in \tau$. If $\{U_\alpha\} \subset \tau$ then $\cup U_\alpha \in \tau$ since for each $x \in \cup U_\alpha$, x belongs to some U_β which implies there exists a $V_\beta \in N$ with $V_\beta + x \subset U_\beta \subset \cup U_\alpha$. Also, if $U_1 \ldots U_n \in \tau$ then $\cap_{i=1}^n U_i \in \tau$ since for each $x \in \cap_{i=1}^n U_i$, there exists a $V_i \in N$ for each i with $V_i + x \subset U_i$. Since N is a filter, $\cap_{i=1}^n V_i \in N$ and $x \in (\cap_{i=1}^n V_i) + x \subset \cap_{i=1}^n U_i$.

Clearly τ has N as its neighborhood system at 0 and if σ is another topology for X that makes (X, +) a topological group and has N as its neighborhood system at 0, it is easily shown that $\sigma = \tau$, for if $U \in \tau$ and $p \in U$, there exists a $V \in N$ with $V + p \subset U$, so U is a neighborhood of p in the topology σ, i.e., $\tau \subset \sigma$. Similarly $\sigma \subset \tau$.

It remains to show that τ itself makes (X, +) into a topological group. For this it suffices to show that $f{:}X^2 \to X$ as defined in Proposition 9.1 ($f(x, y) = x - y$) is continuous. For this let $(p, q) \in X^2$ and let U be a neighborhood of $f(p, q) = p - q$. Then $W = [U - x + y] \in N$. By (1), there exists a $V \in N$ with $V + V \subset W$. By (2), $V^{-1} \in N$ so $A = V \cap V^{-1} \in N$ since N is a filter. Then $A = A^{-1}$ and $A + A \subset V$, so $A + A^{-1} \subset W$. Now $A + p$ is a neighborhood of p. Also, there exists a $B \in N$ with $[p + B - p] \subset A$ by (3). Therefore, $[p + B - p] + q$ is a neighborhood of q, so $[A + p] \times [(p + B - p) + y]$ is a neighborhood of (p, q). Moreover, $f([A + p] \times [(p + B - p) + q]) = \{f(a + p, p + b - p + q) \mid a \in A, b \in B\} \subset \{a_1 + p - p - a_2 + p - q \mid a_1, a_2 \in A\} = [A + A^{-1}] + (x - y) \subset W + (x - y) = U$. Consequently, f is continuous, so the topology τ makes (X, +) into a topological group. ∎

THEOREM 9.4 Let (X, +) be a topological group with topology τ and neighborhood system N at 0. For each $V \in N$ put $\mathcal{U}_V = \{V + x \mid x \in X\}$ and let $\lambda = [\mathcal{U}_V \mid V \in N\}$. Then λ is a basis for a uniformity μ that is compatible with τ. Consequently, every topological group is uniformizable.

Proof: To show λ is a basis for a uniformity, it suffices to show that for each pair $\mathcal{U}_V$, $\mathcal{U}_W \in \lambda$, there exists some $\mathcal{U}_U \in \lambda$ that Δ-refines $\mathcal{U}_V \cap \mathcal{U}_W$. If $\mathcal{U}_V$, $\mathcal{U}_W \in \lambda$, then V, W $\in N$. As shown in the proof of Proposition 9.7, there exists an $A \in N$ with $A + A^{-1} \subset V \cap W$. Then $U = A \cap A^{-1} \in N$ so $S(x, \mathcal{U}_U) = \{z \in X \mid x, z \in U + y \text{ for some } y \in X\} = \{z + X \mid x - y, z - y \in U \text{ for some } y \in X\} \subset \{z \in X \mid (z - y) - (x - y) \in U + U^{-1}\} \subset \{z \in X \mid z - x \in V\} = V + x \in \mathcal{U}_V$. Similarly $S(x, \mathcal{U}_U) \subset W + x \in \mathcal{U}_W$, so $S(x, \mathcal{U}_U) \subset [V + x] \cap [W + x] \in \mathcal{U}_V \cap \mathcal{U}_W$. Therefore, $\mathcal{U}_U$ is a Δ-refinement of $\mathcal{U}_V \cap \mathcal{U}_W$.

To show that μ is compatible with τ, by Proposition 9.7 it suffices to show that $\{S(0, \mathcal{U}) | \mathcal{U} \in \mu\} = N$. For this, first note that if $V \in N$, then $V \subset S(0, \mathcal{U}_V)$ so $S(0, \mathcal{U}_V) \in N$ for each $\mathcal{U}_V \in \lambda$. Since λ is a basis for μ, we see that $\{S(0,\mathcal{U}) | \mathcal{U} \in \mu\} \subset N$. Conversely, if $V \in N$ we can choose $W \in N$ such that $W + W^{-1} \subset V$ so $S(0,\mathcal{U}_W) = \{y \in X | 0, y \in W + x \text{ for some } x \in X\} = \{y \in X | y - x \in W \text{ and } x \in W^{-1}\} = \{y \in X | y \in W + W^{-1}\} \subset V$. Therefore, $N \subset \{S(0, \mathcal{U}) | \mathcal{U} \in \mu\}$ which establishes $\{S(0, \mathcal{U}) | \mathcal{U} \in \mu\} = N$. ∎

By now, it has probably occurred to the reader that the uniformity μ of Theorem 9.4 could just as well have been defined by the basis $\nu = \{\mathcal{W}_V | V \in N\}$ where $\mathcal{W}_V = \{x + V | x \in X\}$, and the argument, with minor notational changes, would still be valid. The uniformity μ is called the **right uniformity** on X whereas the uniformity μ' generated by ν is the **left uniformity** on X. It is left as an exercise (Exercise 3) to show that the uniform spaces (X, μ) and (X, μ') are isogeneous uniform spaces where the collections of homeomorphisms are $R = \{r_x | x \in X\}$ and $L = \{l_x | x \in X\}$ respectively and the isomorphic bases are λ and ν (as defined in Theorem 9.4 and above) respectively. Consequently, if $(X,+)$ is a locally compact topological group, then there exist unique Haar measures m and m' for (X, μ) and (X, μ') respectively. By definition of $\mathcal{U}_V$ and $\mathcal{W}_V$ we then have $m(V + x) = m(V) = m(V + y)$ for each pair $x, y \in X$ and each $V \in N$ (neighborhood system at 0) and similarly, $m'(x + V) = m'(V) = m'(y + V)$.

From Theorem 9.4 it is straightforward to derive a number of useful consequences that are listed in the following proposition and left as an exercise.

PROPOSITION 9.8 If $(X, +)$ is a topological group with neighborhood system N at 0, then:

(1) X has a neighborhood base consisting of symmetric open (closed) sets.

(2) $Int(Y) = \{y \in Y | V + y \subset Y$ for some $V \in N\}$ for each $Y \subset X$.

(3) $Cl(Y) = \cap\{V + Y | V \in N\}$ for each $Y \subset X$.

(4) X is T_1 if and only if $0 = \cap\{V | V \in N\}$.

(5) X is locally compact if and only if there exists a compact $V \in N$.

Let $(X, +)$ be a topological group and let Z be a subgroup of X. For each $x \in X$, the set $Z + x$ is called a **right coset** of Z in X. Similarly $x + Z$ is called a **left coset**. Both the right cosets and the left cosets form partitions of X. To show this for the right cosets, define the relation $p \sim q$ in X by $p - q \in Z$. It is easily verified that $\sim$ is an equivalence relation on X. In fact, $p - q \in Z$ if and only if $p \in Z + q$ so that the equivalence class $[p]$ containing p with respect to $\sim$ is $Z + q$ for each $q \in [p]$. Since $p \in [p]$ we see that $[p] = Z + p$. Let $Q = X/\sim$ be the quotient uniform space defined in Chapter 5. In this section the notation for $X/\sim$ will be X/Z. It should be noted that X/Z may fail to be a topological group if Z is not *normal*. Never-the-less, X/Z will still be referred to as a **quotient** of topological groups, the reference being to the quotient topological and uniform structures rather than a quotient group structure. The natural projection $\pi: X \to X/Z$ is given by $\pi(x) = Z + x$ for each $x \in X$.

Clearly $Z + x = r_x(Z)$ and $x + Z = l_x(Z)$ for each $x \in X$. Also, it is easily seen that the above argument could have been carried out had we chosen to define $p \sim q$ by $-p + q \in Z$. In this case, the partition would consist of the left cosets $\{x + Z \mid x \in X\}$ and we could construct a "left quotient" $\sim/X$ uniform space. If Z is Abelian, clearly $\sim/X = X/\sim$.

PROPOSITION 9.9 The quotient $Q = X/Z$ of a topological group $(X, +)$ with respect to a subgroup Z is T_1 if and only if Z is closed in X.

Proof: Assume Q is T_1. Then the singleton set $\{Z\}$ in Q is closed. Since the canonical projection $\pi: X \to Q$ is continuous it follows that $\pi^{-1}(\{Z\}) = Z$ is closed in X. Conversely, let Z be closed in X. Let $q \in Q$. By definition, q is some right coset $Z + p$ of Z for some $p \in X$. Now $Z + p = r_p(Z)$ and by Proposition 9.2.(2), r_p is a homeomorphism, so $Z + p$ is closed in X. Since $\pi^{-1}(q) = Z + p$ and π is an identification mapping, it follows that $\{q\}$ is a closed subset of Q. Hence Q is T_1. ∎

If H is a normal subgroup of a group G it can be shown (see Exercise 7) that G/H is also a group with respect to the operation + defined on G/H by $(H + x) + (H + y) = H + (x + y)$. In the case G is also a topological group, it is easily verified (see Exercise 8) that $Q = G/H$ is a topological group with respect to the quotient topology, and as such will be referred to as the **quotient topological group** of G with respect to H.

THEOREM 9.5 (N. Howes, 1992) Let G be a topological group and H a subgroup of G. Then the quotient G/H with the quotient topology is an isogeneous uniform space.

Proof: Let $\pi: G \to X = G/H$ be the natural projection defined by $\pi(g) = g + H$ for each $g \in G$. Then π is continuous and onto X. For each $g \in G$ we can define a function $g': X \to X$ by $g'(f + H) = (g + f) + H$ for each $(f + H) \in X$. It is easily seen that for each $g \in G$, g' is one-to-one and onto. Furthermore, if $x, y \in X$, there exists a $g \in G$ with $g'(x) = y$.

We next show that if V is a neighborhood of e (the identity) in G, then for each $g \in G$, if $x, y \in \pi(f + V)$ for some $f \in G$, then $g'(x), g'(y) \in \pi(h + V)$ for some $h \in G$. Now, $x, y \in \pi(f + V)$ implies

$$g'(x), g'(y) \in g'(\pi(f + V)) = \{g'(\pi(f + v)) \mid v \in V\} = \{g + f + v + H \mid v \in V\}.$$

Put $h = g + f$. Then, for some $v_1, v_2 \in V$, $g'(x) = h + v_1 + H$ and $g'(y) = h + v_2 + H$. But then $g'(x), g'(y) \in \pi(h + V)$. Let G' be the collection of all g' such that $g \in G$ and let ν be the collection of all coverings $\pi(\mathcal{V})$ where $\mathcal{V} = \{f + V \mid f \in G$

and V is a neighborhood of e}. Then each $g' \in G'$ is uniformly continuous since $x,y \in \pi(f+V)$ implies $g'(x), g'(y) \in \pi(h+V)$ for some $h \in G$. Furthermore, since G is a group, it is easily shown that each $g' \in G$ is a uniform homeomorphism.

Finally, we show that each $g' \in G'$ is isogeneous with respect to ν. For this, let $x \in X$. By what we have already shown, $g'(S(x, \pi(V)) \subset S(g'(x), \pi(V))$ for each $\pi(V) \in \nu$. Since G is a group, we also have that $(g^{-1})'(S(g'(x), \pi(V))) \subset S((g^{-1})'(g'(x)), \pi(V)) = S(x, \pi(V))$ which implies $S(g'(x), \pi(V)) \subset g'(S(x, \pi(V))$. Hence $g'(S(x, \pi(V)) = S(g'(x),\pi(V))$ for each $g' \in G'$. Therefore, g' is an isomorphism with respect to ν for each $g \in G$. Consequently, ν is an isomorphic basis for a uniformity μ on X. Therefore, (X, μ) is an isogeneous uniform space. ■

THEOREM 9.6 (N. Howes, 1992) Each locally compact isogeneous uniform space is homeomorphic to a quotient of a topological group.

Proof: Let (X, μ) be an isogeneous uniform space and let ν be an isomorphic basis for μ. Let H be the collection of isomorphisms on X with respect to ν. If $f, g \in H$, it is easily shown that $f © g$ and f^{-1} are also isomorphisms. Consequently, H can be extended to a group G with respect to the operation of functional composition. Without loss of generality, we may assume G to be the group of all isomorphisms on X with respect to ν. For each compact $K \subset X$ and $\mathcal{U} \in \nu$, put $[K, \mathcal{U}]$ =

$$\{g \in G \mid \text{there exists a } \mathcal{V} \in \nu \text{ with } S(g(x), \mathcal{V}) \subset S(x, \mathcal{U}) \text{ for each } x \in K\}.$$

Let $\beta = \{[K, \mathcal{U}] \mid K \subset X \text{ is compact and } \mathcal{U} \in \nu\}$. If β is taken as a system of neighborhoods of the identity (mapping) $i \in G$, then G is a Hausdorff topological group with respect to the operation of functional composition. To see this, first note that every member of β contains i. Also, if $[K, \mathcal{U}], [F, \mathcal{V}] \in \beta$, then there exists a $\mathcal{W} \in \nu$ with $\mathcal{W} < \mathcal{U}$ and $\mathcal{W} < \mathcal{V}$, so $[K \cup F, \mathcal{W}] \subset [K,\mathcal{U}] \cap [F,\mathcal{V}]$. Consequently, it will suffice to prove:

(1) For each $[K, \mathcal{U}] \in \beta$, there exists $[F, \mathcal{V}] \in \beta$ with $[F,\mathcal{V}]^{-1} \subset [K, \mathcal{U}]$.
(2) For each $[K, \mathcal{U}] \in \beta$, there exists $[F,\mathcal{V}] \in \beta$ with $[F,\mathcal{V}]^2 \subset [K, \mathcal{U}]$.
(3) For each $[K, \mathcal{U}] \in \beta$ and $g \in G$, there exists $[F,\mathcal{V}] \in \beta$ with $g © [F,\mathcal{V}] © g^{-1} \subset [K, \mathcal{U}]$.
(4) For each $[K, \mathcal{U}] \in \beta$ and $g \in [K, \mathcal{U}]$, there exists $[F,\mathcal{V}] \in \beta$ with $g © [F,\mathcal{V}] \subset [K, \mathcal{U}]$.

To prove (1), let $[K, \mathcal{U}] \in \beta$ and pick $\mathcal{W} \in \nu$ with $\mathcal{W} <^* \mathcal{U}$. Let $g \in [K, \mathcal{W}]$. Then there is a $\mathcal{V} \in \nu$ with $S(g(x),\mathcal{V}) \subset S(x,\mathcal{W})$ for each $x \in K$. Hence, $g^{-1}(x) \in S(x,\mathcal{W})$, so there is a neighborhood $U(g^{-1}(x))$ with compact closure such that $Cl(U(g^{-1}(x))) \subset W_x$ for some $W_x \in \mathcal{W}$ containing x. Since $g^{-1}(K)$ is compact, there exists a finite collection $\{x_1 \ldots x_n\} \subset K$ such that $U = \cup_{i=1}^{n} U(g^{-1}(x_i))$ is a neighborhood of $g^{-1}(K)$. For each j put $G_j = U(g^{-1}(x_j))$. Also for each $j \leq n - 1$, put $F_j = g^{-1}(K) - \cup_{i=j+1}^{n} G_i$ and put $F_n = g^{-1}(K)$. Then each F_j is compact. Choose $Z_1 \in \nu$ such that for each $y \in F_1, S(y, Z_1) \cap Star(X - G_1, Z_1) = \emptyset$. Let k be a positive integer with $k < n$ and assume that for each $m \leq k$, Z_m has been chosen such that for each $y \in F_m$, $S(y, Z_m) \cap Star(X - G_j, Z_m) = \emptyset$ for some positive integer $j \leq m$. Let $H_{m+1} = F_{m+1} \cap Star(X - \cup_{i=1}^{m} G_i, Z_m)$. Now $F_{m+1} - H_{m+1}$ is compact so there is a $Z'_{m+1} \in \nu$ with

$$Star(F_{m+1} - H_{m+1}, Z'_{m+1}) \cap Star(X - G_{m+1}, Z'_{m+1}) = \emptyset.$$

Pick $Z_{m+1} \in \nu$ such that $Z_{m+1} <^* Z'_{m+1}$ and $Z_{m+1} <^* Z_m$. Then for each $y \in F_{m+1}$, $S(y, Z_{m+1}) \cap Star(X - G_j, Z_{m+1}) = \emptyset$ for some positive integer $j \leq m + 1$. Consequently, there exists a Z_n such that for each $y \in g^{-1}(K)$, $S(y, Z_n) \cap Star(X - G_j, Z_n) = \emptyset$ for some positive integer $j \leq n$. Finally, choose $Z \in \nu$ such that $Z < Z_n$ and $Z < \mathcal{W}$. Let $x \in K$. Then $S(g^{-1}(x), Z) \subset G_j \subset W_{x_j}$ for some positive integer $j \leq n$. Hence, there is a $W \in \mathcal{W}$ with $x, x_j \in W$ so $W \cap W_{x_j} \neq \emptyset$. Therefore, $S(g^{-1}(x), Z) \subset S(x, \mathcal{U})$ for each $x \in K$ so $g^{-1} \in [K, \mathcal{U}]$. Thus $[K,\mathcal{W}]^{-1} \subset [K, \mathcal{U}]$ so (1) is proved.

To show (2), let $[K, \mathcal{U}] \in \beta$. Pick $y \in X$ and let $C(y)$ be a compact neighborhood of y. Let $\mathcal{W} \in \nu$ with $S(y,\mathcal{W}) \subset C(y)$. For each $x \in K$, there exists an isomorphism j_x that maps y onto x, so $S(x,\mathcal{W})$ is a neighborhood of x with compact closure. Then there exists a finite collection $\{x_1 \ldots x_n\}$ with $K \subset F = \cup_{i=1}^{n} Cl(S(x_i,\mathcal{W}))$. Hence there is a $\mathcal{V} \in \nu$ with $\mathcal{V} < \mathcal{U}$ and $Star(K,\mathcal{V}) \subset F$. If $f, g \in [F,\mathcal{V}]$, there exists $Z_1, Z_2 \in \nu$ with $S(f(x), Z_1) \subset S(x,\mathcal{V})$ and $S(g(x), Z_2) \subset S(x,\mathcal{V})$ for each $x \in F$. Then for each $x \in K$, $g(x) \in Star(K,\mathcal{V}) \subset F$. Therefore, $S(f(g(x)), Z_1) \subset S(x,\mathcal{V})$ for each $x \in K$, so $f © g \in [K, \mathcal{U}]$. Consequently, $[F,\mathcal{V}]^2 \subset [K, \mathcal{U}]$.

To prove (3), let $[K, \mathcal{U}] \in \beta$ and let $g \in G$. Put $F = g(K)$. Then F is compact. Moreover, $g^{-1} © [F, \mathcal{U}] © g \subset \{g^{-1} © f © g \mid$ there is a $\mathcal{V} \in \nu$ with $S(f(x),\mathcal{V}) \subset S(x, \mathcal{U})$ for each $x \in F\} \subset \{g^{-1} © f © g \mid$ there is a $\mathcal{V} \in \nu$ with $S(g^{-1}(f(g(y))),\mathcal{V}) \subset S(y, \mathcal{U})$ for each $y \in K\} \subset \{h \in G \mid$ there exists a $\mathcal{V} \in \nu$ with $S(h(y),\mathcal{V}) \subset S(y, \mathcal{U})$ for each $y \in K\} = [K, \mathcal{U}]$. Consequently, $g^{-1} © [F, \mathcal{U}] © g \subset [K, \mathcal{U}]$.

To prove (4), first note that if $g \in [K, \mathcal{U}] \in \beta$ then there exists a $\mathcal{V} \in \nu$ with $S(g(x),\mathcal{V}) \subset S(x, \mathcal{U})$ for each $x \in K$. Let $\mathcal{W} \in \nu$ with $\mathcal{W} <^* \mathcal{V}$. If $f \in [K,\mathcal{W}]$, then there exists a $Z \in \nu$ with $S(f(x), Z) \subset S(x,\mathcal{W})$ for each $x \in K$. Therefore, $f(x) \in S(x,\mathcal{W})$ for each $x \in K$ so that $g(f(x)) \in S(g(x),\mathcal{W})$ which implies $S(g(f(x)),\mathcal{W}) \subset S(g(x),\mathcal{V}) \subset S(x, \mathcal{U})$. Therefore, $g © f \in [K, \mathcal{U}]$. Hence $g © [K,\mathcal{W}] \subset [K, \mathcal{U}]$.

We next construct a quotient space from the group G and show that it is homeomorphic with X. For this let $z \in X$ and put $H = \{g \in G \mid g(z) = z\}$. Then H is a subgroup of G. Let $Y = G/H$ be the quotient space of G modulo H. Let $g ©$

$H \in Y$. Then $[g © h](z) = g(z)$ for each $h \in H$, so for each $f \in g © H$, $f(z) = g(z)$. Suppose $y \in X$. Then there exists a $g \in G$ with $g(z) = y$ and hence $f(z) = y$ for each $f \in g © H$. Consequently, there exists a one-to-one correspondence $\phi: Y \to X$ defined by $\phi(g © H) = g(z)$ for each $g © H \in Y$.

Define $\theta: G \to X$ by $\theta(g) = g(z)$ for each $g \in G$. Then $\theta = \phi © \pi$ where π is the quotient mapping $\pi: G \to G/H$. To show ϕ is a homeomorphism of G/H onto X, we must show that ϕ is both continuous and open. Since π is a quotient mapping, it will suffice to show that θ is continuous and open. To show θ is continuous, let $g \in G$. For each $\mathcal{U} \in \nu$, $S(g(z), \mathcal{U})$ is a basic neighborhood of $\theta(g) = g(z)$ in X. Also, $[\{z\}, \mathcal{U}]$ is a basic neighborhood of the identity $i \in G$, so $g © [\{z\}, \mathcal{U}]$ is a basic neighborhood of g in G. Now $\theta(g © [\{z\}, \mathcal{U}]) = \{[g © f](z) \mid f \in [\{z\}, \mathcal{U}]\} = \{g(f(z)) \mid \text{there exists a } \mathcal{V} \in \nu \text{ with } S(f(z),\mathcal{V}) \subset S(z,\mathcal{U})\}$. $S(f(z),\mathcal{V}) \subset S(z, \mathcal{U})$ implies $g(f(z)) \in S(g(z)), \mathcal{U})$. Therefore, $\theta(g © [\{z\},\mathcal{U}]) \subset S(g(z), \mathcal{U})$ so θ is continuous.

To show θ is open, first note that (4) implies each $[K, \mathcal{U}] \in \beta$ is open since if $g \in [K, \mathcal{U}]$, there is an $[F,\mathcal{V}] \in \beta$ with $g © [F,\mathcal{V}] \subset [K, \mathcal{U}]$ so that each $g \in [K, \mathcal{U}]$ has a basic neighborhood $g © [F,\mathcal{V}] \subset [K, \mathcal{U}]$. Next, observe that for each $x \in X$ and $\mathcal{U} \in \nu$ that $[\{x\}, \mathcal{U}](x) = S(x, \mathcal{U})$. To see this, let $y \in S(x, \mathcal{U})$ and let $g \in G$ such that $g(x) = y$. Pick $\mathcal{V} \in \nu$ with $S(y,\mathcal{V}) \subset S(x, \mathcal{U})$. Then $S(x,\mathcal{U}) \subset [\{x\}, \mathcal{U}](x)$ so $[\{x\}, \mathcal{U}](x) = S(x, \mathcal{U})$. Finally, observe that if $U \subset G$ is open and $Y \subset X$, then $U(Y) = \{u(y) \mid u \in U \text{ and } y \in Y\}$ is homeomorphic with $[g © U © g^{-1}](Y)$ for each $g \in G$. To see this define $F: G(X) \to [g © G © g^{-1}](X)$ by $F(h(x)) = [g © h © g^{-1}](x)$ for each $h \in G$ and $x \in X$. Clearly, F is one-to-one and onto. If $h(x) \in G(X)$ and $\mathcal{U} \in \nu$, then $F(S(h(x), \mathcal{U})) = S(F(h(x)), \mathcal{U})$, so F is a homeomorphism. Let $f = F|_{U(Y)}$. Then $f: U(Y) \to [g © U © g^{-1}](Y)$. Consequently, $U(Y)$ and $[g © U © g^{-1}](Y)$ are homeomorphic so that if $U(Y)$ is open in X, so is $[g © U © g^{-1}](Y)$.

Now we show that if $x,y \in X$ and $\mathcal{U} \in \nu$ that $[\{x\}, \mathcal{U}](y)$ is open in X. For this, let $g \in G$ such that $g(x) = y$. It is easily shown that $[\{x\}, \mathcal{U}] = \{f \in G \mid f(x) \in S(x, \mathcal{U})\}$ so $g © [\{x\}, \mathcal{U}] © g^{-1} = \{g © h \mid h(y) \in S(x, \mathcal{U})\} = \{k \in G \mid k(y) \in S(g(x), \mathcal{U})\} = [\{y\}, \mathcal{U}]$. But then $[g © [\{x\}, \mathcal{U}] © g^{-1}](y) = [\{y\}, \mathcal{U}](y) = S(y,\mathcal{U})$ which is open in X. From the proof in the previous paragraph, $[\{x\},\mathcal{U}](y)$ is homeomorphic with $[g © [\{x\}, \mathcal{U}] © g^{-1}](y)$. Hence $[\{x\}, \mathcal{U}](y)$ is open in X.

From this we can show that if $[K, \mathcal{U}] \in \beta$ and $x \in X$ then $[K, \mathcal{U}](x)$ is open in X. For this, pick $\mathcal{V},\mathcal{W} \in \nu$ with $\mathcal{W} <^* \mathcal{V} <^* \mathcal{U}$. Then the covering $\{S(x,\mathcal{W}) \mid x \in K\}$ has a finite subcovering $S(x_1,\mathcal{W}) \ldots S(x_n,\mathcal{W})$. Let $g \in \cap_{i=1}^{n} [\{x_i\},\mathcal{W}]$. Then if $x \in K$, $x \in S(x_j,\mathcal{W})$ for some j, so $g(x) \in S(g(x_j),\mathcal{W})$. Hence, there are $W_1, W_2 \in \mathcal{W}$ with $x, x_j \in W_1$ and $g(x), g(x_j) \in W_2$. Since $g \in [\{x_j\},\mathcal{W}]$, there exists a $W_3 \in \mathcal{W}$ with $x_j, g(x_j) \in W_3$. Therefore, $S(g(x),\mathcal{V}) \subset S(x, \mathcal{U})$. Consequently, $g \in [K, \mathcal{U}]$, so $\cap_{i=1}^{n} [\{x_i\},\mathcal{W}] \subset [K,\mathcal{U}]$. Moreover, $\cap_{i=1}^{n} [\{x_i\},\mathcal{W}](x)$ is open so $[K, \mathcal{U}](x)$ is a neighborhood of x.

Let $y \in [K, \mathcal{U}]$. Then $y = g(x)$ for some $g \in [K, \mathcal{U}]$ which implies there exists an $[F,V] \in \beta$ with $g \circledcirc [F,V] \subset [K, \mathcal{U}]$. Now $g(x) \in [g \circledcirc [F,V]](x)$, since the identity $i \in [F,V]$, and $[g \circledcirc [F,V]](x) \subset [K, \mathcal{U}](x)$. $[g \circledcirc [F,V]](x)$ is a neighborhood of $g(x)$ since $[F,V](x)$ is a neighborhood of x and g is a homeomorphism of X onto itself. Therefore, $y = g(x) \in [g \circledcirc [F,V]](x) \subset [K,\mathcal{U}](x)$ so $[K, \mathcal{U}]$ is a neighborhood of each of its points. Hence $[K, \mathcal{U}]$ is open.

Finally, for each $[K, \mathcal{U}] \in \beta$, $\theta([K, \mathcal{U}]) = [K, \mathcal{U}](z)$ is open. Moreover, if $g \in G$, then $\theta(g \circledcirc [K, \mathcal{U}]) = g([K, \mathcal{U}](z))$ which is open in X since g is a homeomorphism. Therefore, θ is an open mapping. But then ϕ is a homeomorphism, so (X, μ) is homeomorphic to the quotient G/H of the topological group G. ∎

COROLLARY 9.2 The class of locally compact isogeneous uniform spaces is the same as the class of locally compact quotients of topological groups.

Define an isogeneous uniform space (X, μ) to be **strictly isogeneous** if there exists an isometric basis ν for μ such that the group G of all isomophisms with respect to ν contains a subgroup H such that for each $x,y \in X$, there exists only one $i \in H$ with $i(x) = y$. Define (X, μ) to be **perfectly isogeneous** if G contains a commutative subgroup H such that for each $x,y \in X$, there exists an $i \in H$ with $i(x) = y$.

COROLLARY 9.3 The class of locally compact perfectly isogeneous uniform spaces is the same as the class of locally compact abelian topological groups.

Proof: Clearly, a locally compact abelian topological group G is a locally compact perfectly isogeneous uniform space with respect to the basis ν consisting of all coverings of the form $\{g + V \mid g \in G\}$ where V is an open neighborhood of the identity $e \in G$, and the family H of all translations T_g defined by $T_g(x) = g + x$ for all $g \in G$.

To show the converse, let (X, μ) be a locally compact isogeneous uniform space and let ν be an isometric basis for μ such that the group G of all isomorphisms contains a commutative subgroup H such that for each $x,y \in X$, there exists a $j \in H$ with $j(x) = y$. We can now replace G by H in the proof of Theorem 9.6 so that H becomes not only a topological group, but Abelian. Then, just as in the proof of Theorem 9.6, let z be a fixed element of X and put $K = \{h \in H \mid h(z) = z\}$, so that H/K is homeomorphic with X. Let $k \in K$ and let $x \in X$. Pick $j \in H$ such that $j(z) = x$. Then $k(x) = k(j(z)) = j(k(z)) = j(z) = x$ so that $k = i$, the identity mapping on X. Therefore, K consists of the single element i so that $H/K = H$. Consequently, X is homeomorphic with the abelian topological group H. Since X is locally compact, so is H. ∎

COROLLARY 9.4 The class of locally compact strictly isogeneous uniform spaces is the same as the class of locally compact topological groups.

The proof is similar to the proof of the corollary above.

EXERCISES

1. Prove Proposition 9.4.

2. Let (X, +) be a topological group with topology τ and neighborhood system N at 0. For each V $\in N$ put $\mathcal{U}_V = \{(x, y) \in X^2 \mid y - x \in V\}$ and let $B = \{\mathcal{U}_V \mid V \in N\}$. Show that B is the basis for an entourage uniformity $\mathcal{U}$ that generates the topology of τ and that for each $\mathcal{W} \in \mu$ (defined in Theorem 9.4), $\cup\{W\times W \mid W \in \mathcal{W}\}$ is a member of $\mathcal{U}$ and for each V $\in \mathcal{U}$, $\mathcal{V}_V = \{V[x] \mid x \in X\}$ is a member of μ.

3. Let (X, +) be a topological group and let μ and μ' be the right and left uniformities on X respectively. Show that (X, μ) and (X, μ') are isogeneous uniform spaces with respect to the collections $R = \{r_x \mid x \in X\}$ and $L = \{l_x \mid x \in X\}$ of right and left translations respectively and bases λ and ν as defined in Theorem 9.4 and following.

4. Prove Proposition 9.8.

5. Show that a continuous homomorphism h:X $\to$ Y where (X, +) and (Y, +) are topological groups is uniformly continuous with respect to either the right or left uniformities on X and Y respectively.

6. Show that a topological group is pseudo-metrizable if and only if it is first countable and metrizable if and only if it is both first countable and Hausdorff.

7. Show that if H is a normal subgroup of a group G then the quotient G/H is a group with respect to the operation + defined on G/H by (H + x) + (H + y) = H + (x + y) for each pair (H + x), (H + y) $\in$ G/H.

8. Show that if the group G in Exercise 7 is a topological group that G/H is a topological group with respect to the quotient topology.

9. Show that the canonical projection π from a topological group X onto its quotient Q = X/Z over a subgroup Z of X is an open mapping.

10. Show that the quotient topological group **R**/**Z** of the additive topological group **R** of real numbers over its subgroup **Z** of integers is uniformly homeomorphic to the circle group **T**.

RESEARCH PROBLEMS

11. Are isogeneous uniform spaces equivalent to quotients of topological groups without the assumption of local compactness?

12. What is a necessary and sufficient condition for a locally compact uniform space to have a Haar measure?

13. Find a necessary and sufficient condition for a uniform space (not necessarily locally compact) to have a topological group structure that generates the uniformity.

Chapter 10

UNIFORM MEASURES

10.1 Introduction

In 1945, L. Loomis introduced an interesting generalization of the Haar measure to locally compact metric spaces that satisfy a property he called the *congruence axiom* in a paper titled *Abstract congruence and the uniqueness of Haar measure* (Annals of Mathematics, Volume 46). Loomis' concept of *invariance* in this paper did not involve a *translation* (transformation), but rather the property that for a measure μ, $\mu(S_1) = \mu(S_2)$ whenever S_1 and S_2 are compact spheres with the same radius. Because of this distinction, we will refer to these measures as *uniform measures* rather than Haar measures. As we shall see, Loomis' concept, when applied to uniform spaces, generalizes the concept of Haar measure as presented in the last chapter. In this paper, Loomis showed there exists a unique, regular, invariant (uniform) measure on the Lebesgue ring (Borel sets that can be covered by countably many compact sets) on locally compact, metric spaces that satisfy the congruence axiom.

On page 354 of this paper, Loomis mistakenly states that the proofs of the theorems that establish the existence and uniqueness of this uniform measure in locally compact, metric spaces that satisfy the congruence axiom "hold without modification" in locally compact uniform spaces that satisfy the congruence axiom. In 1949, in a paper titled *Haar measure in uniform structures* (Duke Math Journal, Volume 16), Loomis attempted to extend his generalization of Haar measure to locally compact uniform spaces satisfying the congruence axiom using a different approach than in the 1945 paper. Loomis' 1949 proof was not entirely correct either and, strictly speaking, this is still an open problem. However, we will show that a slightly weakened version of Loomis' claimed results still holds on the Baire ring.

In the 1945 paper, Loomis first constructed an outer measure *m*, on the collection of sets that could be covered by countably many compact sets, that had the invariance (uniform) property and then referred to the classical result that in a locally compact, metric space, the members of the Lebesgue ring are all m^*-measurable (we will not prove that result here as our interest is in general proofs for uniform spaces). Thus, by restricting m^* to the Lebesgue ring, he obtained a measure *m* on the Lebesgue ring with the desired properties. In the 1949 paper, Loomis tries to build the measure up from a more elementary set function instead. In the first eight sections of the 1949 paper, Loomis constructs (without the aid of local compactness) a set function λ that is unique (in a sense Loomis defines) on the collection of totally bounded open sets such that whenever S_1 and S_2 are two totally bounded spheres of the "same radius" (generated by the same uniform covering) then $\lambda(S_1)$ is "almost always" equal to $\lambda(S_2)$ in a sense he defines and calls "invariance." Loomis' development contains some errors. The reader can judge their seriousness but should not confuse these gaps in logic with an unsound approach. Loomis claims λ can be extended to a legitimate measure λ^* on the "Borel field generated by the compact sets." But because of the errors and inconsistencies of definitions (e.g., *measure* and *content*) with the classical ones, the classical results to which Loomis appears to be referring (as no proof was given) do not apply.

However, Loomis' insight was keen. About half of the results are correct and the other half can be salvaged with better definitions and modified theorems and proofs to obtain a somewhat weaker result (but more nearly like the result in his original metric space paper). The notion of an *almost uniform measure* is introduced and it is shown that such a measure exists (on the Baire sets) and is unique. Consequently, it is still an open problem whether a uniform measure exists on a locally compact uniform space under Loomis' original assumptions and if so, is it unique?

In the next section Loomis' development of these invariant set functions (that we will refer to as *Loomis contents*) is presented. We will introduce new terminology to clarify the difference between Loomis' definitions and axioms and the classical ones, and in some cases change Loomis' definitions and axioms so they work.

10.2 Prerings and Loomis Contents

In this section, we will find it notationally convenient to use $a \in \mu$ to mean that the covering a belongs to the uniformity μ rather than our usual notation $\mathcal{U} \in \mu$. This is because members of the uniformity μ will frequently be used to index various other collections. If (X, μ) is a uniform space and A and B are subsets of X, Loomis defined $A < B$ to mean there exists a $u \in \mu$ with $Star(A, u) \subset B$. If $a, b \in \mu$, we define $a < b$ to mean there exists a $c \in \mu$ with $Star(S(p,a),c) \subset S(p, b)$ for each $p \in X$. Since $<$ is defined on μ, this could cause confusion with the refinement ordering. Therefore, in this chapter, if we wish to denote the refinement ordering, we will use the "upper case" notation of $\mathcal{U} < \mathcal{V}$ rather than $a < b$.

PROPOSITION 10.1 If (X, μ) is a uniform space then $(\mu,<)$ is a directed set.

Proof: To show $<$ is transitive suppose $a, b, c \in \mu$ with $a < b$ and $b < c$. Then there exists a', $b' \in \mu$ with $Star(S(p, a), a') \subset S(p, b)$ and $Star(S(p, b), b') \subset S(p,c)$ for each $p \in X$. But then $a < c$. To show μ is directed by $<$ note that there exists a $z \in \mu$ with $z <^* a$ and $z <^* b$. Then $Star(S(p, z), z) \subset S(p, a)$ and $Star(S(p, z), z) \subset S(p, b)$ so $z < a$ and $z < b$. ■

PROPOSITION 10.2 If $a, b \in \mu$ with $a < b$ then there is a $c \in \mu$ with $a < c < b$.

Proof: If $a < b$, there exists a $z \in \mu$ with $Star(S(p, a), z) \subset S(p, b)$ for each $p \in X$. Let $x \in \mu$ such that $x <^* z$. Put $c = \{Star(U, x) \mid U \in a\}$. Clearly $c \in \mu$. Let $q \in X$. Then $Star(S(p, a), x) = \cup\{V \in x \mid V \cap S(p, a) \neq \emptyset\} = \cup\{W \in c \mid W \cap U \neq \emptyset$ for some $U \in a$ with $q \in U\} = \cup\{W \in c \mid q \in W\} = S(p, c)$. Hence $a < c$.

To show $c < b$ suppose $r \in Star(S(q, c), x)$. Then there exists $V, W \in x$ and $U \in a$ with $r \in V$, $V \cap W \neq \emptyset \neq V \cap U$ and $q \in U$. Next, there exists a $Z \in z$ with $V \cup W \subset Z$ which implies $r \in Z$ and $Z \cap U \neq \emptyset$, so $r \in Star(S(q, a), z) \subset S(q, b)$. We conclude that $Star(S(q, c), x) \subset S(q, b)$ for each $q \in X$ so $c < b$. ■

Loomis begins by assuming a "uniform structure" μ on a space X that is defined as a collection of coverings μ such that $(\mu, <)$ satisfies Propositions 10.1 and 10.2, i.e., $(\mu, <)$ is directed and satisfies the axiom

A1. If $x, y \in \mu$ with $x < y$, there is a $z \in \mu$ with $x < z < y$.

In addition to axiom A1, Loomis assumes μ satisfies the following additional axioms:

A2. If $p \in S(q, x)$ then $q \in S(p, x)$.
A3. Every y-sphere can be covered by a finite number of x-spheres for each pair $x, y \in \mu$.
A4. The smallest number of x-spheres required to cover a y-sphere is the same for all y-spheres for each pair $x, y \in \mu$.

It is easily shown that every uniformity μ satisfies A1, and A2 is part of the definition of a uniformity. So Loomis' "uniform structure" is a generalization of a uniformity. Axioms A3 and A4 are his uniform space version of the congruence axiom for metric spaces. In this chapter we will restrict ourselves to uniform spaces rather than these more general spaces.

Loomis' problems begin with his axioms. Axioms A3 and A4 rule out most interesting examples of uniform spaces. Either one of them, for instance, rules out the real numbers **R** with the usual metric uniformity. Since in a uniform space, the space itself is an x-sphere for $x = \{X\}$, A3 implies the space is precompact which **R** clearly is not because if it were, its completeness would then imply compactness. Similarly **R** cannot satisfy axiom A4. To see this, consider the uniform covering x consisting of the open interval (-10, 10) and the open intervals $(n\text{-}1, n+1)$ for each integer n with $|n| \geq 10$. Let y be the uniform covering $\{(r\text{-}1, r+1) | r \in \mathbf{R}\}$. The x-sphere about 0 is (-10, 10) while the x-sphere about 12 is (11, 13). Now (11, 13) can be covered by a single y-sphere while (-10, 10) cannot. Surely Loomis did not mean to rule out the most well known uniform space with the archetypical Haar measure when he undertook to generalize the concept of Haar measure to uniform spaces. Consequently, we will assume instead that μ *generates* the uniformity rather than being the uniform structure itself, i.e., we assume that μ is a basis for a uniformity ν. We also need to make adjustments to axioms A3 and A4. We assume they only hold for the basis μ rather than the uniformity ν, i.e., that the x and y mentioned in A3 and A4 are members of μ. Furthermore, this change in assumptions requires a change in the underlying definition of the ordering <. We define A < B to mean that $Star(\text{A}, u) \subset \text{B}$ for some $u \in \mu$ rather than $u \in \nu$. A similar change is needed in the definition of $a < b$ where $a, b \in \mu$.

In this section and the next, the part of Loomis' development that is essentially correct will be presented. The order in which this material is presented has been rearranged somewhat from Loomis' original paper and the terminology has been translated into the terminology that has just been introduced. Also, some of the details omitted in Loomis' original proofs have been included so the reader may be convinced it is all here. It is often in these areas where Loomis omitted proofs altogether that the difficulties are hidden.

PROPOSITION 10.3 For each pair of comparable sets $A_0 < A_1$ there is an interval $\{A_\alpha | \alpha \in [0, 1]\}$ of sets with $Cl(A_\alpha) < Int(A_\beta)$ whenever $\alpha < \beta$.

Proof: Since $A_0 < A_1$, there exists $x, y \in \mu$ with $Star(A_0, x) \subset A_1$ and $y <^* x$. Put $A_{1/2} = Star(A_0, y)$. Then $A_0 < A_{1/2} < A_1$. Let R denote the rationals in [0,1]. Well order R as a sequence $\{r_n\}$ such that $r_1 = 0$, $r_2 = 1/2$ and $r_3 = 1$. Then $A_{r_1} < A_{r_2} < A_{r_3}$. Define A_{r_n} recursively so that $A_{r_m} < A_{r_n}$ whenever $r_m < r_n$. This can be done by letting $a_n = max\{r_k | k < n \text{ and } r_k < r_n\}$ and $b_n = min\{r_k | k < n \text{ and } r_n < r_k\}$. Then $a_n < r_n < b_n$ so $A_{a_n} < A_{b_n}$. As shown above, it is possible to pick A_{r_n} with $A_{a_n} < A_{r_n} < A_{b_n}$. Thus for $\{r_1 \ldots r_n\}$ we have $A_{r_m} < A_{r_n}$ whenever $r_m < r_n$. Clearly, this recursive definition produces a sequence $\{A_{r_n}\}$ of sets indexed by the members of R such that $A_r < A_s$ whenever $r, s \in R$ with $r < s$.

Now for each irrational $\alpha \in [0, 1]$ put $A_\alpha = \cup\{A_r | r \in R \text{ and } r < \alpha\}$. Let $s \in R$ such that $\alpha < s$. Then for each $t \in R$ with $t < \alpha$, $A_t < A_s$ so $A_t \subset A_s$. Consequently, $A_\alpha \subset \cap\{A_s | s \in R \text{ and } \alpha < s\}$. To show that $A_\alpha < A_\beta$ whenever $\alpha < \beta$ are in [0, 1], note that there exists a pair of rationals $r, s \in [0, 1]$ with $\alpha < r < s < \beta$. Now $A_\alpha \subset \cap\{A_t | t \in R \text{ and } \alpha < t\} \subset A_r$ and $A_s \subset \cup\{A_t | t \in R \text{ and } t < \beta\} \subset A_\beta$. Since r and s are rationals, there exists an $x \in \mu$ with $Star(A_r, x) \subset A_s$. Hence $Star(A_\alpha, x) \subset A_\beta$, so $A_\alpha < A_\beta$. Therefore, there exists an interval $\{A_\alpha\}$

of sets with $A_\alpha < A_\beta$ whenever $\alpha < \beta$. But $A_\alpha < A_\beta$ implies $Cl(A_\alpha) < Int(A_\beta)$ by the definition of $<$. ∎

Loomis' definition of a measure ring fails to be a ring in the standard sense. For this reason, we call such an object a *prering* instead. In fact, we adopt the following definitions in order to distinguish Loomis' concepts from the standard ones. Let C be a class of sets and A a collection of members of C. A is said to be a **prering** in C if A is closed under finite unions. A is **hereditary** in C if whenever $\mathrm{A} \in A$ and $\mathrm{B} \in C$ with $\mathrm{B} \subset \mathrm{A}$ then $\mathrm{B} \in A$. A **hereditarily open prering** is a hereditary prering in the class of open sets. A **Loomis content** is a set function λ on a hereditarily open prering P that satisfies the following four properties:

L1. If $\mathrm{U}, \mathrm{V} \in P$ with $\mathrm{U} \subset \mathrm{V}$ then $\lambda(\mathrm{U}) \leq \lambda(\mathrm{V})$.

L2. $\lambda(\mathrm{U} \cup \mathrm{V}) \leq \lambda(Star(\mathrm{U}, x)) + \lambda(Star(\mathrm{V}, x))$ for each $x \in \mu$.

L3. $\lambda(\mathrm{U} \cup \mathrm{V}) = \lambda(\mathrm{U}) + \lambda(\mathrm{V})$ if $Star(\mathrm{U}, z) \cap Star(\mathrm{V}, z) = \varnothing$ for some $z \in \mu$.

Sets $\mathrm{U}, \mathrm{V} \subset \mathrm{X}$ for which $Star(\mathrm{U}, z) \cap Star(\mathrm{V}, z) = \varnothing$ for some $z \in \mu$ are said to be **uniformly separated.** Note that L3 implies $\lambda(\varnothing) = 0$. λ is said to be **left continuous** if it satisfies

L4. $\lambda(\mathrm{U}) = sup\{\lambda(\mathrm{V}) \mid \mathrm{V} < \mathrm{U}\}$.

A subset P' of P is said to be **dense in P with respect to** $<$ if for each $\mathrm{U}, \mathrm{V} \in P$ with $\mathrm{U} < \mathrm{V}$, there is a $\mathrm{W} \in P'$ with $\mathrm{U} < \mathrm{W} < \mathrm{V}$.

THEOREM 10.1 Let P' be dense with respect to $<$ in a hereditarily open prering P and suppose that if U and V are uniformly separated in P' then $U \cup V \in P$. Let λ be a set function defined on P' that satisfies L1 - L4. Then λ has a unique extension to P that is a Loomis content on P.

Proof: For each $\mathrm{U} \in P$ put $\lambda^*(\mathrm{U}) = sup\{\lambda(\mathrm{V}) \mid \mathrm{V} < \mathrm{U} \text{ and } \mathrm{V} \in P'\}$. λ^* is easily seen to be an extension of λ to P. The proof that λ^* satisfies L1 is trivial. To show λ^* satisfies L4, note that if $\mathrm{W} \in P$ with $\mathrm{W} < \mathrm{U}$, then there exists a $\mathrm{V} \in P'$ with $\mathrm{W} < \mathrm{V} < \mathrm{U}$ and by L1, $\lambda^*(\mathrm{W}) < \lambda^*(\mathrm{V})$, so $\lambda^*(\mathrm{U}) = sup\{\lambda^*(\mathrm{W}) \mid \mathrm{W} < \mathrm{U}$ and $\mathrm{W} \in P\}$. Therefore λ^* satisfies L4. To prove L2, let $\mathrm{U},\mathrm{V} \in P$ and let $x \in \mu$. Then

$$\lambda^*(\mathrm{U} \cup \mathrm{V}) = sup\{\lambda(\mathrm{W}) \mid \mathrm{W} < \mathrm{U} \cup \mathrm{V} \text{ and } \mathrm{W} \in P'\}. \tag{10.1}$$

If $\mathrm{W} < \mathrm{U} \cup \mathrm{V}$, there exists an $a \in \mu$ such that $Star(Star(\mathrm{W}, a), a) \subset \mathrm{U} \cup \mathrm{V}$. Put $W_U = \mathrm{W} - Star(\mathrm{X} - \mathrm{U}, a)$ and $W_V = \mathrm{W} - Star(\mathrm{X} - \mathrm{V}, a)$. Then W_U and W_V are uniformly separated, $Star(W_U, a) \subset \mathrm{U}$, $Star(W_V, a) \subset \mathrm{V}$,

$$Star(W_U, a) \cup Star(W_V, a) \subset Star(W, a)$$

and both W_U and W_V are in P'. Hence $[W_U \cup W_V] \in P'$. Therefore,

$$\lambda(W) \leq \lambda(Star(W, a)) = \lambda(Star(W_U, a) \cup Star(W_V, a)) \leq$$

$$\lambda(Star(Star(W_U, a), x)) + \lambda(Star(Star(W_V, a), x)) \leq$$

$$\lambda^*(Star(U, x)) + \lambda^*(Star(V, x))$$

for some $x \in \mu$, so λ^* satisfies L2.

To show L3, let U, V $\in P$ such that for some $z \in \mu$, $Star(U, z) \cap Star(V, x) = \emptyset$. If W < U$\cup$V put W_U = W$\cap$U and W_V = W$\cap$V. Then W_U < U and W_V < V are uniformly separated and belong to P'. Hence $\lambda(W) = \lambda(W_U \cup W_V) = \lambda(W_U) + \lambda(W_V)$. Conversely, for each pair $W_U, W_V \in P'$ with W_U < U and W_V < V we have $W_U \cup W_V$ < U$\cup$V and W_U and W_V are uniformly separated. Therefore, by (10.1)

$$\lambda^*(U \cup V) = sup\{\lambda(W_U) + \lambda(W_V) \mid W_U < U, W_V < V \text{ and } W_U, W_V \in P'\} =$$

$$sup\{\lambda(A) \mid A < U \text{ and } A \in P'\} + sup\{\lambda(B) \mid B < V \text{ and } B \in P'\} = \lambda^*(U) + \lambda^*(V).$$

To show λ^* is unique, let l be another Loomis content on P that is an extension of λ. Then for U $\in P$, $l(U) = sup\{\lambda(W) \mid W < U \text{ and } W \in P\}$. For W $\in P$ with W < U, there exists a V $\in P'$ with W < V < U and since l is an extension of λ, $l(V) = \lambda(V)$. Hence $l(U) = sup\{\lambda(V) \mid V < U \text{ and } V \in P'\} = \lambda^*(U)$. Therefore $l = \lambda^*$ so λ^* is a unique Loomis content on P. ∎

Loomis defined the content λ to be **invariant with respect to** $x \in \mu$ if $\lambda(S(p, x)) = \lambda(S(q, x))$ for each pair $p, q \in X$. It is simply **invariant** if it is invariant with respect to all members $x \in \mu$. In this case, we write $\lambda(y)$ for the common value $\lambda(S(p, y))$ for any $y \in \mu$. We will also use the terminology that some property P **holds for sufficiently small** $x \in \mu$. By this we mean that there exists a $y \in \mu$ for which P holds, and for all $x \in \mu$ with $x < y$, the property continues to hold.

For this section and the next we will restrict our discussion to a Loomis content λ defined on the hereditarily open prering T of totally bounded open sets. A set A is said to have **zero-boundary** (with respect to λ) if for each $\varepsilon > 0$ there exists a closed set F and an open set U $\in T$ with F < A < U and $\lambda(U - F) < \varepsilon$. Let Z denote the zero-boundary sets.

To see that there exist many members of Z, let A and B be sets with A < B. By Proposition 10.3 it is possible to construct an interval $\{A_\alpha\}$ of sets with A_0 = A and A_1 = B such that for each $\alpha < \beta$ in [0, 1], $A_\alpha < A_\beta$. If $\{A_\alpha\} \subset T$ then $\lambda(A_\alpha)$ is an increasing function of α. Hence, from elementary analysis we know that $\lambda(A_\alpha)$ is continuous except (at most) on an increasing sequence $\{x_n\} \subset [0, 1]$. If β is a continuity point of $\lambda(A_\alpha)$, then A_β is a zero-boundary member of T. To see this, let $\varepsilon > 0$. Then there exists a $\delta > 0$ with $|\lambda(A_\beta) - \lambda(A_\alpha)| < \varepsilon$ whenever $\alpha \in (\beta - \delta, \beta + \delta)$ and such that $\lambda(A_\alpha)$ is continuous in $(\beta - \delta, \beta + \delta)$. Choose $r, s \in [0, 1]$ such that $\beta - \delta < r < \beta < s < \beta + \delta$. Then $Cl(A_r) < A_\beta < A_s$ and $\lambda(A_s) - \lambda(A_r) < \varepsilon$. To show that $\lambda(A_s - Cl(A_r)) < \varepsilon$, note that for each $t \in [0,1]$ with $\beta - \delta < t < r$ we have (by L3):

$$\lambda(A_s - Cl(A_r)) + \lambda(A_t) = \lambda[(A_s - Cl(A_r)) \cup A_t] \leq \lambda(A_s)$$

since A_t and $A_s - Cl(A_r)$ are uniformly separated because $A_t < A_r < A_s$. Hence

$$lim_{t\to r}[\lambda(A_s - Cl(A_r)) + \lambda(A_t)] = lim_{t\to r}\lambda[(A_s - Cl(A_r)) \cup A_t] \leq lim_{t\to r}\lambda(A_s).$$

But $\lambda(A_\alpha)$ continuous in $(\beta - \delta, \beta + \delta)$ implies that $lim_{t\to r}\lambda(A_t) = \lambda(A_r)$ so $\lambda(A_s - Cl(A_r)) + \lambda(A_r) \leq \lambda(A_s)$. Therefore, $\lambda(A_s - Cl(A_r)) \leq \lambda(A_s) - \lambda(A_r) < \varepsilon$, so A_β is a zero-boundary member of T.

Conversely, if A is a zero-boundary set with respect to λ, then its interior is a member of an interval of open sets $\{A_\alpha\} \subset T$ at which $\lambda(A_\alpha)$ is continuous. To see this, we can choose a closed set F_1 and an open set $U_1 \in T$ such that $F_1 < Int(\mathrm{A}) \subset Cl(\mathrm{A}) < U_1$ and $\lambda(U_1 - F_1) < 1$. Put $O_1 = Int(F_1)$. Next choose a closed set K_2 and an open set $V_2 \in T$ with $K_2 < Int(\mathrm{A}) \subset Cl(\mathrm{A}) < V_2$ and $\lambda(V_2 - K_2) < 1/2$. Put $F_2 = F_1 \cup K_2$ and $U_2 = U_1 \cap V_2$. Clearly $F_1 < F_2 < Int(\mathrm{A}) \subset Cl(\mathrm{A}) < U_2 < U_1$. Also, let $O_2 = Int(F_2)$. Continuing this process by induction, we get two sequences $\{O_n\}$ and $\{U_n\}$ of members of T such that $O_n < O_{n+1} < Int(\mathrm{A}) \subset Cl(\mathrm{A}) < U_{n+1} < U_n$ for each positive integer n and $\lambda(U_n - Cl(O_n)) < 1/n$. Now we can use Proposition 10.3 to fill in intervals of open sets between the O_n and O_{n+1} and between the U_{n+1} and U_n for each n. Thus the interior of A is a member of an interval of open sets $\{A_\alpha\}$ such that (by construction) each $A_\alpha \in T$ and such that $A_\alpha < A_\beta$ whenever $\alpha < \beta$. Moreover, by construction, it is clear that for each $\varepsilon > 0$, there exists a pair α, β with $A_\alpha < Int(\mathrm{A}) \subset Cl(\mathrm{A}) < A_\beta$ and $\lambda(A_\beta - Cl(A_\alpha)) < \varepsilon$, so $\lambda(A_\alpha)$ is continuous at $Int(\mathrm{A})$.

COROLLARY 10.1 The members of Z are totally bounded (but not necessarily open). λ is uniquely determined on T by its values on the zero-boundary members of T.

Proof: The zero-boundary members of T have the properties of P' in the hypothesis of Theorem 10.1 when we put $P = T$, i.e., they are dense in T, so that λ is uniquely determined on T by its values on the zero-boundary members of T. ■

There is a natural way to extend λ to all members of Z (not necessarily in T). We have seen from the construction above that if $\mathrm{A} \in Z$ then $Int(\mathrm{A})$ is a continuity point of $\lambda(A_\alpha)$ for some interval $\{A_\alpha\} \subset T$ where $A_\alpha < A_\beta$ whenever

$\alpha < \beta$ in [0, 1]. Hence if $Int(A) = A_r$ then $lim_{t \to r}\lambda(A_t) = \lambda(A_r)$ so $sup\{\lambda(O) \mid Int(A) > O \in T\} = \lambda(A_r) = inf\{\lambda(O) \mid Int(A) < O \in T\}$. If $A_r < O$ then $Cl(A) = Cl(A_r) < O$. Note that had we started with some other interval $\{U_\beta\} \subset T$ such that $Int(A) = U_\gamma$ for some $\gamma \in [0, 1]$ that we would get $sup\{\lambda(O) \mid Int(A) > O \in T\} = \lambda(U_\gamma) = inf\{\lambda(O) \mid Int(A) < O \in T\}$ so that $\lambda(U_\gamma) = \lambda(A_r)$. Hence $\lambda(A)$ is well defined. Put $\lambda(A) = \lambda(A_r)$.

For zero-boundary open sets U and V we have $\lambda(U \cup V) = \lambda(U) + \lambda(V)$. To see this, first note that if U and V are zero-boundary open sets then $\lambda(U) = inf\{\lambda(Star(U, x)) \mid x \in \mu\}$ and $\lambda(V) = inf\{\lambda(Star(V, x)) \mid x \in \mu\}$. Then by L2, $\lambda(U \cup V) \le \lambda(Star(U, x)) + \lambda(Star(V, x))$ for each $x \in \mu$. Therefore, $\lambda(U \cup V) = inf_x\lambda(U \cup V) \le inf_x\lambda(Star(U, x)) + inf_x\lambda(Star(V, x)) = \lambda(U) + \lambda(V)$.

THEOREM 10.2 Z is a ring and λ is a finitely additive measure on Z.

Proof: Let $A, B \in Z$. As shown above, there exists intervals $\{A_\alpha\}$ and $\{B_\beta\}$ of members of T with $Int(A) = A_r$ and $Int(B) = B_s$ for some $r, s \in [0,1]$ and A_r and B_s continuity points of $\lambda(A_\alpha)$ and $\lambda(B_\beta)$ respectively. For each $\varepsilon > 0$, there exists $\alpha, \beta, \gamma, \delta \in [0,1]$ with $A_\alpha < A_r < A_\beta$ and $B_\gamma < B_s < B_\delta$ such that both $\lambda(A_\beta - Cl(A_\alpha)) < \varepsilon/2$ and $\lambda(B_\delta - Cl(B_\gamma)) < \varepsilon/2$. Now $Cl(A_\alpha) \cup Cl(B_\gamma) < A_r \cup B_s < A_\beta \cup B_\delta$ and $(A_\beta \cup B_\delta) - (Cl(A_\alpha) \cup Cl(B_\gamma)) \subset (A_\beta - Cl(A_\alpha)) \cup (B_\delta - Cl(B_\gamma))$. Clearly both $A_\beta - Cl(A_\alpha)$ and $B_\delta - Cl(B_\gamma)$ are zero-boundary open sets so by the remarks preceding this theorem, $\lambda[(A_\beta \cup B_\delta) - (Cl(A_\alpha) \cup Cl(B_\gamma))] \le \lambda(A_\beta - Cl(A_\alpha)) + \lambda(B_\delta - Cl(B_\gamma)) < \varepsilon$. It follows that $A_r \cup B_s \in Z$ which implies $A \cup B \in Z$.

Similarly, $Cl(A_\alpha) - Cl(B_\delta) < A_r - B_s < A_\beta - Cl(B_\gamma)$ and $(A_\beta - Cl(B_\gamma)) - (Cl(A_\alpha) - B_\delta) \subset (A_\beta - Cl(A_\alpha)) \cup (B_\delta - Cl(B_\gamma))$, so $\lambda[(A_\beta - Cl(B_\gamma)) - (Cl(A_\alpha) - B_\delta)] < \varepsilon$. Hence $A_r - B_s \in Z$ which implies $A - B \in Z$. Thus Z is a ring. Now if $A \cap B = \varnothing$, then for each $A_\alpha < A_r$ and $B_\beta < B_s$, $Cl(A_\alpha)$ and $Cl(B_\beta)$ are uniformly separated, so by L3, $\lambda(Cl(A_\alpha) \cup Cl(B_\beta)) = \lambda(Cl(A_\alpha)) + \lambda(Cl(B_\beta))$.

Let ϕ:[0, 1] $\to$ [0, 1] be strictly increasing, onto and have $\phi(r) = s$. For each $\alpha \in [0,1]$, put $C_\alpha = B_{\phi(\alpha)}$. Then $\{A_\alpha \cup C_\alpha\}$ is an interval of members of T with $A_\alpha \cup C_\alpha < A_\beta \cup C_\beta$ whenever $\alpha < \beta$. Moreover, for each $\varepsilon > 0$, there exists $\alpha, \beta, \gamma, \delta \in [0, 1]$ with $Cl(A_\alpha) < A_r < A_\beta$, $Cl(C_\gamma) < B_s < C_\delta$ and $\lambda(A_\beta - Cl(A_\alpha)) < \varepsilon/2$ and $\lambda(C_\delta - Cl(C_\gamma)) < \varepsilon/2$. An argument similar to the one above yields $\lambda[(A_\beta \cup C_\delta) - (Cl(A_\alpha) \cup Cl(C_\gamma))] < \varepsilon$, so $A_r \cup B_s$ is a continuity point of $\lambda(A_\alpha \cup C_\alpha)$ which implies $sup_{\alpha<r}\lambda(Cl(A_\alpha) \cup Cl(C_\alpha)) = \lambda(A \cup B)$.

For each $\alpha \in [0, 1]$, $C_\alpha = B_{\phi(\alpha)}$, so if $\alpha < r$ then $A_\alpha < A$ and $C_\alpha < B$ so $Cl(A_\alpha)$ and $Cl(C_\alpha)$ are uniformly separated which implies $\lambda(Cl(A_\alpha) \cup Cl(C_\alpha)) = \lambda(Cl(A_\alpha)) + \lambda(Cl(C_\alpha))$. Hence $sup_{\alpha<r}\lambda(Cl(A_\alpha) \cup Cl(C_\alpha)) = sup_{\alpha<r}[\lambda(Cl(A_\alpha)) + \lambda(Cl(C_\alpha))]$. Since A_r and $C_r = B_s$ are continuity points of $\lambda(A_\alpha)$ and $\lambda(C_\alpha)$ respectively, we have $sup_{\alpha<r}[\lambda(Cl(A_\alpha)) + \lambda(Cl(C_\alpha))] = \lambda(A_r) + \lambda(C_r) = \lambda(A) + \lambda(B)$. Hence $\lambda(A \cup B) = \lambda(A) + \lambda(B)$, so λ is finitely additive on Z. ∎

We can generalize Propositions 8.4 and 8.5 to Loomis contents on T as follows:

PROPOSITION 10.4 Each Loomis content λ on T has the following properties:

L5. If each point of $O \in T$ lies in at least N of the sets $\{O_1 \ldots O_M\} \subset T$ then $\lambda(O) \leq \Sigma_{n=1}^{M} \lambda(Star(O_n, x))/N$ for each $x \in \mu$.

L6. If $\{O_1 \ldots O_M\}$ are subsets of $O \in T$ and no point of O lies in more than N of the O_n, then $\lambda(O) \geq \Sigma_{n=1}^{M} \lambda(O_n)/N$.

The proof is left as an exercise.

EXERCISE

1. Prove Proposition 10.4 Hint: Use Proposition 8.4 and 8.5 on Z and approximate the members O and O_n with members of Z.

10.3 The Haar Functions

We define the **Haar covering function** $h(A, x)$, for totally bounded sets $A \in X$ and $x \in \mu$, to be the smallest number of x-spheres required to cover A. By axiom A4 we have that $h(S(p, y), x) = h(S(q, y), x)$ for each pair $p, q \in X$ and x, $y \in \mu$. We denote this common value by $h(y, x)$. The **Haar function** $H(A, x)$ is the *inf* of the fractions M/N such that for some $y < x$, there exists a collection of M y-spheres covering each point of A at least N times. If $y < x$, clearly $H(A, x) \leq H(A, y)$.

PROPOSITION 10.5 $H(A, x) = inf\{H(A, y) \mid y < x\}$.

Proof: Suppose $H(A, x) \neq inf\{H(A,y) \mid y < x\}$. Then clearly $H(A, x) < inf\{H(A,y) \mid y < x\}$. By the definition of $H(A, x)$, there exists positive integers M and N with $H(A, x) < M/N < inf\{H(A,y) \mid y < x\}$ and M w-spheres covering each point of A at least N times for some $w < x$. By Proposition 10.2, there exists a $y \in \mu$ with $w < y < x$. Then by the definition of $H(A,y)$ we have $H(A,y) \leq M/N$, which is a contradiction. Therefore, $H(A, x) = inf\{H(A,y) \mid y < x\}$. ∎

LEMMA 10.1 If a collection $\mathcal{U}$ of N open r-spheres covers no point more than n times, and if each point covered by $\mathcal{U}$ is covered by at least m spheres from a collection $\mathcal{V}$ of M open r-spheres then $N/n \leq M/m$.

Proof: Let $\mathcal{U}^*$ be a set consisting of m duplicates of each sphere in $\mathcal{U}$ and let $\mathcal{V}^*$ be a set consisting of n duplicates of each sphere in $\mathcal{V}$. If p is the center of a sphere in $\mathcal{U}^*$ then p is the center of a sphere in $\mathcal{U}$ and therefore lies in at least m members of $\mathcal{V}$. Since $\mathcal{V}^*$ was formed by duplicating each sphere in $\mathcal{V}$ n times,

p lies in at least mn formally distinct members of $\mathcal{V}^*$. Since $\mathcal{U}^*$ was formed by duplicating each member of $\mathcal{U}$ m times, there are Nm formally distinct centers of spheres in $\mathcal{U}^*$. Hence there are at least Nm^2n formally distinct members of

$$\mathrm{H} = \{(p,q) \mid S(p,r) \in \mathcal{U}^* \text{ and } p \in S(q,r) \in \mathcal{V}^*\}.$$

On the other hand, if a sphere in $\mathcal{V}^*$ contained more than mn formally distinct members of $\mathcal{U}^*$ then its center, say q, would lie in more than mn formally distinct members of $\mathcal{U}^*$. Since each member of $\mathcal{U}$ was duplicated m times to form $\mathcal{U}^*$, this would imply q was contained in more than n members of $\mathcal{U}$. But this is not possible since $\mathcal{U}$ covers no point more than n times.

Therefore, a sphere in $\mathcal{V}^*$ can contain at most mn formally distinct centers of spheres in $\mathcal{U}^*$. Since there are Mn formally distinct members of H, there can be at most Mmn^2 formally distinct members of H. Thus $Nm^2n \leq Mmn^2$ or $N/n \leq M/m$. ∎

LEMMA 10.2 If $A \in T$ and $r \in \mu$ with $Star(A, r) \subset A^+$, and if λ is a Loomis content on T, invariant with respect to r, then for $r^- < r$,

$$(10.2) \qquad H(A,r) \leq h(A^+,x)/h(r^-,x) \text{ and } \lambda(A)/\lambda(r) \leq h(A^+,x)/h(r^-,x)$$

for all sufficiently small $x \in \mu$.

Proof: Let Σ be a family of $h(A^+, x)$ x-spheres covering A^+. Then Σ covers $S(q,r^-)$ for each $q \in A$ so at least $h(r^-, x)$ members of Σ meet $S(q, r^-)$. If x is sufficiently small that $Star(S(p, r^-), x) \subset S(p, r)$ for each $p \in X$, then at least $h(r^-, x)$ of the centers of members of Σ lie in $S(q, r^-)$. But then q lies in at least $h(r^-, x)$ of the r-spheres concentric with the spheres in Σ. Let Σ^* be the family of r-spheres whose centers are concentric with members of Σ. Then Σ^* is a collection of $h(A^+, x)$ r-spheres covering each point of A at least $h(r^-, x)$ times so the first inequality in (10.2) follows from the definition of $H(A, r)$.

To prove the second inequality in (10.2), assume A is covered by M r^--spheres, say $S_1 \ldots S_M$ that cover each point of A at least N times. Then by L5 of Proposition 10.4, $\lambda(A) \leq \Sigma_{n=1}^{M} \lambda(Star(S_n, x))/N$. Since for each $p \in X$, $Star(S(p, r^-)) \subset S(p, r)$, we have $\lambda(Star(S_n, x)) \leq \lambda(r)$ for each $n = 1 \ldots M$ because λ is invariant with respect to r. Therefore, $\lambda(A) \leq M\lambda(r)N$ which implies $\lambda(A)/\lambda(r) \leq M/N$. By the definition of $H(A, r)$ we get $\lambda(A)/\lambda(r) \leq H(A,r)$. Then the first inequality in (10.2) implies the second. ∎

LEMMA 10.3 If $A \in T$, and $r \in \mu$ with $Star(A^-, r) \subset A$, and λ is a Loomis content on T, invariant with respect to some $r^- < r$, then,

$$(10.3) \qquad H(A, r^-) \geq h(A^-, x)/h(r, x) \text{ and } \lambda(A)/\lambda(r^-) \geq h(A^-, x)/h(r, x)$$

for all sufficiently small $x \in \mu$.

Proof: Let Σ be a family of $h(A^-, x)$ x-spheres covering A^-. If $q \in A$, then at most $h(r, x)$ of the members of Σ can lie in $S(q, r)$, for otherwise these spheres could be replaced by $h(r, x)$ x-spheres covering $S(q, r)$, yielding a covering of A^- by fewer than $h(A^-, x)$ x-spheres which is a contradiction. Thus if x is sufficiently small that $Star(S(p, r^-), x) \subset S(p, r)$ for each $p \in X$, at most $h(r, x)$ of the members of Σ have centers in $S(q, r^-)$ which implies q lies in at most $h(r,x)$ of the r^--spheres concentric with the x-spheres in Σ. Then the first inequality in (10.3) follows from Lemma 10.1 and the definition of $H(A, r^-)$.

To prove the second inequality in (10.3), let Σ^* be the collection of r^--spheres concentric with the x-spheres in Σ. Then Σ^* is a collection of $h(A^-,x)$ r^--spheres that covers each point of A at most $h(r, x)$ times. Since $Star(A^-, r) \subset A$ we have $Star(A^-, r^-) \subset A$ so each member of Σ^* lies in A. Then by L6 of Proposition 10.4, we have

$$\lambda(A) \geq \Sigma_{S \in \Sigma^*} \lambda(S)/h(r, x) = h(A^-, x)\lambda(r^-)/h(r, x)$$

since λ is invariant with respect to r^-. Let $M = h(A^-, x)$ and $N = h(r, x)$. Then $\lambda(A)/\lambda(r^-) \geq M/N$ so by the definition of $H(A, r^-)$, we get $\lambda(A)/\lambda(r^-) \geq H(A, r^-)$. Hence the first inequality in (10.3) implies the second. ∎

LEMMA 10.4 If $A < A^+$ and $B^- < B$ where both A^+ and $B \in T$, then

$$(10.4) \qquad \limsup_r \frac{H(A,r)}{H(B,r)} \leq \liminf_x \frac{h(A^+,x)}{h(B^-,x)} \text{ and}$$

$$(10.5) \qquad \limsup_x \frac{h(B^-,x)}{h(A^+,x)} \leq \liminf_r \frac{H(B,r)}{H(A,r)}.$$

If A and B^- have non-void interiors, then the right members of (10.4) and (10.5) are finite and positive.

Proof: Let $r \in \mu$ such that $Star(A, r) \subset A^+$ and $Star(B^-, r) \subset B$. Then for $r^- < r$, $Star(B^-, r^-) \subset B$. Let $r^{--} < r^-$. Then substituting B for A in the first inequality of (10.3) and B^- for A^-, r^- for r and r^{--} for r^- yields:

$$(10.6) \qquad H(B, r^{--}) \geq h(B^-, x)/h(r^-, x).$$

Dividing the first inequality in (10.2) by (10.6) gives:

$$\text{(10.7)} \qquad \frac{H(A,r)}{H(B,r^{--})} \leq \frac{h(A^+,x)}{h(B^-,x)}$$

for all sufficiently small x. Since (10.7) holds for all $r^{--} < r$, by Proposition 10.5 we get:

$$\text{(10.8)} \qquad \frac{H(A,r)}{H(B,r)} \leq \frac{h(A^+,x)}{h(B^-,x)}$$

for all sufficiently small x. Taking the *limit inferior* of the right hand side of (10.8) yields:

$$\text{(10.9)} \qquad \frac{H(A,r)}{H(B,r)} \leq \liminf_x \frac{h(A^+,x)}{h(B^-,x)}$$

Since (10.9) holds for all sufficiently small r, (10.4) follows. Now, the reciprocal inequality to (10.8) is:

$$\text{(10.10)} \qquad \frac{h(B^-,x)}{h(A^+,x)} \leq \frac{H(B,r)}{H(A,r)}$$

so by an argument similar to the above we get (10.5). ■

In order to use Lemma 10.4 to construct a measure on Z we will need the following result:

LEMMA 10.5 Let $f(\alpha, \beta)$ be a real valued function of two real variables $\alpha, \beta \in [0, 1]$ that increases as α and β increase. Then the discontinuities of f lie on a countable family of decreasing curves (allowing vertical and horizontal segments as curve arcs).

Proof: Since f is increasing in both α and β, a point (α, β) at which f is not continuous has the property that if $\alpha^- < \alpha < \alpha^+$ and $\beta^- < \beta < \beta^+$ then

$$\lim_{(\alpha^-,\beta^-)\to(\alpha,\beta)} f(\alpha^-,\beta^-) < \lim_{(\alpha^+,\beta^+)\to(\alpha,\beta)} f(\alpha^+,\beta^+),$$

and the difference $\Delta f(\alpha, \beta)$ between the right and left sides of this inequality is referred to as the **jump** of f at (α, β). Let E_ε be the set of points (α, β) at which $\Delta f(\alpha, \beta) \geq \varepsilon$. We will call a finite set of points (α_i, β_i) for $i = 1 \ldots n$ an **increasing chain** if $\alpha_i < \alpha_{i+1}$ and $\beta_i < \beta_{i+1}$ for each $i = 1 \ldots n - 1$. If the points (α_i, β_i) are chosen from E_ε then clearly $n \leq [f(1, 1) - f(0, 0)]/\varepsilon$. Let C_1 be the set of points of E_ε which are first points of increasing chains in E_ε but not second points of such chains. Then no pair of points in C_1 can determine a positive slope for otherwise one of them would be an increasing chain in E_ε. A set of points in $[0, 1] \times [0, 1]$, no pair of which determine a positive slope, lies on a decreasing curve (where vertical and horizontal segments are allowed as curve arcs – see Exercise 1).

Consequently, C_1 can be imbedded in a decreasing curve. Similarly, let C_2 be the set of points in E_ε which are second points of increasing chains in E_ε but not third points. Clearly, no pair of points in C_2 determine a positive slope so C_2 can be imbedded in a decreasing curve. Continuing in this manner, we see that E_ε can be imbedded in n decreasing curves for some $n \leq [f(1,1)\text{-}f(0,0)]/\varepsilon$. Now if we choose a decreasing sequence $\{\varepsilon_n\}$ converging to zero, the corresponding sets $\{E_{\varepsilon_n}\}$ include all the points of discontinuities of f and the corresponding finite sets of decreasing curves form the countable collection of such curves in the conclusion of the lemma. ■

LEMMA 10.6 Let $\{B_\beta\} \subset T$ be an interval of sets with non-void interiors. Then there exists an index $\beta_0 \in (0, 1)$ and a unique increasing function $g(\beta)$ that is continuous and equal to 1 at β_0, such that for each interval $\{A_\alpha\} \subset T$ of sets with non-void interiors, there exists a unique, increasing function $f(\alpha)$ with

$$\frac{f(\alpha)}{g(\beta)} = lim_x \frac{h(A_\alpha,x)}{h(B_\beta,x)} = lim_r \frac{H(A_\alpha,r)}{H(B_\beta,r)},$$

where the limits exist at any point (α, β) for which the functions f and g are continuous.

Proof: If $\{A_\alpha\}$ and $\{B_\beta\}$ are intervals of totally bounded open sets with non-void interiors (as obtained by Proposition 10.4), then by Lemma 10.4, for sufficiently small $\varepsilon > 0$,

$$\text{(10.12)} \qquad lim\,sup_x \frac{h(A_{\alpha-\varepsilon},x)}{h(B_{\beta+\varepsilon},x)} \leq lim\,inf_r \frac{H(A_\alpha,r)}{H(B_\beta,r)} \leq lim\,sup_r \frac{H(A_\alpha,r)}{H(B_\beta,r)} \leq$$

$$lim\,inf_x \frac{h(A_{\alpha+\varepsilon},x)}{h(B_{\beta-\varepsilon},x)} \quad \text{and}$$

$$\text{(10.13)} \qquad lim\,sup_r \frac{H(A_{\alpha-\varepsilon},r)}{H(B_{\beta+\varepsilon},r)} \leq lim\,inf_x \frac{h(A_\alpha,x)}{h(B_\beta,x)} \leq lim\,sup_x \frac{h(A_\alpha,x)}{h(B_\beta,r)} \leq$$

$$lim\,inf_r \frac{H(A_{\alpha+\varepsilon},r)}{H(B_{\beta-\varepsilon},r)}.$$

The four functions of (10.12) and (10.13) are all increasing functions as α increases and β decreases. Since (10.12) and (10.13) hold for any sufficiently small $\varepsilon > 0$, they must be identical for any point where any one of them is continuous (and hence identical with respect to their corresponding limits as opposed to their limits superior).

For each $(\alpha, \beta) \in [0, 1] \times [0, 1]$, put $g'(\alpha, \beta) = lim\ sup_x h(A_\alpha, x)/h(B_\beta, x)$. Then the function defined by $f(\alpha, \beta) = g'(\alpha, \beta)$ satisfies the hypothesis of Lemma 10.5, so its set of discontinuities lies on a countable family of decreasing curves which implies the set of discontinuities of g' lie on a countable family F of increasing curves. Then at all points (α, β) of continuity, $g'(\alpha, \beta)$ is the common limit of all four functions in (10.12) and (10.13). Pick β_0 so that none of the curves of F contains a horizontal segment with ordinate β_0. Then (α, β_0) is a continuity point for g' for all α except a countable exceptional set that we denote by E_{β_0}. Then

$$g(\beta) = \frac{g'(\alpha,\beta_0)}{g'(\alpha,\beta)} = lim_{x\in\lambda} \frac{h(B_\beta,x)}{h(B_{\beta_0},x)}$$

is a function only of β, say $g(\beta)$, and is therefore independent of the interval of sets $\{A_\alpha\}$, except for the original choice of β_0. Define f by $f(\alpha) = g(\alpha, \beta_0)$. Then, since β_0 is a continuity point of g,

$$\frac{f(\alpha)}{g(\beta)} = \frac{g(\alpha,\beta_0)}{g(\beta)} = lim_{x\in\lambda} \frac{h(A_\alpha,x)}{h(B_\beta,x)} = g'(\alpha,\beta).$$

Now f depends only on the interval $\{A_\alpha\}$. Furthermore, for any index α, the set A_α can be replaced by either its interior or closure without altering its position in the interval (due to the method of construction in Proposition 10.3), the convergence of the ratios, or the value of $f(\alpha)$. Consequently, any other continuity point of g could have been chosen (which would multiply both f and g by a common factor, leaving the ratio unchanged), and the proof would remain valid. ∎

Let β_0 be the index in the hypothesis of Lemma 10.6 and let P' be the collection of sets A that belong to an interval $\{A_\alpha\} \subset T$ such that if $A = A_\gamma$ then γ is a continuity point of the function $f(\alpha)/g(\beta_0)$ of Lemma 10.6. Put $\lambda(A) = lim_x h(A, x)/h(B_{\beta_0}, x) = lim_r H(A,r)/H(B_{\beta_0}, r)$.

THEOREM 10.3 (L. Loomis, 1949) λ is a Loomis content on T and hence can be extended to a finitely additive measure on Z. Moreover $P' = Z \cap T$.

Proof: We first show that P' is dense in T with respect to $<$ and that if $U, V \in P'$ are uniformly separated then $U \cup V \in P'$. If $U, V \in P'$, there exist intervals $\{A_\alpha\}$, $\{B_\beta\} \subset T$ with $A_\alpha < A_\beta$ and $B_\alpha < B_\beta$ whenever $\alpha < \beta$, and such that $U = A_\theta$ and $V = B_\zeta$ for some θ, ζ that are continuity points of $\lambda(A_\alpha)$ and $\lambda(B_\beta)$ respectively. Let $\phi[0, 1] \to [0, 1]$ be a strictly increasing, onto function with

$\phi(\theta) = \zeta$ and for each $\alpha \in [0, 1]$ put $C_\alpha = B_{\phi(\alpha)}$. Then as shown in the proof of Theorem 10.2, $\{A_\alpha \cup C_\alpha\} \subset T$ is an interval with $A_\alpha \cup C_\alpha < A_\beta \cup C_\beta$ whenever $\alpha < \beta$. To show $U \cup V \in P'$, let $\varepsilon > 0$ and pick $\alpha, \beta, \gamma, \delta \in [0, 1]$ with $A_\alpha < U < A_\beta$ and $B_\gamma < V < B_\delta$ such that $\lambda(A_\beta - Cl(A_\alpha)) < \varepsilon/2$ and $\lambda(B_\delta - Cl(B_\gamma)) < \varepsilon/2$.

Now $U = A_\theta$ and $V = C_\theta$, so there exist $r, s \in [0, 1]$ with $A_r < U < A_s$, $C_r < V < C_s$, $\lambda(A_s - Cl(A_r)) < \varepsilon/2$ and $\lambda(C_s - Cl(C_r)) < \varepsilon/2$. Clearly $A_r \cup C_r < U \cup V < A_s \cup C_s$ and $A_s \cup C_s - Cl(A_r \cup C_r) \subset [A_s - Cl(A_r)] \cup [C_s - Cl(C_r)]$ so $\lambda(A_s \cup C_s - A_r \cup C_r) < \lambda([A_s - Cl(A_r)] \cup [C_s - Cl(C_r)])$. Hence it only remains to show that λ is sub-additive on P' to show that $\lambda(A_s \cup C_s - Cl(A_r \cup C_r)) < \varepsilon$. For this note that for any $A, B \in P'$ that $h(A \cup B, x) \leq h(A, x) + h(B, x)$ for each $x \in \mu$, so $\lambda(A \cup B) \leq \lambda(A) + \lambda(B)$. Consequently, $U \cup V$ is a continuity point of $\lambda(A_\alpha \cup C_\alpha)$ so $U \cup V \in P'$.

To show P' is dense in T, let $A, B \in T$ with $A < B$. By Proposition 10.3, there is an interval $\{A_\alpha\} \subset T$ with $A_0 = A$ and $A_1 = B$ and $A_\alpha < A_\beta$ whenever $\alpha < \beta$. By Lemma 10.6 we can choose a γ such that $A < A_\gamma < B$ and γ is a continuity point of $\lambda(A_\alpha)$. Therefore $A_\gamma \in P'$ so P' is dense in T. That λ satisfies L1 on P' is clear from the definition of λ. L2 and L3 follow from the fact already shown above that for $A, B \in P'$, $h(A \cup B, x) \leq h(A, x) + h(B, x)$ for each $x \in \mu$ and from the fact that $h(A \cup B) = h(A, x) + h(B, x)$ if $Star(A, x) \cap Star(B, x) = \varnothing$. L4 follows from the fact that if $A \in P'$ then $H(A, x) = sup\{H(A, y) \mid y < x\}$ (Proposition 10.6) and the fact that P' is dense in T.

Therefore P' satisfies the conditions of Theorem 10.1 if we set $P = T$, so λ can be uniquely extended to a Loomis content on T. Hence by Theorem 10.2, λ is a finitely additive measure on the ring Z of zero-boundary sets with respect to λ. Since $A \in P'$ implies A is a continuity point of $\lambda(A_\alpha)$ for some $\{A_\alpha\} \subset T$ with $A \in \{A_\alpha\}$, A is a zero-boundary set with respect to λ by the remarks preceding Corollary 10.1. Therefore $A \in Z \cap T$. Conversely, if $A \in Z \cap T$ then A is a continuity point of λ with respect to some interval $\{A_\alpha\} \subset T$ that contains A (also by the remarks preceding Corollary 10.1). Hence $A \in P'$. Therefore, $P' = Z \cap T$. ∎

This concludes the correct part of Loomis' development. In the final two sections, a development of an only marginally weaker version of Loomis' claims will be presented that is closer in spirit to the approach of Loomis' original construction of a uniform measure on a metric space.

EXERCISE

1. Show that a set of points in $[0, 1] \times [0, 1]$, no pair of which determines a positive slope, lie on a decreasing curve (allowing vertical and horizontal segments as curve arcs).

10.4 Invariance and Uniqueness of Loomis Contents and Haar Measures

In this section we first prove the invariance and uniqueness of a Loomis content on its ring Z of zero-boundary sets. We will then use this result to prove the uniqueness of the Haar measure developed in the previous chapter. We have been using μ not only as a basis for the uniformity ν, but also as an index set for defining limits as in the definition of λ in Theorem 10.3. We now want to expand our index set μ to an index set μ^* in such a way that we can have *intervals of indicies* from μ^* analogous to the intervals of sets $\{A_\alpha\}$ such that $Cl(A_\alpha) < Int(A_\beta)$ whenever $\alpha < \beta$ in Proposition 10.3. For any $x, y \in \mu$ with $x < y$, we use axiom A1 to construct a collection $\{y_\alpha\} \subset \mu$ where α ranges over the rationals in $[0, 1]$, $y_0 = x$, $y_1 = y$ and $y_\alpha < y_\beta$ whenever $\alpha < \beta$.

Then for an irrational $\gamma \in [0, 1]$ we let y_γ be the uniform covering defined by $y_\gamma = \{y_\gamma(p) \mid p \in X\}$ where $y_\gamma(p) = \cup_{\alpha<\gamma} S(p, y_\alpha)$. By the argument of the previous section, we have now extended $\{y_\alpha\} \subset \mu$ where α ranges over the rationals in $[0, 1]$ to an interval of coverings $\{y_\alpha\} \subset \nu$, for each $\alpha \in [0, 1]$ such that $y_\alpha < y_\beta$ whenever $\alpha < \beta$. To be precise, we actually need to define what we mean by $y_\alpha < y_\beta$ since this has previously only been defined for $y_\alpha, y_\beta \in \mu$ (i.e., for α, β rationals in $[0, 1]$). By $y_\alpha < y_\beta$ where either α or β is irrational, we mean

> $Star(y_\alpha(p), z) \subset S(p, y_\beta)$ for some $z \in \mu$ and each $p \in X$ if α is irrational,
>
> $Star(S(p, y_\alpha)) \subset y_\beta(p)$ for some $z \in \mu$ and each $p \in X$ if β is irrational, or
>
> $Star(y_\alpha(p), z) \subset y_\beta(p)$ for some $z \in \mu$ and each $p \in X$ if both α and β are irrational.

Define μ^* to be the collection μ together with the coverings $y_\gamma \in \nu$ such that γ is irrational and y_γ is a member of an interval of coverings $\{y_\alpha\} \subset \nu$ where $y_\beta \in \mu$ for each rational index β in $[0, 1]$ and such that $y_\alpha < y_\beta$ whenever $\alpha < \beta$ and for irrational indicies δ, $y_\delta = \{y_\delta(p) \mid p \in X\}$ where $y_\delta(p) = \cup_{\alpha<\delta} S(p, y_\alpha)$. Clearly μ^* is also a basis for ν. It is not necessary for our purposes to prove that μ^* has the properties of μ because we are only going to use μ^* to expand our definition of invariance. We define λ to be **invariant** with respect to y_γ where y_γ in not in μ, if $\lambda(y_\gamma(p)) = \lambda(y_\gamma(q))$ for each pair $p, q \in X$.

Loomis' version of the first theorem we shall prove in this section (Theorem 10.4) was incorrect. It relied on the erroneous assumption that if you select an *arbitrary* interval $\{y_\alpha\} \subset \mu^*$, then for a rational index γ, $y_\gamma \in \mu$. Furthermore, as we will see in the next section, this result is not exactly what is needed to prove uniqueness of the measure that we will eventually construct from λ. The statement of Loomis' version of this theorem (translated into our terminology) was:

The function $\lambda(A) = \lim_x h(A, x)/h(B_0, x)$ *defined by Theorem 10.3 is invariant with respect to all but countably many indicies from any interval of indicies. If m is any other Loomis content on T which is invariant with respect to all except countably many indices from every interval of indicies, then m is a constant multiple of* λ *on Z.*

The way we will prove the "invariance" part of this theorem is by relaxing the "amount" of invariance in the hypothesis to only include those intervals $\{A_\alpha\} \subset \mu^*$ such that $y_\gamma \in \mu$ whenever γ is rational *while still proving* the "uniqueness." Moreover, as we will see in the next section, even this amount of relaxation is not enough (using our approach) to prove the existence of a genuine "invariant" measure on the Baire ring. We will further have to relax our invariance requirement to only include "spheres with compact closures." The restriction to spheres with compact closures (as opposed to totally bounded spheres) was an assumption in Loomis' 1945 paper. Of course it is possible that the above stated "theorem" can still be shown to be true with another argument, but as we will see in the next section we still need the restriction to spheres with compact closures in our approach to extending λ to the Baire sets. Loomis did not include a proof of this in his 1949 paper.

Furthermore, to prove the uniqueness of the genuine measure λ^* that will be constructed in the next section, what is needed is a statement about a smaller ring $Z(G) \subset Z$ consisting of the zero-boundary members of the hereditarily open prering G of open sets with compact closures. For this we need the following terminology.

Let G denote the hereditarily open prering of open sets with compact closures. Then $G \subset T$. The zero-boundary sets, denoted $Z(G)$, with respect to λ (now considered as a Loomis content on $G \subset T$) are those sets A such that for each $\varepsilon > 0$ there exists a closed set F and an open set $U \in G$ such that $F < A < U$ and $\lambda(U - F) < \varepsilon$. It should be intuitively clear that all of Loomis' results documented in the previous two sections hold for λ defined on $Z(G)$. However, to be absolutely sure, one needs to work through all those results replacing T with G and Z with $Z(G)$. We leave this as an exercise for the serious reader.

Next we relax Loomis' definition of invariance as follows: a Loomis content l on G is **invariant on compact spheres with respect to** $x \in \mu$ if $l(S(p,x)) = l(S(q, x))$ for each pair $p, q \in X$ with both $S(p, x)$ and $S(q, x)$ having compact closures. l is simply **invariant on compact spheres** if it is invariant on compact spheres with respect to all $x \in \mu$. In this case, we do not necessarily have a common value $l(x) = l(S(p, x))$ for each $p \in X$ unless we know that $S(p,x)$ has compact closure for each $p \in X$. Again, it should be intuitively clear that Loomis' development recorded in the previous two sections can still be accomplished with this restriction, but to be absolutely sure, one needs to verify it for themselves. For this we note that it suffices to prove Lemmas 10.2 and 10.3 using this definition of invariance on compact spheres with respect to an r

$\in \mu$, since these are the only places where invariance is used in the proofs and the rest of the development follows from these two lemmas.

THEOREM 10.4 The Loomis content λ of Theorem 10.3 (where T is replaced by G) is invariant on compact spheres with respect to all but countably many indices of any interval $\{y_\alpha\} \subset \mu^*$ with $y_\alpha < y_\beta$ whenever $\alpha < \beta$ and such that if γ is rational then $y_\gamma \in \mu$. If l is any other Loomis content on G that is invariant on compact spheres with respect to all but countably many indices of any interval $\{y_\alpha\} \subset \mu^*$, with $y_\alpha < y_\beta$ whenever $\alpha < \beta$ and whose members with rational indices are in μ, then l is a constant multiple of λ on Z(G).

Proof: Let $\{y_\alpha\} \subset \mu^*$ be an interval of indicies with $y_\alpha < y_\beta$ whenever $\alpha < \beta$ such that $y_\gamma \in \mu$ whenever γ is rational. Let $p \in$ X. The function $f(\alpha)/g(\beta_0)$ given by

$$f(\alpha)/g(\beta_0) = lim_x h(y_\alpha(p), x)/h(B_{\beta_0}, x) = \lambda(y_\alpha(p))$$

in Lemma 10.6, is increasing on [0, 1] so it is continuous at all but an increasing sequence $\{r_n\} \subset [0, 1]$. Let $\delta \in [0, 1]$ such that δ is not in $\{r_n\}$. If $y_\delta \in \mu$ then $h(y_\delta(p), x) = h(y_\delta(q), x)$ for each q in X by A4 so $\lambda(y_\delta(p)) = \lambda(y_\delta(q))$ for each q in X. Thus λ is invariant with respect to y_δ. If y_δ is not in μ, δ is still a continuity point of $f(\alpha)/g(\beta_0)$ so there exists an increasing sequence $\{s_n\}$ of rationals with $s_n \to \delta$ and $f(\alpha)/g(\beta_0)$ is continuous at each s_n. Then

$$\lambda(y_\delta(p)) = lim_n \lambda(y_{s_n}(p)) = lim_n \lambda(y_{s_n}(q)) = \lambda(y_\delta(q))$$

because, as shown above, $\lambda(y_{s_n}(p)) = \lambda(y_{s_n}(q))$ for each q in X since each s_n is rational and because δ is a continuity point of $f(\alpha)/g(\beta_0)$. Hence λ is invariant with respect to all but countably many of the y_α's.

Let l be another Loomis content on G that is invariant on compact spheres with respect to all but countably many indices of any interval $\{y_\alpha\} \subset \mu^*$ such that $y_\alpha < y_\beta$ whenever $\alpha < \beta$ and such that $y_\gamma \in \mu$ whenever γ is rational. Let $Z_l(G)$ be the ring of zero-boundary sets with respect to l and $Z_\lambda(G)$ the ring of zero-boundary sets with respect to λ. If r is a member of μ^* - μ, then r^- in Lemma 10.2 can still be chosen from μ and similarly in Lemma 10.3. Therefore, both of these lemmas can be generalized to include the possibility that $r \in \mu^*$.

Since l satisfies L5 and L6 of Proposition 10.4, l satisfies the second inequalities in (10.2) and (10.3), i.e.,

$$l(A)/l(r) \leq h(A^+, x)/h(r^-) \text{ and } l(A)/l(r^-) \geq h(A^-, x)/h(r, x)$$

for all sufficiently small $x \in \mu$, if l is invariant on compact spheres with respect to r. By the first inequalities in (10.2) and (10.3),

$$H(A,r) \le h(A^+,x)/h(r^-,x) \text{ and } H(A,r^-) \ge h(A^-,x)/h(r,x)$$

for all sufficiently small $x \in \mu$. Since $l(A)/l(r)$ and $H(A, r)$ and also $l(A)/l(r^-)$ and $H(A, r^-)$ play the same roles in these fundamental inequalities from which Lemma 10.6 is eventually proved, it is an easy exercise to show that in Lemma 10.6, $f(\alpha)/g(\beta_0)$ can be replaced by

$$lim_r \frac{l(A_\alpha)/l(r)}{l(B_{\beta_0})/l(r)} = \frac{l(A_\alpha)}{l(B_{\beta_0})}$$

since l is invariant on compact spheres with respect to all but countably many indicies from each interval $\{y_\alpha\} \subset \mu^*$ with $y_\alpha < y_\beta$ whenever $\alpha < \beta$ and such that $y_\gamma \in \mu$ whenever γ is rational. But then

$$\lambda(A_\alpha)l(B_{\beta_0}) = l(A_\alpha) \tag{10.14}$$

whenever $A_\alpha \in Z_\lambda(G) \cap Z_l(G)$. Consequently, l is a constant multiple of λ on $Z_\lambda(G) \cap Z_l(G)$. It only remains to show that $Z(G) = Z_\lambda(G) = Z_l(G)$.

For this let $A \in Z_\lambda(G)$. Then $A \in \{A_\alpha\}$ for some interval of sets with $A_\alpha < A_\beta$ whenever $\alpha < \beta$ and if $A = A_\gamma$ then γ is a continuity point of $\lambda(A_\alpha)$. Now $Z(G) = Z_\lambda(G) \cap Z_l(G)$ contains all but an increasing sequence $\{A_{r_n}\} \subset \{A_\alpha\}$ (by the remarks preceding Corollary 10.1 as applied to both $Z_\lambda(G)$ and $Z_l(G)$) and these are the points of discontinuity of either $\lambda(A_\alpha)$ or $l(A_\alpha)$. Let $k > 0$ such that $l(A_\alpha) = k\lambda(A_\alpha)$ for $A_\alpha \in Z(G)$. Let $\varepsilon > 0$. Then there is a $\delta > 0$ such that $(\gamma-\delta, \gamma+\delta) \subset Z(G)$ and for each pair of indicies α, β with $\gamma - \delta < \alpha < \gamma < \beta < \gamma + \delta$, $|k\lambda(A_\beta) - k\lambda(A_\alpha)| < \varepsilon$ since $k\lambda(A_\alpha)$ is continuous at γ. But then $|l(A_\beta) - l(A_\alpha)| < \varepsilon$. By the discussion preceding Corollary 10.1, $l(A_\beta - Cl(A_\alpha)) \le l(A_\beta) - l(A_\beta) < \varepsilon$. Hence A_γ is a zero-boundary set with respect to l so $A \in Z_l(G)$. By a similar argument, if $A \in Z_l(G)$ then $A \in Z_\lambda(G)$. Hence $Z(G) = Z_\lambda(G) = Z_l(G)$. ∎

Theorem 10.4 can now be used to prove uniqueness of the Haar measure of Theorem 9.3.

THEOREM 10.5 Let (X, ν) be a locally compact, isogeneous, uniform space and let μ be an isometric basis for ν. Let m be the Haar measure of Theorem 9.3 and suppose h is another Haar measure on (X, ν). Then h is a constant multiple of m on (X,ν).

Proof: Both m and h are defined on G because all open sets are Borel sets. Since m and h are measures they satisfy L1 - L3. Let $A \in G$. Since m and h are inner regular on open sets, $m(A) = sup\{m(K) | K \subset A \text{ and } K \text{ is compact}\}$ and $h(A) = sup\{h(K) | K \subset A \text{ and } K \text{ is compact}\}$. Clearly $sup\{m(V) | V < A$ and $V \in$

$G\} \le m(A)$ and $sup\{h(V) \mid V < A \text{ and } V \in G\} \le h(A)$, so we can show m and h to satisfy L4 if we show that for a compact $K \subset A$, there exists a $V \in G$ with $K \subset V < A$.

For this put $F = X - A$. For each $z \in K$, there is a uniformly continuous function $f_z : X \to [0, 1]$ such that $f_z(z) = 1$ and $f_z(F) = 0$. Put $U_z = \{x \in X \mid f_z(x) \ge 1/2\}$. Since K is compact, there exists a finite collection $U_{z_1} \ldots U_{z_M}$ that covers K. Put $f = \Sigma_{i=1}^{M} f_i$ and for each $x \in X$ put $g(x) = max\{1, f(x)\}$. Then $g : X \to [0,1]$ is uniformly continuous and $g(K) = 1$ and $g(F) = 0$. Let $\mathcal{W}$ be the uniform covering of [0, 1] consisting of spheres of radius 1/2. Then $\mathcal{U} = g^{-1}(\mathcal{W})$ is a uniform covering of X and $Star(K, \mathcal{U}) \subset A$. Since (X, ν) is a locally compact isogeneous uniform space, it is uniformly locally compact so there exists a $\mathcal{V} \in \mu$ such that each member of $\mathcal{V}$ is open and has compact closure and such that $\mathcal{V} <^* \mathcal{U}$. Since K is compact there is a finite collection $V_1 \ldots V_N \in \mathcal{V}$ that covers K. Put $V = \cup_{n=1}^{N} V_n$. Then $V \in G$ and $Star(V, \mathcal{V}) \subset Star(K, \mathcal{U}) \subset A$, so $K \subset V < A$. Hence both m and h satisfy L4. Therefore, m and h are Loomis contents on G.

To show that the original Loomis content λ exists in this space, we need to show that it satisfies axioms A1 - A4. For this let μ' be an isometric basis for ν. Since (X, ν) is uniformly locally compact, there exists a $\mathcal{V} \in \mu'$ such that each member of $\mathcal{V}$ is compact. Let μ be the collection of members of μ' that refine $\mathcal{V}$. Then μ is also an isometric basis for ν and the uniform coverings in μ have members with compact closures. By Proposition 10.2, μ satisfies A1. By the definition of the basis of a uniformity, μ satisfies A2. Since the members of uniform coverings in μ have compact closures, μ satisfies A3 and since $\mu \subset \mu'$ and μ' is an isometric basis for ν, μ satisfies A4.

Consequently, (X, ν) is a uniform space satisfying Loomis' four axioms, so λ exists in this space. Also, both m and h satisfy the hypothesis of Theorem 10.4. To see this, first note that both m and h are invariant with respect to the members of μ. When we construct μ^* from μ as described in the discussion preceding Theorem 10.4, we do so by adding uniform coverings of the form y_γ where y_γ is a member of an interval $\{y_\alpha\} \subset \nu$ such that $y_\alpha < y_\beta$ whenever $\alpha < \beta$, $y_\alpha \in \mu$ whenever α is rational, and if γ is irrational then $y_\gamma = \{y_\gamma(p) \mid p \in X\}$ where $y_\gamma(p) = \cup_{\alpha<\gamma} S(p, y_\alpha)$. Next we notice that if $\{r_n\} \subset [0,1]$ is a strictly increasing sequence of rationals such that $r_n \to \gamma$, then $y_\gamma(p) = \cup_n S(p, y_{r_n})$. Then by Proposition 8.6,

$$m(y_\gamma(p)) = lim_n m(S(p, y_{r_n})) \text{ and } m(y_\gamma(q)) = lim_n m(S(q, y_{r_n}))$$

for each pair $p, q \in X$. Since m is invariant with respect to each r_n we have $m(y_\gamma(p)) = m(y_\gamma(q))$ for each pair $p, q \in X$. By a similar argument, $h(y_\gamma(p)) = h(y_\gamma(q))$ for each pair $p, q \in X$. Therefore, m and h are invariant with respect to all members of μ^*, so they satisfy the hypothesis of Theorem 10.4. Thus both are constant multiples of λ on Z. But then they are constant multiples of each other on $Z(G)$, i.e., there exists some $k \in \mathbf{R}$ such that $m(Z) = kh(Z)$ for each $Z \in Z(G)$.

By Theorem 10.3, $Z(G) \cap G$ is dense in G with respect to <. Also, as shown above, both m and h are left continuous (L4). Hence for each $A \in G$, $m(A) = sup\{m(V) \mid V < A \text{ and } V \in G\} = sup\{m(W) \mid W < A \text{ and } W \in Z(G) \cap G\} = sup\{kh(W) \mid W < A \text{ and } W \in Z(G) \cap G\} = ksup\{h(V) \mid V < A \text{ and } V \in G\} = kh(A)$. Therefore m is a constant multiple of h on G.

Now let $K \subset X$ be compact. Then $K \subset \cup_{n=1}^{N} V_n$ for some finite collection of open sets V_n with compact closures. Let $U = \cup_{n=1}^{N} V_n$. Then $U \in G$, so $m(U) = kh(U)$. Since K is closed, $U - K \in G$ so $m(U - K) = kh(U - K)$. Since m is a measure, $m(U) = m(K) + m(U - K)$ so $m(K) = kh(U) - kh(U - K) = kh(K)$. Hence $m(K) = kh(K)$ for each compact set K.

Next, let $U \subset X$ be open. Since m and h are inner regular on open sets, there exist ascending sequences $\{K_n\}$ and $\{C_n\}$ of compact subsets of U such that $m(U) = sup\{m(K_n)\}$ and $h(U) = sup\{h(C_n)\}$. For each positive integer n put $H_n = K_n \cup C_n$. Then $\{H_n\}$ is an ascending sequence of compact subsets of U such that $m(U) = sup\{m(H_n)\}$ and $h(U) = sup\{h(H_n)\}$. Consequently,

$$m(U) = sup\{m(H_n)\} = sup\{kh(H_n)\} = ksup\{h(H_n)\} = kh(U).$$

Finally, let E be any Borel set. Since m and h are both outer regular, there exist descending sequences $\{U_n\}$ and $\{V_n\}$ of open sets containing E such that $m(E) = inf\{m(U_n)\}$ and $h(E) = inf\{h(V_n)\}$. For each n put $W_n = U_n \cap V_n$. Then $\{W_n\}$ is a descending sequence of open sets containing E such that $m(E) = inf\{m(W_n)\}$ and $h(E) = inf\{h(W_n)\}$. Hence

$$m(E) = inf\{m(W_n)\} = inf\{kh(W_n)\} = kinf\{h(W_n)\} = kh(E).$$

Therefore $m(E) = kh(E)$ for each Borel set E, so h is a constant multiple of m on (X, ν). ∎

EXERCISE

1. Show that in Lemma 10.6, $f(\alpha)/g(\beta_0)$ can be replaced with $l(A_\alpha)/l(B_{\beta_0})$.

10.5 Local Compactness and Uniform Measures

In the last paragraph of p. 206 in his 1949 paper, Loomis states (without proof) that if we assume our space to be locally compact, the Loomis content λ "has a unique additive extension to the Borel field generated by the ring of totally bounded open sets and M6 (our L6) holds for any sets from this field." The term *Borel field* that Loomis uses should not be confused with our term *Borel algebra.* Although Loomis does not define the term *field* in his paper, it is clear from his proofs of Theorems 2 and 3 that by a field he means a ring (and not a σ-ring) in our terminology. It is also clear that by *additive* he means finitely additive rather than σ-additive. Presumably, this "Borel field" is the smallest ring containing the hereditarily open prering T of totally bounded open sets since Loomis says this "Borel field" is generated by the totally bounded open sets.

How Loomis intended to extend λ to a genuine measure on this "Borel field" is unknown since no proof or reference was given. Had λ been a *Jordan content* (on the compact subsets), one might presume the approach he had in mind was the standard method of Kodaira and von Neumann, referenced in the introduction of this chapter, of extending a Jordan content to a σ-additive measure on the Baire ring B_0. Even though λ is not a Jordan content, it is possible to extend λ to a Jordan content λ^+ (on the compact subsets) in a natural way. However, using Kodaira and von Neumann's method on λ^+ does not solve the problem because the measure m one obtains is an extension of λ^+ on the compact subsets and not necessarily an extension of λ on the non-compact subsets on which the invariance of λ is defined. Consequently, m may not preserve the invariance property.

Loomis may have been aware, or partially aware, of this problem because he states on p. 202 that if our underlying space S is locally totally bounded, then its completion T is locally compact and the Loomis content λ "can be extended to the completely additive measure of Radon on the Borel field . . . . generated by the compact subsets of T. It is not clear at this point in what sense the extended measure is a Haar measure; this question will be considered in § 12." Again, Loomis does not prove this assertion or give a reference. In Section 12 of his paper, he does not deal with the question as promised. Instead, he considers the extension provided by his Theorem 9 that extends a Loomis content into the completion T. He does not deal with the problem that when extending the Loomis content λ to a genuine measure m on the "Borel field," that the extended measure may not be invariant.

Also on p. 202, Loomis uses the term *completely additive* to describe the extended measure. This may indicate that he uses the term *field* inconsistently in various places in the paper, and that what he means by the "Borel field" generated by the compact sets of the completion T, is actually the Baire ring B_0 on T. Also, it should be noted that it is not difficult to show that the smallest ring R, containing the hereditarily open prering T of totally bounded open sets, contains all the compact sets when the space is locally compact. To see this let K be a compact set and for each $p \in$ K, choose an open neighborhood V_p of p with compact closure. Then it is possible to choose a finite collection of points $p_1 \ldots . p_n$ in K such that $V = \cup_{i=1}^{n} V_{p_i}$ belongs to T, so V - K $\in T$. Since $T \subset R$ and R is a ring, K = V - [V - K] $\in R$.

There is a way around the problem of preserving the invariance of λ (such as it is, given by Theorem 10.4) in the Baire measure m if we are willing to restrict our definition of *invariance* to the μ^*-spheres of X with compact closure. We will show how this can be done in this section. It seems surprising that Loomis did not do this because he did make this assumption in his earlier paper on extending the concept of Haar measure to locally compact metric spaces in 1945. We have already seen that Loomis abandoned the compact spheres of his 1945 metric space paper in favor of the totally bounded μ^*-spheres in order to be able to construct the measure λ on the ring Z of its zero-boundary sets. However, it seems that returning to the compact μ^*-spheres for the definition of invariance of the extended Baire measure m is the thing to do because it works (i.e., it preserves the result of Theorem 10.4 in the new measure m for the compact μ^*-spheres).

Had Loomis not attempted to include μ^*-spheres that do not have compact closures in his definition of invariance in the 1949 paper, he would not have had to go to the trouble of introducing the concept of the extension of λ to the completion and introducing the weakened axiom A4′ (p. 205) to try to achieve invariance of an extension in the completion (which he does not accomplish anyway). This is because it turns out that m is already invariant on B_0 with respect to all but countably many members in any interval $\{y_\alpha\} \subset \mu^*$ with $y_\alpha < y_\beta$ whenever $\alpha < \beta$ and $y_\alpha \in \mu$ whenever α is rational, if we rule out members of μ^* whose μ^*-spheres are not compact. This will be shown in what follows.

Based on the previous discussion, we define an **almost uniform measure** to be a measure m on a uniform space (X, ν) such that there exists a basis μ for ν that satisfies axioms A1 - A4 and such that for each pair of points p, q in X, and any interval $\{x_\alpha\} \subset \mu^*$ with $x_\alpha < x_\beta$ whenever $\alpha < \beta$, $x_\gamma \in \mu$ whenever γ is rational, and $Cl(S(p, x_1))$ and $Cl(S(q, x_1))$ are compact, we have $m(S(p, x_\alpha)) = m(S(q, x_\alpha))$ for all but countably many α. We say m is a **uniform measure** if $m(S(p, x)) = m(S(q, x))$ for each x in μ^* such that $S(p, x)$ and $S(q, x)$ have compact closures.

Let K be the collection of compact subsets of X and let θ be a non-negative real valued function on K with the properties:

J1. $\theta(C \cup D) \leq \theta(C) + \theta(D)$ and
J2. if $C \cap D = \emptyset$ then $\theta(C \cup D) = \theta(C) + \theta(D)$ where $C, D \in K$.

Then θ is said to be a **Jordan content** on K. Let λ be as defined in Theorem 10.4. We can extend λ to $K \cup G$ by putting $\lambda^+(A) = \lambda(Int(A))$. Note that if $Int(A) = \emptyset$ that $\lambda^+(A) = 0$ since $\lambda(\emptyset) = 0$. That λ^+ is well defined on $K \cup G$ follows from the fact that $Int(A) \in G$ whenever $A \in K$. That λ^+ is an extension follows from the fact that $Int(U) = U$ if $U \in G$.

PROPOSITION 10.6 λ^+ *is a Jordan content on K.*

Proof: If either C or D has empty interior, the proof is trivial so assume $Int(C) \neq \emptyset \neq Int(D)$. Since $Int(C \cup D) \subset Int(C) \cup Int(D)$, by the definition of h, $h(Int(C \cup D)), x) \leq h(Int(C), x) + h(Int(D), x)$ for each x in μ. Then

$$lim_x \frac{h(Int(C \cup D),x)}{h(B_{\beta_0},x)} \leq lim_x \frac{h(Int(C),x)}{h(B_{\beta_0},x)} + lim_x \frac{h(Int(D),x)}{h(B_{\beta_0},x)}$$

if $Int(C \cup D), Int(C), Int(D) \in Z(G)$. If any of these sets do not belong to $Z(G)$, we still have

$$sup\{lim_x h(Z, x)/h(B_{\beta_0}, x) \mid Z \in Z(G) \text{ and } Z < C \cup D\} \leq$$

$$sup\{lim_x h(Z, x)/h(B_{\beta_0}, x) \mid Z \in Z(G) \text{ and } Z < C\} +$$

$$sup\{lim_x h(Z, x)/h(B_{\beta_0}, x) \mid Z \in Z(G) \text{ and } Z < D\},$$

since $Z(G)$ is a ring, H is left continuous on $Z(G)$ and

$$lim_x h(Z, x)/h(B_{\beta_0}, x) = lim_x H(Z, x)/H(B_{\beta_0}, x).$$

Then by the definition of λ on G, we have $\lambda^+(C \cup D) \leq \lambda^+(C) + \lambda^+(D)$. Now suppose $C \cap D = \emptyset$. It is easily shown (see Exercise 1) that C and D are uniformly separated. Consequently, $Int(C)$ and $Int(D)$ are uniformly separated so $\lambda(Int(C) \cup Int(D)) = \lambda(Int(C)) + \lambda(Int(D))$. But then

$$\lambda^+(C \cup D) = \lambda^+(Cl(Int(C) \cup Int(D))) = \lambda(Int(C) \cup Int(D)) =$$

$$\lambda(Int(C)) + \lambda(Int(D)) = \lambda^+(C) + \lambda^+(D). \blacksquare$$

Next we construct an outer measure λ^* from λ^+ in the following manner. Let τ denote the topology of (X, ν) and let $U \in \tau$. Put $\lambda_*(U) = sup\{\lambda^+(C) \mid C \subset U \text{ and } C \in K\}$. Then for an arbitrary subset A of X put $\lambda^*(A) = inf\{\lambda_*(U) \mid A \subset U \in \tau\}$.

PROPOSITION 10.7 λ^* is an outer measure on X.

Proof: First notice that if A and B are arbitrary subsets of X with $A \subset B$ and if $B \subset U \in \tau$ then $A \subset U \in \tau$. Hence $\lambda^*(A) \leq \lambda^*(B)$. Also notice that if $V \in \tau$, then by definition, $\lambda^*(V) \leq \lambda_*(U)$ for each $U \in \tau$ containing V. Then since $V \subset V \in \tau$, $\lambda^*(V) \leq \lambda_*(V)$. Conversely, if $V \subset U \in \tau$ then $\lambda_*(V) \leq \lambda_*(U)$ so $\lambda_*(V) \leq inf\{\lambda_*(U) \mid V \subset U \in \tau\} = \lambda^*(V)$. Hence $\lambda^*(V) = \lambda_*(V)$ for each open set V.

Next we show that if C is a compact subset of the union $U \cup V$ of two open sets, then there exists compact subsets D and E of U and V respectively such that $C \subset D \cup E$. For this let $W(p)$ be a compact neighborhood of p with $W(p) \subset$

U for each $p \in C \cap U$. Otherwise, for each $p \in C - U$ let $W(p) \subset V$. Then there is finite collection $x_1 \ldots x_N$ such that $C \subset \cup_{n=1}^{N} W(x_n)$. Put $D = \cup\{W(x_n) \mid x_n \in U\}$ and $E = \cup\{W(x_n) \mid x_n \in V - U\}$. Clearly $C \subset D \cup E$, D and E are in K, $D \subset U$ and $E \subset V$.

We are now in a position to show that λ^* is finitely subadditive on open sets. We proceed inductively by first considering two open sets U_1 and U_2. First suppose $\lambda^*(U_1 \cup U_2)$ is finite. Let $\varepsilon > 0$ and let C be a compact subset of $U_1 \cup U_2$ such that $\lambda^+(C) > \lambda^*(U_1 \cup U_2) - \varepsilon$. As we just showed in the preceding paragraph, there exist compact subsets C_1 and C_2 of U_1 and U_2 respectively such that $C \subset C_1 \cup C_2$. Now $\lambda^*(U_1) + \lambda^*(U_2) \geq \lambda^+(C_1) + \lambda^+(C_2) \geq \lambda^+(C_1 \cup C_2) \geq \lambda^+(C) > \lambda^*(U_1 \cup U_2) - \varepsilon$. Since ε was chosen arbitrarily, $\lambda^*(U_1 \cup U_2) \leq \lambda^*(U_1) + \lambda^*(U_2)$. On the other hand, if $\lambda^*(U_1 \cup U_2)$ is infinite, then for any $\alpha \in \mathbf{R}$, there is a $C \in K$ with $C \subset U_1 \cup U_2$ and $\lambda^+(C) > \alpha$. As before, there exist compact subsets C_1 and C_2 of U_1 and U_2 respectively such that $C \subset C_1 \cup C_2$. Then $\lambda^+(C_1) + \lambda^+(C_2) > \alpha$ so $\lambda^*(U_1) + \lambda^*(U_2) > \alpha$. Hence $\lambda^*(U_1) + \lambda^*(U_2)$ is infinite. Therefore, in either case, $\lambda^*(U_1 \cup U_2) \leq \lambda^*(U_1) + \lambda^*(U_2)$.

For any positive integer n such that $\lambda^*(\cup_{i=1}^{n-1} U_i) \leq \Sigma_{i=1}^{n-1} \lambda^*(U_i)$, we can use the proof in the preceding paragraph to show that $\lambda^*(\cup_{i=1}^{n-1} V_i \cup U_n) \leq \Sigma_{i=1}^{n-1} \lambda^*(U_i) + \lambda^*(U_n)$ so we conclude that λ^* is finitely subadditive on open sets by induction.

To show λ^* is countably subadditive on open sets, let $\{U_n\}$ be a sequence of open sets and suppose first that $\lambda^*(\cup U_n) < \infty$. Let $\varepsilon > 0$. Choose a compact $C \subset \cup U_n$ with $\lambda^+(C) > \lambda^*(\cup U_n) - \varepsilon$. Then there is a finite subsequence $\{U_{m_n}\}$ of $\{U_n\}$ that covers C. Then $\lambda^*(\cup U_n) - \varepsilon < \lambda^+(C) \leq \lambda^*(\cup U_{m_n}) \leq \Sigma\lambda^*(U_{m_n}) \leq \Sigma\lambda^*(U_n)$. Since ε was chosen arbitrarily, $\lambda^*(\cup U_n) \leq \Sigma\lambda^*(U_n)$ whenever $\lambda^*(\cup U_n)$ is finite. Now suppose $\lambda^*(\cup U_n)$ is infinite. Then for each $\alpha > 0$ there is a compact $C \subset \cup U_n$ with $\lambda^+(C) > \alpha$. Just as before, there is a finite subsequence $\{U_{m_n}\}$ of $\{U_n\}$ that covers C so $\Sigma\lambda^*(U_{m_n}) > \alpha$. Hence $\Sigma\lambda^*(U_n)$ is also infinite. Therefore λ^* is countably subadditive on open sets.

To conclude the proof we must show that λ^* is countably subadditive. For this let $\{E_n\}$ be a sequence of subsets of X. Note that it is sufficient to show $\lambda^*(\cup E_n) \leq \Sigma\lambda^*(E_n)$ where $\lambda^*(E_n) < \infty$ for each n, since if some $\lambda^*(E_m) = \infty$ the proof is trivial. Let $\varepsilon > 0$. For each positive integer n let U_n be an open set containing E_n such that $\lambda^*(E_n) > \lambda^*(U_n) - \varepsilon/2^n$. Then $\cup E_n \subset \cup U_n$ and $\lambda^*(\cup E_n) \leq \lambda^*(\cup U_n) \leq \Sigma\lambda^*(U_n) < \Sigma_n[\lambda^*(E_n) + \varepsilon/2^n] = \Sigma_n\lambda^*(E_n) + \varepsilon$. Since ε was chosen arbitrarily, this implies $\lambda^*(\cup E_n) \leq \Sigma\lambda^*(E_n)$. ∎

PROPOSITION 10.8 If W is an open set with compact closure then $\lambda^*(W) = \lambda(W)$.

Proof: $\lambda^*(W) = \lambda_*(W) = sup\{\lambda^+(C) \mid C \subset W \text{ and } C \in K\} = sup\{\lambda(Int(C)) \mid C \subset W \text{ and } C \in K\} = sup\{\lambda(U) \mid Cl(U) \subset W \text{ and } U \in \tau\}$ since $Cl(U) \in K$ for each U

$\in \tau$ with $Cl(U) \subset W$. Now for each $U \in \tau$ with $Cl(U) \subset W$ there is a $V \in \tau$ with $Cl(U) \subset V \subset Cl(V) \subset W$ so $\lambda^*(W) = sup\{\lambda(U) \mid U \subset W \text{ and } U \in \tau\} = \lambda(W)$. Therefore $\lambda^*(W) = \lambda(W)$. ∎

COROLLARY 10.2 $\lambda^+(C) \leq \lambda^*(C)$ *for each* $C \in K$.

COROLLARY 10.3 If $x \in \mu^*$ *and* $S(p, x)$ *and* $S(q, x)$ *have compact closures then* $\lambda^*(S(p, x)) = \lambda^*(S(q, x))$ *if and only if* $\lambda(S(p, x)) = \lambda(S(q, x))$.

PROPOSITION 10.9 The open subsets of X are λ^**-measurable.*

Proof: We need to show that for an open set O that $\lambda^*(S) = \lambda^*(O \cap S) + \lambda^*(S\text{-}O)$ for any subset S of X. We start by showing that if $U, V \in \tau$ and $U \cap V = \varnothing$ then $\lambda^*(U \cup V) = \lambda^*(U) + \lambda^*(V)$. For this it is clearly sufficient to consider the case where both $\lambda^*(U)$ and $\lambda^*(V)$ are finite. Let $\varepsilon > 0$ and let C and D be compact subsets of U and V respectively such that $\lambda^*(U) - \lambda^+(C) < \varepsilon$ and $\lambda^*(V) - \lambda^+(D) < \varepsilon$. That this is possible can be seen from $\lambda^*(U) = \lambda_*(U) = sup\{\lambda^+(K) \mid K \subset U \text{ and } K \in K\} \geq \lambda^+(C)$, for any compact $C \subset U$, and similarly for a compact $D \subset V$. Then $C \cap D = \varnothing$ and $\lambda^*(U) + \lambda^*(V) \leq \lambda^+(C) + \varepsilon + \lambda^+(D) + \varepsilon = \lambda^+(C \cup D) + 2\varepsilon \leq \lambda^*(C \cup D) + 2\varepsilon$ (by Corollary 10.2) $\leq \lambda^*(U \cup V) + 2\varepsilon$. Since ε was chosen arbitrarily, $\lambda^*(U) + \lambda^*(V) \leq \lambda^*(U \cup V)$. But then $\lambda^*(U \cup V) = \lambda^*(U) + \lambda^*(V)$.

Next we show that if U and V are open sets then $\lambda^*(V) = \lambda^*(U \cap V) + \lambda^*(V\text{-}U)$. For this let $\varepsilon > 0$. Let C be a compact subset of $U \cap V$ with $\lambda^+(C) > \lambda_*(U \cap V) - \varepsilon = \lambda^*(U \cap V) - \varepsilon$. Let W be an open set such that $C \subset W \subset Cl(W) \subset U \cap V$. Then $V - U \subset V - Cl(W)$, so $\lambda^*(U \cap V) + \lambda^*(V - U) \leq \lambda^*(U \cap V) + \lambda^*(V\text{-}Cl(W)) < \lambda^+(C) + \varepsilon + \lambda^*(V - Cl(W)) \leq \lambda^*(W) + \varepsilon + \lambda^*(V - Cl(W))$. Now W and $V - Cl(W)$ are disjoint open sets so as we just showed in the preceding paragraph, $\lambda^*(W) + \lambda^*(V - Cl(W)) = \lambda^*(W \cup [V - Cl(W)])$. Also, since $Cl(W) \subset V$ we have $W \cup [V - Cl(W)] \subset V$ so $\lambda^*(W \cup [V - Cl(W)]) \leq \lambda^*(V)$. Hence $\lambda^*(U \cap V) + \lambda^*(V - Cl(W)) \leq \lambda^*(V) + \varepsilon$. Since ε was chosen arbitrarily, $\lambda^*(U \cap V) + \lambda^*(V - Cl(W)) \leq \lambda^*(V)$. But then $\lambda^*(U \cap V) + \lambda^*(V - U) \leq \lambda(V)$. Conversely, since λ^* was shown to be subadditive in the proof of Lemma 10.7, $\lambda^*(V) \leq \lambda^*(U \cap V) + \lambda^*(V - U)$. Hence $\lambda^*(V) = \lambda^*(U \cap V) + \lambda^*(V - U)$.

Finally we show that for an open set O that $\lambda^*(S) = \lambda^*(O \cap S) + \lambda^*(S - O)$ for any $S \subset X$. If $\lambda^*(S)$ is infinite then $\lambda^*(O \cap S) + \lambda^*(S - O) \leq \lambda^*(S)$ so assume $\lambda^*(S)$ is finite. Let $\varepsilon > 0$ and choose an open set U containing S such that $\lambda^*(S) > \lambda^*(U) - \varepsilon$. Then $O \cap S \subset O \cap U$ and $S - O \subset U - O$, so $\lambda^*(O \cap S) + \lambda^*(S - O) \leq \lambda^*(O \cap U) + \lambda^*(U - O) = \lambda^*(U)$ as we just showed in the preceding paragraph. Therefore, $\lambda^*(O \cap S) + \lambda^*(S - O) \leq \lambda^*(S) + \varepsilon$. Since ε was chosen arbitrarily, $\lambda^*(O \cap S) + \lambda^*(S - O) \leq \lambda^*(S)$. Conversely, since λ^* is subadditive, $\lambda^*(S) \leq \lambda^*(O \cap S) + \lambda^*(S - O)$, so $\lambda^*(S) = \lambda^*(O \cap S) + \lambda^*(S - O)$. ∎

We have now shown that λ^* is a Borel (and hence Baire) measure on X. This follows from Propositions 10.7 and 10.9 and Theorem 8.4. Next we want to show that λ^* is regular on B_0 (the Baire sets). For this we need a number of results that will eventually enable us to conclude that the mere outer regularity of λ^* on compact sets is sufficient to imply the regularity of λ^* on *all* Baire sets. This is important in our development because the outer regularity of λ^* follows from the definition of λ^* using the result in the first paragraph of the proof of Proposition 10.7. Consequently, this will demonstrate the regularity of λ^* on B_0. The remaining results in this section, up to Theorems 10.10 and 10.11 date back to the 1940s or earlier.

PROPOSITION 10.10 Every Baire set in X is σ-bounded (can be covered by countably many compact sets) and every σ-bounded open set is a Baire set.

Proof: It is easily shown that the class of all σ-bounded sets is a σ-ring. Since each compact set is σ-bounded, this σ-ring contains all compact sets and hence the smallest σ-ring (Proposition 8.3) containing the compact sets, namely, B_0. Hence every Baire set is σ-bounded. The remainder of the proof is left as an exercise. ■

PROPOSITION 10.11 If μ is a Baire measure that is outer regular on compact sets, then μ is outer regular on the difference C - D of any pair of compact sets C and D such that $D \subset C$.

Proof: Since μ is outer regular on C, for each $\varepsilon > 0$ there exists an open Baire set U with $C \subset U$ and $\mu(U) < \mu(C) + \varepsilon$. Put $V = U - D$. Then V is open and $C - D \subset V$. Moreover,

$$\mu(V) - \mu(C - D) = \mu(V - [C - D]) = \mu(U - C) = \mu(U) - \mu(C) < \varepsilon.$$

Hence for each $\varepsilon > 0$ there exists an open $V \in B_0$ that contains C - D such that $\mu(V) < \mu(C - D) + \varepsilon$, so C - D is outer regular. ■

We say that an open set is **bounded** if it is contained in a compact set.

PROPOSITION 10.12 If μ is inner regular on each bounded open set, then μ is inner regular on the difference C - D of any pair of compact sets C and C such that $D \subset C$.

The proof of this proposition is similar to the proof of Proposition 10.11 (remember we are assuming X is locally compact) and will be left as an exercise. Moreover, the proof of the next proposition will be left as an exercise for the same reason.

PROPOSITION 10.13 If μ is inner regular on a finite collection of disjoint sets of finite measure, then μ is inner regular on their union.

THEOREM 10.6 Let μ be a Baire measure and $\{E_n\}$ a sequence of Baire sets. Then the following hold:

(1) If μ is outer regular on each E_n, then μ is outer regular on $\cup E_n$.

(2) If μ is inner regular on each E_n and $\{E_n\}$ is an ascending sequence, then μ is inner regular on $\cup E_n$.

(3) If μ is inner regular on E_n and $\mu(E_n) < \infty$ for each n, then μ is inner regular on $\cap E_n$.

(4) If μ is outer regular on E_n and $\mu(E_n) < \infty$ for each n, and if $\{E_n\}$ is descending, then μ is outer regular on $\cap E_n$.

Proof: To show (1) let $\varepsilon > 0$. For each n there exists an open Baire set U_n with $E_n \subset U_n$ and $\mu(U_n) < \mu(E_n) + \varepsilon/2^n$. Put $U = \cup U_n$ and $E = \cup E_n$. If $\mu(E) = \infty$ then clearly μ is outer regular on E. If $\mu(E) < \infty$, then

$$\mu(U) - \mu(E) \le \mu(\cup_n[U_n - E_n]) \le \Sigma_n \mu(U_n - E_n) < \Sigma_n \varepsilon/2^n = \varepsilon.$$

Since U is also an open Baire set, this shows that μ is outer regular on $\cup E_n = E$.

To show (2) put $E = \cup E_n$. By Proposition 8.6, $\mu(E) = lim_n \mu(E_n)$. Let $\varepsilon > 0$. Choose n such that $\mu(E) - \varepsilon < \mu(E_n)$. Since μ is inner regular on E_n there exists a compact $C \subset E_n$ with $\mu(E) - \varepsilon < \mu(C)$. But then μ is inner regular on $\cup E_n = E$.

To show (3), let $\varepsilon > 0$. For each n there exists a compact $C_n \subset E_n$ with $\mu(E_n) < \mu(C_n) + \varepsilon/2^n$. Put $C = \cap C_n$ and $E = \cap E_n$. Then $C \subset E$ and C is compact. Moreover,

$$\mu(E) - \mu(C) = \mu(\cap_n[E_n - C_n]) \le \Sigma_n \mu(E_n - C_n) < \Sigma_n \varepsilon/2^n = \varepsilon.$$

Since C is compact, this shows μ is inner regular on $\cap E_n$.

To show (4), put $E = \cap E_n$. By Proposition 8.7, $\mu(E) = lim_n \mu(E_n)$. Let $\varepsilon > 0$. Choose n such that $\mu(E_n) < \mu(E) + \varepsilon$. Since μ is outer regular on E_n there exists an open Baire set U such that $\mu(E_n) \le \mu(U) < \mu(E) + \varepsilon$. But then $\mu(E) \le \mu(U) < \mu(E) + \varepsilon$ so μ is outer regular on $\cap E_n = E$. ∎

THEOREM 10.7 A necessary and sufficient condition for a Baire measure μ to be outer regular on compact sets is that it be inner regular on bounded open sets.

Proof: Suppose μ is outer regular on each compact set and let U be a bounded open set. Let $\varepsilon > 0$. Let C be a compact set such that $U \subset C$. Then C - U is compact and hence μ is outer regular on C - U. Therefore, there exists an open Baire set V such that $C - U \subset V$ and $\mu(V) < \mu(C - U) + \varepsilon$. Since $C - V \subset C - (C-U) = U$ we have:

$$\mu(U) - \mu(C - V) = \mu(U - [C - V]) = \mu(U \cap V) \le \mu(V - [C - U]).$$

Since the last term on the right is equal to $\mu(V) - \mu(C - U) < \varepsilon$ we have that $\mu(U) - \varepsilon < \mu(C - V)$. Since C - V is compact we have shown that μ is inner regular on U.

Conversely, suppose μ is inner regular on each bounded open set and let C be compact. Let $\varepsilon > 0$. Since X is locally compact we can find a bounded open set U containing C. Then, since U - C is a bounded open set, there exists a compact set $D \subset U - C$ with $\mu(U - C) < \mu(D) + \varepsilon$. Since $C = U - (U - C) \subset U - D$ we have:

$$\mu(U - D) - \mu(C) = \mu([U - C] - D) = \mu(U - C) - \mu(D) < \varepsilon.$$

Therefore, $\mu(U - D) < \mu(C) + \varepsilon$ and since U - D is a bounded open set, we have shown that μ is outer regular on C. ∎

THEOREM 10.8 The collection R of all finite, disjoint unions of sets of the form C - D where C and D are compact and $D \subset C$ is a ring and the Baire ring B_0 is the smallest σ-ring containing it.

Proof: We first show that B_0 is the smallest σ-ring containing R. Let Σ be the smallest σ-ring containing R (Proposition 8.3). Then Σ contains all compact sets, so by the definition of B_0, $B_0 \subset \Sigma$. But since B_0 is a ring, B_0 contains R so $\Sigma \subset B_0$. Hence $B_0 = \Sigma$.

To show R is a ring let A, B $\in R$. Then $A = \cup_{n=1}^{N} [A_n - C_n]$ for some disjoint collection $\{A_1 \ldots A_N\}$ of compact sets and collection $\{C_1 \ldots C_N\}$ of compact sets such that $C_n \subset A_n$ for each n. Similarly, $B = \cup_{k=1}^{K} [B_k - D_k]$ for some disjoint collection $\{B_1 \ldots B_K\}$ of compact sets and collection $\{D_1 \ldots D_K\}$ of compact sets such that $D_k \subset B_k$ for each k. For each $n = 1 \ldots N$,

$$([A_n - C_n] \cap [B_k - D_k]) \cap ([A_n - C_n] \cap [B_j - D_j]) = \emptyset$$

for each $k \neq j$ since the sets B_k are disjoint. Consequently, the collection of sets of the form $[A_n - C_n] \cap [B_k - D_k]$ for each $k = 1 \ldots K$ is a disjoint collection of the form C - D where C and D are compact and $D \subset C$. But since all the A_n's are disjoint, the collection $[A_n - C_n] \cap [B_k - D_k]$ for each $n = 1 \ldots N$ and $k = 1 \ldots K$ is also disjoint. Now $A \cap B =$

$$(\cup_{n=1}^{N} [A_n - C_n]) \cap (\cup_{k=1}^{K} [B_k - D_k]) = \cup_{n=1}^{N} (\cup_{k=1}^{K} [A_n - C_n] \cap [B_k - D_k]) =$$

$$\cup_{n=1}^{N} (\cup_{k=1}^{K} [A_n \cap B_k] - [C_n \cup D_k])$$

so $A \cap B \in R$. Clearly if $A \cap B = \emptyset$ then $A \cup B \in R$ so R is closed under finite intersections and finite, disjoint unions.

Next we show that if E, F $\in R$ with E $\subset$ F then F - E $\in R$. To do this we start with the simplest case, namely, where F = $C_1 - D_1$ and E = $C_2 - D_2$ for compact sets C_1, D_1, C_2 and D_2 where $D_1 \subset C_1$ and $D_2 \subset C_2$. Now E $\subset$ F implies $[C_2 - D_2] \subset [C_1 - D_1]$ which in turn implies $C_2 \subset C_1$ and $[C_2 - D_2] \cap D_1 = \emptyset$. First suppose $x \in [C_1 - D_1] - [C_2 - D_2]$. Then $x \in [C_1 - D_1]$ and x is not in $[C_2 - D_2]$ which implies $x \in C_1$, x is not in D_1 and ($x \in C_2$ implies $x \in D_2$). Suppose x is not in $[C_1 - C_2 \cup D_1] \cup [D_2 - D_1]$. Then x is not in $[C_1 - C_2 \cup D_2]$ which implies $x \in C_2$ since x is not in D_1 and $x \in C_1$. But then $x \in D_2$ which implies $x \in D_2 - D_1$ which is a contradiction. Hence

$$[C_1 - D_1] - [C_2 - D_2] \subset [C_1 - C_2 \cup D_1] \cup [D_2 - D_1].$$

Conversely, assume $x \in [C_1 - C_2 \cup D_1] \cup [D_2 - D_1]$. Then $x \in [C_1 - C_2 \cup D_1]$ or $x \in [D_2 - D_1]$. If $x \in [C_1 - C_2 \cup D_1]$ then $x \in C_1$ and x is not in $C_2 \cup D_1$ which implies x is not in $C_2 \cup D_1$. Then $x \in [C_1 - D_1]$ and x is not in $[C_2 - D_2]$ because $D_2 \subset C_2$. Therefore, $x \in [C_1 - D_1] - [C_2 - D_2]$. On the other hand, if $x \in D_2 - D_1$ then $x \in D_2 \subset C_2$ and x is not in D_1. But since $C_2 \subset C_1$ we have $x \in C_1$ which implies $x \in [C_1 - D_1]$. Moreover, $x \in D_2$ which implies x is not in $[C_2 - D_2]$. Therefore, $x \in [C_1 - D_1] - [C_2 - D_2]$. Hence

$$[C_1 - D_1] - [C_2 - D_2] = [C_1 - C_2 \cup D_1] \cup [D_2 - D_1].$$

Since $D_2 \subset C_2$ we see that $[C_1 - C_2 \cup D_1] \cap [D_2 - D_1] = \emptyset$ so $[C_1 - C_2 \cup D_1] \cup [D_2 - D_1]$ is a finite, disjoint union of sets of the form C - D where C and D are compact and D $\subset$ C so $[C_1 - D_1] - [C_2 - D_2] \in R$.

Next consider the case where E = H - K for compact sets H and K with K $\subset$ H and F = $\cup_{n=1}^{N} [C_n - D_n]$ for compact sets C_n and D_n with $D_n \subset C_n$ for each n and $[C_m - D_m] \cap [C_n - D_n] = \emptyset$ whenever $m \neq n$. Now E $\subset$ F implies E $\subset \cup_{n=1}^{N} [C_n - D_n]$ which in turn implies E $\subset [C_k - D_k]$ for some k since the $[C_n - D_n]$'s are disjoint. But then as we just showed, $[C_k - D_k]$ - E $\in R$. Since

$$\text{F - E} = (\cup_{n \neq k} [C_n - D_n]) \cup ([C_k - D_k] - \text{E})$$

it is clear that F - E $\in R$.

Finally, we consider the most general case where E = $\cup_{i=1}^{M} [H_i - K_i]$ and F = $\cup_{n=1}^{N} [C_n - D_n]$. Since E $\subset$ F, each $[H_i - K_i] \subset$ F. As just shown, F - $[H_i - K_i] \in R$ for each i. But F - E = $\cap_{i=1}^{M}$(F - $[H_i - K_i]$) and since R is closed under finite intersections, we have F - E $\in R$.

We are now ready to show that R is a ring. Let A, B $\in R$. Now A - B = A - (A$\cap$B). Since R is closed under intersections, A$\cap$B $\in R$ and A$\cap$B $\subset$ A, so by what has been shown above, A - B $\in R$. Moreover, A$\cup$B = [A - B]$\cup$[A$\cap$B] is a disjoint union of members of R and since R is closed under disjoint unions, we have that A$\cup$B $\in R$. Hence R is a ring. ■

To prove the theorem showing that the outer regularity of λ^* on B_0 is sufficient to imply the regularity of λ^* on B_0, we need a set theoretic lemma that we now develop. A sequence $\{E_n\}$ is said to be **monotone** if it is either ascending or descending. The **limit** of $\{E_n\}$ is defined to be $\cup E_n$ if $\{E_n\}$ is ascending or $\cap E_n$ if $\{E_n\}$ is descending. The limit of $\{E_n\}$ is denoted by $lim_n E_n$. A collection of sets M is said to be a **monotone class** if for each monotone sequence $\{E_n\} \subset M$ we have $lim_n E_n \in M$. Clearly the class of all subsets of X is a monotone class and it is easily shown that the intersection of any collection of monotone classes is a monotone class, so there exists a smallest monotone class containing a given collection A of sets. We denote this smallest monotone class containing A by $M(A)$.

LEMMA 10.7 If R is a ring and $\Sigma(R)$ is the smallest σ-ring containing R, then $\Sigma(R) = M(R)$.

Proof: A σ-ring is a monotone class so $M(R) \subset \Sigma(R)$. Conversely, we will show that $M(R)$ is a σ-ring which will imply $\Sigma(R) \subset M(R)$. For this let S $\subset$ X and let M(S) be the collection of all sets T such that S - T, T - S and S$\cup$T all belong to $M(R)$. Then if $\{E_n\}$ is a monotone sequence in M(S) we have S - $lim_n E_n$ = lim_n(S - E_n) $\in M(R)$, $lim_n E_n$ - S = $lim_n(E_n$ - S) $\in M(R)$ and S$\cup$($lim_n E_n$) = lim_n(S$\cup E_n$) $\in M(R)$ so M(S) is also a monotone class if it is not empty.

Now if S, T $\in R$, then since R is a ring, T $\in M$(S). Since this holds for each T $\in R$ it follows that $R \subset M$(S). Since $M(R)$ is the smallest monotone class containing R, we must have $M(R) \subset M$(S). Therefore, if E $\in M(R)$ and S $\in R$ then E $\in M$(S) and hence S $\in M$(E). Since this holds for each S $\in R$ we have $R \subset M$(E) so we also must have $M(R) \subset M$(E). But $M(R) \subset M$(E) for each E $\in M(R)$ implies $M(R)$ is a ring.

To show $M(R)$ is a σ-ring, let $\{S_n\} \subset M(R)$. For each positive integer m put $E_m = \cup_{n=1}^{m} S_n$. Since $M(R)$ is a ring, each E_m is in $M(R)$. Now $\{E_m\}$ is an ascending sequence and $M(R)$ is a monotone class so $\cup S_n = \cup E_m = lim_m E_m \in M(R)$. Hence $M(R)$ is a σ-ring which establishes the lemma. ∎

THEOREM 10.9 Let μ be a Baire measure such that $\mu(C) < \infty$ for each compact C. Then either the outer regularity of μ on each compact set or the inner regularity of μ on each bounded open set is a necessary and sufficient condition for μ to be regular.

Proof: The necessity of either of these conditions is obvious. To demonstrate the sufficiency of either of these conditions, it will suffice to show that μ is regular on each bounded Baire set since each Baire set is σ-bounded by Proposition 10.10 and hence is the union of an ascending sequence of bounded Baire sets. Then by Theorem 10.6, the regularity of μ on the members of this sequence will imply the regularity of μ on the union.

Since both the conditions of the hypothesis are equivalent by Theorem 10.7, to prove the sufficiency, we assume them both. Let E be a bounded Baire set and C a compact set such that $E \subset C$. It is left as an exercise (Exercise 5) to show that $B_0 \cap C = \{B \cap C \mid B \in B_0\}$ is the smallest σ-ring containing all the compact subsets of C. Hence $B_0 \cap C$ is the Baire ring on the compact space C. By the previous theorem applied to the space C, we see that $B_0 \cap C$ is the smallest σ-ring containing the ring R of all finite, disjoint unions of sets of the form A - B where A and B are compact subsets of C with $B \subset A$. By Propositions 10.11 and 10.13 we see that μ is outer regular on every set in R. By Propositions 10.12 and 10.13 we see that μ is inner regular on every set in R.

Let M be the collection of sets S in $B_0 \cap C$ such that μ is both outer regular and inner regular on S. By Theorem 10.6 and the fact that $\mu(E) < \infty$ for each $S \subset C$ we see that M is a monotone class. Also, $R \subset M$, so $M(R) \subset M$. By Lemma 10.7, $M(R) = \Sigma(R)$. But by Theorem 10.8, $\Sigma(R) = B_0 \cap C$. Hence $B_0 \cap C \subset M$ which implies μ is both outer regular and inner regular on each $S \in B_0 \cap S$. In particular, μ is regular on E. Consequently, μ is regular. ∎

THEOREM 10.10 λ^* is a regular, almost uniform Baire measure on X.

Proof: That λ^* is a Borel (and hence Baire) measure on X follows from Propositions 10.7 and 10.9. That λ^* is regular on B_0 follows from the fact that λ^* is outer regular on all Baire sets (see first paragraph of Proposition 10.7), the fact $\lambda^*(C) < \infty$ for compact sets C, and Theorem 10.9. That λ^* is almost uniform follows from Theorem 10.4, Corollary 10.3 and the definition of an almost uniform measure. ∎

THEOREM 10.11 If m is another regular almost uniform Baire measure on X then m is a constant multiple of λ^* on the Baire sets of X.

Proof: If $U \in G$ then $U \in B_0$ so m is defined on G. Clearly m satisfies L1 - L3. To show m satisfies L4 let $U \in G$. Since m is regular there exists an ascending sequence $\{C_n\}$ of compact sets with $m(U) = lim_n m(C_n)$. Now for each n there exists an open $V_n \in G$ with $C_n \subset V_n \subset U$. But then $V_n \in G$ for each n. Moreover, $m(U) = lim_n m(V_n)$ so m satisfies L4. Hence m is a Loomis content on G.

Now by the definition of an almost uniform measure, it is clear that m, considered as a Loomis content on G, is invariant on compact spheres, so by Theorem 10.4, m is a constant multiple of λ on $Z(G)$. Then an argument similar to one of the arguments in Theorem 10.5 can be used to show that m is a constant multiple of λ on G. Since the members of G have compact closures, by Proposition 10.8, $\lambda^*(U) = \lambda(U)$ for each $U \in G$. Therefore, m is a constant multiple of λ^* on G.

Consequently, an argument similar to the last part of the proof of Theorem 10.5 can be used to show that m is a constant multiple of λ^* on the Baire sets of X. ■

EXERCISES

1. Show that if C and D are compact subsets of a locally compact uniform space then C and D are uniformly separated.

2. Complete the proof of Proposition 10.10.

3. Prove Proposition 10.12.

4. Prove Proposition 10.13.

5. Show that the σ-ring $B_0 \cap C = \{B \cap C \mid B \in B_0\}$ for some compact set C is the smallest σ-ring containing all the compact subsets of C. Remember B_0 denotes the Baire ring.

6. Show that $\lambda^*(C) < \infty$ for each compact set C.

RESEARCH PROBLEMS

7. If a locally compact uniform space satisfies axioms A1 - A4, does there exist a regular uniform measure on the Baire ring (as opposed to an almost uniform measure)?

8. What is a necessary and sufficient condition for a locally compact uniform space to have a regular Baire (almost regular Borel) uniform measure?

9. If a locally compact uniform space has a regular Baire (almost regular Borel) uniform measure, is it unique? If not, what is a necessary and sufficient condition for it to be unique?

10. Are there interesting or useful necessary and sufficient conditions for arbitrary (not necessarily locally compact) uniform spaces to have Baire (Borel) uniform measures?

11. Can the construction in this chapter be made to yield a regular Baire (almost regular Borel) uniform measure?

Chapter 11

SPACES OF FUNCTIONS

11.1 L^p-spaces

In this chapter, we will be investigating several types of spaces constructed from functions on a given space X to some other space Y. The spaces X and Y will usually be uniform spaces, but the theory of function spaces is more general than that. Much of this chapter, is independent of uniform spaces. The spaces we will study first will be needed in the next chapter to develop the concept of uniform differentiation. In fact, much of the content of this chapter and the next is needed simply to develop the machinery from classical analysis so that we can discuss uniform differentiation in isogeneous uniform spaces.

We will begin by allowing X and Y to be topological spaces and add special constraints as we proceed. Only when we get to the end of this chapter will we consider abstract uniform function spaces. Some of the function spaces we will consider are special cases of more general classes of spaces, but much of the motivation for studying these spaces seems to have been the function spaces themselves. Examples of this are the Hilbert spaces and Banach spaces that we will introduce later in this chapter. The first function spaces we will consider are the so-called L^p-spaces. To facilitate their study we introduce some results about *convex* functions. A real valued function f defined on an open interval (a, b), where $-\infty \leq a < b \leq \infty$, is said to be **convex** if the inequality

$$f((1-c)x + cy) \leq (1-c)f(x) + cf(y)$$

holds whenever $x, y \in (a, b)$ and $0 \leq c \leq 1$. The intuitive idea behind this inequality is that whenever $x < z < y$, the point $(z, f(z))$ lies on or below the line segment connecting $(x, f(x))$ and $(y, f(y))$ in the plane. This inequality is equivalent to another, namely,

$$\text{(11.1)} \qquad \frac{f(z)-f(x)}{z-x} \leq \frac{f(y)-f(z)}{y-z}$$

whenever $a < x < z < y < b$. Applying the Mean Value Theorem from Calculus to (11.1) shows that a real valued differentiable function f is convex in (a, b) if and only if $a < x < y < b$ implies $f'(x) \leq f'(y)$ or, in other words, if the derivative f' is a monotonically increasing function.

PROPOSITION 11.1 A convex function on (a, b) is continuous on (a,b).

The proof of Proposition 11.1 is left as an exercise. If p and q are positive real numbers such that $pq = p + q$ we say p and q are **conjugate exponents.** Note that this is equivalent to $1/p + 1/q = 1$. Also note that as p approaches 1 this forces q to approach ∞. For this reason we consider 1 and ∞ to be conjugate exponents. In the next theorem, the first inequality is known as *Hölder's Inequality* and the second as *Minkowski's Inequality.*

THEOREM 11.1 Let p and q be conjugate exponents with $1 < p < \infty$. Let μ be a positive measure on X and let f and g be non-negative extended real valued measurable functions on X. Then

$$\int_X fg\,d\mu \leq \{\int_X f^p\,d\mu\}^{1/p}\{\int_X g^q\,d\mu\}^{1/q}$$

and

$$\{\int_X (f+g)^p\,d\mu\}^{1/p} \leq \{\int_X f^p\,d\mu\}^{1/p} + \{\int_X g^p\,d\mu\}^{1/p}.$$

Proof: Put $\alpha = \{\int_X f^p\,d\mu\}^{1/p}$ and $\beta = \{\int_X g^q\,d\mu\}^{1/q}$. If $\alpha = 0$ the $fg = 0$ a. e. on X by Exercise 1 in Section 8.7. Consequently, Hölder's Inequality would hold. Also, if $\alpha > 0$ but $\beta = \infty$, Hölder's Inequality would hold. Therefore, if suffices to consider the case where $0 < \alpha < \infty$ and $0 < \beta < \infty$. Define the functions h and k by $h = f/\alpha$ and $k = g/\beta$. Then

$$\int_X h^p\,d\mu = \frac{1}{\alpha^p}\int_X f^p\,d\mu = \frac{\int_X f^p\,d\mu}{\int_X f^p\,d\mu} = 1.$$

Similarly, $\int_X k^q\,d\mu = 1$. If $x \in X$ such that $0 < h(x) < \infty$ and $0 < k(x) < \infty$, there exist real numbers a and b such that $h(x) = e^{a/p}$ and $k(x) = e^{b/q}$. Then since $1/p + 1/q = 1$, the convexity of the exponential function implies:

$$e^{a/p+b/q} \leq p^{-1}e^a + q^{-1}e^b.$$

To see this use the substitution $(1 - \lambda) = 1/p$ and $\lambda = 1/q$. Then

$$exp((1-\lambda)a + \lambda b) = (1-\lambda)exp(a) + \lambda exp(b) = p^{-1}e^a + q^{-1}e^b.$$

But then $h(x)k(x) \leq p^{-1}h(x)^p + q^{-1}k(x)^q$ for each $x \in X$ with $0 < h(x) < \infty$ and $0 < k(x) < \infty$. Now if $h(x) = 0$ or $h(x) > 0$ and $k(x) = \infty$, it is easily seen that the inequality still holds. Consequently, it holds for each $x \in X$. Integration of both sides then yields

$$\int_X hk\,d\mu \leq p^{-1} + q^{-1} = 1.$$

But then $(1/\alpha\beta)\int_X fg d\mu \le 1$ which implies $\int_X fg d\mu \le \alpha\beta$ which yields Hölder's Inequality.

To prove Minkowski's Inequality we first observe that it suffices to prove it for the case where the left hand side is greater than zero and the right hand side is less than ∞. For this we note that

$$(f+g)^p = f(f+g)^{p-1} + g(f+g)^{p-1} \text{ so}$$

$$\int_X (f+g)^p d\mu = \int_X f(f+g)^{p-1} d\mu + \int_X g(f+g)^{p-1} d\mu.$$

Hölder's Inequality implies that

$$\int_X f(f+g)^{p-1} d\mu \le \{\int_X f^p d\mu\}^{1/p}\{\int_X (f+g)^{(p-1)q} d\mu\}^{1/q} \text{ and}$$

$$\int_X g(f+g)^{p-1} d\mu \le \{\int_X g^p d\mu\}^{1/p}\{\int_X (f+g)^{(p-1)q} d\mu\}^{1/q}.$$

Hence, since $(p-1)q = p$ we have

$$\int_X (f+g)^p d\mu \le [\{\int_X f^p d\mu\}^{1/p} + \{\int_X g^p d\mu\}^{1/p}]\{\int_X (f+g)^p d\mu\}^{1/q}.$$

Dividing both sides of this inequality by $\{\int_X (f+g)^p d\mu\}^{1/q}$ yields

$$\{\int_X (f+g)^p d\mu\}^{1/p} \le \{\int_X f^p d\mu\}^{1/p} + \{\int_X g^p d\mu\}^{1/p}$$

since $1 - 1/q = 1/p$. This is Minkowski's Inequality. ∎

If μ is a positive measure on X, $0 < p < \infty$ and if f is a complex measurable function on X, we define

$$|f|_p = \{\int_X |f|^p d\mu\}^{1/p}$$

and let $L^p(\mu)$ denote the collection of all such functions f such that $|f|_p < \infty$. We say $|f|_p$ is the **L^p-norm** of f. If X is a locally compact isogeneous uniform space and μ is Haar measure on X then we write $L^p(X)$ instead of $L^p(\mu)$.

If μ is a positive measure on X and f:X $\to [0, \infty]$ is a measurable function, put A = $\{a \in \mathbf{R} \mid \mu(f^{-1}((a, \infty])) = 0\}$. If A = $\varnothing$ let $b = \infty$. Otherwise put b = *inf*A. We say that b is the **essential supremum** of f. If f is a complex measurable function on X, we define $|f|_\infty$ to be the essential supremum of f. We denote the collection of all complex measurable functions f such that $|f|_\infty < \infty$ by $L^\infty(\mu)$. The members of $L^\infty(\mu)$ are said to be **essentially bounded** functions. Clearly $|f(x)| \le c$ for almost every $x \in$ X if and only if $c \ge |f|_\infty$.

PROPOSITION 11.2 If μ is a positive measure on X, p and q are conjugate exponents, $1 \le p \le \infty$, and if $f \in L^p(\mu)$ and $g \in L^q(\mu)$, then $fg \in L^1(\mu)$ and $|fg|_1 \le |f|_p |g|_q$.

Proof: If $1 < p < \infty$ then the desired inequality is just Hölder's Inequality applied to f and g. If $p = \infty$, then $|f(x)g(x)| \le |f|_\infty |g(x)|$ for almost all $x \in X$ by the definition of $|f|_\infty$. If $p = \infty$ then $q = 1$, so integrating both sides of this inequality yields:

$$|fg|_1 \le |f|_\infty \int_X |g|\, d\mu = |f|_p |g|_q.$$

Finally, if $p = 1$ then $q = \infty$, so an argument similar to the above again produces the desired inequality. ■

PROPOSITION 11.3 If μ is a positive measure on X, $1 \le p \le \infty$, and $f \in L^p(\mu)$ and $g \in L^p(\mu)$, then $f + g \in L^p(\mu)$ and

$$|f+g|_p \le |f|_p |g|_p.$$

The proof of this proposition is left as an exercise. Proposition 11.3 together with the fact that if $1 \le p \le \infty$ and c is a complex number, implies that $|cf|_p = |c|\,|f|_p$ so $cf \in L^p(\mu)$ which implies $L^p(\mu)$ is a complex vector space.

In fact, $L^p(\mu)$ is also a pseudo-metric space. If we define the distance between f and g by $d(f, g) = |f - g|_p$ for any pair of functions $f, g \in L^p(\mu)$, then it is easily shown that d is a pseudo-metric on $L^p(\mu)$. As shown in Chapter 1, a metric space can be constructed from this pseudo-metric space as follows. Define $f \sim g$ if and only if $d(f, g) = 0$. Clearly $\sim$ is an equivalence relation on $L^p(\mu)$. If $[f]$ and $[g]$ are two equivalence classes containing f and g respectively and if $f' \in [f]$ and $g' \in [g]$ then $d(f, g) = d(f', g')$ so that $d([f], [g])$ is well defined on the equivalence classes of $L^p(\mu)$ with respect to $\sim$. Hence d is a metric space when considered as being defined on the equivalence classes of $L^p(\mu)$ with respect to $\sim$.

Furthermore, this metric space is also a complex vector space since if $f' \in [f]$ and $g' \in [g]$ then $f' + g' \in [f + g]$ and $cf' \in [cf]$. It is customary to regard $L^p(\mu)$ as a metric space even though it is the quotient space $L^p(\mu)/\sim$ that is really the metric space. When this is done, it is with the understanding that $f \in L^p(\mu)$ really means that f is a representative of the equivalence class $[f]$ to which it belongs. The following theorem is especially significant as we will soon discover.

THEOREM 11.2 For a positive measure μ and for $1 \le p \le \infty$, the metric space $L^p(\mu)$ is complete.

Proof: First assume that $p < \infty$. Let $\{f_n\}$ be a Cauchy sequence in $L^p(\mu)$. Then there exists a subsequence $\{f_{n_i}\}$ such that $|f_{n_{i+1}} - f_{n_i}|_p < 2^{-i}$ for each positive integer i. Define the functions g and g_k for each positive integer k as follows: Put

$$g_k(x) = \Sigma_{i=1}^{k} |f_{n_{i+1}}(x) - f_{n_i}(x)| \text{ and } g(x) = \Sigma_{i=1}^{\infty} |f_{n_{i+1}}(x) - f_{n_i}(x)|.$$

Then for each positive integer k,

$$|g_k|_p = [\int_X |\Sigma_{i=1}^{k} |f_{n_{i+1}} - f_{n_i}||^p d\mu]^{1/p} = [\int_X (\Sigma_{i=1}^{k} |f_{n_{i+1}} - f_{n_i}|)^p d\mu]^{1/p}.$$

Applying Minkowski's Inequality inductively to the term on the right produces:

$$|g_k|_p \leq \Sigma_{i=1}^{k} [\int_X |f_{n_{i+1}} - f_{n_i}|^p d\mu]^{1/p} = \Sigma_{i=1}^{k} |f_{n_{i+1}} - f_{n_i}|_p < \Sigma_{i=1}^{k} 2^i < 1.$$

For each $x \in X$ we have:

$$g_k^p(x) = \Sigma_{i=1}^{k} |f_{n_{i+1}}(x) - f_{n_i}(x)|^p \leq \Sigma_{i=1}^{k+1} |f_{n_{i+1}}(x) - f_{n_i}(x)|^p = g_{k+1}^p(x)$$

so $\{g_k^p\}$ is an ascending sequence of real valued functions. Moreover, $g^p(x) = \liminf_k g_k^p(x)$ for each $x \in X$, so by Theorem 8.11,

$$\int_X (\liminf g_k^p) d\mu \leq \liminf \int_X g_k^p d\mu$$

which implies:

$$\int_X g^p d\mu \leq \liminf \int_X (\Sigma_{i=1}^{k} |f_{n_{i+1}} - f_{n_i}|)^p d\mu.$$

Therefore,

$$[\int_X (\Sigma_{i=1}^{\infty} |f_{n_{i+1}} - f_{n_i}|)^p d\mu]^{1/p} \leq [\liminf \int_X (\Sigma_{i=1}^{k} |f_{n_{i+1}} - f_{n_i}|)^p d\mu]^{1/p}$$

so $|g|_p \leq \liminf |g_k|_p \leq 1$. But then $g(x) < \infty$ a.e. so the series

$$f_{n_1}(x) + \Sigma_{i=1}^{\infty} (f_{n_{i+1}}(x) - f_{n_i}(x))$$

converges absolutely a.e. on X. Put $f(x) = f_{n_i}(x) + \Sigma_{i=1}^{\infty} (f_{n_{i+1}}(x) - f_{n_i}(x))$ for each $x \in X$ where the series converges absolutely and put $f(x) = 0$ otherwise. Since $f_{n_i} + \Sigma_{i=1}^{k-1} (f_{n_{i+1}} - f_{n_i}) = f_{n_k}$ for each positive integer k, it is clear that $f(x) = \lim_{i \to \infty} f_{n_i}(x)$ a.e. on X.

Even though f is the pointwise limit a.e. of $\{f_n\}$, it is conceivable that f is not the limit of $\{f_n\}$ with respect to the L^p-norm $|\ |_p$ or that f is not in $L^p(\mu)$. It remains to show that this is not the case. For this let $\varepsilon > 0$. Then there exists a positive integer k such that $|f_m - f_n| < \varepsilon$ whenever $m, n > k$. For each $m > k$,

Theorem 8.11 can be used to show that:

$$\int_X \liminf_{i\to\infty} |f_{n_i}(x) - f_m(x)|^p d\mu \leq \liminf_{i\to\infty} \int_X |f_{n_i}(x) - f_m(x)|^p d\mu \leq \varepsilon^p.$$

But $\liminf_{i\to\infty} |f_{n_i}(x) - f_m(x)|^p = |f(x) - f_m(x)|^p$ so $\int_X |f - f_m|^p d\mu \leq \varepsilon^p$. Hence $|f-f_m|_p \leq \varepsilon^p$ so $f - f_m \in L^p(\mu)$ which implies $f \in L^p(\mu)$ since $f = f_m + (f - f_m)$. Moreover, we have shown that for each $\varepsilon > 0$, there exists a positive integer k such that if $m > k$ then $|f - f_m|_p \leq \varepsilon^p$ so f is the limit of $\{f_n\}$ with respect to the L^p-norm.

The proof that the theorem holds when $p = \infty$ is considerably easier and is left as an exercise. ∎

COROLLARY 11.1 If $\{f_n\}$ is a Cauchy sequence in $L^p(\mu)$, where $1 \leq p \leq \infty$, and if f is the limit of $\{f_n\}$, then $\{f_n\}$ has a subsequence that converges pointwise almost everywhere to f.

The proof of the corollary is contained in the proof of the theorem. Let S_M denote the class of complex valued, simple measurable functions on X whose support has finite measure. Then the members of $L^p(\mu)$ can be approximated by members of S_M for $1 \leq p < \infty$. In fact:

PROPOSITION 11.4 S_M is dense in $L^p(\mu)$ for $1 \leq p < \infty$.

Proof: Clearly $S_M \subset L^p(\mu)$. First suppose $f \geq 0$ is in $L^p(\mu)$ and let $\{s_n\}$ be a sequence of simple measurable functions with $0 \leq s_1 \leq s_2 \leq \ldots \leq f$ and such that $\{s_n(x)\}$ converges to $f(x)$ for each $x \in$ X. That such a sequence exists follows from Theorem 8.7. Since $0 \leq s_n \leq f$ for each positive integer n, we have that $s_n \in L^p(\mu)$ for each n and hence $s_n \in S_M$ for each n.

Now $\int_X |f|^p d\mu < \infty$ since $f \in L^p(\mu)$. But $f \geq 0$ implies $|f|^p = f^p = |f^p|$ so $\int_X |f^p| d\mu < \infty$ which implies $f^p \in L^1(\mu)$. Furthermore, $|f - s_n|^p \leq f^p$ for each n and $\{(f - s_n)^p\}$ converges to 0 for each $x \in$ X. Then by Theorem 8.13,

$$\lim_{n\to\infty} \int_X |(f - s_n)^p - 0| d\mu = 0$$

which implies $\lim_{n\to\infty} \int_X |f - s_n|^p d\mu = 0$ which in turn implies $\lim_{n\to\infty} |f - s_n|_p = 0$ so $\{s_n\}$ converges to f in the L^p-norm. Consequently, S_M is dense in $L^p(\mu)$. ∎

Now let us consider a more specialized class of measure spaces that includes the locally compact isogeneous uniform spaces equipped with Haar measure. Specifically, for the remainder of this section we assume X is locally compact and that μ is a positive, complete, almost regular Borel measure that is finite on compact sets (i.e., the type of measure obtained by applying the Reisz Representation Theorem to a positive linear functional on $C_K(X)$). In such measure spaces we can approximate members of $L^p(\mu)$, for $1 \leq p < \infty$ with sequences of continuous functions in $C_K(X)$. In fact, for such spaces, we will show that $C_K(X)$ is dense in $L^p(\mu)$ for $1 \le p < \infty$, with respect to the L^p-norm.

To do this, we first need a result known in the literature as *Lusin's Theorem.* It is this result that requires the additional assumptions. Fortunately the Haar measures on locally compact uniform spaces satisfy most of the restrictions of classical analysis necessary to build the theory of real analysis on locally compact metric spaces. This is why, once we have Haar measure on this class of uniform spaces we can move the theory of differentiation of a measure to locally compact, isogeneous uniform spaces.

THEOREM 11.3 If f is a complex measurable function on a locally compact space X and μ is a positive, complete, almost regular, Borel measure that is finite on compact sets, and if μ(E) < ∞, and f is zero outside of E, then for each ε > 0 there exists a g ∈ $C_K(X)$ such that the measure of the set on which f and g are not equal is less than ε. Moreover, we can pick g such that sup|g(x)| ≤ sup|f(x)|.

Proof: First assume E is compact and $0 \le f(x) < 1$ for each $x \in$ X. From the proof of Theorem 8.7, there exists a sequence $\{s_n\}$ of simple measurable functions with $0 \le s_1 \le s_2 \le \ldots f$ such that $lim s_n(x) = f(x)$ for each $x \in$ X. Moreover, from the construction in the proof, the s_n have the form

$$s_n = \Sigma_{i=1}^{n2^n} \frac{i-1}{2^n} \chi_{E_{n_i}} + n\chi_{F_n}$$

where $E_{n_i} = \{x \mid (i-1)/2^n \le f(x) < i/2^n\}$ for each positive integer $i = 1 \ldots n2^n$ and $F_n = \{x \mid n \le f(x) \le \infty\}$. Since we assumed $f < 1$ we have $F_n = \emptyset$ for each n and $E_{n_i} = \emptyset$ for $i > 2^n$ so s_n simplifies to

$$s_n = \Sigma_{i=1}^{2^n} \frac{i-1}{2^n} \chi_{E_{n_i}}$$

for each n. Put $\sigma_1 = s_1$ and for each positive integer n let $\sigma_{n+1} = s_{n+1} - s_n$. Then

$$f(x) = \Sigma_{n=1}^{\infty} \sigma_n(x) \text{ for each } x \in X.$$

By inspection, it is easily seen that for each $n > 1$, $s_n - s_{n-1}$ is a simple measurable function that equals 2^{-n} on the set of points where $f - s_{n-1} \ge 2^n$ and 0 otherwise. Consequently, $2^n\sigma_n$ is the characteristic function of this set which we denote by E_n. Then $E_n \subset$ E for each n. Let U be an open set containing E such that Cl(U) is compact. Then for each n, there exists a compact set K_n and an open set U_n with $K_n \subset E_n \subset U_n \subset$ U and $\mu(U_n - K_n) < \varepsilon/2^n$. By Theorem 8.5, there exist functions $g_n \in C_K(X)$ with $0 \le g_n \le 1$ such that $K_n \prec g_n \prec U_n$. Let

$$g(x) = \Sigma_{n=1}^{\infty} g_n(x)/2^n \text{ for each } x \in X.$$

Clearly this series converges uniformly on X. An easy modification to Theorem 1.12 implies g is continuous. Since $\sigma_n = 2^{-n}h_n$ on K_n we have $g = f$ except on $\cup(U_n - K_n)$ which is a set of measure less than ε. Thus we have proved the existence part of the theorem for the case where E is compact and $0 \leq f < 1$.

Consequently, it is easily shown that the existence part holds if E is compact and f is merely a bounded complex measurable function. Furthermore, we can remove our compactness assumption, for if $\mu(E) < \infty$, there exists a compact $K \subset E$ with $\mu(E - K) < \varepsilon$ for any $\varepsilon > 0$. To prove the general case, let f be a complex measurable function and for each positive integer n put $F_n = \{x \mid |f(x)| > n\}$. Then each F_n is measurable and $\cap F_n = \emptyset$. Consequently, $lim_n \mu(F_n) = 0$ by Proposition 8.7. For each positive integer n put $E_n = X - F_n$. Then f is a bounded function on E_n. For each n define f_n by $f_n(x) = f(x)$ if $x \in E_n$ and $f_n(x) = 0$ otherwise. Then f is a bounded measurable function on X and $\mu(\{x \mid f(x) \neq f_n(x)\}) \leq \mu(F_n)$. Now let $\varepsilon > 0$ and pick k with $\mu(F_k) < \varepsilon/2$. Then f_k is a bounded function on X and $\mu(\{x \mid f(x) \neq f_n(x)\}) < \varepsilon/2$. Then there exists a $g \in C_K(X)$ with $\mu(\{x \mid g(x) \neq f_n(x)\}) < \varepsilon/2$ so $\mu(\{x \mid g(x) \neq f(x)\}) < \varepsilon$. This proves the existence part of the proof.

To conclude the proof we first observe that if $sup_x|f(x)| = \infty$ then $sup_x|g(x)| \leq sup_x|f(x)|$ so assume $sup_x|f(x)| < \infty$. Put $b = sup_x|f(x)|$ and let $h(x) = x$ if $|x| \leq b$ and $bx/|x|$ otherwise. Then h is a continuous function from C onto $Cl(S(0, b))$. If $g' \in C_K(X)$ such that $f(x) = g'(x)$ except on a set of measure less than ε and $g = h © g'$ then $f(x) = g(x)$ except on a set of measure less than ε and $sup_x|g(x)| \leq sup_x|f(x)|$. This concludes the proof. ■

THEOREM 11.4 If X is a locally compact space and μ is a positive, complete, almost regular Borel measure on X then $C_K(X)$ is dense in $L^p(\mu)$ for $1 \leq p < \infty$.

Proof: Let S_M be the class of complex valued, simple measurable functions on X whose support has finite measure. If $s \in S_M$ and $\varepsilon > 0$, then by Theorem 11.3, there exists a $g \in C_K(X)$ with $g(x) = s(x)$ except on a set of measure less than ε and $|g| \leq sup_x|s(x)| = |s|_\infty$. Let $E = \{x \mid g(x) \neq s(x)\}$. Then

$$|g - s|_p = [\int_E |g - s|^p d\mu]^{1/p} \leq [\int_E |2|s|_\infty|^p d\mu]^{1/p} = 2|s|_\infty \varepsilon^{1/p}.$$

Therefore, it is possible to find a sequence $\{g_n\} \subset C_K(X)$ such that $\{g_n\}$ converges to s with respect to the L^p-norm metric. But then since S_M is dense in $L^p(\mu)$ by Proposition 11.3, so is $C_K(X)$. ■

Theorems 11.2 and 11.4 together say that for $1 \leq p < \infty$, $L^p(\mu)$ is the completion of $C_K(X)$ with respect to the L^p-norm metric. The case where $p = \infty$ is different because the definition of $L^\infty(\mu)$ is essentially different than the definition of $L^p(\mu)$ for $p < \infty$. Now $C_K(X)$ is a metric space with respect to the L^∞-norm metric. To characterize the completion of $C_K(X)$ with respect to this metric, we need the following definition: A complex valued function f on X is said to **vanish at infinity** if for each $\varepsilon > 0$ there exists a compact set K such that $|f(x)| < \varepsilon$ for each $x \in X - K$. The class of all continuous, complex valued functions that vanish at infinity is denoted by $C_\infty(X)$. On $C_K(X)$ the L^∞-norm coincides with another norm called the **supremum norm** that is defined by $|f| = sup_x|f(x)|$.

THEOREM 11.5 If X is a locally compact space then $C_\infty(X)$ is the completion of $C_K(X)$ with respect to the L^∞-norm and the supremum norm.

Proof: It is left as an exercise to show that $C_\infty(X)$ is, indeed, a metric space with respect to the L^∞-norm. We need to show that $C_K(X)$ is dense in $C_\infty(X)$ and that $C_\infty(X)$ is complete. That $C_K(X)$ is dense in $C_\infty(X)$ follows from the fact that if $f \in C_\infty(X)$ and $\varepsilon > 0$, then there exists a compact set K with $|f(x)| < \varepsilon$ outside K and an $h \in C_K(X)$ with $0 \le h \le 1$ and $h(K) = 1$. If we put $g = fh$ then clearly $g \in C_K(X)$ and $|f - g|_\infty = sup_x|f(x) - g(x)| < \varepsilon$.

To show $C_\infty(X)$ is complete let $\{f_n\}$ be a Cauchy sequence in $C_\infty(X)$. For each $\varepsilon > 0$, there exists a positive integer N such that if $m, n > N$ then $|f_m - f_n| < \varepsilon$ so $sup_x|f_m(x) - f_n(x)| < \varepsilon$. Then for each $x \in X$, the sequence $\{f_n(x)\}$ is a Cauchy sequence in $\mathbb{C}$ and hence converges to some point $f(x) \in \mathbb{C}$. This pointwise limit function defined by $f(x) = lim_n f_n(x)$ is well defined. Moreover, $\{f_n\}$ converges uniformly to f since for each $\varepsilon > 0$ there exists a positive integer N such that if $m, n > N$ then $sup_x|f_m(x) - f_n(x)| < \varepsilon/2$. Also, for each $x \in X$, there exists a $k > N$ with $|f_k(x) - f(x)| < \varepsilon/2$. But then if $m > N$, $|f_m(x) - f_k(x)| < \varepsilon/2$ so $|f_m(x) - f(x)| < \varepsilon$ for each $x \in X$. Now an easy modification of Theorem 8.12 shows that f is continuous.

It remains to show that $f \in C_\infty(X)$. For this let $\varepsilon > 0$. Then there exists n with $|f - f_n|_\infty < \varepsilon/2$ and there exists a compact set K such that $|f_n(x)| < \varepsilon/2$ outside K. But then $|f(x)| < \varepsilon$ outside K so $f \in C_\infty(X)$. ∎

EXERCISES

1. Prove Proposition 11.1.

2. Prove Proposition 11.3.

3. Show that d defined by $d(f, g) = |f - g|_p$ is a pseudo-metric on $L^p(\mu)$.

4. Prove Theorem 11.2 for the case $p = \infty$.

5. (Jensen's Inequality) If μ is a positive measure on a σ-algebra M in a space

X such that $\mu(X) = 1$ and if g is a real valued function in $L^1(\mu)$ with $a < g(x) < b$ for each $x \in X$, and if f is convex on (a, b), show that:

$$f\left(\int_X g\,d\mu\right) \leq \int_X (f \circledcirc g)\,d\mu.$$

6. If $\alpha_i > 0$ for each $i = 1 \ldots n$ such that $\Sigma\alpha_i = 1$ and if x_i is a real number for each $i = 1 \ldots n$, show that $e(\Sigma\alpha_i x_i) \leq \Sigma\alpha_i e^{x_i}$ where e is the exponential function.

7. Show that if X is a locally compact space and μ is a positive, complete, almost regular Borel measure on X and if the distance $d(f, g)$ between two functions $f, g \in C_K(X)$ is defined by $d(f, g) = \int_X |f - g|\,d\mu$ then $(C_K(X), d)$ is a metric space and the completion of $(C_K(X), d)$ is precisely the class of Lebesgue integrable functions on X. [Recall that the Lebesgue integrable functions were defined in Exercise 1 of Section 8.6.]

8. Show that if $p < q$ that for some measures μ, $L^p(\mu) \subset L^q(\mu)$ whereas for other measures $L^q(\mu) \subset L^p(\mu)$, and that there are some measures such that neither $L^p(\mu)$ or $L^q(\mu)$ contains the other. What are the conditions on μ for which these situations occur?

11.2 The Space $L^2(\mu)$ and Hilbert Spaces

The space $L^2(\mu)$ is known as the space of **square integrable** functions. It plays a major role in modern physics and in many other mathematical applications. In fact, it is the mathematical model that underlies the wave interpretation of quantum mechanics, when X is Euclidean space and μ is Lebesgue measure on X. $L^2(\mu)$ is a special case of a more general class of spaces called the *Hilbert spaces*. Other Hilbert spaces also play important roles in quantum mechanics, in fact, they are the mathematical models behind all quantum phenomenon. We will need some results about Hilbert spaces for our development of uniform differentiation in the next chapter. Hilbert spaces derive their name from David Hilbert who published a series of six papers between 1904 and 1910 titled *Grundzuge einer allgemeinen Theorie der linearen Integralgleichungen I - VI* in Nachr. Akad. Wiss. Gottingen Math.- Phys. Kl. that involved these spaces. They were republished in book form by Teubner Verlagsgesellschaft, Leipzig in 1912 and reprinted by Chelsea Publishing Co., New York in 1952. We now define these spaces.

A complex vector space H is called an **inner product space** if for each pair of vectors $u, v \in H$ there is a complex number (u, v) called the **inner product** or sometimes the *scalar product* or *dot product* that satisfies the following axioms:

(1) $(u, v) = (v, u)^*$ (the * representing complex conjugation).
(2) $(u + v, w) = (u, w) + (v, w)$ for $u, v, w \in H$.
(3) $(cu, v) = c(u, v)$ for $u, v \in H$ and $c \in \mathbf{C}$.
(4) $(u, u) \geq 0$ for each $u \in H$.
(5) $(u, u) = 0$ if and only if u is the zero vector in H.

There are a number of observations we can make about these axioms. First, (3) implies $(0, x) = 0$ for each $x \in H$ and (1) and (3) together imply $(x, cy) = c^*(x, y)$ for each pair $x, y \in H$ and $c \in \mathbf{C}$. Next we observe that (2) and (3) together imply that for each $y \in H$, the mapping defined by $\lambda(x) = (x, y)$ for each $x \in H$ is a linear functional on H. (1) and (2) can be combined to show that $(x, y + z) = (x, y) + (x, z)$ for $x, y, z \in H$. Finally, by (4) we can define a norm $|x|$ for each $x \in H$ by $|x| = (x, x)^{1/2}$ so that $|x|^2 = (x, x)$.

PROPOSITION 11.5 (Schwarz Inequality) For each $x, y \in H$, $|(x, y)| \leq |x|\,|y|$ where the norm on the left is the modulus of the complex number (x,y).

Proof: If $x = 0$ or $y = 0$ then $|(x, y)| \leq |x|\,|y|$ so assume $x \neq 0 \neq y$. Let α be an arbitrary complex number. Then $(x + \alpha y, x + \alpha y) \geq 0$ and $(x + \alpha y, x + \alpha y) = |x|^2 + |\alpha|^2|y|^2 + \alpha(y, x) + \alpha^*(x, y)$ and $\alpha(y, x) + \alpha^*(x, y) = 2Re(\alpha(y, x))$ so

$$|x|^2 + |\alpha|^2|y|^3 + 2Re(\alpha(y, x)) \geq 0.$$

Now each complex number α can be represented by $\alpha = re^{it}$ for some real number $r \geq 0$ and some complex number e^{it} for some real number t. Recall that $|e^{it}| = 1$. Similarly, $(y, x) = |(y, x)|e^{is}$ for some real number s. Hence $Re(\alpha(y,x)) = Re(re^{it}|(y, x)|e^{is}) = Re(r|(y, x)|e^{i(s+t)}) = r|(y, x)|Re(e^{i(s+t)})$ and $Re(e^{i(s+t)}) \leq 1$. Consequently,

$$|x|^2 + |e^{it}|^2|r|^2|y|^2 + 2r|(y, x)|Re(e^{i(s+t)}) \geq 0$$

so $|x|^2 + |r|^2|y|^2 + 2r|(y, x)| \geq 0$ for each real number $r \geq 0$. Put $r = -|x|/|y|$. Then substituting this value for r in the previous inequality yields $|x|\,|y| \leq |(x, y)|$. ∎

An immediate result of the Schwarz inequality is the so called *triangle inequality*: $|x + y| \leq |x| + |y|$. It follows from the observation that $|x + y|^2 = (x + y, x + y) = (x, x) + (x, y) + (y, x) + (y, y) \leq |x|^2 + 2Re((x, y)) + |y|^2 \leq |x|^2 + 2|(x, y)| + |y|^2 \leq |x|^2 + 2|x|\,|y| + |y|^2 = (|x| + |y|)^2$. Consequently, if we define the distance $d(u, v)$ between two vectors $u, v \in H$ to be $|u - v|$, it is easily shown that d satisfies the axioms of a metric space. In particular, if $u, v, w \in H$ we see that $d(u, v) \leq d(u, w) + d(w, v)$ follows from $|u - v| \leq |u - w| + |w - v|$. If the metric space (H, d) is complete, we call H a **Hilbert space**. We Now observe that if μ is a positive measure, then $L^2(\mu)$ is a Hilbert space.

For this we define an inner product on $L^2(\mu)$ by $(f, g) = \int_X fg^* d\mu$. Since $g \in L^2(\mu)$ implies $g^* \in L^2(\mu)$ and the exponents $p = 2 = q$ are conjugate exponents, Proposition 11.2 implies $fg^* \in L^1(\mu)$ so (f, g) is well defined on H. Now we observe that if we define $|f|^2 = (f, f)$ or equivalently, $|f| = (f, f)^{1/2}$ then we have

$$|f| = [\int_X ff^* d\mu]^{1/2} = [\int_X |f|^2 d\mu]^{1/2} = |f|_2$$

so the L^2-norm $| \ |_2$ on $L^2(\mu)$ is equivalent to the inner product norm $| \ |$ on $L^2(\mu)$. Since μ was assumed to be positive, by Theorem 11.2 we know $L^2(\mu)$ is complete with respect to the L^2-norm, so $L^2(\mu)$ is complete with respect to $| \ |$ and therefore is a Hilbert space.

PROPOSITION 11.6 For a given $y \in H$, the mappings defined by $f(x) = (x,y)$, $g(x) = (y, x)$, $h(x) = |x|$ for each $x \in X$ are uniformly continuous functions on H.

Proof: To show f is uniformly continuous let $\varepsilon > 0$ and put $\delta = \varepsilon/|y|$. Then if $|x_1 - x_2| < \delta$ we have by Proposition 11.5 that:

$$|f(x_1) - f(x_2)| = |(x_1, y) - (x_2, y)| = |(x_1 - x_2, y)| \leq |x_1 - x_2||y| < \varepsilon.$$

Therefore, f is uniformly continuous on H. A similar argument shows that g is also uniformly continuous on H. To show h is uniformly continuous, let $\varepsilon > 0$ and put $\delta = \varepsilon$. By the triangle inequality, if $|x_1 - x_2| < \delta$, then $|x_1| \leq |x_1 - x_2| + |x_2|$, so $|x_1| - |x_2| \leq |x_1 - x_2|$. Similarly, $|x_2| - |x_1| = |x_2 - x_1| = |x_1 - x_2|$, so $||x_1| - |x_2|| \leq |x_1 - x_2|$. But then

$$|h(x_1) - h(x_2)| = ||x_1| - |x_2|| \leq |x_1 - x_2| < \delta = \varepsilon$$

so that h is also uniformly continuous on H. ■

Recall that a subset S of a vector space V is called a *subspace* of V if S is a vector space with respect to the addition and scalar multiplication operations defined on V. A necessary and sufficient condition for $S \subset V$ to be a subspace is that $u + v \in S$ and $cu \in S$ for each $u, v \in S$ and $c \in \mathbf{C}$. If H is an inner product space, a **closed subspace** of H is a subspace that is closed with respect to the metric topology generated by the inner product norm. A set E in a vector space is said to be **convex** if for each $u, v \in E$ and $0 \leq x \leq 1$, the vector $(1 - x)u + xv$ is contained in E. One can visualize this property of convexity by imagining a "line segment" being traced out between u and v as x goes from 0 to 1, and that all the vectors on this "line segment" are contained in E.

If $(u, v) = 0$ for some $u, v \in H$ we say u is **orthogonal** to v and denote this by $u \perp v$. Let $u^\perp$ be the collection of $v \in H$ which are orthogonal to u. If $v, w \in u^\perp$ then $(u, v + w) = (u, v) + (u, w) = 0$ and if $c \in \mathbf{C}$ then $(u, cv) = c(u, v) = 0$ so v

$+ w \in u^\perp$ and $cv \in u^\perp$. Hence $u^\perp$ is a subspace of H. Now $u^\perp$ is the set of vectors $x \in H$ where the continuous function $g(x) = (u, x) = 0$, so $u^\perp$ is a closed set in H. If S is any subspace of H, let $S^\perp$ denote the collection of all $v \in H$ that are orthogonal to every $u \in S$. Clearly $S^\perp = \cap\{u^\perp | u \in S\}$. Since each $u^\perp$ is a closed subspace of H, so is $S^\perp$.

PROPOSITION 11.7 Each nonempty, closed, convex set in a Hilbert space has a smallest element with respect to the inner product norm.

Proof: Let E be a nonempty, closed, convex subset of the Hilbert space H. Put $b = inf\{|v| \,|\, v \in \text{E}\}$. Then there exists a sequence $\{v_n\} \subset \text{E}$ such that $\{|v_n|\}$ converges to b. For any pair $x, y \in H$, $|x + y|^2 + |x - y|^2 = (x + y, x + y) + (x-y,x-y) = 2|x|^2 + 2|y|^2$. If we apply this identity to $x/2$ and $y/2$ we get $(1/4)|x-y|^2 = |x|^2/2 + |y|^2/2 - |(x + y)/2|^2$. Since E is convex, $(x + y)/2 \in \text{E}$ which implies $|(x + y)/2| \geq b$, so

$$(11.2) \qquad |x - y|^2 \leq 2|x|^2 + 2|y|^2 - 4b^2$$

for each pair $x, y \in H$. If we replace x and y in this inequality by v_m and v_n we see that as $m, n \to \infty$, the right side of (11.2) tends to zero. Hence $|v_m - v_n| \to 0$ as $m, n \to \infty$ which implies $\{v_n\}$ is a Cauchy sequence in $\text{E} \subset H$. Since H is complete, $\{v_n\}$ converges to some $v \in H$, and since E is closed, $v \in \text{E}$. Also, since the norm function $h(x) = |x|$ is continuous on H by Proposition 11.6, we have $|v| = lim_n |v_n| = b$. Consequently, there exists a $v \in \text{E}$ of smallest norm.

It remains to show that v is unique. If u is another member of H such that $|u| = b = |v|$, then the inequality (11.2) implies $|u - v|^2 \leq 0$ so $u = v$. Therefore, E has a smallest element with respect to the inner product norm. ▪

Let S be a closed subspace of the Hilbert space H and let $u \in H$. Then the set $u + S = \{u + v | v \in S\}$ is closed and convex. To see that $u + S$ is closed, let w be a limit point of $u + S$. Then there exists a sequence $\{u + v_n\}$ in $u + S$ that converges to w, so $\{u + v_n\}$ is Cauchy. If $\varepsilon > 0$, there exists a positive integer N such that if $m, n > N$ then $|u + v_m - u - v_n| < \varepsilon$ which implies $|v_m - v_n| < \varepsilon$, so $\{v_n\}$ is Cauchy in S. Since S is a closed subspace of the complete space H, S is complete which implies $\{v_n\}$ converges to some $v \in S$. Then if $\varepsilon > 0$, there exists a positive integer N such that if $n > N$ then $|v - v_n| < \varepsilon$ which implies $|(u+v) - (u + v_n)| < \varepsilon$, so $\{u + v_n\}$ converges to $u + v \in u + S$. But since limits are unique in metric spaces, $u + v = w$ so $w \in u + S$. Hence $u + S$ is closed.

To see $u + S$ is convex, let $u + v$ and $u + w \in u + S$ and let $0 \leq \lambda \leq 1$. Then $(1 - \lambda)(u + v) + \lambda(u + w) = u + (1 - \lambda)v + \lambda w \in u + S$. Therefore, $u + S$ is convex. Consequently, we can apply Proposition 11.7 to $u + S$ and get a smallest element $p_{S^\perp}(u)$ in $u + S$ with respect to the inner product norm. Next, put $p_S(u) = u - p_{S^\perp}(u)$. Then p_S and $p_{S^\perp}$ are functions on H. Since $p_{S^\perp}(u) \in u +$

S, $p_S(u) \in S$. The function p_S is called the **orthogonal projection** of H onto S. The function $p_{S^\perp}$ is called the orthogonal projection of H onto $S^\perp$.

For this later definition to make sense, we need to show that $p_{S^\perp}(u) \in S^\perp$. For this let $x = p_{S^\perp}(u)$. Then by the definition of the orthogonal projection onto $S^\perp$ we have, for each $y \in S$ with $|y| = 1$,

$$|x|^2 \leq |x - cy|^2 = (x - cy, x - cy) = |x|^2 - c(y, x) - c^*(x, y) + |c|^2$$

for each scalar c. If we substitute $c = (x, y)$ into this inequality, we get $0 \leq -|(x,y)|^2$ which implies $(x, y) = 0$ for each $y \in S$ with $|y| = 1$. But then $(x, y) = 0$ for any $y \in S$. Therefore, $x = p_{S^\perp}(u) \in S^\perp$.

THEOREM 11.6 If S is a closed subspace of the Hilbert space H, then the orthogonal projections p_S and $p_{S^\perp}$ of H onto S and $S^\perp$ respectively have the following properties:

(1) $u = p_S(u) + p_{S^\perp}(u)$ for each $u \in H$.
(2) The orthogonal projections are unique.
(3) The orthogonal projections are linear.
(4) If $u \in S$ then $p_S(u) = u$ and $p_{S^\perp}(u) = 0$.
(5) If $u \in S^\perp$ then $p_S(u) = 0$ and $p_{S^\perp}(u) = u$.
(6) $|u - p_S(u)| = inf\{|u - v| \mid v \in S\}$ for each $u \in H$.
(7) $|u|^2 = |p_S(u)|^2 + |p_{S^\perp}(u)|^2$ for each $u \in H$.

Proof: (1) follows immediately from the definition of $p_S(u)$. To show (2), first note that $S \cap S^\perp = \{0\}$. This is because if $x \in S \cap S^\perp$ then $(x, x) = 0$ which implies $x = 0$. Next let $u = v + w$ where $v \in S$ and $w \in S^\perp$. Then $p_S(u) + p_{S^\perp}(u) = u = v + w$ which implies $p_S(u) - v = w - p_{S^\perp}(u)$. Since the left side of this equation is in S while the right side is in $S^\perp$, we conclude that both sides are the zero vector. Therefore, $v = p_S(u)$ and $w = p_{S^\perp}(u)$, so the orthogonal projections are unique.

To show (3), let $u, v \in H$ and $c, d \in \mathbf{C}$. Then by (1), $p_S(cu + dv) + p_{S^\perp}(cu + dv) = cu + dv = c[p_S(u) + p_{S^\perp}(u)] + d[p_S(v) + p_{S^\perp}(v)]$ so $p_S(cu + dv) - cp_S(u) - dp_S(v) = cp_{S^\perp}(u) + dp_{S^\perp}(v) - p_{S^\perp}(cu + dv)$. Again, the left side of this equation is in S while the right side is in $S^\perp$ so we conclude both sides are the zero vector. Therefore, $p_S(cu + dv) = cp_S(u) + dp_S(v)$ and $p_{S^\perp}(cu + dv) = cp_{S^\perp}(u) + dp_{S^\perp}(v)$ so the orthogonal projections are linear.

To show (4) and (5), note that if $u \in S$, then (1) implies $u - p_S(u) = p_{S^\perp}(u)$ so the left hand side of this equation is in S while the right hand side is in $S^\perp$. Again, we conclude both sides are the zero vector, so $p_S(u) = u$ and $p_{S^\perp}(u) = 0$. This proves (4). A similar argument proves (5).

To show (6), note that by definition of $p_{S^\perp}(u)$ we have $|u - p_S(u)| = |p_{S^\perp}(u)| = inf\{|u + v| \mid v \in S\} = inf\{|u - v| \mid v \in S\}$. This proves (6). To show (7), observe that $|u|^2 = (u, u) = (p_S(u) + p_{S^\perp}(u), p_s(u) + p_{S^\perp}(u)) = |p_S(u)|^2 +$

$(p_S(u), p_{S^\perp}(u)) + (p_{S^\perp}(u), p_S(u)) + |p_{S^\perp}(u)|^2$. Since $(p_S(u), p_{S^\perp}(u)) = 0 = (p_{S^\perp}(u), p_S(u))$ we have that $|u|^2 = |p_S(u)|^2 + |p_{S^\perp}(u)|^2$ which proves (7). ∎

COROLLARY 11.2 If $S \neq H$ then there exists a $u \in H$ such that $u \perp S$ and $u \neq 0$.

Proof: Pick $v \in H - S$. Put $u = p_{S^\perp}(v)$. Then $v \neq p_S(v)$ so $u \neq 0$. But $u \perp S$ since $u \in S^\perp$. ∎

In Proposition 11.6 we saw that the function $f(x) = (x, y)$ for a fixed $y \in H$ is uniformly continuous. Since $(x, y) \in \mathbf{C}$ and since the definition of the inner product causes f to be linear, we see that f is a continuous linear functional. The Riesz Representation Theorem (Chapter 8) showed that positive linear functionals on $C_K(X)$ could be represented as positive measures on X. It is therefore natural to ask if continuous linear functionals on a Hilbert space can be represented as inner product functions with respect to a given vector. The answer is affirmative as the next theorem shows.

THEOREM 11.7 If λ is a continuous linear functional on a Hilbert space H, then there exists a unique $v \in H$ such that $\lambda(u) = (u, v)$ for each $u \in H$.

Proof: If $\lambda(u) = 0$ for each $u \in H$ put $v = 0$. Otherwise put $K = \{u \mid \lambda(u) = 0\}$. Since λ is linear, K is a subspace of H and since λ is continuous, K is closed. Since λ is not identically zero, Theorem 11.6 shows that $K^\perp \neq \{0\}$. Choose $w \in K^\perp$ such that $w \neq 0$. Then w is not in K so $\lambda(w) \neq 0$. Put $c = \lambda(w)/|w|^2$ and let $v = c^*w$. Then $v \in K^\perp$, $v \neq 0$ and $\lambda(v) = \lambda(c^*w) = c^*\lambda(w) = c^*c|w|^2 = cc^*(w, w) = (c^*w, c^*w) = (v, v)$. For any $u \in H$ put $x = u - \lambda(u)v/\lambda(v)$. Then $\lambda(x) = \lambda(u) - \lambda(u)\lambda(v)/\lambda(v) = 0$ so $x \in K$ which implies $(x, v) = 0$. Now $(u, v) = (x+\lambda(u)v/\lambda(v),v) = (x, v) + (\lambda(u)v/\lambda(v), v) = 0 + [\lambda(u)/\lambda(v)](v, v) = [\lambda(u)/\lambda(v)]\lambda(v) = \lambda(u)$. Consequently, $\lambda(u) = (u, v)$ for each $u \in H$.

To show that v is unique, let w be another vector in H such that $\lambda(u) = (u,w)$ for each $u \in H$. Put $z = v - w$. Then $(u, z) = (u, v - w) = (u, v) - (u, w) = \lambda(u) - \lambda(u) = 0$ for each $u \in H$. But then $(z, z) = 0$ which implies $z = 0$ so $v = w$. ∎

Recall how a basis is defined in a vector space V. First we define a **linear combination** of vectors $v_1 \dots v_n \in V$ to be a vector sum of the form $c_1v_1 + \dots + c_nv_n$ for some $c_1 \dots c_n \in \mathbf{C}$. We define the vectors $v_1 \dots v_n$ to be **linearly independent** if $c_1v_1 + \dots + c_nv_n = 0$ implies $c_i = 0$ for any $c_1 \dots c_n \in \mathbf{C}$. A subset S of V is said to be linearly independent if every finite subset of S is linearly independent. The set $V(S)$ of all linear combinations of finite subsets of S is clearly a vector space. In fact, it is easily seen to be the smallest subspace of V containing S. $V(S)$ is called the **span** of S or the subspace spanned by S. If $V(S) = V$ then S is called a spanning subset of V or we say S **spans** V. Finally, we define a **basis** of a vector space V to be a linearly independent subset that spans V.

PROPOSITION 11.8 A linearly independent subset of a vector space is a basis if and only if it is maximal.

Proof: Let S be a maximal linearly independent subset of a vector space V. Assume $V(S) \neq V$ for otherwise S would be a basis for V. Then $0 \in V(S)$ so there exists a $u \in V - V(S)$ with $u \neq 0$. Let $v_1 \ldots v_n \in S$ and suppose $c_1v_1 + \ldots + c_nv_n + c_{n+1}u = 0$ for $c_1 \ldots c_{n+1} \in \mathbf{C}$. If $c_{n+1} \neq 0$ then $u = \Sigma_{i=1}^n c_iv_i/c_{n+1}$ which implies $u \in V(S)$ which is a contradiction. Therefore, $c_{n+1} = 0$ which implies $c_1v_1 + \ldots + c_nv_n = 0$ so $c_i = 0$ for each $i = 1 \ldots n$. But then $S \cup \{u\}$ is linearly independent which implies S is not maximal which is a contradiction. Therefore, S is a basis for V.

Conversely, assume $\{u_\alpha\}$ is a basis for V and suppose $\{u_\alpha\}$ is not maximal. Then there exists a $u \in H$ with $u \neq 0$ and $\{u_\alpha\} \cup \{u\}$ linearly independent. Since $\{u_\alpha\}$ spans V, there exists a finite subcollection $u_1 \ldots u_n$ of $\{u_\alpha\}$ and a finite collection $c_1 \ldots c_n \in \mathbf{C}$ with $u = c_1u_1 + \ldots + c_nu_n$. Then $c_1u_1 + \ldots + c_nu_n + (-1)u = 0$ but $(-1) \neq 0$, so $\{u_\alpha\} \cup \{u\}$ is not linearly independent after all. ∎

A set of vectors $\{u_\alpha\}$ in a Hilbert space H is said to be **orthogonal** if $(u_\alpha, u_\beta) = 0$ if $\alpha \neq \beta$. $\{u_\alpha\}$ is said to be **normalized** if $|u_\alpha| = 1$ for each α. If $\{u_\alpha\}$ is both orthogonal and normalized it is said to be **orthonormal**. Clearly, $\{u_\alpha\}$ is orthonormal if and only if $(u_\alpha, u_\beta) = 1$ if $\alpha = \beta$ and 0 otherwise.

PROPOSITION 11.9 If $u_1 \ldots u_n$ is an orthonormal set and $v = c_1u_1 + \ldots + c_nu_n$, then $c_i = (v, u_i)$ for each $i = 1 \ldots n$ and $|v|^2 = \Sigma_{i=1}^n |c_i|^2$.

Proof: For each $i = 1 \ldots n$, $(v, u_i) = (c_1u_1 + \ldots + c_nu_n, u_i) = c_i(u_i, u_i) = c_i$ since $u_1 \ldots u_n$ is an orthonormal set. Also, $|v|^2 = (v, v) = (c_1u_1 + \ldots + c_nu_n, c_1u_1 + \ldots + c_nu_n) = c_1c_1{}^*(u_1, u_1) + \ldots + c_nc_n{}^*(u_n, u_n) = \Sigma_{i=1}^n |c_i|^2$. ∎

COROLLARY 11.3 An orthonormal set is linearly independent.

Proof: Let $\{u_\alpha\}$ be an orthonormal set and let $u_1 \ldots u_n \in \{u_\alpha\}$. Suppose $c_1u_1 + \ldots + c_nu_n = 0$ for $c_1 \ldots c_n \in \mathbf{C}$. By Proposition 11.9, $c_i = (0, u_i) = 0$ for each $i = 1 \ldots n$. Consequently, $\{u_\alpha\}$ is linearly independent. ∎

THEOREM 11.8 Each Hilbert space has an orthonormal basis.

Proof: If H is a non-trivial Hilbert space, then there exists a $u \in H$ with $|u| = 1$. Then $\{u\}$ is an orthonormal subset of H. Let P be the collection of all orthonormal subsets of H containing $\{u\}$, partially ordered by set inclusion. Since $\{u\} \in P$, $P \neq \varnothing$ so by the Hausdorff Maximal Principle (an equivalent form of the Axiom of Choice), there exists a maximal linearly ordered subcollection Q of P. Clearly $\{u\} \in Q$, so $Q \neq \emptyset$. Let $R = \cup Q$. If $v, w \in R$ then $v \in A \in A$ and $w \in B \in Q$ for some A, B $\in Q$. Since Q is linearly ordered by inclusion, either $v, w \in$ A or $v, w \in$ B. Since both A and B are orthonormal subsets of H, $(v, w) = 1$ if $v = w$ and 0 otherwise. Therefore, R is orthonormal.

Suppose R is not a maximal orthonormal set. Then there exists an orthonormal set S containing R with $S - R \neq \emptyset$. Now S is not in Q and S contains each member of Q so we can adjoin S to Q and still have a linearly ordered set with respect to inclusion which implies Q is not maximal which is a contradiction. Therefore, R is a maximal orthonormal set. By Corollary 11.3, R is linearly independent. Suppose R is not a maximal linearly independent set. Then there exists a linearly independent set T with $R \subset T$ and $T - R \neq \emptyset$. Let $x \in T - R$. Let V be the subspace of H spanned by R. Then x is not in V which implies $y = p_{V^\perp}(x) \neq 0$ by Theorem 11.7, so $(y, v) = 0$ for each $v \in R$. Put $z = y/|y|$ which implies $|z| = 1$ and $(z, v) = 0$ for each $v \in R$. Therefore, z can be adjoined to R to obtain an orthonormal set in H containing R as a proper subset which is a contradiction. Consequently, R is a maximal linearly independent set so R is a basis. Since R is orthonormal, R is an orthonormal basis. ∎

LEMMA 11.1 If V is a closed subspace of the Hilbert space H and if $u \in H - V$, then the subspace W spanned by $V \cup \{u\}$ is closed.

Proof: Suppose v is a limit point of W. Then $v = lim_n(v_n + c_n u)$ where $\{v_n\} \subset V$ and $\{c_n\} \subset \mathbf{C}$. Consequently, there exists a $b < \infty$ such that $|v_n = c_n u| < b$ for each positive integer n. Assume $\{c_n\}$ has no convergent subsequence. Since closed and bounded subsets in $\mathbf{C}$ are compact, this implies $lim_n|c_n| = \infty$. But then $|v_n + c_n u|/|c_n| < b/|c_n|$ for each n and $lim_n b/|c_n| = 0$. Therefore, $lim_n|v_n/c_n + u| = 0$, so $lim_n(v_n/c_n) = -u$ which implies $-u \in V$ since V is closed which is a contradiction.

Hence we may assume $\{c_n\}$ has a convergent subsequence $\{c_{m_n}\}$ that converges to some $c \in \mathbf{C}$. Now $v = lim_n(v_{m_n} + c_{m_n}u) = lim_n v_{m_n} + cu$ which implies $\{v_{m_n}\}$ converges to $v - cu$. Since $\{v_{m_n}\} \subset V$ and V is closed, $[v - cu] \in V$ and $v = [v - cu + cu] \in W$. Therefore, W is closed. ∎

THEOREM 11.9 If $u_1 \ldots u_n$ is an orthonormal set of vectors in the Hilbert space H and $v \in H$, then $|v - \Sigma_{i=1}^n (v, u_i)u_i| \leq |v - \Sigma_{i=1}^n c_i u_i|$ for all $c_1 \ldots c_n \in \mathbf{C}$ and equality holds if and only if $c_i = (v, u_i)$ for each $i = 1 \ldots n$. The vector $\Sigma_{i=1}^n (v, u_i)u_i$ is the orthogonal projection of v onto the subspace W generated by $\{u_1 \ldots u_n\}$ and if d is the distance from v to W then $|v|^2 = d^2 + \Sigma_{i=1}^n |(v, u_i)|^2$.

Proof: The subspace $\{0\}$ of H is obviously closed. By Lemma 11.1, the subspace of H spanned by $\{u_1\}$ is closed since it is the subspace spanned by $\{0\} \cup \{u_1\}$. Proceeding inductively, it is clear that W is closed. Then by

Theorem 11.6, $p_W(v)$ is an element of W such that $|v - p_W(v)| \leq |v - w|$ for each $w \in W$ and since the mappings p_W and $p_{W^\perp}$ are unique, $p_W(v)$ is unique. Consequently, $p_W(v)$ has the property that $|v - p_W(v)| \leq |v - \Sigma_{i=1}^n c_i u_i|$ for each collection $c_1 \ldots c_n$ in C.

Let $p_W(v) = \Sigma_{i=1}^n a_i u_i$ for some $a_1 \ldots a_n \in$ C. By Proposition 11.9, $a_i = (p_W(v), u_i)$ for each $i = 1 \ldots n$. Now $v = p_W(v) + p_{S^\perp}(v)$ so $(v, u_i) = (p_W(v), u_i) + (p_{W^\perp}(v), u_i) = a_i + 0$ since $p_{W^\perp}(v)$ is orthogonal to all the u_i. Hence $p_W(v) = \Sigma_{i=1}^n (v, u_i)u_i$. Therefore, $|v - \Sigma_{i=1}^n (v, u_i)u_i| \leq |v - \Sigma_{i=1}^n c_i u_i|$ for all $c_1 \ldots c_n \in$ C and equality holds if and only if $c_i = (v, u_i)$ for each $i = 1 \ldots n$.

Finally, the distance d from v to W is the minimum value of $|v - w|$ such that $w \in W$. Therefore, $d = |v - p_W(v)|$ so $d^2 = (v - p_W(v), v - p_W(v)) = (v - p_W(v), v) - (v - p_W(v), p_W(v))$. Now $(v - p_W(v), p_W(v)) = 0$ so $d^2 = (v - p_W(v), v)$. To see this, note that for each $i = 1 \ldots n$, $(v - p_W(v), u_i) = (v - \Sigma_{i=1}^n (v, u_i)u_i, u_i) = (v, u_i) - (v, u_i)(u_i, u_i) = 0$. Therefore, $(v - p_W(v), p_W(v)) = (v - p_W(v), \Sigma_{i=1}^n (v,u_i)u_i) = \Sigma_{i=1}^n (v, u_i)(v - p_W(v), u_i) = 0$. Consequently, $d^2 = (v - p_W(v), v) = (v - \Sigma_{i=1}^n (v,u_i)u_i, v) = |v|^2 - \Sigma_{i=1}^n (v, u_i)(u_i, v) = |v|^2 + \Sigma_{i=1}^n |(v, u_i)|^2$ which implies $|v|^2 = d^2 + \Sigma_{i=1}^n |(v, u_i)|^2$. ∎

Let $\{u_\alpha\}$ be an orthonormal set in the Hilbert space H. For each vector $v \in H$ put $v_\alpha = (v, u_\alpha)$ for each α. By the symbol $\Sigma_\alpha |v_\alpha|^2$ is meant the supremum of the set of finite sums of the form $|v_1|^2 + \ldots + |v_n|^2$ where $v_i = (v, u_i)$ for each finite collection $u_1 \ldots u_n \in \{u_\alpha\}$. With this notation we can state and prove the following classical result:

THEOREM 11.10 (Bessel's Inequality) $\Sigma_\alpha |v_\alpha|^2 \leq |v|^2$.

Proof: For any finite collection $u_1 \ldots u_n \in \{u_\alpha\}$, Theorem 11.9 gives $\Sigma_{i=1}^n |(v,u_i)|^2 = |v|^2 - d^2$ where d is the distance from v to the subspace W spanned by the vectors $u_1 \ldots u_n$. Since $d \geq 0$, this means $\Sigma_{i=1}^n |(v, u_i)|^2 \leq |v|^2$ and hence the supremum of such finite sums is less than or equal to $|v|^2$. ∎

Sums of the form $\Sigma_\alpha c_\alpha$ where $0 \leq c_\alpha \leq \infty$ for each α and where the summation is defined as the supremum of finite sums of the form $c_1 + \ldots + c_n$ where $c_1 \ldots c_n \in \{c_\alpha\}$ are especially interesting in Hilbert spaces because they are used in the characterization of the structure of Hilbert spaces. We will now develop this characterization. Let X be a set. For each E $\in$ X put μ(E) = ∞ if E is infinite and μ(E) = card(E) if E is finite. It is easily seen that μ is a measure on the σ-algebra of all subsets of X. The measure μ is called the **counting measure** on X. Let f:X $\to$ C. Then it is easily seen that $\Sigma_{x \in X} |f(x)|$, where the summation is the supremum of the finite sums of the form $|f(x_1)| + \ldots + |f(x_n)|$ where $x_1 \ldots x_n \in$ X, is the Lebesgue integral of $|f|$ with respect to the counting measure on X. We use the notation l^2(X) to denote the L^2-space $L^2(\mu)$ where μ is the counting measure on X.

In particular, if $\{u_\alpha \mid \alpha \in A\}$ is an orthonormal basis in H and for some $v \in H$, v':A $\to$ C is defined by $v'(\alpha) = v_\alpha = (v, u_\alpha)$ for each $\alpha \in A$, then it is a consequence of Bessel's Inequality that $v' \in l^2(A)$ since:

$$|v'|_2 = [\Sigma_\alpha |v'(\alpha)|^2]^{1/2} = [\Sigma_\alpha |(v, u_\alpha)|^2]^{1/2} \leq [|v'|^2]^{1/2} < \infty.$$

The importance of the Riesz-Fischer Theorem that we will prove next is that if $\{u_\alpha \mid \alpha \in A\}$ is an orthonormal set of vectors in H then $f \in l^2(A)$ implies that f is of the form v':A $\to$ C for some vector $v \in H$. Before we prove the Riesz-Fischer Theorem, we first observe that Bessel's Inequality implies an even stronger statement about a function v':A $\to$ C defined by $v'(\alpha) = (v, u_\alpha)$ for each $\alpha \in A$. Bessel's Inequality implies the set of $\alpha \in A$ for which $v'(\alpha) \neq 0$ is at most countable.

To see this, suppose the set of $\alpha \in A$ for which $v'(\alpha) \neq 0$ is uncountable. For each positive integer n put $E_n = \{\alpha \in A \mid |v'(\alpha)| > 1/n\}$. Then for some positive integer m, E_m must be uncountable. But then there exists a finite subset S of E_m with $\Sigma_{\alpha \in S} |v'(\alpha)|^2 > |v|^2$ which is a contradiction.

THEOREM 11.11 (Riesz-Fischer) If $\{u_\alpha \mid \alpha \in A\}$ is an orthonormal set in a Hilbert space H and if $f \in l^2(A)$, then $f = v'$ for some $v \in H$.

Proof: For each positive integer n put $E_n = \{\alpha \in A \mid |f(\alpha)| > 1/n\}$. Then each E_n must be finite, for otherwise there would exist a finite collection $\alpha_1 \ldots \alpha_n \in$ A such that $\Sigma_{i=1}^n |f(\alpha_i)|^2 > |f|_2$. For each positive integer n let $v_n = \Sigma_{\alpha \in E_n} |f(\alpha)| u_\alpha$. Then each v_n is in H. For each n define v_n':A $\to$ C by $v_n'(\alpha) = (v_n, u_\alpha)$ for each $\alpha \in A$. Then for a given $\beta \in A$, $v_n'(\beta) = (\Sigma_{\alpha \in E_n} |f(\alpha)| u_\alpha, u_\beta) = \Sigma_{\alpha \in E_n} |f(\alpha)| (u_\alpha, u_\beta) = |f(\beta)|$ if $\beta \in E_n$ and 0 otherwise. Therefore, $v_n' = |f| \chi_{E_n}$. Consequently, $|f - v_n'| \leq |f|^2$ which implies $|f - v_n'| \leq |f|$.

Since $E_m \subset E_n$ if $m < n$, it is clear that $lim_n v_n' = f$, so by Theorem 8.13, $lim_n |f - v_n'| = 0$ which implies $lim_n |f - v_n'| = 0$ since we can choose an N large enough so that $n > N$ implies $[\Sigma_{\alpha \in E_n} |f - v_n'|^2]^{1/2} < \Sigma_{\alpha \in E_n} |f - v_n'|$ and therefore $|f - v_n'|_2 = [\Sigma_{\alpha \in E_n} |f - v_n'|^2]^{1/2} < \Sigma_{\alpha \in E_n} |f - v_n'| = |f - v_n'|_1$ for $n > N$.

Then since $lim_n |f - v_n'|_2 = 0$, $\{v_n'\}$ is a Cauchy sequence in $l^2(A)$. Now for each n, $v_n = \Sigma_{\alpha \in E_n} |f(\alpha)| u_\alpha$ and since E_n is finite we can apply Proposition 11.9 to get $|f(\alpha)| = (v_n, u_\alpha)$ for each $\alpha \in E_n$. If m and n are positive integers with $m < n$, then $E_m \subset E_n$. Put $E(n, m) - E_n - E_m$. Then

$$|v_n' - v_m'| = [\Sigma_{\alpha \in E(n,m)} |(v_n, u_\alpha) - (v_m, u_\alpha)|^2]^{1/2} = [\Sigma_{\alpha \in E(n,m)} ||f(\alpha)| - 0|^2]^{1/2} =$$

$$(\Sigma_{\alpha \in E(n,m)} |f(\alpha)| u_\alpha, \Sigma_{\alpha \in E(n,m)} |f(\alpha)| u_\alpha)^{1/2} = (v_n - v_m, v_n - v_m)^{1/2} = |v_n - v_m|,$$

so $|v_n - v_m| = |v_n' - v_m'|_2$ which implies $\{v_n\}$ is Cauchy in H and therefore converges to some $v \in H$.

Then $v':A \to C$ defined by $v'(\alpha) = (v, u_\alpha)$ for each α is the desired function. To see this, for a fixed $\beta \in A$, the function $g(x) = (x, u_\beta)$ for each $x \in H$ is uniformly continuous (Proposition 11.6). Therefore, since $\{v_n\}$ converges to v, $\{g(v_n)\}$ converges to $g(v)$. Hence $lim_n(v_n, u_\beta) = (v, u_\beta)$. Then for each $\alpha \in A$, $v'(\alpha) = (v, u_\alpha) = lim_n(v_n, u_\alpha) = lim_n v_n'(\alpha) = f(\alpha)$ since $\{v_n'\}$ converges to f. Consequently, $f = v'$. ∎

PROPOSITION 11.10 Let $\{u_\alpha\}$ be an orthonormal set in the Hilbert space H. Then $\{u_\alpha\}$ is a basis for H if and only if the set S of finite linear combinations of members of $\{u_\alpha\}$ is dense in H.

Proof: Assume $\{u_\alpha\}$ is a basis for H. By Proposition 11.8, $\{u_\alpha\}$ is a maximal linear independent subset of H. Suppose $\{u_\alpha\}$ is not a maximal orthonormal set in H. Then there exists a $u \in H$ with $|u| = 1$ and $\{u_\alpha\}\cup\{u\}$ orthonormal. By Corollary 11.3, $\{u_\alpha\}\cup\{u\}$ is linearly independent which is a contradiction. Hence $\{u_\alpha\}$ is a maximal orthonormal set in H. Now assume S is not dense in H. Then there exists a $v \in H - Cl(S)$ which implies there exists a $d > 0$ such that $|v - u| > d$ for each $u \in Cl(S)$. Now $v = p_S(v) + p_{S^\perp}(v)$ and by the remarks preceding Theorem 11.6, $w = p_{S^\perp}(v)/|p_{S^\perp}(v)|$ is in $S^\perp$. Since $|w| = 1$, $\{u_\alpha\}\cup\{w\}$ is orthonormal which is a contradiction. Therefore, S is dense in H.

Conversely, assume S is dense in H and suppose $\{u_\alpha\}$ is not a basis for H. Then $\{u_\alpha\}$ is not a maximal orthonormal set in H, so there exists a $u \in H$ with $|u| = 1$ and $u \perp u_\alpha$ for each α. Now there exists a sequence $\{u_n\} \subset S$ such that $lim_n u_n = u$. Clearly $u \perp u_n$ for each positive integer n. Since the function $f:H \to C$ defined by $f(v) = (u, v)$ for each $v \in H$ is uniformly continuous, $\{f(u_n)\}$ must converge to $f(u) = (u, u) = 1$. But this is impossible since $(u, u_n) = 0$ for each n. Hence $\{u_\alpha\}$ is a basis for H. ∎

The property that S is dense in H has some very interesting ramifications. In fact, it leads to a representation of all Hilbert spaces H as $l^2(A)$ where $card(A) = card(\{u_\alpha\})$ where $\{u_\alpha\}$ is a basis for H. To show this is the case, we need the following lemma.

LEMMA 11.2 Let $\{u_\alpha | \alpha \in A\}$ be an orthonormal set in the Hilbert space H and let S be the collection of finite linear combinations of members of $\{u_\alpha\}$. For each pair $x, y \in H$ let $x', y' \in l^2(A)$ be defined by $x'(\alpha) = (x, u_\alpha)$ and $y'(\alpha) = (y, u_\alpha)$ for each $\alpha \in A$. Then S is dense in H if and only if $(x, y) = (x', y')$ for each pair $x, y \in H$.

Proof: First assume S is dense in H. Let $v \in H$. Now choose $\varepsilon > 0$. Then there exists a finite collection $u_1 \ldots u_n \in \{u_\alpha\}$ and $c_1 \ldots c_n \in C$ such that $w = c_1u_1 + \ldots + c_nu_n$ is within ε of v. By Theorem 11.9, $|v - \Sigma_{i=1}^n (v, u_i)u_i| \le |v - w| < \varepsilon$. Put $z = \Sigma_{i=1}^n (v, u_i)u_i$. Then $|v - z| < \varepsilon$ which implies $|v| < |z| + \varepsilon$ which in turn implies $(|v| - \varepsilon)^2 < |z|^2 = |(v, u_1)|^2 + \ldots + |(v, u_n)|^2 \le \Sigma_\alpha |v'(\alpha)|^2$ by Proposition 11.9. But $\Sigma_\alpha |v'(\alpha)|^2 = \Sigma_\alpha v'(\alpha)[v'(\alpha)]^* = (v', v')$. Hence, for each v

$\in H$, $(v, v) = |v|^2 \le (v', v')$. By Bessel's Inequality, $(v', v') \le (v, v)$ so $(v, v) = (v', v')$ for each $v \in H$.

Now let $u, v \in H$. Then for each $c \in \mathbf{C}$ we have $(u + cv, u + cv) = (u' + cv', u' + cv')$ which implies $(u, cv) + (cv, u) = (u', cv') + (cv', u')$ or $c^*(u, v) + c(v, u) = c^*(u', v') + c(v', u')$. Since this holds for each $c \in \mathbf{C}$, it holds for $c = 1$ and $c = i$. When $c = 1$ we have $(u, v) + (v, u) = (u', v') + (v', u')$. When $c = i$ we have $(u,v) - (v, u) = (u', v') - (v', u')$. Adding these two equations yields: $2(u, v) = 2(u',v')$ which implies $(u, v) = (u', v')$.

Conversely, assume $(x, y) = (x', y')$ for each pair $x, y \in H$. Suppose S is not dense in H. Pick $u \in H - Cl(S)$ and let $w = p_{S^\perp}(u)$. Clearly $w \ne 0$ and $(w, u_\alpha) = 0$ for each α. Put $x = w = y$. Then $(x, y) = |w|^2 > 0$ but $(x', y') = \Sigma_\alpha(w,u_\alpha)(w,u_\alpha)^* = 0$ so $(x, y) \ne (x', y')$ which is a contradiction. Hence S must be dense in H. ∎

Recall from Section 1.5 that two metric spaces X and M are isomorphic if there exists a uniform homeomorphism f:X → M that preserves distance. Two Hilbert spaces H_1 and H_2 are said to be **isomorphic** if there is an isomorphism $f:H_1 \to H_2$ that is also a linear transformation, i.e., one that preserves sums and scalar products. Such a mapping is called a **Hilbert space isomorphism**.

THEOREM 11.12 If $\{u_\alpha | \alpha \in A\}$ is an orthonormal basis for the Hilbert space H and if for each $x \in H$, x' is the element of $l^2(A)$ defined by $x'(\alpha) = (x, u_\alpha)$ for each α, then the mapping $\lambda:H \to l^2(A)$ defined by $\lambda(x) = x'$ for each $x \in H$ is a Hilbert space isomorphism of H onto $l^2(A)$.

Proof: Let $u, v \in H$ and $c, d \in \mathbf{C}$. Then $\lambda(cu + dv) = (cu + dv)'$. Then for each $\alpha \in$ A, $(cu + dv)'(\alpha) = (cu + dv, u_\alpha) = c(u, u_\alpha) + d(v, u_\alpha) = cu'(\alpha) + dv'(\alpha) = c\lambda(v) + d\lambda(v)$ so λ is a linear transformation from H into l^2(A). That λ is onto follows from the Riesz-Fischer Theorem.

Suppose $u \ne v$ but $u' = v'$. Then for each $\alpha \in$ A, $(u, u_\alpha) = (v, u_\alpha)$. Now $(u\text{-}v)' \in l^2$(A) and $(u - v)'(\alpha) = (u - v, u_\alpha) = (u, u_\alpha) - (v, u_\alpha) = 0$ for each α so $(u\text{-}v)'$ is the zero element of l^2(A). Since $u \ne v$, $|u - v| > 0$ which implies $w = (u\text{-}v)/|u - v|$ is a unit vector in H such that $w \perp u_\alpha$ for each α, so $\{u_\alpha\}\cup\{w\}$ is an orthonormal set in H. But then $\{u_\alpha\}$ is not maximal which is a contradiction. Hence $u - v = 0$ so $u = v$. Therefore, λ is one-to-one.

Since $(u, u) = (u', u')$ for each $u \in H$, λ preserves inner products and hence distance. Therefore, λ is a metric space isomorphism between H and l^2(A). Since λ is a linear transformation, it is a Hilbert space isomorphism. ∎

EXERCISES

1. Show that the vector space C^*(X) of all real valued continuous functions on

X = [0, 1] is an inner product space with respect to $(f, g) = \int_X fg^* dx$ (where dx denotes integration with respect to Lebesgue measure) but not a Hilbert space.

2. Show that if S is a closed subspace of the Hilbert space H then $(S^{\perp})^{\perp} = S$.

3. Show that a Hilbert space is separable if and only if it contains a countable orthonormal basis.

4. Let $\{u_n\}$ be an orthonormal sequence in the Hilbert space H. Let S be the set of all $v \in H$ with $v = \Sigma_{n=1}^{\infty} c_n u_n$ where $|c_n| \leq 1/n$. Then S is isomomorphic with the Hilbert cube and is an example of a closed and bounded subset of H that is not compact.

5. Show that no Hilbert space containing an orthonormal sequence is locally compact.

6. Show that for each pair of Hilbert spaces, one of them is isomorphic to a subspace of the other.

7. Let u be a member of the Hilbert space H and let S be a closed linear subspace of H. Show that $min\{|u - v| \mid v \in S\} = max\{|(u, w)| \mid w \in S^{\perp} \text{ and } |w| = 1\}$.

THE TRIGONOMETRIC SYSTEM

Let T be the unit circle in the complex plane. If $F:T \to \mathbf{C}$ is any function on T then the function f defined by $f(x) = F(e^{ix})$ for each $x \in \mathbf{R}$ is a periodic function of $\mathbf{R}$ of period 2π, i.e., $f(x + 2\pi) = f(x)$ for each $x \in \mathbf{R}$. Conversely, if $f:\mathbf{R} \to \mathbf{C}$ is periodic with period 2π, it is easily seen that there is a function $F:T \to \mathbf{C}$ such that $f(x) = F(e^{ix})$ for each $x \in \mathbf{R}$. Therefore, we can identify the complex valued functions on T with the complex valued 2π periodic functions on $\mathbf{R}$. Define $L^p(T)$, where $1 \leq p \leq \infty$, to be the class of all complex valued, Lebesgue measurable, 2π periodic functions on $\mathbf{R}$ for which

$$|f|_p = [1/2\pi \int_{[-\pi,\pi)} |f(x)|^p dx]^{1/p} < \infty.$$

In other words, $L^p(T) = L^p(\mu)$ where μ is Lebesgue measure on $[-\pi, \pi)$ divided by $1/2\pi$. The factor $1/2\pi$ is not essential here but it simplifies the following development. For instance, with this definition, the L^p-norm $|\ |_p$ of the constant function 1 is 1.

8. For each pair $f, g \in L^2(T)$ define $(f, g) = 1/2\pi \int_{-\pi}^{\pi} f(x)g(x)^* dx$ for each $x \in [-\pi,\pi)$. Show that (f, g) defines an inner product on $L^2(T)$ and that $|f|_2^2 = (f, f)$ for each $f \in L^2(T)$.

9. A **trigonometric polynomial** is a finite sum of the form $p(x) = a_0 + \Sigma_{n=1}^{N}[a_n cosnx + b_n sinnx]$ for each $x \in \mathbf{R}$ where the a_n's and b_n's are complex numbers. As is well known from calculus, for each $x \in \mathbf{R}$ we have $e^{ix} = cosx + isinx$. Show this implies that $p(x)$ can be written as $p(x) = \Sigma_{n=-N}^{N} c_n e^{inx}$ for some suitable c_n's $\in \mathbf{C}$ for $n = -N \ldots N$.

10. For each integer n (not necessarily positive) put $u_n(x) = e^{inx}$ for each $x \in \mathbf{R}$. Then for each n, $u_n \in L^2(T)$. Show that for each pair of integers m, n that $(u_m, u_n) = 1$ if $m = n$ and 0 otherwise. Consequently, $\{u_n \mid n \in Z\}$ is an orthonormal set in $L^2(T)$.

11. Show that for each positive integer n it is possible to choose c_n such that if $p_n(x) = 2^{-n}c_n(1 + cosx)^n$ for each $x \in \mathbf{R}$ then the sequence of functions $\{p_n\}$ converges uniformly on $[-\pi, -\varepsilon] \cup [\varepsilon, \pi)$ for each $\varepsilon > 0$.

12. Show that $\{p_n\}$ is a sequence of trigonometric polynomials such that

(1) $p_n(x) \geq 0$ for each $x \in \mathbf{R}$.

(2) $1/2\pi \int_{-\pi}^{\pi} p_n(x)dx = 1$.

(3) If $\delta_n(\varepsilon) = sup\{p_n(x) \mid x \in [-\pi, -\varepsilon] \cup [\varepsilon, \pi)\}$ then $lim_n \delta_n(\varepsilon) = 0$ for each $\varepsilon > 0$.

13. Let $C(T)$ denote the class of all continuous, complex valued, 2π-periodic functions on $\mathbf{R}$ with norm $|f|_\infty = sup_x |f(x)|$. For each $f \in C(T)$ we can associate a sequence of functions $\{P_n\}$ defined by

$$P_n(y) = 1/2\pi \int_{[-\pi,\pi)} f(y - x)p_n(x)dx.$$

Show that $\{P_n\}$ is a sequence of trigonometric polynomials such that for each positive integer n,

$$P_n(y) = 1/2\pi \int_{[-\pi,\pi)} f(x)p_n(y - x)dx.$$

14. Let $f \in C(T)$. Since f is uniformly continuous on $[-\pi, \pi]$, if $\varepsilon > 0$, there exists a $\delta > 0$ such that $|f(x) - f(y)| < \varepsilon$ whenever $|x - y| < \delta$. By (2) of Exercise 12, it follows that

$$P_n(y) - f(y) = 1/2\pi \int_{[-\pi,\pi)} [f(y - x) - f(y)]p_n(x)dx.$$

Show that $lim_{n \to \infty} |p_n - f|_\infty = 0$, and hence for each $f \in C(T)$ and $\varepsilon > 0$ there exists a trigonometric polynomial P such that $|f(x) - P(x)| < \varepsilon$ for each $x \in \mathbf{R}$.

15. Show that $\{u_n \mid n \in Z\}$ where u_n is defined by $u_n(x) = e^{inx}$ for each $x \in \mathbf{R}$ is a basis for $L^2(T)$.

11.3 The Spaces $L^p(\mu)$ and Banach Spaces

Hilbert space norms are fairly specialized, being based on the concept of an inner product. This is what accounts for there being so few of them, essentially one for each cardinal. The L^p-spaces are a special case of a more general class of spaces called the *Banach spaces.* These spaces derive their name from Stefan Banach who published a series of papers between 1922 and 1932 in several journals that culminated in the now famous *Theorie des Operations lineaires* (Monografje Matematyczne, Volume 1, Warsaw, 1932). In the special case $p = 2$, $L^2(\mu)$ is both a Hilbert space and a Banach space. As we shall see from the definition of Banach spaces, this is because all Hilbert spaces are Banach spaces.

A **normed linear space** X is a vector space over the complex field **C** (i.e., an abelian group in which multiplication of group members by complex numbers, called scalar multiplication, is defined that satisfies the distributive laws) in which a non-negative real number $|x|$ (called the norm of x) is associated with each $x \in X$ that satisfies the following properties:

B1. $|x| = 0$ if and only if $x = 0$.
B2. $|x + y| \leq |x| + |y|$ for each pair $x,y \in X$.
B3. $|\alpha x| = |\alpha|\,|x|$ for each $x \in X$ and $\alpha \in \mathbf{C}$.

Here, $|\alpha|$ denotes the modulus of the complex number α. A metric can be defined in a normed linear space in the following way. Define $d{:}X \times X \to [0,\infty)$ by $d(x,y) = |x - y|$ for each pair $x, y \in X$. That d is actually a metric is left as an exercise. If the metric space (X, d) is complete, X is said to be a **Banach space.** The simplest Banach space is merely **C** itself, normed by absolute value, i.e., $|x|$ is simply the absolute value (modulus) of x for each $x \in \mathbf{C}$. One can also discuss *real* Banach spaces by restricting the field of scalars to **R**.

The topology induced on X by d is called the **norm topology** and the set $S(0,1) = \{x \in X \mid |x| \leq 1\}$ is the **closed unit sphere** in X. A mapping T of a normed linear space X into a normed linear space Y is said to be a **linear transformation** if $T(x + y) = T(x) + T(y)$ and $T(\alpha x) = \alpha T(x)$ for each pair $x,y \in X$ and $\alpha \in \mathbf{C}$ (linear transformations are sometimes called vector space homomorphisms – the definition does not depend on the norm, only on the vector spaces X and Y). Linear transformations are also commonly referred to as **linear operators.** In the special case the space Y is the Banach space **C**, T is referred to as a **linear functional.** The **kernel** of a linear operator is the set of all elements in X that get mapped onto the zero element of Y, i.e., $Ker(T) = \{x \in X \mid T(x) = 0\}$. The proof of the following proposition is left as an exercise.

PROPOSITION 11.11 The kernel of a linear operator $T{:}X \to Y$ from the linear space X into a linear space Y is a linear subspace of X.

A linear transformation T is said to be **bounded** if there exists a real number α such that $|T(x)| \le \alpha|x|$ for each $x \in X$. The smallest α with this property is called the **norm** of T and is denoted by $|T|$. It is easily seen that $|T| = sup\{|T(x)|/|x| \mid x \in X \text{ and } x \neq 0\}$. Since $|T(\alpha x)| = |\alpha T(x)| = |\alpha||T(x)|$ for each $x \in X$ and $\alpha \in \mathbf{C}$, we could restrict ourselves to *unit vectors* (i.e., $x \in X$ such that $|x| = 1$) in the definition of the norm of T. In this case, we would have $|T| = sup\{|T(x)|/|x| \mid x \in X \text{ and } |x| = 1\}$. A bounded linear transformation T maps the closed unit sphere in X into the closed sphere $S(0,|T|) = \{y \in Y| \; |y - 0| \le |T|\}$. To see this, let $x \in S(0,1)$ in X. Then $|x| \le 1$. Since $|T(x)| \le |T||x|$, we have $|T(x)|/|x| \le |T|$. Since $|x| \le 1$ this implies $|T(x)| \le |T|$ which in turn implies $|T(x) - 0| \le |T|$ so $T(x) \in S(0,|T|)$.

THEOREM 11.13 Let $T:X \to Y$ be a linear operator from a normed linear space X into a normed linear space Y. Then the following statements are equivalent:

(1) T is bounded.

(2) T is uniformly continuous.

(3) T is continuous at some point of X.

Proof: If T is bounded, $|T(x) - T(y)| = |T(x - y)| \le |T||x - y|$ for each pair $x,y \in X$. Then if $\varepsilon > 0$ and $|x - y| < \varepsilon/|T|$, we have $|T(x) - T(y)| \le |T|(\varepsilon/|T|) = \varepsilon$ so T is uniformly continuous. Thus (1) $\to$ (2). That (2) $\to$ (3) is clear.

To show (3) $\to$ (1), assume T is continuous at $z \in X$ and suppose $\varepsilon > 0$. Then there exists a $\delta > 0$ such that $|x - z| < \delta$ implies $|T(x) - T(z)| < \varepsilon$. Then $|x| \le \delta$ implies $|z + x - x| \le \delta$ which in turn implies $|T(z + x) - T(z)| < \varepsilon$. But $|T(z + x) - T(z)| = |T(x)|$ so $|x| < \delta$ which implies $|T(x)| < \varepsilon$. Hence $|T(x)|/\delta > |x|/\varepsilon$ which implies $|T(x)|/|x| > \delta/\varepsilon$ which in turn implies $|T(x)|/|x| < \varepsilon/\delta$. Therefore, $|x| < \delta$ implies $|T(x)|/|x| < \varepsilon/\delta$. Now let $y \in X$ and let $\alpha \in \mathbf{C}$ such that $|\alpha y| < \delta$. Then $|T(\alpha y)|/|\alpha y| < \varepsilon/\delta$ implies $|\alpha||T(y)|/|\alpha||y| < \varepsilon/\delta$ so $|T(y)|/|y| < \varepsilon/\delta$. Thus T is bounded by ε/δ. ▪

Let $L(X, Y)$ denote the set of all bounded linear operators of a normed linear space X into a normed linear space Y. Define *addition* and *scalar multiplication* on $L(X, Y)$ as follows: if $S, T \in L(X, Y)$ and $\alpha \in \mathbf{C}$, then $S + T$ is defined by $(S + T)(x) = S(x) + T(x)$ for each $x \in X$ and αT is defined by $(\alpha T)(x) = \alpha T(x)$ for each $x \in X$. The following lemma is left as an exercise.

LEMMA 11.3 $L(X, Y)$ is a normed linear space with respect to $|\;|$.

THEOREM 11.14 If Y is a Banach space then so is $L(X, Y)$.

Proof: Let $\{T_n\}$ be a Cauchy sequence in $L(X, Y)$. For each $x \in X$, $\{T_n(x)\}$ is a sequence in Y. For $x_0 \in X$, let $\varepsilon > 0$. Since $\{T_n\}$ is Cauchy, there exists a positive integer N such that $m, n > N$ implies $|T_m - T_n| < \varepsilon/|x_0|$. Now $|T_m(x_0) - T_n(x_0)| \le |T_m - T_n||x_0| < \varepsilon$ so $\{T_n(x_0)\}$ is Cauchy in Y and therefore converges to some $x_0' \in Y$. Define $T:X \to Y$ by $T(x) = x'$ for each $x \in X$. Then $\{T_n(x)\}$ converges to $T(x)$ for each $x \in X$.

To show $T \in L(X, Y)$, we must show T is linear and bounded. To show T is linear, let $x, y \in X$ and $\alpha \in \mathbf{C}$. Then $T(x + y) = (x + y)'$ and $T(\alpha x) = (\alpha x)'$. Let $\varepsilon > 0$. Since $\{T_n(x)\}$ converges to $T(x)$ for each $x \in X$, there is a positive integer N such that $n > N$ implies $|T_n(x) - x'| < \varepsilon/2$ and $|T_n(y) - y'| < \varepsilon/2$. Now $|T_n(x+y) - (x' + y')| = |T_n(x) - x' + T_n(y) - y'| \le |T_n(x) - x'| + |T_n(y) - y'| < \varepsilon$. Therefore, $n > N$ implies $|T_n(x + y) - (x' + y')| < \varepsilon$ so $\{T_n(x + y)\}$ converges to $(x' + y')$. Hence $(x + y)' = (x' + y')$ so $T(x + y) = T(x) + T(y)$. Also, there exists a positive integer M such that $n > M$ implies $|T_n(x) - x'| < \varepsilon/|\alpha|$ which in turn implies $|T_n(\alpha x) - \alpha x'| < \varepsilon$. Therefore, $\{T_n(\alpha x)\}$ converges to $\alpha x'$ so $(\alpha x)' = \alpha x'$ which implies $T(\alpha x) = \alpha T(x)$. Consequently, T is linear.

To show that T is bounded, let $\varepsilon > 0$. Since $\{T_n\}$ is Cauchy, there exists a positive integer M such that $m, n > M$ implies $|T_m - T_n| < \varepsilon$. Pick $m > M$. Then $|T_m| < \infty$. For each $n > M$, $|T_n| = |T_n - 0| \le |T_n - T_m| + |T_m - 0| < \varepsilon + |T_m|$. Consequently, there exists a bound β such that $|T_n| < \beta$ for each n. Let $x \in X$ such that $|x| = 1$. Then there exists a positive integer N such that $n > N$ implies $|T_n(x) - T(x)| < \varepsilon$. Now $|T_n(x)| \le |T_n||x| < \beta$ and $|T(x)| \le |T(x) - T_n(x)| + |T_n(x) - 0| < \varepsilon + \beta$. Therefore, $|T| = sup\{|T(x)| \mid |x| = 1\} < \varepsilon + \beta$ so T is bounded.

Consequently, $T \in L(X, Y)$. It remains to show that $\{T_n\}$ converges to T with respect to the norm $|\ |$. For this, let $\varepsilon > 0$. Then there exists a positive integer M such that $m, n > M$ implies $|T_m - T_n| < \varepsilon/4$. Suppose there is no $n > M$ with $|T_n - T| < \varepsilon/2$. Then for each $n > M$, $|T_n - T| \ge \varepsilon/2$. Let $m > M$. Then $sup\{|T_m(x) - T(x)| \mid |x| = 1\} \ge \varepsilon/2$ which implies there exists an $x \in X$ with $|x| = 1$ such that $|T_m(x) - T(x)| \ge \varepsilon/2$. Now, for each $n > M$, $|T_m(x) - T_n(x)| \le |T_m - T_n||x| < \varepsilon/4$. Therefore, for each $n > M$, $|T_n(x) - T(x)| \ge \varepsilon/4$ for otherwise, $|T_m(x) - T(x)| \le |T_m(x) - T_n(x)| + |T_n(x) - T(x)| < \varepsilon/2$ which is a contradiction. Hence $\{T_n(x)\}$ does not converge to $T(x)$ which is a contradiction. Therefore, there exists a $k > M$ with $|T_k - T| < \varepsilon/2$. But then for each $n > M$, $|T_n - T| \le |T_n - T_k| + |T_k - T| < \varepsilon$ so $\{T_n\}$ converges to T. ∎

$L(X, Y)$, where Y is a Banach space, is our first example of a Banach space whose elements are functions. There are many more function spaces that are Banach spaces. It is easily shown that for each $1 \le p \le \infty$ that $L^p(\mu)$ is a Banach space with respect to the L^p-norm. If Z is a dense linear subspace of X and Y is a Banach space, then by Theorem 1.16, if $T \in L(Z, Y)$, T has a unique extension to a member T' of $L(X, Y)$.

THEOREM 11.15 $|T'| = |T|$.

Proof: Clearly, $|T| \le |T'|$ from the definition of the $|\ |$-norm. Suppose $|T| < |T'|$. Then there exists an $x \in X - Z$ with $|x| = 1$ such that $|T'(x)| > |T|$. Since Z is dense in X, there exists a sequence $\{z_n\} \subset Z$ that converges to x. For each positive integer n, put $w_n = \alpha_n z_n$ where $\alpha_n = 1/|z_n|$. Then $\{w_n\}$ is a sequence of *unit vectors* in Z that also converges to x. This follows from the fact that $|x - w_n| \le |x - z_n| + |z_n - w_n|$ and both terms on the right hand side of this inequality can be made arbitrarily small if n is large enough. $|z_n - w_n| = |(1 - \alpha_n)z_n| = |1 - \alpha_n||z_n| = |(|z_n| - 1)|$ and $|z_n| \le |z_n - x| + |x - 0| = |z_n - x| + 1$. This last term converges to 1 as $n \to \infty$ so $\{|z_n|\}$ converges to 1 which implies $\{|z_n - w_n|\}$ converges to 0.

Therefore, $|T'(w_n)| = |T(w_n)| \le |T|$ for each positive integer n. Since T' is continuous, $\{T'(w_n)\}$ converges to $T'(x)$. But this is impossible since $|T'(x)| > |T|$ and $|T'(w_n)| \le |T|$ for each n. We conclude that $|T'| = |T|$. ∎

The complex numbers **C**, normed by their *absolute value* form a Banach space. $L(X, \mathbf{C})$ is called the **dual space** of X and is denoted by X^*. $L(X, \mathbf{C})$ is a Banach space by Theorem 11.14. The interplay between X and its dual space X^* is the basis of much of that field of mathematics known as *functional analysis*. To explore this interplay, one needs the *Hahn-Banach Theorem*. The Hahn-Banach Theorem essentially says that if $Y = \mathbf{C}$, then we can drop the assumption that Z be dense in X. Linear transformations from Z into **C** can then be extended to X in such a way that Theorem 11.15 still holds.

By a **real linear functional**, we mean a linear operator from a real vector space (vector space over the real field). Let f be a complex linear functional on a linear space X. Then for each $x \in X$, $f(x) = u(x) + iv(x)$ for some real valued functions u and v on X. Since X is a vector space over the complex field, it is clearly a vector space over the real field as well. It is easily seen that the linearity of f implies the linearity of u and v, i.e., u and v are real linear functionals.

PROPOSITION 11.12 If X is a linear space and f is a linear functional on X then:

(1) If u is the real part of f then $f(x) = u(x) - iu(ix)$ for each $x \in X$.

(2) If u is a real linear functional on X and $f(x) = u(x) - iu(x)$ for each $x \in X$, then f is a (complex) linear functional on X.

(3) If X is a normed linear space and $f(x) = u(x) - iu(ix)$ for each $x \in X$, then $|f| = |u|$.

Proof: If $\alpha, \beta \in \mathbf{R}$ and $\gamma = \alpha + i\beta$, then the real part of $i\gamma$ is $-\beta$. Therefore, $\gamma - Re(\gamma) - iRe(i\gamma)$ for each $\gamma \in \mathbf{C}$. Then (1) follows with $\gamma = f(x)$. To show (2), it is clear that $f(x + y) = f(x) + f(y)$ and that $f(\alpha x) = \alpha f(x)$ for each $\alpha \in \mathbf{R}$. But we must also show this second equation for $\alpha \in \mathbf{C}$. It will suffice to show it for $\alpha = i$. For this note that $f(ix) = u(ix) - iu(-x) = u(ix) + iu(x) = if(x)$.

To show (3) note that $|u(x)| \le |f(x)|$ for each $x \in X$ which implies $|u| \le |f|$. Let $x \in X$. Put $\beta = f(x)/|f(x)|$. Then $|\beta| = 1$ and $\beta|f(x)| = f(x)$. Put $\alpha = \beta^{-1}$. Then $|\alpha| = 1$ and $\alpha f(x) = |f(x)|$. Hence $|f(x)| = f(\alpha x)$ which implies

$f(\alpha x)$ is real so $f(\alpha x) = u(\alpha x) \le |u|\,|\alpha x|$. Then $|f| = sup\{|f(x)| \mid |x| = 1\} \le |u|$. Therefore, $|f| = |u|$. ∎

One of the most important theorems in the theory of Banach spaces is the *Hahn-Banach Theorem*. It allows us to extend bounded linear functionals on subspaces of a normed linear space in such a way that the norm is preserved.

THEOREM 11.16 (S. Banach, H Hahn, 1932) If Y is a subspace of a normed linear space X and if f is a bounded linear functional on Y, then f can be extended to a bounded linear functional F on X such that $|F| = |f|$.

Proof: We first prove the theorem assuming the field of scalars to be real, i.e., we assume X is a real normed linear space and f is a real bounded linear functional on $Y \subset X$. If $|f| = 0$ then clearly, the desired extension is $F = 0$. If $|f| > 0$, we may assume, without loss of generality, that $|f| = 1$ since if $|f| \neq 1$, there exists an $\alpha \in \mathbf{R}$ with $|\alpha|\,|f| = 1$ which implies $|\alpha f| = 1$. We can then prove the theorem for αf and simply divide the extension F by $|\alpha|$. Assuming $|f| = 1$, pick $z \in X - Y$ and let N be the subspace of X spanned by z and Y (i.e., $N = \{x + \alpha z \mid x \in Y \text{ and } \alpha \in \mathbf{R}\}$). Define $f_N: N \to \mathbf{R}$ by $f_N(x + \alpha z) = f(x) + \alpha\lambda$ for any fixed $\lambda \in \mathbf{R}$. It is left as an exercise to show that f_N is a linear functional on N such that f_N restricted to Y is f. We next show that it is possible to pick λ such that $|f_N| = 1$. For this, first note that by the definition of $|\ |$, that $|f| \le |f_N|$. Also,

$$|f_N| = sup\{\frac{|f(x)+\alpha\lambda|}{|x+\alpha z|} \mid x \in Y \text{ and } \alpha \in \mathbf{R}\}$$

so that if $|f(x) + \alpha\lambda| \le |x + \alpha z|$ for each $x \in Y$ and $\alpha \in \mathbf{R}$, then $|f_N| \le 1$. Hence $|f(x) + \alpha\lambda| + |x + \alpha z|$ for each $x \in Y$ and $\alpha \in \mathbf{R}$ which implies $|f_N| = 1$. Therefore, it suffices to show that λ can be chosen such that $|f(x) + \alpha\lambda| \le |x+\alpha z|$ for each $x \in Y$ and $\alpha \in \mathbf{R}$. This can be shown if we choose λ such that

$$\frac{|f(x)+\alpha\lambda|}{|\alpha|} \le \frac{|x+\alpha z|}{|\alpha|} \text{ for each } x \in Y \text{ and } \alpha \in \mathbf{R}.$$

But this is equivalent to showing that we can find a λ such that $|f(x/\alpha) + \lambda| \le |x/\alpha + z|$ which in turn is equivalent to finding a $w \in Y$ such that $|f(w) - \lambda| \le |w - z|$ in view of the substitution $w = -x/\alpha$. For each $w \in Y$ put $\gamma(w) = f(w) - |w - z|$ and $\beta(w) = f(w) + |w - z|$. If $\gamma(w) \le \lambda \le \beta(w)$, then $|f(w) - \lambda| \le |w - z|$. Hence $|f_N| = 1$ if $\gamma(w) \le \lambda \le \beta(w)$ for each $w \in Y$.

To show this, let $I(w)$ be the interval $[\gamma(w), \beta(w)]$ for each $w \in Y$. Then $|f_N| = 1$ if $\cap\{I(w) \mid w \in Y\} \neq \varnothing$, or equivalently if $\gamma(w) \le \beta(v)$ for each pair $w,v \in Y$. Now $\gamma(w) - \beta(w) = f(w - v) - |w - z| - |z - v|$. Since $f(w - v) \le |f(w - v)| \le |f|\,|w - v| = |w - v| \le |w - z| + |z - v|$, $\gamma(w) - \beta(v) \le 0$ so $\gamma(w) \le \beta(v)$ for each pair $w,v \in Y$. Hence $\cap\{I(w)\,|\,w \in Y\} \neq \emptyset$. Pick $\lambda \in \cap\{I(w)\,|\,w \in Y\}$. Then $\gamma(w) \leq \lambda \leq \beta(w)$ for each $w \in Y$ so $|f_N| = 1$.

We have shown that f can be extended to N such that the extension f_N has the same norm as f. Let P be the collection of all subspaces N such that $Y \subset N$ and there exists a real linear extension f_N of f on N with $|f_N| = 1$. Define a partial order $\leq$ on P by $N \leq S$ if $N \subset S$ and $f_N(x) = f_S(x)$ for each $x \in N$. By the Hausdorff Maximal Principle (an equivalent form of the Axiom of Choice), there exists a maximal totally ordered subcollection Q of P. Put $A = \cup\{N\,|\,N \in Q\}$. It is easily shown that A is a subspace of X. For each $x \in A$, define $F(x) = f_N(x)$ such that $x \in N$ for some $N \in Q$. Clearly, F is well defined since $f_N(x) = f_S(x)$ if $N \subset S$ and Q is totally ordered. It is left as an exercise to show that F is a linear functional on A.

To show that $|F| = 1$, note first that $|f| \leq |F|$ so that $1 \leq |F|$. Suppose $|F| > 1$. Then there exists an $x \in A$ such that $|F(x)| > 1$. But $x \in A$ implies $x \in N$ for some $N \in Q$ so $|f_N(x)| > 1$ which is a contradiction since $|f_N| = 1$. It is easily seen that $A = X$, for otherwise, there would be $y \in X - A$ and as we have already seen, the space Y spanned by y and A would be a subspace of X larger than A and F could be extended to a linear functional G on Y such that $|G| = 1$. But then $Y \in Q$ which implies $Y \subset A$ which is a contradiction. Therefore, F is the desired extension of f to X.

To prove the theorem for a complex bounded linear functional f on a subspace Y of a complex normed linear space X, let u be the real part of f. As shown above, u can be extended to a real linear functional U on X such that $|U| = |u|$. Define $F{:}X \to \mathbf{C}$ by $F(x) = U(x) - iU(ix)$ for each $x \in X$. By Proposition 11.12, F is a complex linear functional on X such that $|F| = |U| = |u| = |f|$. Moreover, if $x \in Y$ then $F(x) = U(x) - iU(ix) = u(x) - iu(ix) = f(x)$ so F is an extension of f. ∎

COROLLARY 11.4 If Y is a linear subspace of a normed linear space X and u is a vector in $X - Y$ with distance $d(u, Y) > 0$ from u to Y, then there exists a bounded (continuous) linear functional $f{:}X \to \mathbf{C}$ such that $|f| = 1$, $f(Y) = 0$ and $f(u) = d(u, Y)$.

Proof: Let $K = \{x \in X\,|\,|x| = 0\}$. Then $Z = Y + K = \{y + x\,|\,y \in Y \text{ and } x \in K\}$ is a linear subspace of X. Since $Y \subset Z$ it follows that $d(u, Z) \leq d(u, Y)$. Let $\varepsilon > 0$. From the definition of $d(u, Z)$, there exists a vector $y + x \in Z$ with $y \in Y$ and $x \in K$ such that $|u - (y + x)| < d(u, Z) + \varepsilon$. Hence

$$d(u, Y) \leq |u - y| \leq |u - (y + x)| + |(y + x) - y| =$$

$$|u - (y + x)| + |x| = |u - (y + x)| + 0 < d(u, Z) + \varepsilon.$$

Since this holds for each $\varepsilon > 0$, we have $d(u, Y) \leq d(u, Z)$. But then $d(u, Y) =$

$d(u, Z)$. Let V denote the linear subspace of X spanned by u and Z, i.e., $V = \{cu + z \mid c \in \mathbf{C} \text{ and } z \in Z\}$. Define $g{:}V \to \mathbf{C}$ by $g(cu + z) = cd(u, Y)$. Clearly g is a linear mapping.

To show that g is a bounded linear functional let $cu + z$ be an arbitrary vector in V. If $c = 0$ then $g(cu + z) = 0$. If $c \neq 0$, put $w = -z/c$. Then $w \in Z$ and $|cu + z| = c||u + z/c| = |c||u - w| \geq cd(u, Z) = |c|d(u, Y) > 0$. Hence $|g(cu+z)| = |c|d(u, Y) \leq |cu + z|$ so that g is bounded. This also shows that $|g| \leq 1$.

To show $|g| = 1$, let $r \in (0, 1)$. Then there exists a $z \in Z$ such that $|u - z| < d(u, Z)/r$ which implies $d(u, Z) > |u - z|r$. Now $|g(u - z)| = d(u, Z) > r|u - z|$ so $|g| > r$. Since this is true for each $r \in (0, 1)$ we have $|g| \geq 1$. But then $|g| = 1$. By the Hahn-Banach Theorem, g can be extended to a bounded linear functional $f{:}X \to \mathbf{C}$ such that $|f| = |g| = 1$. Since f is an extension of g and $Y \subset V$ we have $f(Y) = g(Y) = 0$. Finally, $f(u) = g(u) = d(u, Y)$. ∎

COROLLARY 11.5 For each vector u in a normed linear space X with $|u| > 0$, *there exists a continuous linear functional* $f{:}X \to \mathbf{C}$ *with* $|f| = 1$ *and* $f(u) = |u|$.

Returning now to the dual space X^* of a normed linear space X, we note that it is possible to form the **second dual space** of X or *bidual space* or the *second conjugate space* $(X^*)^* = L(X^*, \mathbf{C})$. $(X^*)^*$ is often denoted by X^{**}. Again by Theorem 11.14, X^{**} is a Banach space. Define a function $\Omega{:}X \to X^{**}$ as follows. If $x \in X$, put $\Omega(x) = \Phi_x$ where Φ_x denotes the functional $\Phi_x{:}X^* \to \mathbf{C}$ such that $\Phi_x(\phi) = \phi(x)$ for each $\phi \in X^*$. Since $\phi \in X^*$ implies $\phi(x) \in \mathbf{C}$ the function Ω is well defined.

To verify that Ω is a linear function, let $u, v \in X$ and $c, d \in \mathbf{C}$. Then $\Omega(cu+dv) = \Phi_{(cu+dv)}$. Let $\phi \in X^*$. Then $\Phi_{(cu+dv)}(\phi) = \phi(cu + dv) = c\phi(u) + d\phi(v) = c\Phi_u(\phi) + d\Phi_v(\phi) = [c\Phi_u + d\Phi_v](\phi)$. Hence $\Phi_{(cu+dv)} = c\Phi_u + d\Phi_v = c\Omega(u) + d\Omega(v)$. But then $\Omega(cu + dv) = c\Omega(u) + d\Omega(v)$ so Ω is linear.

To show Ω is bounded (continuous) recall that $|\Omega| = sup\{|\Omega(x)|/|x| \mid x \in X \text{ and } |x| = 1\}$. Let $y \in X$ with $|y| = 1$. Then $|\Omega(y)| = |\Phi_y| = sup\{\Phi_y(\phi)|/|\phi| \mid \phi \in X^* \text{ and } |\phi| = 1\}$. Let $\psi \in X^*$ such that $|\psi| = 1$. Then $|\Phi_y(\psi)| = |\psi(y)| \leq |\psi||y| = |y|$. Therefore, $|\Phi_y| \leq |y| = 1$ which implies $|\Omega(x)| \leq 1$ for each x with $|x| = 1$. Hence Ω is a bounded (continuous) linear operator and $|\Omega| \leq 1$.

Now if $x \in X$ with $|x| = 0$, then by the discussion above, $|\Omega(x)| = 0$. If $x \in X$ with $|x| > 0$ then by Corollary 11.5 there exists a $\phi \in X$ with $|\phi| = 1$ and $\phi(x) = |x|$. But then $|\Phi_x(\phi)| = |\phi(x)| = |x| = |x||\phi|$ which implies $|\Phi_x(\phi)|/|\phi| = |x|$ for each $\phi \in X^*$ with $|\phi| = 1$. Therefore, $|\Omega(x)| = |x|$. Consequently, $|\Omega| = 1$. Since $|\Omega(x)| = |x|$ for each x, Ω preserves distances so if $x \neq y$ then $\Omega(x) \neq \Omega(y)$ so Ω is one-to-one. Hence Ω is an isometric imbedding of X into X^{**}. We record this as:

THEOREM 11.17 For a normed linear space X the mapping $\Omega{:}X \to X^{**}$ *defined by* $\Omega(x) = \Phi_x$ *where* Φ_x *is the member of* X^{**} *defined by* $\Phi_x(\phi) = \phi(x)$ *for each* $\phi \in X^*$ *is an isometric imbedding of X into* X^{**} *such that* Ω *is linear and* $|\Omega| = 1$.

If Ω is an onto mapping, Ω is said to be a *Banach space isomorphism* of X onto X^{**}. In general, a continuous linear operator $f{:}X \to Y$ between Banach spaces X and Y is said to be a **Banach space isomorphism** if it is a metric space isomorphism. Clearly if Ω is onto then X must have been a Banach space to start with. If Ω is onto, X is said to be a **reflexive** normed linear space. Clearly, reflexive normed linear spaces are Banach spaces. Examples of reflexive normed linear spaces are the Euclidean space $\mathbf{R}^n$. An example of a Banach space that is not reflexive is $C([0, 1])$ where $|f|$ is defined to be $sup\{\,|f(x)|\,|\,x \in [0, 1]\}$ for each $f \in C([0, 1])$.

In addition to the Hahn-Banach Theorem, there are three other theorems of great significance in the theory of Banach spaces. These are the *open mapping theorem*, the *closed graph theorem* and the *uniform boundedness theorem*. Together, these four theorems are considered the cornerstones of functional analysis. Two of the theorems, the open mapping theorem and the uniform boundedness theorem are consequences of a classical theorem by R. Baire that has many ramifications in modern analysis and topology. Baire's Theorem was proven for the real numbers in 1889. It was F. Hausdorff who proved it for complete metric spaces in *Grundzuge der Mengenlehre* published in Leipzig in 1914. It is Hausdorff's version of the theorem that we now present.

THEOREM 11.18 (R. Baire, 1889, F. Hausdorff, 1914) The intersection of a countable family of dense open sets in a complete metric (or pseudo-metric) space X is dense in X.

Proof: Let $\{U_n\}$ be a sequence of dense open sets in X and put $U = \cap U_n$. To show that U is dense in X, we need to demonstrate that for each open set $V \subset X$ that $U \cap V \neq \varnothing$. Since U_1 is dense in X, $U_1 \cap V \neq \varnothing$. Pick $x_1 \in U_1 \cap V$. Then there exists an $r_1 < 1$ with $r_1 > 0$ such that $Cl(S(x_1, r_1)) \subset U_1 \cap V$. Next, since U_2 is dense in X, the open set $S(x_1, r_1) \cap U_2$ is not empty. Pick $x_2 \in S(x_1, r_1) \cap U_2$. Then there exists an r_2 with $0 < r_2 < 1/2$ such that $Cl(S(x_2, r_2)) \subset S(x_1, r_1) \cap U_2$. Continuing this process inductively, we obtain a sequence $\{x_n\} \subset X$ and a sequence $\{r_n\} \subset \mathbf{R}$ such that for each n, $r_n < 2^{-n}$ and $S(x_n, r_n) \cap U_n$ contains $Cl(S(x_{n+1}, r_{n+1}))$.

To show $\{x_n\}$ is Cauchy, let $\varepsilon > 0$. Select a positive integer m such that $2^{-m+1} < \varepsilon$. Then for each positive integer $k > m$ we have $x_k \in S(x_m, r_m)$. Hence for each pair of positive integers $i, j > m$ we have $d(x_i, x_j) \leq d(x_i, x_m) + d(x_m, x_j)$

$< r_m + r_m < 2^{-m} + 2^{-m} = 2^{-m+1} < \varepsilon$. Therefore, $\{x_n\}$ is Cauchy and hence converges to some point $x \in X$ since X is complete. To show $x \in U \cap V$ let m be any positive integer. Then

$$x_n \in Cl(S(x_{m+1}, r_{m+1})) \subset S(x_m, r_m) \cap U_m \subset V \cap U_m$$

for each positive integer $n > m$. But then $x \in Cl(S(x_{m+1}, r_{m+1})) \subset V \cap U_m$ for each positive integer m. Hence $x \in V \cap U$. ∎

A subset E of a topological space X is said to be **nowhere dense** in X if each nonempty open set U contains a nonempty open set V that contains no point of E. The integers are an example of a nowhere dense subset of **R**. A space X is said to be of the **first category** (having nothing to do with the concept of category introduced in Chapter 5) if it is the union of a countable collection of nowhere dense subsets. X is said to be of the **second category** if it is not of the first category.

PROPOSITION 11.13 For a topological space X the following are equivalent:

(1) E is nowhere dense in X.

(2) $Int(Cl(E)) = \varnothing$.

(3) $X - Cl(E)$ is dense in X.

Proof: To show (1) → (2) assume $Int(Cl(E)) \neq \varnothing$. Since E is nowhere dense in X there is a nonempty open set $V \subset Int(Cl(E))$ that contains no point of E which is impossible. Therefore, (1) → (2). To show (2) → (3) assume $D = X - Cl(E)$ is not dense in X. Then $U = X - Cl(D)$ is a nonempty open set in $Cl(E)$. But then $U \subset Int(Cl(E))$ which implies $Int(Cl(E)) = \varnothing$ which is a contradiction. To show (3) → (1) assume E is not nowhere dense in X. Then there exists a nonempty open set $U \subset X$ such that every nonempty open subset of U contains a point of E. Hence $U \subset Cl(E)$ which implies $X - Cl(E)$ is not dense in X which is a contradiction. ∎

THEOREM 11.19 A complete metric (pseudo-metric) space X is of the second category.

Proof: Let $\{E_n\}$ be a countable collection of nowhere dense subsets of X. By Proposition 11.13, $X - Cl(E_n)$ is a dense open subset of the complete metric space X. By Theorem 11.18, $\cap(X - Cl(E_n)) \neq \varnothing$. Then there exists a point $p \in X - Cl(E_n))$ for each n which implies p is not in E_n for each n so $\cup E_n \neq X$. Therefore, X is of the second category. ∎

LEMMA 11.4 For any continuous linear operator $f: X \to Y$ of a Banach space X onto a Banach space Y, there exists a $\delta > 0$ such that the image of the open unit sphere in X contains each $y \in Y$ with $|y| < \delta$.

Proof: Let $S = \{x \in X \mid |x| < 1\}$ be the unit sphere in X. For each integer $n \geq 0$ put $S_n = S(0, 2^{-n})$. Then $S = S_0$. Clearly the sequence $\{nS_1\}$, where $nS_1 = \{nx \mid x \in S_1\}$, covers X. Since f is onto, $Y = \cup f(nS_1) = \cup nf(S_1)$. By Theorem 11.19, Y is of the second category. It is easily shown that for each n, the function $n:Y \to Y$, defined by $n(y) = ny$ for each $y \in Y$, is a homeomorphism. Consequently, S_1 cannot be nowhere dense in Y. Therefore, there exists a $v \in Y$ and a $\delta > 0$ such that $S(v, 2\delta) \subset Cl(f(S_1))$. But then

$$S(0, 2\delta) \subset Cl(f(S_1)) - v \subset 2Cl(f(S_1)) = Cl(f(S)).$$

Since f is linear, this implies $S(0, \delta/2^{n-1}) \subset Cl(f(S_n))$ for each nonnegative integer n.

Let $y \in Y$ with $|y| < \delta$. Since $y \in Cl(f(S_1))$, there exists a vector $x_1 \in S_1$ such that $|y - f(x_1)| < \delta/2$. Then $y - f(x_1) \in Cl(f(S_2))$ and therefore there exists a vector $x_2 \in S_2$ with $|y - f(x_1) - f(x_2)| < \delta/4$. Continuing this process inductively, we obtain a sequence $\{x_n\} \subset X$ with $x_n \in S_n$ for each n and such that $|y - \Sigma_{i=1}^n f(x_i)| < \delta/2^n$ for each n. But this means that the sequence $\{\sigma_n\}$ of partial sums $\sigma_n = \Sigma_{i=1}^n f(x_i)$ converges to y.

Since $x_n \in S_n$ for each n, $|x_n| < 2^{-n}$. If we let $z_n = x_1 + \ldots + x_n$ be the nth partial sum of the series $\Sigma_{i=1}^\infty x_i$, we see that $|z_n| < 2^{-n}$ for each n which implies $\{z_n\}$ is a Cauchy sequence and hence converges to some $z \in X$. For each n, $\sigma_n = f(z_n)$ so $\{f(z_n)\}$ converges to y. Since f is continuous $f(z) = y$. It remains to show that $z \in S$. Since the metric $d(x, y) = |x - y|$ is continuous in X, we have $|z| = d(0, z) = lim_n d(0, z) = lim_n |z_n| = \Sigma_{n=1}^\infty |x_n| < \Sigma_{n=1}^\infty 2^{-n} = 1$. Therefore, $|z| < 1$ so $z \in S$. Hence S contains each $y \in Y$ such that $|y| < \delta$. ∎

THEOREM 11.20 (Open Mapping Theorem) Each continuous linear operator $f:X \to Y$ from a Banach space X onto a Banach space Y is an open mapping.

Proof: Let U be open in X. To show that $f(U)$ is open in Y, let $v \in f(U)$. Then there exists a $u \in U$ with $f(u) = v$. Since U is open there exists an $\varepsilon > 0$ such that $S(u, \varepsilon) \subset U$. By Lemma 11.4 there is a $\delta > 0$ such that $S(0, \delta) \subset f(S(0, 1))$. Let $y \in S(v, \delta\varepsilon)$. Since $|(y - v)/\varepsilon| = 1/\varepsilon |y - v| < 1/\varepsilon(\delta\varepsilon) = \delta$, there exists a $w \in S(0, 1) \subset X$ with $f(w) = (y - v)/\varepsilon$. Let $x = \varepsilon w + u$. Then $|x - u| = |\varepsilon w| = \varepsilon |w| < \varepsilon$ so $x \in U$. Since $f(x) = f(\varepsilon w + u) = \varepsilon f(w) + f(u) = (y - v) + v = y$ it follows that $y \in f(U)$. Hence $S(v, \delta\varepsilon) \subset f(U)$ so f is open. ∎

COROLLARY 11.6 A one-to-one, onto, bounded linear operator $f:X \to Y$ from a Banach space X to a Banach space Y is a homeomorphism, and the inverse $f^{-1}:Y \to X$ is also a continuous linear operator.

COROLLARY 11.7 Let X be a Banach space with norm $|\ |$. Let $|\ |'$ be another norm for X that makes X a Banach space, such that there exists a $\delta > 0$ with $|x| \leq \delta|x|'$ for each $x \in X$. Then there exists an $\varepsilon > 0$ such that $|x|' \leq \varepsilon|x|$ for each $x \in X$.

Proof: Let X' denote the Banach space X with the norm $|\ |'$. Then the identity operator $i:X' \to X$ is clearly linear. Since $|x| \leq \delta|x|'$ for each $x \in X$, i is a bounded linear operator. Since i is a one-to-one, onto, continuous linear operator it is a homeomorphism by Corollary 11.6 and $i^{-1}:X \to X'$ is also a bounded linear operator. Therefore, there exists an $\varepsilon > 0$ such that $|x|' \leq \varepsilon|x|$ for each $x \in X$. ■

LEMMA 11.5 The graph $\Gamma = \{(x, y) \in X \times Y | f(x) = y\}$ of an operator $f:X \to Y$ from a Banach space X into a Banach space Y is a closed subset of $X \times Y$ if and only if for any sequence $\{x_n\} \subset X$ that converges to some $x \in X$ such that $\{f(x_n)\}$ converges to $y \in Y$, we must have $f(x) = y$.

Proof: Assume Γ is closed and suppose $\{x_n\}$ is a sequence in X converging to a point x. Further, suppose $\{f(x_n)\}$ converges to $y \in Y$. Then $\{x_n\}$ is Cauchy in X and $\{f(x_n)\}$ is Cauchy in Y. Hence the product sequence $\{(x_n, f(x_n))\} \subset \Gamma$ is Cauchy in $X \times Y$ by Exercise 3 of Section 5.3. Since X and Y are complete, so is $X \times Y$ (Exercise 4, Section 5.3). Therefore, $\{(x_n, f(x_n))\}$ converges to a point $(a,b) \in \Gamma$ since Γ is closed. But then $x = a$ and $y = b$ (Exercise 2, Section 5.3). Suppose $f(x) \neq y$. Then (x, y) is not in Γ which implies there exists open sets $U \subset X$ and $V \subset Y$ such that $(x, y) \in U \times V$ and $U \times V \cap \Gamma = \emptyset$. But $\{(x_n, f(x_n))\} \subset \Gamma$ converges to (x, y) which is a contradiction. Hence $f(x) = y$.

Conversely, assume that for each sequence $\{x_n\} \subset X$ that converges to some $x \in X$ such that $\{f(x_n)\}$ converges to some $y \in Y$, we have $f(x) = y$. Suppose Γ is not closed. Then there exists a point $(x, y) \in X \times Y$ and a sequence $\{(x_n, f(x_n))\} \subset \Gamma$ such that $\{(x_n, f(x_n))\}$ converges to (x, y) but (x, y) is not in Γ. Since $\{(x_n, f(x_n))\}$ converges to (x, y), $\{x_n\}$ must converge to x and $\{f(x_n)\}$ must converge to y by Exercise 2 of Section 5.3. But then $f(x) = y$ which implies $(x,y) \in \Gamma$ which is a contradiction. Hence Γ is closed. ■

THEOREM 11.21 (Closed Graph Theorem) The graph Γ of a linear operator $f:X \to Y$ from a Banach space X into a Banach space Y is closed if and only if f is continuous.

Proof: Assume Γ is closed. Define $|x|' = |x| + |f(x)|$ for each $x \in X$. To show that $|\ |'$ is a norm in X first note that if $x = 0$ then $|x|' = 0$ since the linearity of f implies $f(0) = 0$. On the other hand, if $|x|' = 0$ then $|x| = -|f(x)|$ which implies $|x| = 0 = |f(x)|$ since $|x|$ and $|f(x)|$ are both nonnegative. But then x

$= 0$ since $|\ |$ is a norm on X. Therefore, $|\ |'$ satisfies B1. That $|\ |'$ satisfies B2 and B3 is an easy exercise. Consequently, X is a normed linear space with respect to $|\ |'$. Let X' denote the linear space X equipped with the norm $|\ |'$.

To show that X' is complete let $\{x_n\} \subset X$ be a Cauchy sequence with respect to $|\ |'$. Then for a given $\varepsilon > 0$ there exists a positive integer N such that for $m, n > N$, $|x_m - x_n| + |f(x_m) - f(x_n)| < \varepsilon$. But then $|x_m - x_n| < \varepsilon$ and $|f(x_m) - f(x_n)| < \varepsilon$ for each pair $m, n > N$. Hence $\{x_n\}$ is Cauchy in X and $\{f(x_n)\}$ is Cauchy in Y which implies $\{x_n\}$ converges to some $x \in X$ and $\{f(x_n)\}$ converges to some $y \in Y$. By Lemma 11.5, $f(x) = y$.

To show that $\{x_n\}$ converges to x with respect to $|\ |'$, let $\varepsilon > 0$. Then there exists a positive integer N such that for each $n > N$, $|x_n - x| < \varepsilon/2$ and $|f(x_n) - f(x)| < \varepsilon/2$. Therefore, $|x_n - x|' = |x_n - x| + |f(x_n) - f(x)| < \varepsilon/2 + \varepsilon/2 = \varepsilon$. Hence $\{x_n\}$ converges to x with respect to $|\ |'$. Therefore, X' is complete. Since $|x| \le 1|x|'$ for each $x \in X$, by Corollary 11.7, there exists a $\delta > 0$ such that $|x|' \le \delta|x|$ for each $x \in X$. But then $|f(x)| \le |x| + |f(x)| = |x|' \le \delta|x|$ which implies f is a bounded linear operator and therefore continuous.

Conversely, if f is continuous, then if $\{x_n\}$ is any sequence in X that converges to some $x \in X$, $\{f(x_n)\}$ converges to $f(x) \in Y$ by Proposition 1.12. But then by Lemma 11.5, Γ is closed. ∎

LEMMA 11.6 (Uniform Boundedness Principle) If X is a metric (pseudo-metric) space and $F \subset C(X)$ such that for each $x \in X$ there exists a real number b_x with $|f(x)| \le b_x$ for each $f \in F$, then there exists a nonempty open set $U \subset X$ and a real number b such that $|f(x)| \le b$ for each $f \in F$ and $x \in U$.

Proof: For each positive integer n and each $f \in F$, $f^{-1}[-n, n]$ is a closed subset of X since f is continuous. Therefore, the intersection $E_n = \cap_{f \in F} f^{-1}[-n, n]$ is a closed subset of X. By hypothesis, for each $x \in$ X there exists a positive integer n_x such that $|f(x)| \le n_x$ for each $f \in F$. Therefore, $x \in E_{n_x}$. But then $X = \cup E_n$. By Theorem 11.19, X is of the second category, so one of the E_n's, say E_m, is not nowhere dense in X. By Proposition 11.13, the open set $U = Int(Cl(E_m)) \neq \varnothing$. Since E_m is closed, $U = Int(E_m)$. Consequently, $|f(x)| \le m$ for each $f \in F$ and $x \in$ U. Put $b = m$. Then b is the bound we were seeking. ∎

THEOREM 11.22 (Uniform Boundedness Theorem) If F is a family of continuous linear operators from a Banach space X into a Banach space Y such that for each $x \in X$ there exists a real number b_x with $|f(x)| \le b_x$ for each $f \in F$, then there exists a real number b such that $|f| \le b$ for each $f \in F$.

Proof: For each $g \in F$ define a function $\Omega_g : X \to \mathbf{R}$ by $\Omega_g(x) = |g(x)|$ for each $x \in X$. Put $G = \{\Omega_g \mid g \in F\}$. Then $G \subset C(X)$ and $|\Omega_g(x)| \le b_x$ for each $x \in X$ and $\Omega_g \in G$. By the previous lemma, there exists a nonempty open set $U \subset X$

and a $\delta > 0$ such that $|\Omega_g(x)| \leq \delta$ for each $\Omega_g \in G$ and every $x \in U$. Let $u \in U$. Since U is open there exists an $\varepsilon > 0$ such that $Cl(S(u, \varepsilon)) \subset U$.

Let $b = (\delta + b_u)/\varepsilon$. To show $|f| \leq b$ for each $f \in F$ let $f \in F$ and $x \in X$ with $|x| = 1$. Then $|(\varepsilon x + u) - u| = |\varepsilon x| = \varepsilon|x| = \varepsilon$. Hence $(\varepsilon x + u) \in U$. Therefore, $|f(x)| = |f(\varepsilon^{-1}\varepsilon x)| = \varepsilon^{-1}|f(\varepsilon x)| = \varepsilon^{-1}|f(\varepsilon x + u - u)| \leq \varepsilon^{-1}(|f(\varepsilon x+u)| + |f(u)|) \leq \varepsilon^{-1}(\delta + b_u) = b$. Consequently, $|f| = sup\{|f(x)| \,|\, |x| = 1\} \leq b$ for each $f \in F$. ∎

COROLLARY 11.8 (Banach-Steinhaus Theorem) If $\{f_n\}$ is a sequence of continuous linear operators from a Banach space X into a normed linear space Y that converges pointwise to an operator $f{:}X \to Y$, then f is also a continuous linear operator.

Proof: Since limits are unique in Y, it is easily seen that f is a linear operator. Let $x \in X$. Since $\{f_n(x)\}$ converges to $f(x)$, it follows that $b_x = sup\{|f_n(x)|\} < \infty$. By the Uniform Boundedness Theorem, there exists a $b \in \mathbf{R}$ such that $|f_n| \leq b$ for each n. Hence, for each $x \in X$, $|f(x)| = lim_n|f_n(x)| \leq b|x|$ since the distance function $d(x, y) = |x - y|$ is continuous in Y. But then f is bounded and therefore continuous. ∎

EXERCISES

1. Prove Proposition 11.11.

2. Prove Lemma 11.3.

3. Let $I = [0, 1]$ and let $f \in L^p(I)$ for $p > 1$. Show that there exists a $g \in L^q(I)$, where p and q are conjugate exponents, such that $|f|_p = |g|_q$ and $f(\phi) = \int_0^1 \phi(x)g(x)dx$ for each $\phi \in L^p(I)$.

4. Show that the composition $f \circledcirc g$ of two continuous linear operators of normed linear spaces is a continuous linear operator with $|f \circledcirc g| \leq |f||g|$.

THE DUAL SPACE

5. Let $f{:}X \to Y$ be a continuous linear operator from a normed linear space X into a normed linear space Y. Let X^* and Y^* denote the dual spaces of X and Y respectively. Define a function $f^*{:}Y^* \to X^*$ as follows. If $\phi \in Y^*$, let $f^*(\phi) = \phi^* \in X^*$ where ϕ^* is defined by $\phi^*(x) = \phi(f(x))$ for each $x \in X$. Show that f^* is a continuous linear operator with $|f^*| = |f|$. f^* is called the **dual** operator or *adjoint operator* of f.

6. A sequence $\{x_n\}$ in a normed linear space X is said to **converge weakly** to a

vector $x \in X$ if for each $f \in X^*$, $\{f(x_n)\}$ converges to $f(x)$. Show that a convergent sequence is weakly convergent.

7. Let τ denote the metric topology of a normed linear space X. Let σ be the collection of subsets U of X such that no sequence in $X - U$ weakly converges to a vector $u \in U$. Show that σ is a topology for X and that $\sigma \subset \tau$. Also show that σ is the coarsest topology in X that causes all members of X^* to be continuous. σ is called the **weak topology** of X.

8. By the **weak * topology** in the dual space X^* of a normed linear space X is meant the coarsest topology in X^* that causes the members of X^{**} to be continuous. Let $\Omega: X \to X^{**}$ denote the operator of Theorem 11.17. Show that the closed unit sphere in X^* is compact in the weak * topology of X^*. Hence, in a reflexive normed linear space the closed unit sphere is compact in the weak topology.

9. Let $f: X \to Y$ be a linear operator from a Banach space X into a Banach space Y. Show that f is continuous if and only if $g \circledcirc f \in X^*$ for each $g \in Y^*$.

QUOTIENT NORMED LINEAR SPACES

10. Let X be a normed linear space and Y a closed subspace of X. Show that the set $Q = X/Y$ is a linear space and that $|g| = inf\{|x| \, | x \in q\}$ defines a norm on Q called the **quotient norm**. The normed linear space Q with the norm $|g|$ is called the **quotient normed linear space** of X over Y. Show that it has the following properties:

(1) If X is complete, so is Q.
(2) The canonical projection $p: X \to Q$ is an open continuous linear operator with $|p| \leq 1$.
(3) If X is complete, p maps the open unit sphere in X onto the open unit sphere in Q.
(4) If both Y and Q are complete, then so is X.

11. Let $Y = C([0, 1])$ and let X be the linear subspace of Y consisting of the members of Y possessing continuous derivatives. Show that the operator $f: X \to Y$ defined by $f(g) = g'$ for each $g \in X$ has a closed graph but that f is not continuous. Consequently, X is not complete.

FOURIER SERIES

For any $f \in L^1(T)$ where T is the unit circle in **C** (see Exercises 8 through 15 of Section 11.2). The **Fourier coefficients** of f are defined by $c_n(f) = 1/2\pi\int_{-\pi}^{\pi} f(t)e^{-int}dt$ for each integer n (not necessarily positive). The **Fourier series** of f is $\Sigma_{-\infty}^{\infty} c_n(f)e^{int}$ and its partial sums are $\sigma_N(t) = \Sigma_{-N}^{N} c_n(f)e^{int}$ for each nonnegative integer N. Since $L^2(T) \subset L^1(T)$ it is possible to compute the Fourier coefficients and the Fourier series for each $f \in L^2(T)$.

12. Let Z denote the integers. Show that if $\{z_n \mid n \in Z\} \subset \mathbf{C}$ such that $\Sigma_{-\infty}^{\infty} |z_n|^2 < \infty$ that there exists an $f \in L^2(T)$ such that $z_n = c_n(f)$ for each $n \in Z$. [Hint: Use the Riesz-Fischer Theorem.]

13. Let $f, g \in L^2(T)$. Show that $\Sigma_{-\infty}^{\infty} c_n(f) c_n(g)^* = 1/2\pi \int_{-\pi}^{\pi} f(t) g(t)^* dt$.

14. If σ_N denotes the Nth partial sum of the Fourier series of $f \in L^2(T)$, show that $lim_{N\to\infty} |f - \sigma_N|_2 = 0$ so that every $f \in L^2(T)$ is the L^2-limit of the partial sums of its Fourier series; i.e., the Fourier series of f converges to f in the L^2-norm.

15. Let $c_f: Z \to \mathbf{R}$ be defined by $c_f(n) = c_n(f)$ for each $n \in Z$. Let $\Omega: L^2(T) \to l^2(Z)$ be defined by $\Omega(f) = c_f$ for each $f \in L^2(T)$. Show that Ω is a Hilbert space isomorphism of $L^2(T)$ onto $l^2(Z)$.

16. For each nonnegative integer N put $\phi_N(t) = \Sigma_{-N}^{N} e^{int}$ for each $t \in T$. Show that $\phi_N \in L^1(T)$ for each nonnegative integer N and the partial sum $\sigma_N(x)$ of the Fourier series of $f \in C(T) \subset L^1(T)$ is given by $\sigma_N(x) = 1/2\pi \int_{-\pi}^{\pi} f(t)\phi_N(x - t) dt$.

17. $C(T)$ is a Banach space with respect to the supremum norm $|f|_\infty$. For each $f \in C(T)$ let $\sigma_N^f(x)$ denote the Nth partial sum of the Fourier series of f. Let $f \in C(T)$. For each positive integer N put $\lambda_N(f) = \sigma_N^f(0)$. Show that for each positive integer N, λ_N is a bounded linear functional on $C(T)$ and $|\lambda_N| \leq |\phi_N|_1$.

18. Show that $lim_N |\phi_N|_1 = \infty$.

19. Show that $|\lambda_N| \geq |\phi_N|_1$ for each positive integer N so that $|\lambda_N| = |\phi_N|_1$ and $lim_N |\lambda_N| = \infty$.

20. Put $\sigma_\infty^f(x) = sup_n |\sigma_n^f(x)| = |\Sigma_{-\infty}^{\infty} c_n(f) e^{inx}|$. Show that for each $x \in T$ there is a set $D_x \subset C(T)$ that is a dense G_δ in $C(T)$ such that $\sigma_\infty^f(x) = \infty$ for each $f \in D_x$. In other words, the Fourier series of f diverges for each $f \in D_x$. [Hint: Use the Banach-Steinhaus Theorem.]

21. Show that there exists a $D \subset C(T)$ that is a dense G_δ in $C(T)$ such that for each $f \in D$, the set $T_f = \{x \in T \mid \sigma_\infty^f(x) = \infty\}$ is a dense G_δ in $\mathbf{R}$.

22. Show that for each $f \in L^1(T)$, $c_n(f)$ converges to 0 as $|n|$ approaches ∞.

11.4 Uniform Function Spaces

So far, the spaces of functions we have been considering have been metric spaces. This is because the L^p-norms for $1 \leq p \leq \infty$ define metrics. In general, spaces of functions may not be metric. In the case of spaces of functions from a space X into a uniform space (Y, ν), it is possible to define uniformities on the space $F(X, Y)$ of functions from X to Y that generate various topologies on $F(X,Y)$. Often, we will denote $F(X, Y)$ simply by F when the domain and range spaces are known. These uniformities on function spaces allow us to consider questions of *nearness* of functions and questions of uniform convergence of nets $\{f_\alpha\}$ of functions. In particular we are often interested in when the uniform limit of a net of continuous functions $\{g_\alpha\}$ is itself continuous. In general, it is possible to consider spaces of functions that are not uniformizable. While we will be interested in the topologies of the spaces of functions we now consider, we will not pursue this more general investigation.

Probably the most simple uniformity we can define on the space of functions F from a space X into a uniform space (Y, ν) is the product uniformity on $\Pi_{x\in X}(Y, \nu)$ relativized to $F = Y^X$. That F can be considered as a subspace of $\Pi_{x\in X}(Y, \nu)$ is readily seen by observing that for a given $f \in F$ that for each $x \in$ X, $f(x) \in$ Y and hence we can identify f with the point $\phi \in \Pi_{x\in X}(Y, \nu)$ such that $\pi_x(\phi) = f(x)$ for each $x \in$ X where π_x denotes the projection of $\Pi_{x\in X}(Y, \nu)$ onto the xth coordinate subspace (Y, ν) [i.e., the xth copy of (Y,ν)]. We will refer to this uniformity on F as the **product uniformity** on F and denote it by π. π is also called the **uniformity of pointwise convergence** because a net $\{f_\alpha\}$ in F is Cauchy if and only if $\{f_\alpha(x)\}$ is Cauchy in (Y, ν) for each $x \in$ X and $\{f_\alpha(x)\}$ converges to some $f \in F$ if and only if $\{f_\alpha(x)\}$ converges for each $x \in$ X. The topology generated by π is referred to as the **topology of pointwise convergence** or simply the **product topology** on F. We will abuse the terminology and also denote the topology of pointwise convergence by π. If Z $\subset$ X we define F[Z] to be the set $\{f(x) \mid x \in$ Z and $f \in F\} \subset$ Y. If $z \in$ X we denote $F[\{z\}]$ by $F[z]$. Clearly $\pi_x(F) = F[x]$ for each $x \in$ X. The following theorem about the pointwise topology for F is central to the study of topologies on spaces of functions.

THEOREM 11.23 A necessary and sufficient condition for a family G $\subset$ F, of functions from a space X into a uniform space (Y, ν) to be compact with respect to the topology of pointwise convergence is that G be a closed subset of F and for each $x \in$ X, G[x] has a compact closure in Y.

Proof: The family of functions G is contained in $\Pi_{x\in X}Cl(G[x]) = \Pi_{x\in X}Cl(\pi_x(G))$. To prove the sufficiency, assume G is closed in F and for each $x \in$ X, $G[x]$ has a compact closure in Y. Then $\Pi_{x\in X}Cl(G[x])$ is compact by the Tychonoff Product Theorem and since G is closed in F, it is closed in $\Pi_{x\in X}Cl(G[x])$. Hence G is compact. To prove the necessity, assume G is compact. Since $\Pi_{x\in X}(Y, \nu)$ is Hausdorff, so is F. Since G is compact in F, G is closed. Furthermore, the set $G[x]$ is closed and compact for each $x \in X$ since $G[x] = \pi_x(G)$ and π_x is continuous. ∎

Theorem 11.23 can now be used to prove things about finer topologies for G in the following way. If τ is a finer topology than π on G and if (G, τ) is compact then $\tau = \pi$. To see this note that the identity mapping $i:(G, \tau) \to (G, \pi)$ is continuous. Let $g \in G$ and assume U is an open set in G containing g. Put K $= G -$ U. Then K is compact since G is compact with respect to τ, so $i(\mathrm{K}) = \mathrm{K}$ is compact in G with respect to π. For each $k \in \mathrm{K}$ there exists open sets U_k, V_k with respect to π such that $g \in U_k$, $k \in V_k$ and $U_k \cap V_k = \varnothing$. Then there exists a finite collection $k_1 \ldots k_n \in \mathrm{K}$ such that $\cup_{i=1}^n V_{k_i}$ covers K so $\mathrm{W} = \cap_{i=1}^n U_{k_i}$ is open with respect to π, contains g and lies entirely in U. Hence i^{-1} is continuous so i is a homeomorphism. Therefore, $\tau = \pi$.

Consequently, Theorem 11.23 can be used to determine the compactness or non-compactness of (G, τ). Clearly if G either fails to be closed in F with respect to π or if there exists an x such that $G[x]$ does not have compact closure in Y, then by Theorem 11.23, (G, τ) cannot be compact. Conversely, if G is closed in F with respect to π and $G[x]$ has compact closure in Y for each $x \in X$ then (G, τ) is compact provided convergence of a net $\{g_\alpha\} \subset G$ that converges with respect to π implies $\{g_\alpha\}$ converges with respect to τ.

If A is a subset of X it is sometimes useful to be able to talk about pointwise convergence on A. For each $f \in F$ the restriction $f_A = f|_A$ is a member of Y^A. We can define a *restriction mapping* $\rho: F \to Y^A$ by $\rho(f) = f_A$ for each $f \in F$. We then define the **uniformity of pointwise convergence on** A to be the smallest uniformity on F that makes ρ uniformly continuous. We denote the uniformity of pointwise convergence on A by π_A. The topology generated by π_A is called the **topology of pointwise convergence on** A. We can prove the following results about π and π_A for each $\mathrm{A} \subset \mathrm{X}$.

PROPOSITION 11.14 Let X be a space and let $F = Y^X$ where (Y, ν) is a uniform space. Let $A \subset X$ and let π and π_A denote the topologies of pointwise convergence on X and A respectively. Then

(1) A subbase for π on F is the family of all sets of the form $\{f \in F \mid f(x) \in U\}$ for some point $x \in X$ and open set $U \subset Y$.

(2) A subbase for π_A on F is the family of all sets of the form $\{f \in F \mid f(x) \in U\}$ for some point $x \in A$ and open set $U \subset Y$.

(3) π_A is the smallest topology for F that makes the restriction mapping $\rho: F \to Y^A$ continuous.

Proof: To prove (1) note that sets of the form $p_{\alpha_1}^{-1}(W_1) \cap \ldots p_{\alpha_n}^{-1}(W_n)$ for some open sets $W_1 \ldots W_n$ and projections $p_{\alpha_1} \ldots p_{\alpha_n}$ (where for each i, p_{α_i} is the projection of F onto the α_i^{th} coordinate subspace of $Y^X = F$), form a basis for π since π is the product topology on F. Now for each i, $p_{\alpha_i}^{-1}(W_i) = \{f \in F \mid p_{\alpha_i}(f) \in$

$W_i\} = \{f \in F \mid f(\alpha_i) \in W_i\}$. Hence the sets of the form $\{f \in F \mid f(x) \in \mathrm{U}\}$ for some point $x \in \mathrm{X}$ and open set $\mathrm{U} \subset \mathrm{Y}$ forms a subbase for π on F.

To prove (2), first note that by (1) above, sets of the form $\{f \in Y^A \mid f(y) \in \mathrm{V}\}$ for some point $y \in \mathrm{A}$ and open set $\mathrm{V} \subset \mathrm{Y}$ form a subbase for the π topology on Y^A. Since π_A is a uniformity that makes the restriction mapping $\rho: F \to Y^A$ uniformly continuous, each set of the form $\rho^{-1}(\{f \in Y^A \mid f(y) \in \mathrm{V}\})$ for some $y \in \mathrm{A}$ and open set $\mathrm{V} \subset \mathrm{Y}$ belongs to the π_A topology on F.

$$\text{Now } \rho^{-1}(\{f \in Y^A \mid f(y) \in \mathrm{V}\}) = \{f \in F \mid f(y) \in \mathrm{V}\},$$

so sets of the form $\{f \in F \mid f(y) \in \mathrm{V}\}$ for some $y \in \mathrm{A}$ and open $\mathrm{V} \subset \mathrm{Y}$ belong to the π_A topology on F. To show these sets form a subbase for the π_A topology on F we need to show that for each open U in π_A and $g \in \mathrm{U}$, there exists a finite collection of sets $\{f \in F \mid f(y_1) \in V_1\} \ldots \{f \in F \mid f(y_n) \in V_n\}$ for some points $y_1 \ldots y_n \in \mathrm{A}$ and open sets $V_1 \ldots V_n$ contained in Y, such that $g \in \cap_{i=1}^n \{f \in F \mid f(y_i) \in V_i\} \subset \mathrm{U}$.

Since $\mathrm{U} \in \pi_A$ on F, there exists a $\mathcal{U} \in \pi_A$ (the uniformity) such that $g \in S(g, \mathcal{U}) \subset \mathrm{U}$. Since π_A is the smallest uniformity on F that makes ρ uniformly continuous, there exists a finite collection of open coverings $\mathcal{V}_1 \ldots \mathcal{V}_n \in \pi$ on Y^A such that $\mathcal{W} = \cap_{i=1}^n \rho^{-1}(\mathcal{V}_i) < \mathcal{U}$. Then $g \in S(g, \mathcal{W}) \subset \mathrm{U}$. Now $S(g, \mathcal{W}) = \{f \in F \mid f, g \in \mathrm{W}$ for some $\mathrm{W} \in \mathcal{W}\} = \{f \in F \mid f, g \in \cap_{i=1}^n \rho^{-1}(V_i)$ for some $V_1 \ldots V_n$ that are open in Y$\}$. Therefore, $S(g, \mathcal{W}) = \cap_{i=1}^n \{f \in F \mid f, g \in \rho^{-1}(V_i)\}$ for some $V_1 \ldots V_n \subset \mathrm{Y}$ which implies $S(g, \mathcal{W}) = \cap_{i=1}^n \{f \in F \mid f(y_i), g(y_i) \in V_i\}$ for some points $y_1 \ldots y_n \in \mathrm{A}$ and open sets $V_1 \ldots V_n$ in Y. Therefore, sets of the form $\{f \in F \mid f(y) \in \mathrm{V}\}$ for some $y \in \mathrm{A}$ and open $\mathrm{V} \subset \mathrm{Y}$ form a subbase for the π_A topology on F.

To prove (3), note that since the π_A topology on F is generated by the π_A uniformity on F and the π_A uniform is the smallest that makes ρ uniformly continuous, it is clear that the π_A topology on F makes ρ continuous. Let τ be another topology on F that makes ρ continuous. Then for each $y \in \mathrm{A}$ and open set $\mathrm{V} \subset \mathrm{Y}$, the set $\{f \in Y^A \mid f(y) \in \mathrm{V}\}$ is π-open in Y^A which implies

$$\rho^{-1}(\{f \in Y^A \mid f(y) \in \mathrm{V}\}) = \{f \in F \mid f(y) \in \mathrm{V}\} \in \tau.$$

But then τ contains a subbase for π_A which implies π_A is coarser than τ. Hence π_A is the smallest topology for F that makes ρ continuous. ∎

From the definition of the product uniformity in Chapter 5 and by Exercise 5.2, it is seen that the coverings of the form $\pi_\alpha^{-1}(\mathcal{V})$ (where $\mathcal{V} \in \nu$ and π_α denotes the projection of F onto the αth coordinate subspace of $Y^X = F$) determine a subbase for the π uniformity on F. For a particular $\alpha \in \mathrm{X}$ and $\mathcal{V} \in \nu$ we have $\pi_\alpha(\mathcal{V}) = \{\pi_\alpha^{-1}(\mathrm{V}) \mid \mathrm{V} \in \mathcal{V}\} = \{\{f \in F \mid f(\alpha) \in \mathrm{V}\} \mid \mathrm{V} \in \mathcal{V}\}$. Hence for $g \in F$ we have $S(g, \pi_\alpha^{-1}(\mathcal{V})) = \{f \in F \mid f(\alpha), g(\alpha) \in \mathrm{V}$ for some $\mathrm{V} \in \mathcal{V}\}$.

PROPOSITION 11.15 Let X be a space and let $F = Y^X$ where (Y, ν) is a uniform space. Let $A \subset X$ and let π_A denote the uniformity of pointwise convergence on A on F. Then

(1) Coverings of the form $\mathcal{V}_\alpha = \{\{f \in F \mid f(\alpha) \in V\} \mid V \in \mathcal{V}\}$ where $\alpha \in A$ and $\mathcal{V} \in \nu$ form a subbase for the π_A uniformity on F.

(2) A net $\{f_\beta\} \subset F$ is Cauchy with respect to π_A if and only if $\{f_\beta(x)\}$ is Cauchy in Y for each $x \in A$.

(3) If (Y, ν) is complete and $\rho(F)$ is closed in Y^A with respect to the π topology on Y^A, then F is complete with respect to the π_A uniformity.

Proof: By the definition of the π_A uniformity on F, sets of the form $\rho^{-1}(\mathcal{V})$ where $\mathcal{V}$ is a member of the π uniformity on Y^A form a subbase for the π_A uniformity on F. For a particular $\mathcal{V}$ in the π uniformity on Y^A we have that $\mathcal{V}$ is refined by some $\cap_{i=1}^{n}\pi_{\alpha_i}^{-1}(\mathcal{V}_i)$ for some $\alpha_1 \ldots \alpha_n \in$ A and $\mathcal{V}_1 \ldots \mathcal{V}_n \in \nu$. Hence $\rho^{-1}(\mathcal{V})$ is refined by $\cap_{i=1}^{n}\rho^{-1}(\pi_{\alpha_i}^{-1}(\mathcal{V}_i))$. For each $i = 1 \ldots n$,

$$\rho^{-1}(\pi_{\alpha_i}^{-1}(\mathcal{V}_i)) = \rho^{-1}(\{\pi_{\alpha_i}^{-1}(\mathrm{V}) \mid \mathrm{V} \in \mathcal{V}\}) =$$

$$\rho^{-1}(\{\{f \in Y^A \mid f(\alpha_i) \in \mathrm{V}\} \mid \mathrm{V} \in \mathcal{V}_i\}) = \{\{f \in F \mid f(\alpha_i) \in \mathrm{V} \mid \mathrm{V} \in \mathcal{V}_i\}.$$

Hence coverings of the form $\mathcal{V}_\alpha = \{\{f \in F \mid f(\alpha) \in \mathrm{V}\} \mid \mathrm{V} \in \mathcal{V}\}$ where $\alpha \in$ A and $\mathcal{V} \in \nu$ form a subbase for the π_A uniformity on F.

To prove (2) first assume $\{f_\beta\}$ is Cauchy in F with respect to π_A. Then for each covering of the form $\mathcal{V}_\alpha = \{\{f \in F \mid f(\alpha) \in \mathrm{V}\} \mid \mathrm{V} \in \mathcal{V}\}$ there exists a residual set R such that $\{f_\beta \mid \beta \in \mathrm{R}\} \subset \{f \in F \mid f(\alpha) \in \mathrm{V}\}$ for some $\mathrm{V} \in \mathcal{V}$. But then $\{f_\beta(\alpha) \mid \beta \in \mathrm{R}\} \subset \mathrm{V}$ so $\{f_\beta(\alpha)\}$ is Cauchy in (Y, ν). Hence $\{f_\beta(x)\}$ is Cauchy in (Y, ν) for each $x \in$ A. Conversely, assume $\{f_\beta(x)\}$ is Cauchy in (Y,ν) for each $x \in$ A. Let $\mathcal{V}_\alpha$ be a subbasic member of π_A. For some $\mathrm{V} \in \mathcal{V}$ there exists a residual set R such that $\{f_\beta(\alpha) \mid \beta \in \mathrm{R}\} \subset \mathrm{V}$ since $\{f_\beta(\alpha)\}$ is Cauchy in (Y, ν). But then $\{f_\beta \mid \beta \in \mathrm{R}\} \subset \{f \in F \mid f(\alpha) \in \mathrm{V}\}$ so $\{f_\beta\}$ is Cauchy with respect to π_A.

To prove (3), assume (Y, ν) is complete and $\rho(F)$ is closed in Y^A with respect to π. Let $\{f_\beta\}$ be a Cauchy net with respect to π_A. Then $\{\rho(f_\beta)\}$ is a Cauchy net in Y^A with respect to π since ρ is uniformly continuous. Since $\rho(F)$ is closed in Y^A with respect to π and Y^A is complete with respect to π, since π is the product uniformity on Y^A, $\{\rho(f_\beta)\}$ converges to some $f \in \rho(F)$ with respect to π. Let $g \in F$ such that $\rho(g) = f$. Let $\mathcal{V} \in \pi_A$. Then there exists some $\alpha_1 \ldots \alpha_n \in$ A and $\mathcal{W}_1 \ldots \mathcal{W}_n \in \pi$ such that $\mathcal{W} = \cap_{i=1}^{n}\rho^{-1}(\pi_{\alpha_i}^{-1}(\mathcal{W}_i)) < \mathcal{V}$. Now $S(g,\mathcal{W}) \subset S(g, \mathcal{V})$ and since $\{\rho(f_\beta)\}$ converges to f, there exists a residual set R with $\{\rho(f_\beta) \mid \beta \in \mathrm{R}\} \subset \pi_{\alpha_i}^{-1}(W_i)$ for some $W_i \in \mathcal{W}$ with $f \in \pi_{\alpha_i}^{-1}(W_i)$ for each i = 1 ... n. But then $\{\rho(f_\beta) \mid \beta \in \mathrm{R}\} \subset \cap_{i=1}^{n}\pi_{\alpha_i}^{-1}(W_i)$ which implies $\{f_\beta \mid \beta \in \mathrm{R}\} \subset \cap_{i=1}^{n}\rho^{-1}(\pi_{\alpha_i}^{-1}(W_i))$. Moreover, $g \in \cap_{i=1}^{n}\rho^{-1}(\pi_{\alpha_i}^{-1}(W_i)) \subset S(g, \mathcal{W})$. Hence $\{f_\beta \mid \beta$

$\in$ R$\} \subset S(g, \mathcal{V})$. Therefore, $\{f_\beta\}$ converges to g with respect to π_A so F is complete with respect to π_A. ∎

While the π uniformity for F is one of the most simple, it is not the most interesting. In fact, the π topology, when Y = **R**, is so coarse that the characteristic functions of finite sets converge to the unit function. For each finite F $\subset$ X let χ_F denote the characteristic function of F. Since the family of all finite subsets F of X is directed by set inclusion, the family $\{\chi_F\}$ is a net. Moreover, $\{\chi_F\}$ converges to the unit function 1 defined by $1(x) = 1$ for each $x \in$ X. To see this, observe that for each $x \in$ X, $\{x\}$ is a finite set so if F is a finite subset of X containing x then $\chi_F(x) = 1$. Hence by Proposition 11.14.(1) $\{\chi_F\}$ is eventually in each subbasic open set containing 1. A topology for which the characteristic functions of finite sets get arbitrarily *near* the unit function is too coarse for investigating the problems we will be interested in throughout the remainder of this section.

A more interesting uniformity for F is the *uniformity of uniform convergence* which we will denote as the *uc* uniformity. The *uc* uniformity is also independent of the topology of X, but it has the interesting (and useful) property that the family of continuous functions C from X to Y is closed in F. In other words, with the *uc* uniformity, the uniform limit of continuous functions is also continuous. The *uc* uniformity will be the finest uniformity we will consider for F.

Let $\mathcal{V} \in \nu$. For each $f \in F$ define the **ball about** f **of radius** $\mathcal{V}$, denoted $B(f, \mathcal{V})$ to be the set $\{g \in F \mid$ for each $x \in$ X, $g(x), f(x) \in$ V for some V $\in \mathcal{V}\}$. Put $\mathcal{V}_F = \{B(f, \mathcal{V}) \mid f \in F\}$. Let *uc* denote the family of all coverings of F that are refined by coverings of the form $\mathcal{V}_F$ where $\mathcal{V} \in \nu$. If $\mathcal{V}_F \in uc$ then $S(f, \mathcal{V}_F) = \{g \in F \mid f, g \in B(h, \mathcal{V})$ for some $h \in F\} = \{g \in F \mid$ for each $x \in$ X, $f(x), h(x) \in V_1$ and $g(x), h(x) \in V_2$ for some $V_1, V_2 \in \mathcal{V}$, for some $h \in F\} = \{g \in F \mid$ for each $x \in$ X, $f(x), g(x) \in S(h(x), \mathcal{V})$ for some $h \in F\}$. Since $h(x)$ ranges over all of Y as h ranges over F we have that $S(f, \mathcal{V}_F) = B(f, \mathcal{V}^\Delta)$ where $\mathcal{V}^\Delta = \{S(y,\mathcal{V}) \mid y \in$ Y$\}$ (i.e., $\mathcal{V}^\Delta$ is the covering made from spheres with respect to $\mathcal{V}$).

If $\mathcal{U}_F$ and $\mathcal{V}_F$ are members of *uc* then clearly there exists a $\mathcal{W} \in \nu$ such that $\mathcal{W}^{\Delta *} < \mathcal{U} \wedge \mathcal{V}$ (just pick $\mathcal{W}$ such that $\mathcal{W}^{**} < \mathcal{U} \wedge \mathcal{V}$). Now

$$\mathcal{W}_F^* = \{Star(B(f,\mathcal{W}),\mathcal{W}_F) \mid f \in F\} =$$

$$\{Star(\{g \in F \mid \text{for each } x \in \text{X}, g(x), f(x) \in \text{W for some W} \in \mathcal{W}\}, \mathcal{W}_F) \mid f \in F\} =$$

$$\{\{h \in F \mid \text{for each } x \in \text{X}, h(x), g(x) \in W_1 \text{ and } g(x), f(x) \in W_2 \text{ for some } g \in F \text{ and some } W_1, W_2 \in \mathcal{W}\} \mid f \in F\} =$$

$$\{\{h \in F \mid \text{for each } x \in \text{X}, h(x) \in Star(S(f(x), \mathcal{W}))\} \mid f \in F\}.$$

Since $f(x)$ ranges over all of Y as f ranges over F we have:

$$\mathcal{W}_F^* = \{\{h \in F \mid \text{for each } x \in X, h(x), f(x) \in W \text{ for some } W \in \mathcal{W}^{\Delta *}\} \mid f \in F\}.$$

Therefore, $\mathcal{W}_F^* = \{B(f, \mathcal{W}^{\Delta *}) \mid f \in F\}$. Since $\mathcal{W}^{\Delta *} < \mathcal{U} \cap \mathcal{V}$ we have that $B(f,\mathcal{W}^{\Delta *}) \subset B(f,\mathcal{U})$ and $B(f, \mathcal{W}^{\Delta *}) \subset B(f, \mathcal{V})$ for each $f \in F$, so $\mathcal{W}_F^* < \mathcal{U}_F$ and $\mathcal{W}_F^* < \mathcal{V}_F$. Hence $\mathcal{W}_F^* < \mathcal{U}_F \cap \mathcal{V}_F$, so *uc* is a uniformity for *F*. We call *uc* the **uniformity of uniform convergence** and the topology generated by *uc* the **topology of uniform convergence** or simply the *uc* topology.

PROPOSITION 11.16 uc is finer than π.

Proof: For a subbasic uniform covering $\pi_\alpha^{-1}(\mathcal{V})$ in π and each $f \in F$, the sphere $S(f,\pi_\alpha^{-1}(\mathcal{V})) = \{g \in F \mid f(\alpha), g(\alpha) \in V \text{ for some } V \in \mathcal{V}\}$ contains the set $\{g \in F \mid \text{for each } x \in X, f(x), g(x) \in V \text{ for some } V \in \mathcal{V}\} = B(f, \mathcal{V})$. Therefore, $\mathcal{V}_F < \pi_\alpha^{-1}(\mathcal{V})^\Delta$. Then for a subbasic covering $\pi_\alpha^{-1}(\mathcal{U})$ in π, there exists a $\mathcal{V} \in \nu$ such that $\mathcal{V}^* < \mathcal{U}$ so $\pi_\alpha^{-1}(\mathcal{V}) <^* \pi_\alpha^{-1}(\mathcal{U})$ and hence $\mathcal{V}_F < \pi_\alpha^{-1}(\mathcal{V})^\Delta < \pi_\alpha^{-1}(\mathcal{U})$ so $\pi_\alpha^{-1}(\mathcal{U}) \in uc$. Since *uc* contains a subbase for π it contains π. ∎

PROPOSITION 11.17 Let X be a space and F the family of functions from X to the uniform space (Y, ν). *Then the uc uniformity has the following properties:*

(1) *uc is generated by the family of all pseudo-metrics of the form* $d(f, g) = sup\{d(f(x), g(x)) \mid x \in X\}$ *where d is a bounded member of the family of pseudo-metrics that generates* ν.

(2) *A net* $\{f_\alpha\}$ *in F converges uniformly to* $f \in F$ *if and only if it is Cauchy with respect to uc and* $\{f_\alpha(x)\}$ *converges to* $f(x)$ *for each* $x \in X$.

(3) *If* (Y, ν) *is complete then* (F, uc) *is complete.*

Proof: Clearly the family of all coverings of the form $\mathcal{V}(d, r) = \{S_d(x, r) \mid x \in X\}$, where $r > 0$ and $S_d(x, r)$ denotes the *r*-sphere about *x* with respect to a bounded member *d* of the family of pseudo-metrics that generates ν, forms a basis for ν. Now $\mathcal{V}(d, r)_F = \{B(f,\mathcal{V}(d, r)) \mid f \in F\} = \{\{g \in F \mid \text{for each } x \in X, g(x), f(x) \in S_d(y, r) \text{ for some } y \in Y\} \mid f \in F\}$. Now $d(f, g) = sup\{d(f(x), g(x)) \mid x \in X\}$ so if $d(f, g) \leq r$ then for each $x \in X$, $f(x), g(x) \in S_d(f(x), r)$ so $\mathcal{V}(d, r)_F$ contains $\{\{g \in F \mid d(g, f) \leq r\} \mid f \in F\} = \{S_d(f, r) \mid f \in F\}$ Therefore, $\{S_d(f, r) \mid f \in F\} < \mathcal{V}(d, r)_F$ so the family of pseudo-metrics of the form $d(f, g) = sup\{d(f(x),g(x)) \mid x \in X\}$ where *d* is a bounded member of the family of pseudo-metrics that generates ν, generates the *uc* uniformity.

To prove (2) assume $\{f_\alpha\}$ is a net in *F* that converges uniformly to some $f \in F$. Then for each $\mathcal{V} \in \nu$ there exists a residual set R such that for each $\alpha \in R$, and $x \in X$, $f_\alpha(x), f(x) \in V(\alpha, x)$ for some $V(\alpha, x) \in \mathcal{V}$. Therefore, $f_\alpha \in B(f, \mathcal{V})$ for each $\alpha \in R$. Since $S(f, \mathcal{V}_F) = B(f, \mathcal{V}^\Delta)$, $f_\alpha \in S(f, \mathcal{V}_F)$ for each $\alpha \in R$. Hence $\{f_\alpha\}$ is eventually in $\mathcal{V}_F$ for each $\mathcal{V} \in \nu$ so $\{f_\alpha\}$ is Cauchy with respect to *uc*. Moreover, since for each $\mathcal{V} \in \nu$, there exists a residual R such that for each $\alpha \in R$, $f_\alpha(x), f(x) \in V$ for some $V \in \mathcal{V}$, we have that $\{f_\alpha(x)\}$ is eventually in $S(f(x), \mathcal{V})$ for each $\mathcal{V} \in \nu$ so $\{f_\alpha(x)\}$ converges to $f(x)$ for each $x \in X$.

Conversely, assume $\{f_\alpha\}$ is Cauchy with respect to *uc* and that for each $x \in X$, $\{f_\alpha(x)\}$ converges to $f(x)$. Let $\mathcal{V} \in \nu$. Pick $\mathcal{U} \in \nu$ such that $\mathcal{U}^{**} < \mathcal{V}$. Since $\{f_\alpha\}$ is Cauchy with respect to *uc*, there exists a residual set R such that $\{f_\alpha \mid \alpha \in R\} \subset B(g, \mathcal{U})$ for some $g \in F$. Then for each $x \in X$ and $\alpha \in R$, $f_\alpha(x)$, $g(x) \in U(\alpha, x)$ for some $U(\alpha, x) \in \mathcal{U}$. Therefore, $f_\alpha(x), g(x) \in S(g(x), \mathcal{U})$ for each $\alpha \in R$ and $x \in X$. Now for each $x \in X$, $\{f_\alpha(x)\}$ converges to $f(x)$ so there exists a $W_x \in \mathcal{U}$ containing $f(x)$ and a residual R_x such that $\{f_\alpha(x) \mid \alpha \in R_x\} \subset W_x$. Now $R_x \cap R \neq \varnothing$ so pick $\beta \in R_x \cap R$. Then $f_\beta(x) \in W_x$ and $f_\beta(x), g(x) \in U(\beta, x)$. But $f(x) \in W_x$ and $f_\alpha(x) \in S(g(x), \mathcal{U})$ for each $\alpha \in R$ and $x \in X$. Therefore, $f_\alpha(x), f(x) \in S(g(x), \mathcal{U}) \cup W_x$ for each $x \in X$ which implies $f_\alpha(x), f(x) \in Star(S(g(x), \mathcal{U}))$ for each $x \in X$. But $Star(S(g(x), \mathcal{U})) \subset V_x$ for some $V_x \in \mathcal{V}$ since $\mathcal{U}^{**} < \mathcal{V}$. Then for each $\alpha \in R$ and $x \in X$, $f(x), f_\alpha(x) \in V_x$ for some $V_x \in \mathcal{V}$ so $\{f_\alpha\}$ converges uniformly to f.

To prove (3), let $\{f_\alpha\}$ be a Cauchy net in F with respect to *uc*. Let $\mathcal{V} \in \nu$. Then $\mathcal{V}_F$ is a basic open covering in *uc*, so there exists a $g \in F$ and a residual set R such that $\{f_\alpha \mid \alpha \in R\} \subset B(g, \mathcal{V})$. Then for each $\alpha \in R$ and $x \in X$,

$$f_\alpha(x), g(x) \in V(\alpha, x) \text{ for some } V(\alpha, x) \in \mathcal{V}.$$

Fix $y \in X$. Then $f_\alpha(y) \in S(g(y), \mathcal{V})$ which implies $\{f_\alpha(y)\}$ is Cauchy in Y and therefore converges to some $p_x \in Y$ for each $x \in X$. Define $f \in F$ by $f(x) = p_x$ for each $x \in X$. Then by (2) above, $\{f_\alpha\}$ converges to f since $\{f_\alpha\}$ is Cauchy with respect to *uc* and $\{f_\alpha(x)\}$ converges to $f(x)$ for each $x \in X$. ∎

Notice that in (F, uc), convergence of a net $\{f_\alpha\}$ to $f \in F$ implies uniform converge of $\{f_\alpha\}$ to f. This is because if $\{f_\alpha\}$ converges to f with respect to *uc* then $\{f_\alpha\}$ is Cauchy with respect to *uc* and since *uc* is finer than π by Proposition 11.16, $\{f_\alpha\}$ converges to f with respect to π. But then $\{f_\alpha(x)\}$ converges to $f(x)$ for each $x \in X$ so by Proposition 11.17.(2), $\{f_\alpha\}$ converges uniformly to f with respect to *uc*.

THEOREM 11.24 Let X be a space and let $F = Y^X$ where (Y, ν) is a uniform space. Let $C \subset F$ be the family of continuous functions from X to Y. Then C is closed in F with respect to the uc topology and hence (C, uc) is complete if (Y, ν) is complete.

Note that by the previous remarks, this theorem asserts that the uniform limit of a net of continuous functions from X to Y is itself continuous.

Proof: Let $G = F - C$ and let $g \in G$. Then there exists an $x_0 \in X$ such that g is not continuous at x_0. Therefore, there exists a $\mathcal{V} \in \nu$ such that $g^{-1}(S(g(x_0), \mathcal{V}))$ is not a neighborhood of x_0. Choose $\mathcal{U} \in \nu$ such that $\mathcal{U}^5 < \mathcal{V}$ (where $\mathcal{U}^5 = \mathcal{U}^{*****}$). Let $f \in S(g, \mathcal{U}_F) = B(g, \mathcal{U}^\Delta)$. Then for each $x \in X$, $f(x), g(x) \in$

$S(y,\mathcal{U})$ for some $y \in Y$. Hence $f(x) \in Star(S(g(x), \mathcal{U}))$ and $g(x) \in Star(S(f(x),\mathcal{U}))$ for each $x \in X$. Consequently, if $Z \subset X$ then for each $z \in Z$, $g(z) \in Star(S(f(z),\mathcal{U}))$ so $z \in g^{-1}(Star(S(f(z), \mathcal{U}), \mathcal{U})$ which implies

$$Z \subset g^{-1}(Star(f[Z], \mathcal{U}^*)) \text{ for each } Z \subset X.$$

Substituting $f^{-1}(S(f(x), \mathcal{U}))$ for Z we get

$$f^{-1}(S(f(x), \mathcal{U})) \subset g^{-1}(Star(f[f^{-1}(S(f(x),\mathcal{U}))], \mathcal{U}^*)) = g^{-1}(Star(S(f(x), \mathcal{U}), \mathcal{U}^*)).$$

Since $g(x) \in Star(S(f(x), \mathcal{U}))$ we have that $S(f(x), \mathcal{U}) \subset Star(S(g(x), \mathcal{U}), \mathcal{U}^*)$. Hence $f^{-1}(S(f(x), \mathcal{U})) \subset g^{-1}(Star(Star(S(g(x), \mathcal{U}), \mathcal{U}^*, \mathcal{U}^*)) \subset g^{-1}(S(g(x), \mathcal{V}))$ since $\mathcal{U}^5 < \mathcal{V}$. But then $f^{-1}(S(f(x_0), \mathcal{U}))$ cannot be a neighborhood of x_0 since $g^{-1}(S(g(x_0), \mathcal{V})$ is not a neighborhood of x_0 so f is not continuous at x_0. Consequently, $S(g, \mathcal{U}_F) \subset G$ so G is open which implies C is closed. ∎

Let τ be a topology for the function space $F(X, Y)$ where (Y, ν) is a uniform space. Then we can consider the product space $F \times X$ and the mapping $e{:}F \times X \to Y$ defined by $e(f, x) = f(x) \in Y$, for each $(f, x) \in F \times X$. The topology τ is said to be **jointly continuous** if the mapping e is continuous. Usually the topology π of pointwise convergence is not jointly continuous (see Exercise 1). However, if we restrict our attention to the continuous members of F that we denote by C, then the discrete topology is always jointly continuous, i.e., $e{:}C \times X \to Y$ is continuous (see Exercise 2). Consequently, there is a finest topology on a family of functions that is jointly continuous. The opposite problem of finding a coarsest topology on a family of functions that is jointly continuous also has a solution sometimes as we will see a little later. However, in general, there is no coarsest topology for a family of functions that is jointly continuous.

If $A \subset X$, a topology for F is **jointly continuous on** A if $e_A = e\mid_A$ is continuous on $F \times A$. If $\mathcal{A}$ is a family of subsets of X it may turn out that a topology for F is jointly continuous on each member of $A \in \mathcal{A}$. It will be shown that the *uc* topology is jointly continuous on $C \subset F$ and that some interesting uniformities can be constructed on F by considering the uniformity of uniform convergence on each member $A \in \mathcal{A}$. These uniformities then generate topologies that are jointly continuous on each $A \in \mathcal{A}$. Joint continuity will play an important role in studying the so-called *uniformity of uniform convergence on compacta* on F, which we shall now begin.

PROPOSITION 11.18 Let X be a space and let C be the family of continuous functions from X to a uniform space (Y, ν). Then the uc topology on C is jointly continuous.

Proof: To show the continuity of $e{:}C \times X \to Y$ at some point $(f, x) \in C \times X$, let $\mathcal{V} \in \nu$. It suffices to show there exists a basic open set $W \times U \subset C \times X$

containing (f, x) such that $e(W \times U) \subset S(f(x), \mathcal{V})$. For this, first observe that

$$e^{-1}(S(f(x), \mathcal{V}) = \{(g, z) \in C \times X \mid g(z) \in S(f(x), \mathcal{V})\}.$$

Pick $\mathcal{U} \in \nu$ such that $\mathcal{U}^{**} < \mathcal{V}$. Put $U = f^{-1}(S(f(x), \mathcal{U}))$. U is open in X since f is continuous and $x \in U$. Also, $W = B(f, \mathcal{U})$ is open in C and contains f. Therefore, $W \times U$ is a basic open set in $C \times X$ containing (f, x). Let $(g, z) \in W \times U$. Then $g \in B(f, \mathcal{U})$ and $z \in f^{-1}(S(f(x), \mathcal{U}))$ so $f(z), f(x) \in U_x$ for some $U_x \in \mathcal{U}$ and for each $y \in X$, $f(y), g(y) \in U_y$ for some $U_y \in \mathcal{U}$. Hence $f(z), g(z) \in U_z$ for some $U_z \in \mathcal{U}$. Then $f(x), g(z) \in U_x \cup U_z$ so $g(z) \in Star(S(f(x), \mathcal{U})) \subset S(f(x), \mathcal{V})$. Hence $e(W \times U) \subset S(f(x), \mathcal{V})$ so e is continuous at (f, x). But then e is continuous at each $(g, z) \in C \times X$ so uc is jointly continuous on C. ■

Let us now consider some uniformities for F constructed by considering the uc uniformity restricted to some subsets A of X belonging to a family A. For each $A \in A$ let $\rho_A : F \to Y^A$ be defined by $\rho_A(f) = f_A = f|_A$. Then define the uniformity $uc|A$ to be the coarsest uniformity on F such that for each $A \in A$, ρ_A is uniformly continuous with respect to $uc|A$ and the uc uniformity on $F_A = Y^A$.

PROPOSITION 11.19 A subbase for the $uc|A$ uniformity on F is the family of all coverings of the form $\mathcal{V}_F(E) = \{B_E(f, \mathcal{V}) \mid f \in F_E\}$ where $E \in A$ and $B_E(f, \mathcal{V}) = \{g \in F \mid$ for each $x \in E$, $g(x), f(x) \in V$ for some $V \subset \mathcal{V}\}$ and $F_E = Y^E$.

The proof of Proposition 11.19 is left as an exercise. We are now in a position to state the central structural theorem of this section. Its proof is also left as an exercise.

THEOREM 11.25 Let X be a space and let F be the family of functions from X to a uniform space (Y, ν). Let C_A be the family of all functions in F that are continuous on each $E \in A$ where A is a family of subsets of X that covers X. Then

(1) $uc|A$ is finer than π on F but coarser than uc.
(2) A net $\{f_\alpha\}$ in F converges to $f \in F$ with respect to $uc|A$ if and only if it is Cauchy with respect to $uc|A$ and converges to f with respect to π.
(3) If (Y, ν) is complete then $(F, uc|A)$ is complete.
(4) C_A is closed in F with respect to $uc|A$, so if (Y, ν) is complete then so is $(C_A, uc|A)$.
(5) $uc|A$ on C_A is jointly continuous on each $E \in A$.

If S is the family of singleton sets $\{x\}$ such that $x \in X$ then $uc|S = \pi$ and the family $C \subset F$ of continuous functions is generally not complete with respect to π. Consequently, Theorem 11.25 does not allow us to say anything about C. However, if the continuity of each $f \in F$ on each member of A implies the continuity of f on X then $C = C_A$ so by Theorem 11.25.(4) $(C, uc|A)$ is complete if (Y, ν) is complete.

We now turn our attention to the collection K of compact subsets of X. $uc|K$ is called the uniformity of **uniform convergence on compacta**. We will denote $uc|K$ by κ. The topology generated by κ is called the **topology of compact convergence** or simply the κ topology. It turns out that on the family C of all continuous functions from X to Y that the κ topology is identical to an important function space topology that is not dependent on the concept of a uniform structure on the function space. This topology is known as the **compact-open** topology which we now define. If K is a compact set in X and U is open in Y we write U(K) to denote the set $\{f \in F | f(K) \subset U\}$. The family of all sets of the form U(K) where U is open in Y and K is compact in X is a subbase for the compact-open topology.

THEOREM 11.26 Let X be a space and let C be the family of all continuous functions from X to a uniform space (Y, ν). Then κ is the compact-open topology on C.

Proof: Let $f \in C$ and let $\mathcal{V}_C(K)$ be a subbasic covering in κ for some compact set $K \subset X$ and some $\mathcal{V} \in \nu$. Then $\mathcal{V}_C(K) = \{B_K(g, \mathcal{V}) | g \in C_K\}$ where in this case C_K denotes $C(K, Y)$ [rather than the family of continuous functions with compact support]. Then $S(f, \mathcal{V}_C(K)) = \{g \in C | g, f \in B_K(h, \mathcal{V})$ for some $h \in C\}$ $= \{g \in C|$ for each $x \in K$, $g(x), h(x) \in V_1$ and $f(x), h(x) \in V_2$ for some $V_1, V_2 \in \mathcal{V}$ and some $h \in C\} = \{g \in C|$for each $x \in K$, $g(x) \in Star(S(f(x), \mathcal{V}), \mathcal{V})\} = \{g \in C | g(K) \subset Star(S(f(x), \mathcal{V}), \mathcal{V})\}$. Put $W = Star(S(f(x), \mathcal{V}), \mathcal{V})$. Then $S(f, \mathcal{V}_C(K)) = W(K)$ so $S(f, \mathcal{V}_C(K))$ is open with respect to the compact-open topology on C. Therefore, the compact-open topology is finer than the κ topology.

Conversely, if K is compact in X and U is open in Y then U(K) is a subbasic member of the compact-open topology on C. Let $f \in C$ such that $f \in U(K)$. Then $f(K)$ is a compact subset of U so there exists a $\mathcal{V} \in \nu$ such that $Star(f[K], \mathcal{V}) \subset U$. Let $\mathcal{U} \in \nu$ such that $\mathcal{U}^* < \mathcal{V}$. If $g \in S(f, \mathcal{U}_C(K))$ then $f, g \in B_K(h, \mathcal{U})$ for some $h \in C_K$. Then as we saw above, $g(x) \in Star(S(f(x), \mathcal{U}))$ for each $x \in K$ so $g[K] \subset Star(S(f(x), \mathcal{U}), \mathcal{U}) \subset S(f(x), \mathcal{V})$ for each $x \in K$ which implies $g[K] \subset Star(f[K], \mathcal{V}) \subset U$. Therefore, $S(f, \mathcal{U}_C(K)) \subset U(K)$ so U(K) is open in the κ topology. Hence, κ is finer than the compact-open topology, so κ is the compact-open topology on C. ∎

A topology for F is said to be **jointly continuous on compacta** if it is jointly continuous on each compact subset of X. In a paper titled *On topologies for function spaces*, published in the Bulletin of the American Mathematical Society in 1945(Vol. 51, pp. 429-432), R. H. Fox showed that the compact-open topology on C is coarser than each jointly continuous topology for C and that the compact-open topology on C is jointly continuous itself if X is locally compact.

THEOREM 11.27 Let X be a space and let $F = Y^X$ for a uniform space (Y, ν). Each topology for F that is jointly continuous on compacta is finer than the compact-open topology on F. If X is regular or Hausdorff and C_K is the family of functions $f \in F$ such that f is continuous on each compact subset of X, then the compact-open topology on C_K is jointly continuous on compacta.

Proof: Let τ be a topology for F that is jointly continuous on compacta. Let U(K) be a subbasic member of the compact-open topology on F for some open $U \subset Y$ and compact $K \subset X$. Let e be the mapping on $F \times X$ to Y defined by $e(f,x) = f(x)$. Put $W = [F \times X] \cap e^{-1}(U)$. Then W is open in $F \times K$ since τ is jointly continuous on compacta. If $f \in U(K)$ then $\{f\} \times K \subset W$. Since $\{f\} \times K$ is compact, there exists a τ-open set $V \subset F$ containing f such that $V \times K \subset e^{-1}(U)$ (see Exercise 5). If $g \in V$ then $g[K] \subset U$ so $g \in U(K)$. Therefore, $V \subset U(K)$ so for each $f \in U(K)$, there exists a $V \in \tau$ such that $f \in V \subset U(K)$. Hence U(K) is τ-open so τ is finer than the compact-open topology on F.

Next assume X is either regular or Hausdorff and let K be a compact subset of X. Let $(f, x) \in C_K \times K$. Then $f \in C_K$ and $x \in K$. Let U be an open set in Y containing $f(x) = e(f, x)$. Since f is continuous on K, there exists an open set V in X containing x such that $f(V) \subset U$. Since X is either regular or Hausdorff, there exists a compact neighborhood W of x with $f(W) \subset U$. Then $U(W) \times W$ is a neighborhood of (f, x) in $C_K \times K$ and $e(U(W) \times W) \subset U$. Hence e is continuous at $x \in K$. But then e is continuous on K so the compact-open topology on C_K is jointly continuous on K. Therefore, the compact-open topology on C_K is jointly continuous on compacta. ∎

COROLLARY 11.9 If X is a regular locally compact space and C is the family of continuous functions from X to a uniform space (Y, ν), then the κ topology (compact-open topology) is the coarsest jointly continuous topology on C.

THEOREM 11.28 Let X be a regular of Hausdorff space and let C_K be the family of all functions from X to a uniform space (Y, ν) that are continuous on each compact subset of X. Let κ and π denote the compact-open and the product topologies respectively. Then a subfamily E of C_K is κ-compact in C_K if and only if

(1) E is closed in C_K with respect to κ.
(2) E[x] has a compact closure in Y for each $x \in X$.
(3) The π topology on the π-closure of E in F is jointly continuous on compacta.

Proof: First assume E is κ-compact in C_K. Then E is closed with respect to κ in

C_K so (1) holds. Let $x \in X$. Then $E[x] = \{f(x) \mid f \in E\} = \{\pi_x(f) \mid f \in E\} = \pi_x(E)$. Since the xth projection is continuous with respect to the product topology, and hence continuous with respect to κ, $E[x]$ is compact so (2) holds. By the remarks following Theorem 11.23, since E is compact, the κ and π topologies arc identical on E. Consequently, E is closed in F with respect to π. Thus the π-closure of E is E. By Theorem 11.27, κ is jointly continuous on compacta on E so π is jointly continuous on compacta on E. Hence π is jointly continuous on compacta on the π-closure of E so (3) holds.

Conversely, assume (1), (2) and (3) hold. Let E^* denote the closure of E in F with respect to π. By (2), $Cl_Y(E[x])$ is compact for each $x \in X$ so $\Pi_{x\in X}Cl_Y(E[x])$ is compact. Since $E^* \subset \Pi_{x\in X}Cl_Y(E[x])$, E^* is compact with respect to π. By (3), π is jointly continuous on compacta on E^*. Let $f \in E^*$ and let K be a compact subset of X. Then $e{:}E^* \times K \to Y$ is continuous at each point (f, x) such that $x \in K$. Now $e\mid_{f\times K}{:}\{f\} \times K \to Y$ is defined by $e\mid_{f\times K}(f, x) = e(f,x) = f(x)$ so we can identify $e\mid_{f\times K}$ with f. Since e is continuous on $\{f\} \times K$, f is continuous on K. Therefore, f is continuous on each compact subset of X, so $E^* \subset C_K$. By Theorem 11.27, since π is jointly continuous on compacta on E^*, π is finer than κ on E^* so π is identical with κ on E^*. But then E^* is compact with respect to κ. By (1), E is closed in C_K with respect to κ so E is closed in C_K with respect to π. Hence $E = E^*$ so E is compact with respect to κ.
∎

We now turn our attention to a concept that can replace condition (3) in Theorem 11.28, and which allows us to strengthen the conclusion of the theorem to the function space C rather than C_K to get a uniform version of the classical *Ascoli Theorem.* This concept is called *equicontinuity* and it applies to a *family* of mappings as does the concept of uniform convergence. A family of mappings E from a space X into a uniform space (Y, ν) is said to be **equicontinuous at a point** $x \in X$ if for each $\mathcal{V} \in \nu$ there exists an open set $U \subset X$ containing x such that $f(U) \subset Star(f(x), \mathcal{V})$ for each $f \in E$. In other words, for each $\mathcal{V} \in \nu$ there exists a neighborhood U of x whose image under any $f \in E$ is $\mathcal{V}^{\Delta}$-small. E is said to be an **equicontinuous** family if E is equicontinuous at each point $x \in X$. Clearly, equicontinuity implies continuity for each member of the family. We divide the proof of the Ascoli Theorem into three lemmas that are of significant interest themselves.

LEMMA 11.7 If E is equicontinuous at x, then the closure of E with respect to π in F is also equicontinuous at x. Hence if E is an equicontinuous family, then so is the π-closure in F.

Proof: Let E^* be the closure of E in F with respect to π. Let $\mathcal{V} \in \nu$ and let $\mathcal{U} \in \nu$ such that $\mathcal{U}^* < \mathcal{V}$. Since E is equicontinuous at x, there exists an open $U \subset X$ containing x such that $f(U) \subset S(f(x), \mathcal{U})$ for each $f \in E$. Let H be the family of all members $h \in F$ such that $h(U) \subset Cl(S(h(x), \mathcal{U}))$. Clearly $E \subset H$. Now

$$H = \{h \in F \mid \text{for each } u \in U, h(u) \in Cl(S(h(x), \mathcal{U}))\} =$$
$$\cap_{u \in U}\{h \in F \mid h(u) \in Cl(S(h(x), \mathcal{U}))\}.$$

Since for each $u \in$ U, $\{h \in F \mid h(u) \in Cl(S(h(x), \mathcal{U}))\}$ is closed with respect to π, it is clear that H is closed in F with respect to π. But $E \subset H$ so $E^* \subset H$, so for each $g \in E^*$, g(U) $\subset Cl(S(g(x), \mathcal{U})) \subset S(g(x), \mathcal{V})$. Hence E^* is equicontinuous at x. ∎

LEMMA 11.8 If E is equicontinuous then π is jointly continuous on E. Hence π is identical with κ on E.

Proof: First we want to show that π is jointly continuous on E. For this we need to show that $e{:}E \times \text{X} \to \text{Y}$ defined by $e(f, x) = f(x)$ is continuous at each point of $E \times$ X. Let $(f, x) \in E \times$ X and let $\mathcal{V} \in \nu$. Pick $\mathcal{U} \in \nu$ such that $\mathcal{U}^* < \mathcal{V}$. Let W be an open set in X containing x such that g(W) $\subset S(g(x), \mathcal{U})$ for each $g \in E$. Put U$(f, x) = \{g \in E \mid g(x) \in S(f(x), \mathcal{U})\}$. Then U$(f, x)$ is a subbasic open set in π. Now U$(f, x) \times$ W contains (f, x) in $E \times$ X and if $(g, y) \in$ U$(f, x) \times$ W then $g(x) \in S(f(x), \mathcal{U})$ since $g \in$ U(f, x) and $g(y) \in S(g(x), \mathcal{U})$ since $y \in$ W. But then $g(y) \in Star(S(f(x), \mathcal{U}), \mathcal{U}) \subset S(f(x), \mathcal{V})$ so $e(g, y) \in S(e(f, x), \mathcal{V})$ for each $(g, y) \in$ U$(f,x) \times$ W. Hence π is jointly continuous on E. Then by Theorem 11.27, the π topology on E is finer than the compact-open topology on E so π is identical with the compact-open topology on E. Since the members of E are all continuous, by Theorem 11.26, $\pi = \kappa$ on E. ∎

COROLLARY 11.10 If E is an equicontinuous family of functions then E is compact with respect to κ if it is compact with respect to π.

LEMMA 11.9 If E is a family of functions from a space X to a uniform space (Y,ν) and E is compact with respect to some jointly continuous topology for E, then E is equicontinuous.

Proof: Let $x \in$ X and $\mathcal{V} \in \nu$. Pick $\mathcal{U} \in \nu$ with $\mathcal{U}^* < \mathcal{V}$. Since τ is jointly continuous on E, for each $f \in E$ there is a τ-open set U_f containing f and an open set W $\subset$ X containing x with $e(U_f \times \text{W}) \subset S(f(x), \mathcal{U})$. If $g \in U_f$ and $y \in$ W, then $g(x)$ and $g(y)$ are in $S(f(x), \mathcal{U})$ which implies $g(y) \in Star(S(g(x),\mathcal{U}),\mathcal{U}) \subset S(g(x), \mathcal{V})$ for each $y \in$ W so $g(\text{W}) \subset S(g(x), \mathcal{V})$ for each $g \in U_f$. Since E is compact with respect to τ, there is a finite collection $U_{f_1} \ldots U_{f_n} \in \tau$ covering E and open sets $W_1 \ldots W_n \subset$ X, each containing x such that for each $g \in U_{f_i}$, $g(W_i) \subset S(g(x), \mathcal{V})$. Put U $= \cap_{i=1}^{n} W_i$. Then U is open in X and contains x. Moreover, g(U) $\subset S(g(x), \mathcal{V})$ for each $g \in E$. Thus, E is equicontinuous. ∎

THEOREM 11.29 (Ascoli) Let C be the family of all continuous functions from a locally compact space X to a uniform space (Y, ν) and let κ be the topology of uniform convergence on compacta. Then a subfamily E of C is compact with respect to κ if and only if

(1) E is closed in C with respect to κ,

(2) E[x] has compact closure in Y for each x ∈ X and

(3) E is equicontinuous.

Proof: Since X is locally compact, X is uniformizable and hence regular. Also, local compactness implies $C = C_K$ (those members of F that are continuous on compact subsets). Then if E is compact in C with respect to κ, (1) and (2) hold by Theorem 11.28, and the π topology on the π-closure of E in F is jointly continuous on compacta. Then as shown in the last part of the proof of Lemma 11.8, π = κ on E. Hence κ is jointly continuous on compacta. But since X is locally compact, κ is jointly continuous on E by Lemma 11.9, since E is compact with respect to κ, E is equicontinuous. Hence (3) holds.

Conversely, assume (1), (2) and (3) hold. By (3), E is equicontinuous so by Lemma 11.7, the closure E^* of E with respect to π in F is equicontinuous. Then by Lemma 11.8, π is jointly continuous on E^*. Hence π is jointly continuous on compacta on E^* so (1), (2) and (3) of Theorem 11.28 hold which implies if $E \subset C_K$ then E is compact in C_K with respect to κ. Since X is locally compact, $C = C_K$ so $E \subset C_K$ and hence E is compact in C with respect to κ. ∎

EXERCISES

1. Let X = [0, 1] = Y and let $F = Y^X$. Let $f \in F$ be defined by $f(x) = 0$ if x is rational and $f(x) = 1$ otherwise. Then the function $e{:}F \times X \to Y$ defined by $e(f,x) = f(x) \in Y$ is zero at $(f, 1/2) \in F \times X$. Show that e is not continuous at the point $(f, 1/2)$.

2. Let X and Y be spaces and let $G \subset Y^X$. Let δ be the discrete topology on G. Show that δ is jointly continuous on G.

3. Prove Proposition 11.19.

4. Prove Theorem 11.25.

5. If X and Y are topological spaces and H and K are subsets of X and Y respectively, and if U is open in X × Y and contains H × K, then there exists open sets V and W containing H and K respectively such that V × W ⊂ U.

THE π TOPOLOGY ON THE DUAL SPACE

6. Show that the weak* topology on the dual space X* of X in Exercise 8 of Section 11.3 is the topology π of pointwise convergence.

7. Show that the unit sphere in X* is compact in the weak* topology.

Chapter 12

UNIFORM DIFFERENTIATION

12.1 Complex Measures

At the time of writing this chapter, very little is known about uniform differentiation. It is an area where all the good theorems remain to be proven. It is also an area where we know the approximate form we would like for some of these yet unproven theorems to take. This is because we expect the uniform derivative to be equivalent to the Radon-Nikodym derivative when it exists. In this chapter we develop both the concepts of the Radon-Nikodym derivative and the uniform derivative. However, the assumptions we make about the spaces on which the uniform derivatives are defined, and perhaps about the uniform derivative itself, are probably unnecessarily restrictive. This is because we do not know how to prove the equivalence of the uniform derivative and the Radon-Nikodym derivative at the present time without them.

The concept of the Radon-Nikodym derivative is more general than the concept of the uniform derivative. The Radon-Nikodym derivative is established by a nonconstructive existence proof and consequently we know almost nothing about its structure. On the other hand, the uniform derivative is constructed in a more familiar manner as the limit of ratios of the measure of uniform neighborhoods of a point to the Haar measure of these uniform neighborhoods. Consequently, to even discuss the concept of uniform differentiation we need assumptions strong enough to guarantee the existence of Haar measure. The Radon-Nikodym derivative exists in a more general setting than that required to assure the existence of Haar measure.

Both the Radon-Nikodym derivative and the uniform derivative of a measure can be defined for complex valued measures as well as for the positive measures that we have been considering up until now. In this section we will first introduce the concept of a complex valued measure and then develop some tools for dealing with them. For our purposes we define a **complex measure** to be a countably additive, complex valued function on a σ-algebra. We will not develop the concept of complex valued measures on rings and σ-rings.

Let M be a σ-algebra in a set X. A countable collection $\{E_n\} \subset M$ is said to be a **partition** of E if $E_m \cap E_n = \emptyset$ whenever $m \neq n$ and if $E = \cup E_n$. If μ is a complex measure on M then $\mu(E) = \Sigma_n \mu(E_n)$ for each partition $\{E_n\}$ of E. Consequently, the series $\Sigma_n \mu(E_n)$ must converge if μ is a complex measure.

Since the union $\cup E_n$ of the sequence $\{E_n\}$ is not changed by a permutation of the subscripts, each rearrangement of the series must also converge. From a well known result about infinite series that can be found in several advanced calculus texts, the series must converge absolutely.

Let λ be a positive measure on M. λ is said to **dominate** μ if $|\mu(E)| \leq \lambda(E)$ for each $E \in M$. Then for the partition $\{E_n\}$ of E, $\lambda(E) = \Sigma_n\lambda(E_n) \geq \Sigma_n|\mu(E_n)|$ so $\lambda(E) \geq sup\{\Sigma_n|\mu(E_n)| \mid \{E_n\}$ is a partition of E$\}$. We define a set function $|\mu|$ on M by $|\mu|(E) = sup\{\Sigma_n|\mu(E_n)| \mid \{E_n\}$ is a partition of E$\}$. This function $|\mu|:M \to [0, \infty)$ is called the **total variation** of μ.

PROPOSITION 12.1 The total variation of a complex measure μ is a positive measure. It is the smallest positive measure that dominates μ.

Proof: Let $\{E_n\}$ be a partition of $E \in M$. For each n let $r_n \in \mathbf{R}$ such that $r_n < |\mu|(E_n)$. Then each E_n has a partition $\{E_{n_m}\}$ such that $\Sigma_m|\mu(E_{n_m})| > r_n$. Since $\{E_{n_m} \mid n, m$ are positive integers$\}$ is a partition of E, we have $\Sigma_n r_n \leq \Sigma_{n,m}|\mu(E_{n_m})| \leq |\mu|(E)$. Then $sup\{\Sigma_n r_n \mid r_n < |\mu|(E_n)$ for each $n\} = \Sigma_n|\mu|(E_n)$ and $sup\{\Sigma_n r_n \mid r_n < |\mu|(E_n)$ for each $n\} \leq \Sigma_{n,m}|\mu(E_{n_m})|$ so $\Sigma_n|\mu|(E_n) \leq |\mu|(E)$.

To prove the opposite inequality, let $\{A_m\}$ be another partition of E. Then for a fixed n, $\{E_n \cap A_m\}$ is a partition of E_n and for a fixed m, $\{E_n \cap A_m\}$ is a partition of A_m. Therefore, $\Sigma_m|\mu(A_m)| = \Sigma_m|\Sigma_n\mu(E_n \cap A_m)| \leq \Sigma_m\Sigma_n|\mu(E_n \cap A_m)| \leq \Sigma_n|\mu|(E_n)$. Since this holds for each partition $\{A_m\}$ of E, we have $|\mu|(E) \leq \Sigma_n|\mu|(E_n)$. Consequently, $\Sigma_n|\mu|(E_n) = |\mu|(E)$ so μ is countably additive. That $|\mu|$ satisfies the other properties of a positive measure are easily demonstrated. That $|\mu|$ is the smallest positive measure that dominates μ follows from the fact that if λ is another positive measure that dominates μ, then $\lambda(E) = \Sigma_n\lambda(E_n) \geq \Sigma_n|\mu(E_n)|$ for each partition $\{E_n\}$ of E, so $\lambda(E) \geq |\mu|(E)$ for each $E \in M$. ■

LEMMA 12.1 If $c_1 \ldots c_n \in \mathbf{C}$, there exists a subset of $\{1 \ldots n\}$ such that $|\Sigma_{i\in S}c_i| \geq 1/6\Sigma_{i=1}^n|c_i|$.

Proof: Put $r = \Sigma_{i=1}^n|c_i|$. C is the union of four "diagonal quadrants" bounded by the lines $y = x$ and $y = -x$. Let Q_1 denote the one defined by $|y| \leq x$ for each $z = (x, y)$ in Q_1. Now one of these quadrants, say Q, has the property that the sum of the $|c_i|$'s for which $c_i \in Q$ is at least $r/4$. Assume first that $Q = Q_1$. For $c \in Q_1$, $\text{Re}(c) \geq |c|^{-1/2}$. If S is the set of all i such that $c_i \in Q_1$, it then follows that:

$$|\Sigma_{i\in S}c_i| \geq \Sigma_{i\in S}Re(c_i) \geq \Sigma_{i\in S}|c_i|^{-1/2} \geq r/4\sqrt{2} \geq r/6.$$

This proves the lemma for $Q = Q_1$. If Q is one of the other quadrants, a similar argument will work by replacing the formula $Re(c) \geq |c|^{-1/2}$ with the appropriate formula for that quadrant. ■

PROPOSITION 12.2 If μ is a complex measure on X then $|\mu|(X) < \infty$.

Proof: Assume $|\mu|(X) = \infty$. Then for each $r > 0$ there exists a partition $\{X_n\}$ of X such that $\Sigma_n |\mu(X_n)| > r$. Now $\mu(X) = c$ for some $c \in$ C so $|\mu(X)| < \infty$. Put $s = 6(|\mu(X)| + 1)$. Then there exists a partition $\{E_n\}$ of X with $\Sigma_{n=1}^{m} |\mu(E_n)| > s$ for some positive integer m. By Lemma 12.1, if we put $c_i = \mu(E_i)$ for $i = 1 \ldots m$ there exists a subset S of $\{1 \ldots m\}$ with $|\Sigma_{n \in S}\mu(E_n)| \geq 1/6\Sigma_{n=1}^{m} |\mu(E_n)|$. Put H $= \cup_{n \in S} E_n$ and K = X - H. Then $H \cap K = \emptyset$, $X = H \cup K$ and $|\mu(H)| = |\mu(\cup_{n \in S} E_n)| = |\Sigma_{n \in S}\mu(E_n)| \geq 1/6\Sigma_{n=1}^{m} |\mu(E_n)| > s/6 = |\mu(X)| + 1 \geq 1$, so $|\mu(H)| \geq 1$.

For any pair $a, b \in$ C, it is easily seen that $|a - b| \geq |b| - |a|$. Hence $|\mu(K)| = |\mu(X) - \mu(H)| \geq |\mu(H)| - |\mu(X)| > s/6 - |\mu(X)| = 1$. Since $|\mu|(X) = |\mu|(H) + |\mu|(K)$ by Proposition 12.1, we have $|\mu|(H) = \infty$ or $|\mu|(K) = \infty$. Consequently, X is the union of two disjoint sets H and K such that for one of the sets, say H, $|\mu|(H) = \infty$ and for the other, $|\mu(K)| \geq 1$. Put H_1 = H and K_1 = K. A similar argument can be used to show that H_1 can be partitioned into two disjoint sets H_2 and K_2 such that $|\mu|(H_2) = \infty$ and $|\mu(K_2)| \geq 1$. Continuing inductively, we obtain a sequence $\{K_n\}$ of subsets of X such that $K_m \cap K_n = \emptyset$ if $m \neq n$ and $|\mu(K_n)| \geq 1$ for each n. Put $Y = \cup K_n$. Then $\mu(Y) = \Sigma_n \mu(K_n)$. But $\Sigma_n \mu(K_n)$ cannot converge absolutely since $|\mu(K_n)| \geq 1$ for each n, which is a contradiction. Therefore, $|\mu|(X) < \infty$. ■

Proposition 2.12 shows that complex measures are *bounded* in the sense that their range lies in a sphere of finite radius. To see this let μ be a complex measure on a σ-algebra *M* and E ∈ *M*. Then $|\mu(E)| \leq |\mu|(E) \leq |\mu|(X)$. This fact has many interesting and useful consequences, among which is the the following: if μ and λ are complex measures on the same σ-algebra *M* and if $c \in$ C, we define $\mu + \lambda$ and $c\mu$ on *M* as follows: for each E ∈ *M* put $[\mu + \lambda](E) = \mu(E) + \lambda(E)$ and $[c\mu](E) = c\mu(E)$. Clearly $\mu + \lambda$ and $c\mu$ are also complex measures on *M*. Therefore, the collection of complex measures on *M* is a vector space. In fact, if we put $|\mu| = |\mu|(X)$, it is easily verified that the collection of complex measures on *M* is a normed linear space.

Now assume that μ is a positive measure on *M* and λ is either a complex measure or a positive measure on *M*. λ is said to be **absolutely continuous** with respect to μ, denoted by $\lambda << \mu$, if $\lambda(E) = 0$ for each E ∈ *M* for which $\mu(E) = 0$. If there exists a Y ∈ *M* such that $\mu(E) = \mu(E \cap Y)$ for each E ∈ *M*, we say μ is **concentrated** on Y. In other words, μ is concentrated on Y if and only if $\mu(E) = 0$ whenever $E \cap Y = \emptyset$. If μ and λ are any measures on *M* such that there exists disjoint sets H and K with λ concentrated on H and μ concentrated on K, then μ and λ are said to be **orthogonal** or to be *mutually singular*, denoted $\mu \perp \lambda$.

PROPOSITION 12.3 If λ, μ and σ are measures on a σ-algebra M and if σ is positive, then:

(1) If μ is concentrated on $E \in M$, then so is $|\mu|$.
(2) If $\mu \perp \lambda$ then $|\mu| \perp |\lambda|$.
(3) If $\mu << \sigma$ then $|\mu| << \sigma$.
(4) If $\mu \perp \sigma$ and $\lambda \perp \sigma$, then $\mu + \lambda \perp \sigma$.
(5) If $\mu << \sigma$ and $\lambda << \sigma$, then $\mu + \lambda << \sigma$.
(6) If $\mu << \sigma$ and $\lambda \perp \sigma$, then $\mu \perp \lambda$.
(7) If $\mu << \sigma$ and $\mu \perp \sigma$, then $\mu = 0$.

The proof of this proposition is left as an exercise.

EXERCISES

1. Prove Proposition 12.3.

2. Let μ be a *real valued measure* on a σ-algebra M and define $\mu^+(E) = 1/2[|\mu|(E) + \mu(E)]$ and $\mu^-(E) = 1/2[|\mu|(E) - \mu(E)]$ for each E in M. Show that both m^+ and μ^- are bounded positive measures on M.

3. Show that $\mu = \mu^+ - \mu^-$ and that $|\mu| = \mu^+ + \mu^-$. Expressing μ as $\mu^+ - \mu^-$ is known as the **Jordan decomposition** of μ. Also, μ^+ and μ^- are said to be the **positive and negative variations** of μ respectively.

4. Let λ and μ be measures on a σ-algebra M such that μ is positive and λ is complex. Show that λ is absolutely continuous with respect to μ if and only if for each $\varepsilon > 0$ there exists a $\delta > 0$ such that $|\lambda(E)| < \varepsilon$ for each $E \in M$ with $\mu(E) < \delta$.

5. Let $M(X)$ be the family of all complex regular Borel measures on a locally compact Hausdorff space (Note: a complex measure μ is said to be **regular** if $|\mu|$ is regular). Show that $M(X)$ is a Banach space with norm $|\mu| = |\mu|(X)$.

12.2 The Radon-Nikodym Derivative

Let M be a σ-algebra in a topological space X and let μ be a positive measure on M. If f:X $\to [0, \infty]$ or f:X $\to$ **C** we can define a measure λ on M by $\lambda(E) = \int_E f d\mu$. In the case f:X $\to [0, \infty]$, Theorem 8.12 assures us that λ is a positive measure. In the case f:X $\to$ **C**, let $E \in M$ such that $E = \cup E_n$ for some sequence $\{E_n\} \subset M$ for which $E_m \cap E_n = \varnothing$ whenever $m \neq n$. For each positive integer n put $f_n = f © \chi_{E_n}$. Then $f_n(x) = f(x)$ for each $x \in E_n$ and $f_n(x) = 0$ otherwise. Clearly $f(x) = \Sigma_n f_n(x)$ for each $x \in$ E, so by Theorem 8.10, $\lambda(E) = \int_E f d\mu = \Sigma_n \int_E f_n d\mu = \Sigma_n \int_{E_n} f d\mu = \Sigma_n \lambda(E_n)$ so λ is countably additive and hence a complex measure on M. λ is called the measure **generated** by f and μ. Clearly $\lambda << \mu$. In the event that μ is also σ-finite, the converse of this situation is true, i.e., if σ is a positive or a complex measure on M that is absolutely continuous with respect to μ ($\sigma << \mu$) then there exists a unique $g \in L^1(\mu)$ such that $\sigma(E) = \int_E g\,d\mu$ for each $E \in M$. This function $g: X \to [0, \infty]$ or $g: X \to \mathbf{C}$, depending on whether σ is positive or complex is called the **Radon-Nikodym Derivative** of σ with respect to μ and is often denoted by $d\sigma/d\mu$ or, since $\int_E d\sigma = \int_E g\,d\mu$ for each $E \in M$, by $d\sigma = g\,d\mu$. The proof of the existence and the uniqueness of the Radon-Nikodym derivative will constitute the remainder of this section. The following theorem establishes the existence and uniqueness when λ is positive and bounded or complex. Exercise 4 extends Theorem 2.1 to the case where λ is positive and σ-finite.

THEOREM 12.1 (J. Radon and O. Nikodym, 1930) If μ is a positive σ-finite measure on a σ-algebra M in a space X and λ is a positive bounded or complex measure on M that is absolutely continuous with respect to μ, then there exists a unique function f in $L^1(\mu)$, where $f: X \to [0, \infty)$ or $f: X \to \mathbf{C}$, depending on whether λ is positive or complex, such that $\lambda(E) = \int_E f\,d\mu$ for each $E \in M$.

Proof: We will first prove the theorem for the special case where μ and λ are both positive bounded measures on M. Let $\sigma = \lambda + \mu$. Then σ is also a positive bounded measure on M. If $E \in M$ and $f = \chi_E$ then $\int_X f\,d\sigma = \int_X \chi_E\,d\sigma = \int_X \chi_E\,d(\lambda+\mu) = (\lambda + \mu)(E) = \lambda(E) + \mu(E) = \int_X \chi_E\,d\lambda + \int_X \chi_E\,d\mu = \int_X f\,d\lambda + \int_X f\,d\mu$. Hence

$$\int_X f\,d\sigma = \int_X f\,d\lambda + \int_X f\,d\mu \tag{12.1}$$

whenever f is a characteristic function. If s is a simple measurable function then the range of s is $\{a_1 \ldots a_n\}$ for some finite subset of $[0, \infty)$. For each $i = 1 \ldots n$ put $E_i = \{x \in X \mid s(x) = a_i\}$. Then $E_i \in M$ for each i. Clearly $s = \Sigma_{i=1}^n a_i \chi_{E_i}$ so each simple measurable function is a linear combination of characteristic functions. Consequently, (12.1) holds for simple measurable functions. Let $f: X \to [0, \infty]$ be measurable. By the definition of $\int_X f\,d\sigma$ and the fact that the supremum of the sum of the two sets in $[0, \infty]$ is the sum of the supremums of the two sets, it follows that (12.1) also holds for all nonnegative measurable functions.

If $f \in L^2(\sigma)$, the Schwarz inequality (Chapter 11) gives $|(f, 1)| \leq |f|_2 |1|_2$. Now $|f|_2 = [\int_X |f|^2 d\sigma]^{1/2}$ and $|1|_2 = [\int_X |1|^2 d\sigma]^{1/2} = [\sigma(X)]^{1/2}$. Also, $|(f, 1)| = |\int_X f\,d\sigma|$. Hence $|\int_X f\,d\lambda| \leq |\int_X f\,d\sigma| \leq [\int_X |f|^2 d\sigma]^{1/2} [\sigma(X)]^{1/2}$. Since $\sigma(X) < \infty$ and $|f|_2 < \infty$ it is clear that $|\int_X f\,d\lambda| < \infty$. Hence the mapping $\Lambda: L^2(\sigma) \to \mathbf{C}$ defined by $\Lambda(f) = \int_X f\,d\lambda$ is a bounded linear functional on $L^2(\sigma)$.

By Theorem 11.13, Λ is continuous so by Theorem 11.7, there exists a unique $g \in L^2(\sigma)$ such that $\Lambda(f) = (f, g^*)$ for each $f \in L^2(\sigma)$, or in other words, $\int_X f\,d\lambda = \int_X fg\,d\sigma$ for each $f \in L^2(\sigma)$. Let $E \in M$ with $\sigma(E) > 0$. Then

(12.2) $\lambda(E) = \int_X \chi_E d\lambda = \int_X \chi_E g d\sigma = \int_E g d\sigma.$

Since $0 \le \lambda \le \sigma$, $0 \le \lambda(E)/\sigma(E) \le 1$ so $0 \le [1/\sigma(E)]\int_E g d\sigma \le 1$. Consequently, $g(x) \in [0, 1]$ for almost all x (with respect to σ) by Exercise 5 of Section 8.7. So we may assume $g(x) \in [0, 1]$ for all $x \in X$ without affecting (12.2). Let $E \in M$. By (12.2), $\lambda(E) = \int_E g d\sigma$ so $\int_E d\lambda = \int_E g d\sigma$ which implies $\int_X \chi_E(1 - g)d\lambda = \int_X \chi_E g d\mu$. Hence

(12.3) $\int_X (1 - g)f d\lambda = \int_X f g d\mu$

whenever $f = \chi_E$ for some $E \in M$. Then by an argument similar to the one above, (12.3) holds for any nonnegative measurable function.

Let $H = \{x \mid g(x) < 1\}$ and $K = \{x \mid g(x) = 1\}$. Then $H \cap K = \varnothing$. Put $\lambda_H(E) = \lambda(H \cap E)$ and $\lambda_K(E) = \lambda(K \cap E)$ for each $E \in M$. Then λ_H and λ_K are both positive measures on M. If we put $f = \chi_K$ in (12.3) we get $\int_K (1 - g)d\lambda = \int_K g d\mu$ which implies $\int_K d\lambda - \int_K g d\lambda = \int_K g d\mu$, so $\lambda(K) - \lambda(K) = \mu(K)$. Hence $\mu(K) = 0$, so $\lambda(K) = 0$. But then $\sigma(K) = 0$ and $\lambda_H(E) = \lambda(E)$ for each $E \in M$ so $\lambda = \lambda_H$. Consequently, we may assume $g(x) \in [0, 1)$ for all $x \in X$ without affecting (12.3).

Let $E \in M$ and let n be a positive integer. If we let $f = \chi_E(1 + g + \ldots + g^n)$ in (12.3) we get:

(12.4) $\int_E (1 - g^{n+1})d\lambda = \int_E (1 + g + \ldots + g^n) g d\mu.$

At each point $x \in X$, $g^{n+1}(x)$ converges monotonically to 0 (as $n \to \infty$) so the left side of (12.4) converges to $\lambda(E)$. Since $g(x) < 1$ for each $x \in X$, $(1 + g + \ldots + g^n)g = \Sigma_{m=1}^{n+1}[g(x)]^m$ converges monotonically to some $f(x)$ for each $x \in X$. Then $f: X \to \mathbf{R}$ is nonnegative and measurable by Proposition 8.18. Consequently, the right hand side of (12.4) converges to $\int_E f d\mu$ by Theorem 8.9. Clearly the left hand side converges to $\int_E d\lambda$. Therefore, $\lambda(E) = \int_E f d\mu$ for each $E \in M$. Taking $E = X$ shows that $f \in L^1(\mu)$ since λ is bounded. This concludes the proof for the case where both μ and λ are positive and bounded.

Next let μ be σ-finite and let λ be a positive bounded measure. Since μ is σ-finite, there exists a countable collection $\{E_n\}$ of subsets of X such that $\cup E_n = X$ and $\mu(E_n) < \infty$ for each positive integer n. Put $X_1 = E_1$ and for each positive integer $n > 1$ put $X_n = E_n - \cup_{i=1}^{n-1} X_i$. Then $\{X_n\}$ is a decomposition of X into disjoint measurable sets such that $\cup X_n = X$ and $\mu(X_n) < \infty$ for each positive integer n. Since $\lambda(X) < \infty$, $\lambda(X_n) < \infty$ for each n, so we can apply the result we just proved to each X_n to get a function $f_n \in L^1(\mu)$ such that $f_n = 0$ outside X_n and $\lambda(E \cap X_n) = \int_{E \cap X_n} f_n d\mu$. Now put $f(x) = \Sigma_{n=1}^{\infty} f_n$. Then $f: X \to \mathbf{R}$ is nonnegative and measurable. Now $\lambda(E) = \lambda(E \cap X) = \lambda(E \cap [\cup X_n]) = \lambda(\cup [E \cap X_n]) = \Sigma_n \lambda(E \cap X_n) = \Sigma_n \int_{E \cap X_n} f_n d\mu = \Sigma_n \int_E f_n d\mu$. For each positive integer m, $\Sigma_{n=1}^{m} \int_E f_n d\mu = \int_E (\Sigma_{n=1}^{m} f_n) d\mu$ and by Theorem 8.9, $\int_E (\Sigma_{n=1}^{m} f_n) d\mu$ converges to $\int_E f d\mu$ as $m \to \infty$. Hence $\Sigma_{n=1}^{m} \int_E f_n d\mu$ converges to $\int_E f d\mu$ as $m \to \infty$.

But $\Sigma_{n=1}^{m}\int_E f_n d\mu$ converges to $\Sigma_n\int_E f_n d\mu$ so $\Sigma_n\int_E f_n d\mu = \int_E f d\mu$. Therefore, $\lambda(E) = \int_E f d\mu$. Again, taking $E = X$ shows that $f \in L^1(\mu)$ since $\lambda(E) < \infty$. This concludes the proof for the case where μ is σ-finite and λ is a positive bounded measure.

Finally assume μ is σ-finite and that λ is complex. Then $\lambda = \lambda_1 + i\lambda_2$ for some real valued measures λ_1 and λ_2 on M. We can then apply the above results to the positive and negative variations of λ_1 and λ_2 (see Exercises 2 and 3 of Section 12.1) to obtain the desired result.

To show that f is unique, let g be another member of $L^1(\mu)$ such that $\lambda(E) = \int_E g d\mu$ for each $E \in M$. First assume f:X $\to [0, \infty)$ and g:X $\to [0, \infty)$. Now $f - g$ is a measurable function. Let $A = \{x \in X \mid [f - g](x) < 0\}$ and $B = \{x \in X \mid [f\text{-}g](x) > 0\}$. Clearly both A and B are measurable and $A \cup B = \{x \in X \mid f(x) \neq g(x)\}$. Suppose $\lambda(A) \neq 0$. For each positive integer n put $A_n = \{x \in A \mid -1/2^n \leq [f - g](x) < 1/2^n\}$ and let $A_0 = \{x \in X \mid [f - g](x) < -1/2\}$. Then $A = \cup_{n=0}^{\infty} A_n$, $A_m \cap A_n = \varnothing$ if $m \neq n$, and each A_n is measurable. Therefore, there exists some nonnegative integer k such that $\lambda(A_k) \neq 0$ which implies $g(x) - f(x) > 1/2^k$ for each $x \in A_k$ so $\int_{A_k} f d\lambda \neq \int_{A_k} g d\lambda$ which is a contradiction. Hence $\lambda(A) = 0$. Similarly $\lambda(B) = 0$. Therefore, $\{x \in X \mid f(x) \neq g(x)\}$ has zero measure, so $f = g$ almost everywhere.

Next assume f:X $\to$ **C** and g:X $\to$ **C**. Then $f = f_1 + if_2$ for two real valued functions f_1 and f_2 and $g = g_1 + ig_2$ for two real valued functions g_1 and g_2. Then $\int_E f_1 d\mu = \int_E g_1 d\mu$ and $\int_E f_2 d\mu = \int_E g_2 d\mu$ for each $E \in M$. Now the argument above used to show $f = g$ almost everywhere when f:X $\to [0, \infty)$ and g:X $\to [0,\infty)$ can easily be modified to show $f_1 = g_1$ almost everywhere when f_1:X $\to$ **R** and g_1:X $\to$ **R**. Therefore, $f_1 = g_1$ almost everywhere. Similarly $f_2 = g_2$ almost everywhere. Hence $f = g$ almost everywhere, so f is unique in $L^1(\mu)$. ∎

The existence of the Radon-Nikodym derivative is extremely significant and we will devote a good deal of space in this chapter to considering its consequences. For instance, it allows us to show that if p and q are conjugate exponents and μ is a positive σ-finite measure, then $L^q(\mu)$ is the *dual space* of $L^p(\mu)$. It also allows us to extend the Reisz Representation Theorem to bounded linear functionals Λ on $C_\infty(X)$ to obtain a representation of the form $\Lambda(f) = \int_X f d\mu$ for a unique regular, complex measure μ, for each $f \in C_\infty(X)$. Furthermore, it will be used to prove the Hahn Decomposition Theorem in the next section. However, as important as the Radon-Nikodym derivative is, its existence proof gives us no insight into the intrinsic nature or structure of this derivative. These results we have just mentioned only give us insight into its *behavior*.

Later in this chapter we will use the Haar measure on an isogeneous uniform space to construct a derivative in the more familiar manner as a limiting process of ratios of a measure μ with respect to the Haar measure m for suitably well behaved measures μ, to produce a *uniform derivative* $d\mu/dm$. We will show that for these measures μ, the uniform derivative is the Radon-Nikodym derivative.

We now turn our attention to showing that for a positive σ-finite measure μ and conjugate exponents p and q that $L^q(\mu)$ is the dual space of $L^p(\mu)$. If we let μ be any positive measure, $1 \le p < \infty$, and q is an exponent conjugate to p, then the Hölder inequality (Theorem 11.1) implies that if $g \in L^q(\mu)$ and if Λ is defined by $\Lambda(f) = \int_X fg d\mu$ for each $f \in L^p(\mu)$ then Λ is a bounded linear functional on $L^p(\mu)$ of norm $|\Lambda|$ at most $|g|_q$. To see this, note that Hölder's inequality is simply $|fg|_1 \le |f|_p |g|_q$. Then $|\Lambda| = sup\{|\int_X fg d\mu|/|f|_p \,|\, f \in L^p(\mu) \text{ and } f \ne 0\} \le sup\{\int_X |fg| d\mu/|f|_p \,|\, f \in L^p(\mu) \text{ and } f \ne 0\} = sup\{|fg|_1/|f|_p \,|\, f \in L^p(\mu) \text{ and } f \ne 0\} \le sup\{|f|_p |g|_q/|f|_p \,|\, f \in L^p(\mu) \text{ and } f \ne 0\} = |g|_q$. Hence $|\Lambda| \le |g|_q$.

Now if all bounded linear functionals on $L^p(\mu)$ have a unique representation of this form, then we can define a one-to-one, onto mapping $\Phi: L^p(\mu)^* \to L^q(\mu)$ by $\Phi(\Lambda) = g$ where g is the unique member of $L^q(\mu)$ such that $\Lambda(f) = \int_X fg d\mu$ for each $f \in L^p(\mu)$. Clearly, Φ is a linear operator. If it were the case that $|\Lambda| = |g|_q$ where $\Phi(\Lambda) = g$ then Φ would be distance preserving (norm preserving) and consequently a Banach space isomorphism, so $L^q(\mu)$ would be the dual space $L^p(\mu)^*$ of $L^p(\mu)$. This is indeed the case for $1 < p < \infty$. However, there are cases where it fails when $p = 1$ or $p = \infty$. But if μ is a σ-finite, positive measure, then it holds for $p = 1$.

THEOREM 12.2 If μ is a σ-finite, positive measure on X, $1 \le p < \infty$, and Λ is a bounded linear functional on $L^p(\mu)$, then there exists a unique $g \in L^q(\mu)$, where p and q are conjugate exponents, such that $\Lambda(f) = \int_X fg d\mu$ for each $f \in L^p(\mu)$ and such that $|\Lambda| = |g|_q$.

Proof: The uniqueness of such a g is easily shown using the technique of Theorem 12.1, so it only remains to show that such a g exists and that $|\Lambda| = |g|_q$. We first prove it for the case when μ is bounded, i.e., $\mu(X) < \infty$. As shown in the preceding discussion, Hölder's inequality implies that $|\Lambda| \le |g|_q$. If $|\Lambda| = 0$ then the theorem holds with $g = 0$ so suppose $|\Lambda| > 0$. For any measurable set $E \subset X$ put $\lambda(E) = \Lambda(\chi_E)$ where χ_E is the characteristic function of E. Since Λ is linear, λ is additive. Suppose $\{E_n\}$ is a sequence of pairwise disjoint measurable sets and $\cup E_n = E$. For each positive integer m put $F_m = \cup_{n=1}^m E_n$. Now $|\chi_E - \chi_{F_m}| = [\mu(E - F_m)]^{1/p}$ and the right side of this equation converges to zero as $m \to \infty$.

Since Λ is bounded, Λ is continuous and since $[\mu(E - F_m)]^{1/p} \to 0$ as $m \to \infty$, $\{\chi_{F_m}\}$ converges to χ_E in the L^p-norm, $\{\Lambda(\chi_{F_m})\}$ converges to $\Lambda(\chi_E)$ and hence $\{\lambda(F_m)\}$ converges to $\lambda(E)$. Consequently, λ is a complex measure. If $\mu(E) = 0$ then $\lambda(E) = \Lambda(\chi_E) = \int_X \chi_E g d\mu = \int_E g d\mu = 0$ so λ is absolutely continuous with respect to μ. Then the existence of the Radon-Nikodym derivative assures us of the existence of a $g \in L^1(\mu)$ such that for each measurable set $E \subset X$,

(12.5) $$\Lambda(\chi_E) = \lambda(E) = \int_E g\,d\mu = \int_X \chi_E g\,d\mu.$$

Because of the linearity of Λ it is easily shown that whenever f is a simple measurable function that:

(12.6) $$\Lambda(f) = \int_X fg\,d\mu.$$

If $f \in L^\infty(\mu)$ then there exists a $b \geq 0$ such that $|f| \leq b$ almost everywhere on X. Let $B = \{x \in X \mid |f(x)| \leq b\}$. Then $\mu(X - B) = 0$ and $|f|$ is bounded on B. It is not difficult to show that the method of proof in Theorem 8.7 yields a uniformly convergent sequence $\{s_n\}$ of simple measurable functions on B that converges to $|f|$. By Proposition 8.17 there exists a measurable function a on X such that $|a| = 1$ and $f = a|f|$. Since $\{s_n\}$ converges uniformly to $|f|$, $\{as_n\}$ converges uniformly to f. Now the uniform convergence of $\{as_n\}$ implies $|as_n - f|_p$ converges to zero as $n \to \infty$. Since Λ is continuous on $L^p(\mu)$ and $\mu(X) < \infty$ implies $L^\infty(\mu) \subset L^p(\mu)$, $f \in L^p(\mu)$. But then $\{\Lambda(as_n)\}$ converges to $\Lambda(f)$ on B as $n \to \infty$. Hence $\Lambda(f) = lim_n\Lambda(as_n) = lim_n\int_X as_n g\,d\mu = lim_n\int_B as_n g\,d\mu$. Now $|as_n|(x) \leq |f|(x)$ for each $x \in B$ and $\int_B |f|\,d\mu < \infty$. Therefore, by Theorem 8.13, $lim_n\int_B as_n g\,d\mu = \int_B fg\,d\mu = \int_X fg\,d\mu$. Hence $\Lambda(f) = \int_X fg\,d\mu$ for each $f \in L^\infty(\mu)$.

Next we want to show that $g \in L^q(\mu)$ and that $|\Lambda| = |g|_q$. We can do this by showing that $|g|_q \leq |\Lambda|$ because, as previously shown, the Hölder inequality implies $|\Lambda| \leq |g|_q$ so we would then have $|\Lambda| = |g|_q$ which in turn implies $g \in L^q(\mu)$ since Λ is bounded. For this there are two cases to consider.

Case 1: ($p = 1$). Now $p = 1$ implies $q = \infty$ and by (12.5), for each measurable set E, $|\int_E g\,d\mu| = |\Lambda(\chi_E)| \leq |\Lambda|\,|\chi_E|_1 \leq |\Lambda|\mu(E) = \int_E |\Lambda|\,d\mu$ which implies $|\Lambda| = [1/\mu(E)]\int_E |\Lambda|\,d\mu$. Then by Exercise 5 of Section 8.7, $g(x) \leq |\Lambda|$ almost everywhere on X. Hence $|g|_q = |g|_\infty \leq |\Lambda|$.

Case 2: ($p > 1$). By Proposition 8.7, there exists a measurable function a on X such that $|a| = 1$ and $g = a|g|$. For each positive integer n put $E_n = \{x \in X \mid |g(x)| \leq n\}$ and let $h_n = (1/a)\chi_{E_n}|g|^{q-1}$. Then for each $x \in E_n$ we have $|h_n(x)|^p = |(1/a(x))\chi_{E_n}(x)|g(x)|^{q-1}|^p = |(1/a(x))|^p|g(x)|^{pq-p}$. Since p and q are conjugate exponents, $pq = p + q$ so $|h_n(x)|^p = |g(x)|^q$. Hence $|h_n|^p = |g|^q$ on E_n. Clearly $h_n \in L^\infty(\mu)$ for each n so (12.6) yields $\int_{E_n}|g|^q d\mu = \int_X |g|\chi_{E_n}|g|^{q-1}d\mu = \int_X (g/a)\chi_{E_n}|g|^{q-1}d\mu = \int_X h_n g\,d\mu = \Lambda(h_n) \leq |\Lambda|\,|h_n|_p = |\Lambda|[\int_X |h_n|^p d\mu]^{1/p} = |\Lambda|[\int_{E_n}|g|^q d\mu]^{1/p}$.

Now $\int_{E_n}|g|^q d\mu \leq |\Lambda|[\int_{E_n}|g|^q d\mu]^{1/p}$ so $[\int_{E_n}|g|^q d\mu]^{1-1/p} \leq |\Lambda|$ which implies $[\int_{E_n}|g|^q d\mu]^{1/q} \leq |\Lambda|$ so $\int_{E_n}|g|^q d\mu \leq |\Lambda|^q$ which implies $\int_X \chi_{E_n}|g|^q d\mu \leq |\Lambda|^q$. Then $lim_n\chi_{E_n}|g|^q = |g|^q$ and $\chi_{E_m}|g|^q \leq \chi_{E_n}|g|^q$ if $m < n$ so by Theorem 8.9, $\int_X |g|^q d\mu = lim_n\int_X \chi_{E_n}|g|^q d\mu \leq |\Lambda|^q$ so $\int_X |g|^q d\mu \leq |\Lambda|^q$. But then $|g|^q \leq |\Lambda|$. Consequently, $g \in L^q(\mu)$ and $|\Lambda| = |g|_q$. It only remains to show that (12.6) holds for each $f \in L^p(\mu)$.

Both $|\Lambda|(f)$ and $\int_X fgd\mu$ are continuous on $L^p(\mu)$ since if $f \in L^p(\mu)$, $g \in L^q(\mu)$ implies $fg \in L^1(\mu)$ and $|fg|_1 \le |f|_p |g|_q$ (Proposition 11.2) so $|\int_X fgd\mu| \le \int_X |fg| d\mu \le |f|_p |g|_q = |\Lambda| |f|_p = \Lambda(f)$. Hence $\int_X fgd\mu$ is a bounded linear functional on $L^p(\mu)$ and therefore continuous. Now $L^\infty(\mu)$ is a dense subset of $L^p(\mu)$. To see this recall that by Proposition 11.4, the class S_M of simple measurable functions is dense in $L^p(\mu)$. Clearly $S_M \subset L^\infty(\mu)$. Since $\mu(X) < \infty$ implies $L^\infty(\mu) \subset L^p(\mu)$ we see that $L^\infty(\mu)$ is dense in $L^p(\mu)$. Since $\Lambda(f)$ and $\int_X fgd\mu$ are continuous and agree on the dense subset $L^\infty(\mu)$, they must agree on $L^p(\mu)$. Therefore, (12.6) holds for each $f \in L^p(\mu)$. This concludes the proof for the case where $\mu(X) < \infty$.

Next suppose μ is σ-finite and $f \in L^p(\mu)$. Then as previously observed in the proof of the Radon-Nikodym Theorem, there exists a sequence $\{E_n\}$ of disjoint sets such that $\mu(E_n) < \infty$ for each n. For each positive integer n put $A_n = \cup_{i=1}^n E_i$. Now for each n, $|\chi_{E_n} f|_p \le |f|_p$ so the mapping $\Phi_n: L^p(\mu) \to \mathbf{C}$ defined by $\Phi_n(f) = \Lambda(\chi_{E_n} f)$ for each $f \in L^p(\mu)$ is a bounded linear functional of norm at most $|\Lambda|$ on X. Then since $\mu(E_n) < \infty$ for each n, we can apply the above proof to Φ_n restricted to E_n to obtain a function g_n on E_n with $\Phi_n(\chi_{E_n} f) = \int_{E_n} fg_n d\mu$ for each $f \in L^p(\mu)$.

Now extend each g_n to X by setting $g_n(x) = 0$ for each x not in E_n and put $g = \Sigma_n g_n$. Also, for each n let L_n^p and L_n^q denote $L^p(\mu)$ and $L^q(\mu)$ on A_n respectively. Since for each i, $g_i(x) = 0$ for each x that is not in E_i, $\Lambda(\chi_{A_n} f) = \int_{A_n} f(g_1 + \ldots + g_n) d\mu$ for each $f \in L^p(\mu)$ and since $\mu(A_n) < \infty$ for each n, the theorem is true for each A_n, so $(g_1 + \ldots + g_n) |_{A_n}$ is the unique member of L_n^q such that $\Lambda_n(f) = \int_{A_n} f(g_1 + \ldots + g_n) d\mu$ where Λ_n denotes Λ restricted to L_n^p. Consequently, $|g_1 + \ldots + g_n|_q = |\Lambda_n| \le |\Lambda|$.

For each positive integer n put $h_n = \Sigma_{i=1}^n g_i$. Then $|h_n|_q = |\Lambda_n|$ for each n. By Theorem 8.11, $\int_X (\liminf_{n\to\infty} |h_n|^q) d\mu \le \liminf_{n\to\infty} \int_X |h_n|^q d\mu$ which implies $\int_X |g|^q d\mu \le \liminf_{n\to\infty} (|\Lambda_n|^q) \le \liminf_{n\to\infty} (|\Lambda|^q) = |\Lambda|^q$. Hence $[\int_X |g|^q d\mu]^{1/q} \le |\Lambda|$ so $g \in L^q(\mu)$.

Now let $f \in L^p(\mu)$. By Proposition 11.2, $fg \in L^1(\mu)$ so $\int_X |fg| d\mu < \infty$. Then by Theorem 8.13, since $[fg](x) = \lim_n [fh_n](x)$ and $|fh_n(x)| \le |fg(x)|$ for each $x \in X$, we have $\int_X fgd\mu = \lim_n \int_X fh_n d\mu = \lim_n \int_{A_n} fh_n d\mu = \lim_n \Lambda_n(f) = \lim_n \Lambda(\chi_{A_n} f)$. Since $\{\chi_{A_n} f\}$ converges to f in $L^p(\mu)$ and Λ is continuous, $\{\Lambda(\chi_{A_n} f)\}$ converges to $\Lambda(f)$. Therefore, $\Lambda(f) = \int_X fgd\mu$. That $|\Lambda| = |g|_q$ follows from the fact that $\lim_n h_n = g$ and $|h_n|_q \le |\Lambda|$ for each n. ∎

EXERCISES

1. Let λ denote Lebesgue measure on (0, 1) and let μ be the counting measure on the σ-algebra of Lebesgue measurable sets in (0, 1). Show that λ is absolutely continuous with respect to μ but there does not exist an $f \in L^1(\mu)$

such that $\lambda(E) = \int_E f d\mu$ for each Lebesgue measurable set E. In other words, the Radon-Nikodym derivative may not exist if μ is not σ-finite.

2. Show that if $1 < p < \infty$, then Theorem 12.2 still holds even if μ is not σ-finite, i.e., if q is an exponent conjugate to p then $L^q(\mu)$ is still the dual space of $L^p(\mu)$ even if μ is not σ-finite.

3. Let λ, μ be two measures on M with $\lambda << \mu$ such that the Radon-Nikodym derivative g exists. Show that for each $E \in M$ and measurable function $f{:}X \to \mathbf{R}$ with respect to μ, $\int_E f d\lambda = \int_E fg d\mu$.

4. Show that the Radon-Nikodym Theorem (12.1) can be extended to the case where both μ and λ are positive σ-finite measures.

12.3 Decompositions of Measures and Complex Integration

We have already encountered a decomposition of a real valued measure μ, namely, the *Jordan decomposition* of μ defined in Exercise 2 and 3 of Section 12.1. This decomposition was helpful in proving the existence of the Radon-Nikodym derivative in the case where λ was a complex measure, absolutely continuous with respect to μ and μ was positive and σ-finite. In this section we introduce other useful decompositions, called the *Lebesgue decomposition* and the *polar decomposition* and also prove an important theorem about the Jordan decomposition called the *Hahn Decomposition Theorem.* All the major results in this section depend on the existence of the Radon-Nikodym derivative. The Lebesgue decomposition is introduced by means of the following theorem.

THEOREM 12.3 Let μ be a positive σ-finite measure on a σ-algebra M in a space X and let λ be a positive bounded measure or a complex measure on M that is absolutely continuous with respect to μ. Then there exists a unique pair of measures λ_1 and λ_2 on M such that $\lambda = \lambda_1 + \lambda_2$ and such that $\lambda_1 \perp \mu$, $\lambda_2 << \mu$ and $\lambda_1 \perp \lambda_2$.

Proof: We first prove the theorem for the case where μ and λ are positive bounded measures on M. We can use the first part of the proof of the Radon-Nikodym Theorem to obtain a unique function $g \in L^2(\mu)$ with $g(x) \in [0, 1]$ for each x such that equations (12.2) and (12.3) hold. Then, as in the proof of the Radon-Nikodym Theorem, let $H = \{x \in X \mid g(x) < 1\}$ and $K = \{x \in X \mid g(x) = 1\}$ and let $\lambda_1(E) = \lambda(K \cap E)$ and $\lambda_2(E) = \lambda(H \cap E)$ for each $E \in M$. Then λ_1 and λ_2 are both positive measures on M.

If we take $f = \chi_K$ in (12.3) we get $\int_K (1 - g) d\lambda = \int_K g d\mu$ which implies $\int_K d\lambda - \int_K d\lambda = \int_K d\mu$ so $\mu(K) = 0$. Therefore, μ is concentrated on H. But clearly λ_1 is concentrated on K and $H \cap K = \varnothing$. Hence $\lambda_1 \perp \mu$. As shown in the proof of the Radon-Nikodym Theorem, since g is bounded, (12.3) holds if we replace f by

$(1+g+\ldots+g^n)\chi_E$ for each positive integer n and each E $\in M$, we get (12.4). At each point of K, $g(x) = 1$ so $1 - g^{n+1}(x) = 0$. On the other hand, at each point x of H, $\{g^{n+1}(x)\}$ converges to zero monotonically. Consequently, the left hand side of (12.4) converges to $\lambda(\text{H}\cap\text{E}) = \lambda_2(\text{E})$. Also, as shown in the proof of the Radon-Nikodym Theorem, the right side of (12.4) converges to $\int_E f d\mu$ and $n \to \infty$ where f is the Radon-Nikodym derivative, so $\lambda_2(\text{E}) = \int_E f d\mu$ for each E $\in M$. Therefore, $\mu(\text{E}) = 0$ implies $\lambda_2(\text{E}) = 0$ so $\lambda_2 << \mu$. It remains to show that λ_1 and λ_2 are unique since clearly $\lambda_1 \perp \lambda_2$.

But the uniqueness of λ_1 and λ_2 is easily seen. If λ_1' and λ_2' are measures such that $\lambda = \lambda_1' + \lambda_2'$ and such that $\lambda_1' \perp \mu$ and $\lambda_2' << \mu$, then $\lambda_1' - \lambda_1 = \lambda_2' - \lambda_2$ are measures such that $\lambda_1' - \lambda_1 \perp \mu$ and $\lambda_2' - \lambda_2 << \mu$. But then it is easily shown that $\lambda_1' - \lambda_1 = 0$ and $\lambda_2' - \lambda_2 = 0$.

Next we prove the case where μ is a positive σ-finite measure but λ remains positive and bounded. Then, as shown in the proof of the Radon-Nikodym Theorem there is a disjoint sequence $\{E_n\} \subset M$ with $\text{X} = \cup E_n$ such that $\mu(E_n) < \infty$ for each n. Since $\lambda(\text{X}) < \infty$ we have $\lambda(E_n) < \infty$ for each n, so the above result holds for each n. Therefore, there exists sequences $\{\lambda_n^1\}$ and $\{\lambda_n^2\}$ of measures such that for each n, λ_n^1 and λ_n^2 are positive measures concentrated on E_n with $\lambda(E_n \cap \text{A}) = \lambda_n^1(E_n \cap \text{A}) + \lambda_n^2(E_n \cap \text{A})$ for each A $\in M$, $\lambda_n^1 \perp \mu$, $\lambda_n^2 << \mu$ and $\lambda_n^1 \perp \lambda_n^2$. Put $\lambda_1 = \Sigma_n \lambda_n^1$ and $\lambda_2 = \Sigma_n \lambda_n^2$. Then for each A $\in M$, $\lambda_1(\text{A}) + \lambda_2(\text{A}) = \Sigma_n \lambda_n^1(\text{A} \cap E_n) + \Sigma_n \lambda_n^2(\text{A} \cap E_n) = \Sigma_n \lambda(\text{A} \cap E_n) = \lambda(\text{A})$. If $\mu(\text{A}) = 0$ then $\lambda_n^2(\text{A}) = 0$ for each n so $\lambda_2(\text{A}) = \Sigma_n \lambda_n^2(\text{A}) = 0$. Therefore, $\lambda_2 << \mu$. Since $\lambda_n^1 \perp \mu$ for each n and μ is concentrated on H, then λ_n^1 is concentrated on K for each n which implies λ_1 is concentrated on K so $\lambda_1 \perp \mu$. Finally, since λ_n^1 is concentrated on K for each n, λ_n^2 is concentrated on H for each n which implies λ_2 is concentrated on H so $\lambda_1 \perp \lambda_2$. This proves the case where μ is a positive σ-finite measure but λ remains positive and bounded since the uniqueness proof is the same as for the previous case.

To complete the proof, assume λ is complex and μ is positive and σ-finite. Then $\lambda = \nu + i\sigma$ for some real valued measures ν and σ. Moreover, $\nu = \nu^+ - \nu^-$ and $\sigma = \sigma^+ - \sigma^-$ (see Exercises 2 and 3 of Section 12.1) where ν^+, ν^-, σ^+ and σ^- are all positive bounded measures. Then applying the above result to these measures, we obtain the desired decomposition of λ into a unique pair of measures λ_1 and λ_2 on M such that $\lambda = \lambda_1 + \lambda_2$, $\lambda_1 \perp \lambda_2$, $\lambda_1 \perp \mu$ and $\lambda_2 << \mu$. ∎

The pair λ_1 and λ_2 in Theorem 12.3 is known as the **Lebesgue decomposition** of λ. Another useful decomposition of a complex measure is the so called *polar decomposition.* Like the Lebesgue decomposition we introduce it by means of a theorem.

THEOREM 12.4 If μ is a complex valued measure on a σ-algebra M in X, then there exists a complex measurable function h on X with $|h(x)| = 1$ for each $x \in$ X and such that $\mu(E) = \int_E h d|\mu|$ for each $E \in M$.

Proof: Clearly $\mu << |\mu|$ so there exists a Radon-Nikodym derivative $f \in L^1(|\mu|)$ of μ with respect to $|\mu|$ such that $\mu(E) = \int_E fd|\mu|$ for each $E \in M$. We would like to show that $|f(x)| = 1$ for each $x \in X$. Let $a \in \mathbf{R}$ and put $A = \{x \in X \,|\, |f(x)| < a\}$. Then $|\mu|(A) = sup\{\Sigma_n |\mu(E_n)| \,|\, \{E_n\}$ is a partition of $A\}$. Let $\{A_n\}$ be some partition of A. Then $\Sigma_n |\mu(A_n)| = \Sigma_n |\int_{A_n} fd|\mu|| \le \Sigma_n a|\mu|(A_n) = a|\mu|(A)$ so $|\mu|(A) \le a|\mu|(A)$. If $a < 1$ then $|\mu|(A) = 0$ so $|f(x)| \ge 1$ almost everywhere on X.

Now if $E \in M$ with $|\mu|(E) > 0$ then $|[1/|\mu|(E)]\int_E fd|\mu|| = |[1/|\mu|(E)]\mu(E)| = |\mu(E)|/|\mu|(E) \le 1$. Then by Exercise 5 of Section 8.7, we have that $|f(x)| \le 1$ almost everywhere on X. But then $|f(x)| = 1$ almost everywhere on X. If we define $h:X \to C$ by $h(x) = f(x)$ if $|f(x)| = 1$ and $h(x) = 1$ otherwise then $\int_E fd|\mu| = \int_E hd|\mu|$ for each $E \in M$, so $h \in L^1(|\mu|)$, $|h(x)| = 1$ for each $x \in X$ and $\mu(E) = \int_E hd|\mu|$ for each $E \in M$. ▪

The function h of Theorem 12.4 and the total variation $|\mu|$ are said to be the **polar decomposition** of μ. This is because when we define integration with respect to a complex measure, which we will do next, we will have $\mu(E) = \int_E d\mu$ for each $E \in M$ so $\int_E d\mu = \int_E hd|\mu|$. We use the notation $d\mu = hd|\mu|$ to denote $\int_E d\mu = \int_E hd|\mu|$ for each $E \in M$. It is this notation that gives intuitive meaning to the term *polar decomposition*. Notice that the polar decomposition is different in nature than the Jordan and Lebesgue decompositions where $\lambda = \lambda_1 + \lambda_2$. We cannot write $\mu = h|\mu|$ meaningfully. It is only meaningful in the sense that $d\mu = hd|\mu|$.

We defined integration with respect to a positive measure in Section 8.6 as the supremum of finite sums of products of complex numbers a_i with measures of sets E_i such that $0 \le \Sigma_i a_i \chi_{A_i} \le f$. That definition of integration was extended to complex valued functions in Exercise 1 of Section 8.6. Since then we have been able to avoid the notion of integration with respect to a complex valued measure. But we now have a mechanism for defining integration with respect to a complex measure and a need for this idea in the next section. If μ and λ are complex measures then so is $\mu + \lambda$. Using our intuition about what integration with respect to a complex measure should mean and our experience with integration with respect to positive measures we would expect the following identities to hold for each $E \in M$:

$$\mu(E) = \int_E d\mu \tag{12.7}$$

$$\int_E fd(\mu + \lambda) = \int_E fd\mu + \int_E fd\lambda \tag{12.8}$$

for each complex measurable function $f \in C_\infty(X)$. That (12.8) holds for the special case where μ and λ are positive measures is an easy exercise (Exercise 1).

By Theorem 12.4, there exists a complex measurable function h with $|h| = 1$ such that $\mu(E) = \int_E hd|\mu|$. Since we want (12.7) to hold, we expect our definition of integration with respect to a complex measure to yield $\mu(E) = \int_E 1 d\mu = \int_E 1 h d|\mu|$. If $f = 1$ (the unit function) then $\int_E f d\mu = \int_E f h d|\mu|$. Using this formula for motivation, we define the integral of g with respect to μ by

(12.9) $$\int_E g d\mu = \int_E g h d|\mu|$$

for each complex measurable function $g \in C_\infty(X)$ and for each $E \in M$. Let us now see if the definition (12.9) implies (12.7) and (12.8). That (12.9) implies (12.7) is immediate since by Theorem 12.4, $\mu(E) = \int_E h d|\mu|$ so by (12.9), $\mu(E) = \int_E d\mu$. That (12.9) implies (12.8) takes more work.

To show (12.9) implies (12.8), first note that by (12.9), $\int_X \chi_E d\mu = \int_X \chi_E h d|\mu| = \int_E h d|\mu| = \mu(E)$. Therefore, if μ and λ are complex measures and $\sigma = \mu + \lambda$, then for each $E \in M$ we have

$$\int_X \chi_E d\sigma = \sigma(E) = \mu(E) + \lambda(E) = \int_X \chi_E d\mu + \int_X \chi_E d\lambda$$

so (12.8) holds if $f = \chi_E$ for some $E \in M$. Next suppose f is a nonnegative bounded measurable function. By Theorem 8.7 there exists a sequence $\{s_n\}$ of simple measurable functions with $0 \le s_n \le f$ for each positive integer n and $s_n \le s_m$ if $n < m$ such that $\{s_n(x)\}$ converges to $f(x)$ for each $x \in X$. For each s_n, $s_n = \Sigma_{i=1}^m a_i \chi_{A_i}$ for some positive integer m and some nonnegative real numbers $a_1 \ldots a_m$ and characteristic functions $\chi_{A_1} \ldots \chi_{A_m}$ for measurable sets $A_1 \ldots A_m$. Let k be a complex measurable function with $|k| = 1$ such that $d\sigma = k d|\sigma|$. Then

$$\int_X s_n d\sigma = \int_X s_n k d|\sigma| = \int_X \Sigma_{i=1}^m a_i \chi_{A_i} k d|\sigma| = \Sigma_{i=1}^m a_i \int_X \chi_{A_i} k d|\sigma| =$$

$$\Sigma_{i=1}^m a_i \int_X \chi_{A_i} d\sigma = \Sigma_{i=1}^m a_i \int_X \chi_{A_i} d\mu + \Sigma_{i=1}^m \int_X \chi_{A_i} d\lambda =$$

$$\int_X \Sigma_{i=1}^m a_i \chi_{A_i} d\mu + \int_X \Sigma_{i=1}^m a_i \chi_{A_i} d\lambda = \int_X s_n d\mu + \int_X s_n d\lambda.$$

So (12.8) holds for $E = X$ and f a simple measurable function. Now let p be a complex measurable function with $|p| = 1$ such that $d\lambda = p d|\lambda|$. Then if $E \in M$ we have

$$\int_E s_n d\sigma = \int_E s_n k d|\sigma| = \int_X \chi_E s_n k d|\sigma| = \int_X \chi_E s_n d\sigma.$$

Since s_n is a simple measurable function, $\chi_E s_n$ is also a simple measurable function. Therefore,

$$\int_X \chi_E s_n d\sigma = \int_X \chi_E s_n d\mu + \int_X \chi_E s_n d\lambda = \int_E s_n h d|\mu| +$$

$$\int_E s_n p d|\lambda| = \int_E s_n d\mu + \int_E s_n d\lambda.$$

Consequently, (12.8) holds for the simple measurable functions. Define Λ on

$C_\infty(X)$ by $\Lambda(g) = \int_X g\,d\sigma$. By (12.9) $\Lambda(g) = \int_X gk\,d|\sigma|$ so Λ is clearly a linear functional on $C_\infty(X)$. Since $|g|_\infty < \infty$ and $|\sigma|(X) < \infty$ we have

$$|\Lambda(g)| \le \int_X |gk|\,d|\sigma| = \int_X |g|\,d|\sigma| \le |g|_\infty |\sigma|(X)$$

so Λ is bounded and hence continuous by Theorem 11.13. Similarly define the continuous linear functionals Ω and Φ on $C_\infty(X)$ by $\Omega(g) = \int_X g\,d\mu$ and $\Phi(g) = \int_X g\,d\lambda$ for each $g \in C_\infty(X)$. Since $\{s_n\}$ converges to f we have $\{\Omega(s_n)\}$ converges to $\Omega(f)$, $\{\Phi(s_n)\}$ converges to $\Phi(f)$ and $\{\Lambda(s_n)\}$ converges to $\Lambda(f)$. Now for each positive integer n, $\Lambda(s_n) = \int_X s_n\,d\sigma = \int_X s_n\,d\mu + \int_X s_n\,d\lambda$ so $\{\Lambda(s_n)\} = \{\int_X s_n\,d\mu + \int_X s_n\,d\lambda\} = \{\Omega(s_n) + \Phi(s_n)\}$. Clearly, $\{\Omega(s_n) + \Phi(s_n)\}$ converges to $\Omega(f) + \Phi(f) = \int_X f\,d\mu + \int_X f\,d\lambda$ and $\{\Lambda(s_n)\}$ converges to $\int_X f\,d\sigma$. Since limits are unique, $\int_X f\,d\sigma = \int_X f\,d\mu + \int_X f\,d\lambda$. Therefore, (12.8) holds when f is a nonnegative bounded measurable function.

It is a straightforward exercise (Exercise 2) to show that (12.8) holds for complex bounded measurable functions. Now if μ is a complex measure then $\mu = \mu_1 + i\mu_2$ for some pair μ_1 and μ_2 of real measures. If $E \in M$ then $i\mu_2(E) = i\int_E d\mu_2$. Consequently, $\int_E d\mu = \mu(E) = \mu_1(E) + i\mu_2(E) = \int_E d\mu_1 + i\int_E d\mu_2$ so

$$\int_E d(\mu_1 + i\mu_2) = \int_E d\mu_1 + i\int_E d\mu_2 \tag{12.10}$$

for each $E \in M$. This concludes our discussion of integration with respect to complex measures. In the next section we will use these results to significantly generalize the Riesz Representation Theorem.

PROPOSITION 12.4 If $g \in L^1(\mu)$, where μ is a positive measure on a σ-algebra M in X, and if $\lambda(E) = \int_E g\,d\mu$ for each $E \in M$, then $|\lambda|(E) = \int_E |g|\,d\mu$ for each $E \in M$.

Proof: Let $A = \{x \in X \mid g(x) = 0\}$ and $B = \{x \in X \mid g(x) \ne 0\}$. Then $\lambda(A) = 0 = \int_A |g|\,d\mu$. Since $\lambda(A) = 0$ implies $|\lambda|(A) = 0$ we have $|\lambda|(A) = \int_A |g|\,d\mu$.

Next let $E \in M$. Then $|\lambda|(E\cap B) = 0$ implies $\lambda(E\cap B) = 0$ so $\int_{E\cap B} g\,d\mu = 0$. It can be shown that $\int_{E\cap B} g\,d\mu = 0$ implies $\mu(E\cap B) = 0$. To see this note that

$$\int_{E\cap B} g\,d\mu = \int_{E\cap B} g_1^+\,d\mu - \int_{E\cap B} g_1^-\,d\mu + i\int_{E\cap B} g_2^+\,d\mu - i\int_{E\cap B} g_2^-\,d\mu$$

for some nonnegative functions g_1^+, g_1^-, g_2^+ and g_2^-. Let $C_1 = \{x \in X \mid g_1^+(x) > 0\}$, $C_2 = \{x \in X \mid g_1^-(x) > 0\}$, $C_3 = \{x \in X \mid g_2^+(x) > 0\}$ and $C_4 = \{x \in X \mid g_2^-(x) > 0\}$. Clearly $B = C_1 \cup C_2 \cup C_3 \cup C_4$. Suppose $\mu(E\cap B) > 0$. Then $\mu(E\cap C_i) > 0$ for some $i = 1 \ldots 4$. Assume $\mu(E\cap C_1) > 0$. Then $g_1^-(x) = 0$ for each $x \in E\cap C_1$ so $\int_{C_1} g_1\,d\mu = \int_{C_1\cap E} g_1^+\,d\mu > 0$. But $E\cap C_1 \subset E\cap B$ so $0 = \lambda(E\cap C_1) = \int_{E\cap C_1} g\,d\mu$, and $\int_{E\cap C_1} g\,d\mu \ne 0$ since $\int_{E\cap C_1} g_1\,d\mu > 0$ which is a contradiction. Hence $\mu(E\cap C_1) = 0$.

Similarly $\mu(E\cap C_2) = 0 = \mu(E\cap C_3) = \mu(E\cap C_4)$. But then $\mu(B\cap E) = 0$. Therefore, $|\lambda|(E\cap B) = 0$ which implies $\mu(E\cap B) = 0$ so $\mu << |\lambda|$ on B. Hence the Radon-Nikodym derivative $d\mu/d|\lambda|$ exists on B since $|\lambda|(X) < \infty$ by Proposition 12.3. By Theorem 12.4 there exists an $h \in L^1(\mu)$ with $|h| = 1$ such that $\lambda(E) = \int_E hd|\lambda|$ for each $E \in M$ and by hypothesis, $\lambda(E) = \int_E gd\mu$ so $\int_E hd|\lambda| = \int_E gd\mu$ for each $E \in M$. But then $\int_{E\cap B} hd|\lambda| = \int_{E\cap B} g[d\mu/d|\lambda|]d|\lambda|$ for each $E \in M$ by Exercise 3 of Section 12.2. Thus $h = gd\mu/d|\lambda|$ almost everywhere with respect to $|\lambda|$ on B by Exercise 2 of Section 8.7. Therefore, $\int_{E\cap B}|h|d|\lambda| = \int_{E\cap B}|g|[d\mu/d|\lambda|]d|\lambda|$ since μ and $|\lambda|$ are both positive, so $|d\mu/d|\lambda|| = d\mu/d|\lambda|$. But then $|\lambda|(E\cap B) = \int_{E\cap B} d|\lambda| = \int_{E\cap B}|g|d\mu$ by Exercise 3 of Section 12.2 and the fact that $|h| = 1$. Consequently, $|\lambda|(E) = \int_E |g|d\mu$ for each $E \in M$. ∎

We conclude this section with the so called *Hahn Decomposition Theorem*. This theorem is a statement about the Jordan decomposition of a real valued measure μ on a σ-algebra M in a space X. It says that X can be decomposed into two disjoint sets A and B such that μ^+ is concentrated on A and μ^- is concentrated on B.

THEOREM 12.5 (Hahn Decomposition Theorem) If μ is a real valued measure on a σ-algebra M on a space X, there exists disjoint sets A and B with $A\cup B = X$ such that $\mu^+(E) = \mu(E\cap A)$ and $\mu^-(E) = -\mu(E\cap B)$ for each $E \in M$.

Proof: By Theorem 12.4, there exists a complex measurable function f on X with $|f| = 1$ and $\mu(E) = \int_E fd|\mu|$ for each $E \in M$. Since μ is real f must be real almost everywhere. Define h on X by $h(x) = f(x)$ if $f(x)$ is real and $h(x) = 1$ otherwise. Then h is a real measurable function with $|h| = 1$ and $\mu(E) = \int_E hd|\mu|$ for each $E \in M$. Put $H = \{x \in X | h(x) = 1\}$ and $K = \{x \in X | h(x) = -1\}$. Clearly $H\cup K = X$ and $H\cap K = \varnothing$. Now $\mu^+ = (|\mu| + \mu)/2$ and $(1 + h)/2 = h$ on H and zero on K so for each $E \in M$ we have:

$$\mu^+(E) = \int_E(1/2)d|\mu| + \int_E(1/2)d\mu = \int_E[(1+h)/2]d|\mu| =$$

$$\int_{E\cap H}[(1+h)/2]d|\mu| + \int_{E\cap K}[(1+h)/2]d|\mu| = \int_{E\cap H}hd|\mu| = \mu(E\cap H).$$

Hence $\mu^+(E) = \mu(E\cap H)$. Now $\mu(E) = \mu(E\cap H) + \mu(E\cap K)$ and $\mu(E) = \mu^+(E) - \mu^-(E)$ so $\mu(E\cap H) + \mu(E\cap K) = \mu^+(E) - \mu^-(E)$. Since $\mu^+(E) = \mu(E\cap H)$ we have $\mu^-(E) = -\mu(E\cap K)$. ∎

EXERCISES

1. Show that the identity (12.8) holds when μ and λ are positive measures, without reference to complex measures.

2. Show that the identity (12.8) holds when μ and λ are complex measures, based on the fact that (12.8) holds when f is a nonnegative bounded measurable function.

3. Use a similar technique to the one used to prove the identity (12.8) to show that if $c \in \mathbf{C}$ and μ is a complex measure, then for each $f \in C_\infty(X)$, $\int_E f d(c\mu) = c\int_E f d\mu$ for each measurable set E.

12.4 The Riesz Representation Theorem

One of the more important applications of the Radon-Nikodym Theorem is to generalize the version of the Riesz Representation Theorem presented as a series of exercises in Section 8.8, to include bounded linear functionals on $C_\infty(X)$ as opposed to the positive linear functionals on $C_K(X)$ as given in Section 8.8. For this we will find the following lemma useful.

LEMMA 12.2 Let Ω be a bounded linear functional on $C_\infty(X)$. Then there exists a positive linear functional Λ on $C_K(X)$ with $|\Omega(f)| \le \Lambda(|f|) \le |f|_\infty$ for each $f \in C_K(X)$.

Proof: We first define Λ on the subset $C_K^+(X)$ consisting of all nonnegative real valued members of $C_K(X)$. For this put $\Lambda(f) = sup\{|\Omega(g)| \mid g \in C_K(X)$ and $|g| \le f\}$ for each $f \in C_K^+(X)$. Then $\Lambda(f) \ge 0$ for each $f \in C_K^+(X)$ and $|\Omega(f)| \le \Lambda(|f|) \le |f|_\infty$. Moreover, if $f, g \in C_K^+(X)$ with $f \le g$ then $\Lambda(f) \le \Lambda(g)$ and if $\alpha > 0$ then $\Lambda(\alpha f) = \alpha\Lambda(f)$. We want to show that if $f, g \in C_K^+(X)$ that $\Lambda(f + g) = \Lambda(f) + \Lambda(g)$.

For this let $f, g \in C_K^+(X)$ and let $\varepsilon > 0$. Then there exists $h, k \in C_K(X)$ with $|h| \le f$, $|k| \le g$ and $\Lambda(f) \le |\Omega(h)| + \varepsilon$ and $\Lambda(g) \le |\Omega(k)| + \varepsilon$. Moreover, there are complex numbers α and β with $|\alpha| = 1 = |\beta|$ such that $\alpha\Omega(h) = |\Omega(h)|$ and $\beta\Omega(k) = |\Omega(k)|$. Then $\Lambda(f) + \Lambda(g) \le |\Omega(h)| + |\Omega(k)| + 2\varepsilon = \alpha\Omega(h) + \beta\Omega(k) + 2\varepsilon = \Omega(\alpha h + \beta k) + 2\varepsilon$. Clearly $\Omega(\alpha h + \beta k) \ge 0$ so $\Omega(\alpha h + \beta k) = |\Omega(\alpha h+\beta k)|$. Therefore, $\Lambda(f) + \Lambda(g) \le |\Omega(\alpha h + \beta k)| + 2\varepsilon \le \Lambda(|\alpha h| + |\beta k|) + 2\varepsilon \le \Lambda(f + g) + 2\varepsilon$. Since this inequality holds for each $\varepsilon > 0$ we have $\Lambda(f) + \Lambda(g) \le \Lambda(f + g)$.

To show $\Lambda(f + g) \le \Lambda(f) + \Lambda(g)$, let $h \in C_K(X)$ with $|h| \le f + g$. Put $U = \{x \in X \mid f(x) + g(x) > 0\}$ and $F = X - U$. Define s and t on X by $s(x) = f(x)h(x)/[f+g](x)$ and $t(x) = g(x)h(x)/[f + g](x)$ for each $x \in U$ and $s(x) = 0 = t(x)$ for each $x \in F$. Clearly s and t are continuous on U and $h = s + t$. Moreover, $|s(x)| \le |h(x)|$ and $|t(x)| \le |h(x)|$ for each $x \in X$ so s and t are also continuous on F since $h(F) = 0$ and h is continuous on X. Therefore, $s, t \in C_K(X)$. Since $|s| \le f$ and $|t| \le g$ we have: $|\Omega(h)| = |\Omega(s) + \Omega(t)| \le |\Omega(s)| + |\Omega(t)| \le \Lambda(f) + \Lambda(g)$. Since this holds for each $h \in C_K(X)$ with $|h| \le f + g$ we have $\Lambda(f + g) \le \Lambda(f) + \Lambda(g)$. Hence $\Lambda(f + g) = \Lambda(f) + \Lambda(g)$ for each pair $f, g \in C_K^+(X)$.

To extend Λ to the real valued members of $C_K(X)$ note that if f is a real valued member of $C_K(X)$ that $f^+ = (|f| + f)/2$ and $f^- = (|f| - f)/2$ so that both f^+ and f^- are in $C_K^+(X)$. Define $\Lambda(f) = \Lambda(f^+) - \Lambda(f^-)$. To extend Λ to $C_K(X)$ let $g \in C_K(X)$. Then $g = h + ik$ for some real valued members h and k of $C_K(X)$. Define $\Lambda(g) = \Lambda(h) + i\Lambda(k)$. Now Λ is defined for all $g \in C_K(X)$.

It remains to show that Λ is linear on $C_K(X)$ and that $|\Omega(f)| \leq \Lambda(|f|) \leq |f|_\infty$ for each $f \in C_K(X)$. We leave this as an exercise (Exercise 1). ■

A complex Borel (Baire) measure μ on X is said to be **regular** or **almost regular** if $|\mu|$ is regular or almost regular, respectively. The linear operator Λ defined on $C_\infty(X)$ by $\Lambda(f) = \int_E f d\mu$ is clearly bounded as shown in the last section and therefore continuous. Moreover, $|\Lambda| \leq |\mu|(X)$ since $|\Lambda| = sup\{|\int_X f d\mu| \,|\, f \in C_\infty(X) \text{ and } |f|_\infty = 1\} \leq sup\{\int_X |f| d|\mu| \,|\, f \in C_\infty(X) \text{ and } |f|_\infty = 1\} = \int_X d|\mu| = |\mu|(X)$. Our generalization of the Riesz Representation Theorem shows that all bounded linear functionals on $C_\infty(X)$ are formed this way with respect to regular complex measures.

THEOREM 12.6 (Riesz Representation Theorem) Let Ω be a bounded linear functional on $C_\infty(X)$ where X is locally compact and Hausdorff. Then there exists a unique complex regular Borel measure μ on X such that $\Omega(f) = \int_X f d\mu$ for each $f \in C_\infty(X)$ and such that $|\mu|(X) = |\Omega|$.

Proof: Since $C_K(X)$ is a dense subspace of $C_\infty(X)$ with respect to the supremum norm (Theorem 11.15) and each bounded linear functional Λ in $C_K(X)$ has a unique extension to a bounded linear functional Φ on $C_\infty(X)$ with the same norm (Theorem 11.6), it suffices to prove the Theorem for $C_K(X)$.

For each $f \in C_K(X)$ put $\Phi(f) = \Omega(f)/|\Omega|$. Then $|\Phi| = 1$ and Φ is a linear functional on $C_K(X)$. By Lemma 12.2, there exists a positive linear functional Λ on $C_K(X)$ with $|\Phi(f)| \leq \Lambda(|f|) \leq |f|_\infty$ for each $f \in C_K(X)$. By Theorem 8.14 there exists a positive almost regular Borel measure λ such that $\Lambda(f) = \int_X f d\lambda$ for each $f \in C_K(X)$. From the proof of Theorem 8.14 (see Exercise 1, Section 8.8) $\lambda(X) = sup\{\Lambda(f) \,|\, f \prec X\} = sup\{\Lambda(f) \,|\, f \in C_K(X) \text{ and } f(X) \subset [0, 1]\}$. Since for each $f \in C_K(X)$ with $|f|_\infty \leq 1$ we have $|\Lambda(f)| \leq 1$ it is clear that $\lambda(X) \leq 1$. Hence λ is regular. Moreover, $|\Phi(f)| \leq \Lambda(|f|) = \int_X |f| d\lambda = |f|_1$ for each $f \in C_K(X)$ where $|\ |_1$ denotes the $L^1(\lambda)$ norm.

Now $|\Phi|_1 = sup\{|\Phi(f)| \,|\, f \in C_K(X) \text{ with } |f|_1 = 1\} \leq 1$ so the norm of Φ is at most 1 with respect to the $L^1(\lambda)$ norm on $C_K(X)$. By Theorem 11.4, $C_K(X)$ is dense in $L^1(\lambda)$ and by Theorem 11.6 there exists a norm preserving extension to a linear functional Φ' on $L^1(\lambda)$. By Theorem 12.2 there exists a unique Borel measurable function g with $|g|_\infty = |\Phi'|$ such that

$$\Phi'(f) = \int_X fg d\lambda \tag{12.11}$$

for each $f \in L^1(\lambda)$. Consequently, $|g|_\infty \leq 1$ so $|g| \leq 1$. Let $f \in C_K(X)$ with

$|f|_\infty = 1$. Then $|f(x)| \leq 1$ for each $x \in X$ so $\int_X |g|d\lambda \geq \int_X |fg|d\lambda \geq |\int_X fgd\lambda|$ for each $f \in C_K(X)$ with $|f|_\infty = 1$. Therefore, $\int_X |g|d\lambda \geq sup\{|\int_X fgd\lambda| \mid f \in C_K(X)$ with $|f|_\infty = 1\} = sup\{|\Phi(f)| \mid f \in C_K(X)$ with $|f|_\infty = 1\} = |\Phi|$. Since $|\Phi| = 1$ (with respect to the $|\ |_\infty$ norm on $C_K(X)$) we have $\int_X |g|d\lambda \geq 1$. But since $\lambda(X) \leq 1$ and $|g| \leq 1$, this can only happen if $\lambda(X) = 1$ and $|g| = 1$ almost everywhere with respect to λ. Redefine g so that $|g(x)| = 1$ for each $x \in X$. Clearly we can do this without disturbing the results so far. Define μ' on X by $\mu'(E) = \int_E gd|\lambda|$ for each Borel set $E \subset X$. Then μ' is a complex measure and by Theorem 12.4, $|\mu'|(E) = \int_E |g|d|\lambda| = \int_E d\lambda$ for each Borel set E. But then $|\mu'|(X) = \lambda(X) = 1 = |\Phi|$. Now (12.11) and (12.9) yield

$$\Phi(f) = \int_X fgd\lambda = \int_X fgd|\lambda| = \int_X fd\mu' \tag{12.12}$$

for each $f \in C_K(X)$. Then $\Omega(f) = \Phi(f)|\Omega| = \int_X f|\Omega|d\mu'$. Put $\mu(E) = |\Omega|\mu'(E)$ for each Borel set $E \subset X$. By Exercise 3 of Section 12.3 we have

$$\Omega(f) = |\Omega|\Phi(f) = |\Omega|\int_X fd\mu' = |\Omega|\int_X fd(|\Omega|\mu') = \int_X fd\mu \tag{12.13}$$

for each $f \in C_K(X)$, so the bounded linear functional Ω on $C_K(X)$ is represented by the complex measure μ in the sense that $\Omega(f) = \int_X fd\mu$ for each $f \in C_K(X)$. Moreover, $|\mu|(X) = sup\{\Sigma_{n=1}^{\infty} |\mu(E_n)| \mid \{E_n\}$ is a partition of $X\} = sup\{|\Omega|\Sigma_{n=1}^{\infty} |\mu'(E_n)| \mid \{E_n\}$ is a partition of $X\} = |\Omega||\mu'|(X) = |\Omega|$. Hence $|\Omega| = |\mu|(X)$.

Since the measure λ was regular, it is not difficult to show that μ is regular, but we leave that as an exercise (Exercise 2). It remains to show the uniqueness of μ. For this let μ and ν be two Borel measures on X such that for each $f \in C_K(X)$, $\int_X fd\mu = \Omega(f) = \int_X fd\nu$. Then $\int_X fd\sigma = 0$ for each $f \in C_K(X)$ where $\sigma = \mu - \nu$. Let h be a Borel measurable function on X with $|h| = 1$ such that $d\sigma = hd|\sigma|$. Then $h^* \in L^1(|\sigma|)$. Since $C_K(X)$ is dense in $L^1(|\sigma|)$ we can find a sequence $\{f_n\} \subset C_K(X)$ such that $|h^* - f_n|$ converges to 0 as $n \to \infty$. Now for each positive integer n, $-\int_X f_n d\sigma = 0$ so $-\int_X f_n hd|\sigma| = 0$. Therefore,

$$|\sigma|(X) = \int_X |h|d|\sigma| = \int_X h^*hd|\sigma| - \int_X f_n hd|\sigma| = \int_X (h^* - f_n)hd|\sigma|.$$

Consequently, $|\sigma|(X) = ||\sigma|(X)| \leq \int_X |h^* - f_n|d\sigma = |h^* - f_n|_1$. Since $|h^* - f_n|_1$ converges to 0 as $n \to \infty$ we see that $|\sigma|(X) = 0$. But then $\sigma(X) = 0$ so $\sigma(E) = 0$ for each Borel set $E \subset X$. Therefore, $\mu(E) = \nu(E)$ for each Borel set E so $\mu = \nu$. ■

EXERCISES

1. Show that the functional Λ constructed in Lemma 12.2 is linear and that $|\Omega(f)| \leq \Lambda(|f|) \leq |f|_\infty$ for each $f \in C_K(X)$.

2. Show that the measure constructed in Theorem 12.6 is regular.

12.5 Uniform Derivatives of Measures

On the real line **R** there is a natural way to think about the *derivative* of a complex valued Borel measure μ at a point $x \in \mathbf{R}$. Let m denote Lebesgue measure on **R**. Define $f: \mathbf{R} \to \mathbf{C}$ by $f(x) = \mu((-\infty, x))$ for each $x \in \mathbf{R}$. Then it is possible to prove the following:

THEOREM 12.7 f is differentiable at x with derivative $D \in \mathbf{C}$ *if and only if for each* $\varepsilon > 0$ *there exists a* $\delta > 0$ *with* $|\mu(S)/m(S) - D| < \varepsilon$ *whenever S is an open interval containing x with* $m(S) < \delta$.

This theorem, whose proof we leave as an exercise, suggests defining the derivative of a complex Borel measure μ on more general spaces, on which Haar measure m exists, as a limit of quotients of the form $\mu(U)/m(U)$ for some suitably restricted class of open sets U, that contain some given point x, as these sets shrink uniformly to the point x. In this more general setting, Haar measure replaces Lebesgue measure used in Theorem 12.7. Constructing such a generalization will be the program of this section.

For this let (X, λ) be a locally compact, isogeneous uniform space with isometric basis ν. Let μ be a complex Borel measure on X and let m denote Haar measure on X. If $p \in X$ and $D \in \mathbf{C}$ such that whenever $\varepsilon > 0$, there exists a $b \in \nu$ with

(12.14) $$\left| \frac{\mu(S(p,a))}{m(S(p,a))} - D \right| < \varepsilon \text{ for each } a \in \nu \text{ with } a < b$$

then we say μ is **differentiable at** p with respect to m and write $d\mu/dp = D$. $d\mu/dp$ is called the **derivative of** μ **at** p (with respect to m). Clearly, if μ is differentiable at p with derivative D, the net $\{x_a \mid a \in \nu\}$ defined by $x_a = \mu(S(p,a))/m(S(p,a))$ converges to D and the convergence of $\{x_a\}$ is equivalent to (12.14). We write

(12.15) $$d\mu/dp = lim_a \frac{\mu(S(p,a))}{m(S(p,a))}$$

to denote this. The derivative $d\mu/dx$ of a complex Borel measure is not required to exist at all points of X by the definition, nor is it required to be well behaved where it exists. If the derivative exists almost everywhere and if the function $f(x) = d\mu/dx$ is Borel measurable on the subset where it exists, then we say μ is a **differentiable measure** on X.

As we saw in Section 12.2, the Radon-Nikodym derivative is a function $h \in L^1(m)$. If in addition to being differentiable, the measure μ is such that $d\mu/dx$

$\in L^1(m)$, we say μ is **L^1-differentiable.** If in addition to being L^1-differentiable, the measure μ is such, that the ratios in (12.15) converge uniformly to some $g \in L^1(m)$, then we say μ is **uniformly differentiable** and that g is the **uniform derivative** of μ with respect to m.

THEOREM 12.8 (N. Howes, 1994) Let X be a locally compact, σ-compact, isogeneous uniform space with isometric basis ν and Haar measure m. Let μ be a complex, uniformly differentiable measure on X that is absolutely continuous with respect to m. Then dμ/dx coincides with the Radon-Nikodym derivative f almost everywhere with respect to m on X.

Proof: We first prove the theorem for the case where μ is real. By Theorem 12.1, there exists a unique Radon-Nikodym derivative $f \in L^1(m)$ with $\mu(E) = \int_E f dm$ for each Borel set $E \subset X$. By hypothesis there exists a $g \in L^1(m)$ such that the sequence of functions $\{\mu(S(x, a))/m(S(x, a))\}$ converges uniformly to $g(x)$. We need to show that $f = g$ almost everywhere with respect to m. Now for each Borel set $E \subset X$, $\int_E g dm = \int_E [lim_a \mu(S(x, a))/m(S(x, a))]dm$. We begin by showing that for Borel sets E with $m(E) < \infty$ the limit can be moved out from under the integral sign, i.e., that

$$\text{(12.16)} \qquad \int_E g dm = lim_a \int_E [\mu(S(x, a))/m(S(x, a))]dm.$$

For this let $\varepsilon > 0$. Then since locally compact isogeneous uniform spaces are uniformly locally compact, there exists a $b \in \nu$ such that $Cl(S(x, b))$ is compact for each $x \in X$ and with $|\mu(S(x, a))/m(S(x, a)) - g(x)| < \varepsilon$ for each $a \in \nu$ with $a < b$ and for each $x \in X$. Now if $m(E) = 0$, (12.16) holds. In fact, if $m(E) = 0$, then $\int_E f dm = \int_E g dm$. So suppose $m(E) > 0$. Then

$$\text{(12.17)} \qquad \left|\int_E \left[\frac{\mu(S(x,a))}{m(S(x,a))} - g(x)\right]dm\right| \le \int_E \left|\frac{\mu(S(x,a))}{m(S(x,a))} - g(x)\right| dm < \int_E \varepsilon dm = \varepsilon m(E).$$

Consequently, $|\int_E [\mu(S(x, a))/m(S(x, a))]dm - \int_E g dm| < \varepsilon m(E)$ for each $\varepsilon > 0$ so (12.16) holds. Next we show that for $m(E) < \infty$ the right hand side of (12.16) is equal to $\mu(E)$, i.e.,

$$\text{(12.18)} \qquad \mu(E) = \int_E g dm.$$

Now $lim_{a<b} \int_E [\mu(S(x, a))/m(S(x, a))]dm = lim_{a<b} \int_E [1/m(S(x,a))] \int_{S(x,a)} f dm dm$. We would like to be able to change the order of integration in the right hand side of this identity, i.e., we would like to have for each $a \in \nu$ with $a < b$

$$\text{(12.19)} \qquad \int_E [1/m(S(x,a))] \int_{S(x,a)} f dm dm = \int_{S(x,a))} [1/m(S(x, a))] \int_E f dm dm.$$

Since m is a σ-finite measure (due to the σ-compactness of X), since the function $\phi(x) = 1/m(S(x, a))$ is a constant function (due to the translation invariance of m) so that $\phi^*(x) = \int_E \phi dm < \infty$, and since $\int_{S(x,a)} \phi^* dm < \infty$, then by a classical theorem from analysis known as *Fubini's Theorem*, the order of integration can be changed in this case, i.e., (12.19) holds. Fubini's Theorem is developed in a series of exercises at the end of this section. Now the right hand side of (12.19) is merely $\mu(E)$ since $\int_E f dm = \mu(E)$ (which can then be moved outside the remaining integral sign) and $\int_{S(x,a)} [1/m(S(x, a))] dm = 1$ since for a given a, $1/m(S(x, a))$ is a constant. But then the limit of the right hand side of (12.19), with respect to $a \in \mathrm{V}$, is simply $\mu(E)$ so by (12.16) and (12.19), (12.18) holds. Therefore, for Borel sets $E \subset X$ with $m(E) < \infty$ we have $\int_E f dm = \mu(E) = \int_E g dm$.

In particular, if K is a compact set then $m(K) < \infty$ (since Haar measure as developed in Chapter 9 was derived from the Riesz Representation Theorem). Since X is σ-compact, m is regular and there exists a sequence $\{K_n\}$ of compact sets such that $lim_n m(K_n) = m(X)$. For each positive integer n, $\int_{K_n} g dm = \int_X \chi_{K_n} g dm$ and $\int_{K_n} f dm = \int_X \chi_{K_n} f dm$. Clearly $lim_n \{\chi_{K_n} g\} = g$ and $lim_n \{\chi_{K_n} f\} = f$. Also, $|\chi_{K_n} g| \le |g|$ and $|\chi_{K_n} f| \le |f|$ so by Theorem 8.13 we have:

(12.20) $$lim_n \int_{K_n} g dm = lim_n \int_X \chi_{K_n} g dm = \int_X g dm \quad \text{and}$$

$$lim_n \int_{K_n} f dm = lim_n \int_X \chi_{K_n} f dm = \int_X f dm.$$

Consequently, $\int_X f dm = lim_n \int_{K_n} f dm = lim_n \int_{K_n} g dm = \int_X g dm$ so $f = g$ almost everywhere with respect to m.

It remains to prove the theorem for the case where μ is a complex measure. But this is now easy since if μ is complex then $\mu = \mu_1 + i\mu_2$ for some pair μ_1 and μ_2 of real measures. Since μ is absolutely continuous with respect to m, so are μ_1 and μ_2 so by Theorem 12.1 there exists unique functions $f_1, f_2 \in L^1(m)$ such that for each Borel measurable set $E \subset X$, $\mu_1(E) = \int_E f_1 dm$ and $\mu_2(E) = \int_E f_2 dm$. Also, there exists a $g \in L^1(m)$ such that $\{\mu(S(x, a))/m(S(x, a))\}$ converges uniformly to g. Now $g = g_1 + ig_2$ for some real valued functions g_1, $g_2 \in L^1(m)$ and $\{\mu(S(x, a))/m(S(x, a))\} = \{\mu_1(S(x, a))/m(S(x, a)) + i\mu_2(S(x,a))/m(S(x, a))\}$ and the only way for this net to converge uniformly to $g_1 + ig_2$ is for $\{\mu_1(S(x, a))/m(S(x, a))\}$ to converge uniformly to g_1 and $\{\mu_2(S(x,a))/m(S(x, a))\}$ to converge uniformly to g_2. But then by what has already been shown, $f_1 = g_1$ almost everywhere and $f_2 = g_2$ almost everywhere with respect to m. Consequently, $f = g$ almost everywhere with respect to m. ∎

If $f \in L^1(m)$ and $\mu(E) = \int_E f dm$ for each measurable $E \subset X$, it is consistent with the terminology of calculus to refer to μ as the **indefinite integral** of f. Using this terminology, we can say that Theorem 12.8 states that *the uniform derivative of an indefinite integral is the integrand almost everywhere*. We can also say that μ *is the indefinite integral of its uniform derivative*.

EXERCISES

1. Prove Theorem 12.7.

FUBINI'S THEOREM

Let (X, M) and (Y, N) be measurable spaces. By $M \times N$ we mean the smallest σ-algebra in $X \times Y$ that contains each set of the form $A \times B$ where $A \in M$ and $B \in N$. If $E \subset X \times Y$ and $x \in X$ and $y \in Y$, we define the **x-section**, E_x and the **y-section**, E^y as follows:

$$E_x = \{y \in Y \mid (x, y) \in E\} \text{ and } E^y = \{x \in X \mid (x, y) \in E\}.$$

2. Show that if $E \in M \times N$, then $E_x \in M$ and $E^y \in N$ for each $x \in X$ and $y \in Y$.

3. Show that $M \times N$ is the smallest monotone class that contains all sets E that are finite unions of sets of the form $A \times B$ where $A \in M$ and $B \in N$.

For each function f on $X \times Y$ and $x \in X$, we associate the function f_x on Y defined by $f_x(y) = f(x, y)$. Similarly, for each $y \in Y$ we defined the function f^y on X by $f^y(x) = f(x, y)$.

4. Show that if f is an $M \times N$-measurable function on $X \times Y$ then for each $x \in X$, f_x is an N-measurable function and for each $y \in Y$, f^y is an M-measurable function.

5. Let (X, M, μ) and (Y, N, ν) be σ-finite measure spaces, and let $E \in M \times N$. If $f(x) = \nu(E_x)$ for each $x \in X$ and $g(y) = \mu(E^y)$ for each $y \in Y$, show that f is M-measurable, g is N-measurable and $\int_X f d\mu = \int_X g d\nu$.

Let (X, M, μ) and (Y, N, ν) be σ-finite measure spaces. For each $E \in M \times N$ define $[\mu \times \nu](E) = \int_X \nu(E_x) d\mu = \int_Y \mu(E^y) d\nu$. Then $\mu \times \nu$ is said to be the **product** of μ and ν, or the **product measure** on $M \times N$.

6. [Fubini] Show that if (X, M, μ) and (Y, N, ν) are σ-finite measure spaces, and if f is $M \times N$-measurable on $X \times Y$, then:
   (a) If $0 \le f(x) \le \infty$ for each x and $\phi(x) = \int_Y f_x d\nu$ and $\gamma(y) = \int_X f^y d\mu$ for each $x \in X$ and $y \in Y$, then ϕ is M-measurable, γ is N-measurable and $\int_X \phi d\mu = \int_{X \times Y} f d(\mu \times \nu) = \int_Y \gamma d\nu$.
   (b) If f is complex valued and if $\phi^*(x) = \int_Y |f|_x d\nu$ and $\int_X \phi^* d\mu < \infty$ then $f \in L^1(\mu \times \nu)$, $f_x \in L^1(\nu)$ almost everywhere on X, $f^y \in L^1(\mu)$ almost everywhere on Y, $\phi \in L^1(\mu)$ almost everywhere on X, $\gamma \in L^1(\nu)$ almost everywhere on Y and $\int_X \phi d\mu = \int_{X \times Y} f d(\mu \times \nu) = \int_Y \gamma d\nu$.

RESEARCH PROBLEMS

7. Can we relax the definition of uniform differentiability? In other words, if the measure μ is only L^1 differentiable, does Theorem 12.8 still hold?

8. Show that for Euclidean space $\mathbf{R}^n$, Theorem 12.8 still holds if the measure μ is only L^1-differentiable. What about the general case when X is a metric space?

9. Is it possible to relax the requirements on the space X in Theorem 12.8 and still have the theorem hold?

INDEX

Abelian group, 273
absolute continuity, 372
absolutely closed, 42
abstract category, 147
accumulation point, complete, 74
additive set function, 235
Alexandrov, P., 73
algebra, 230
algebra, measure, 231
almost Cauchy, 121
almost everywhere (a.e.), 256
almost paracompact, 189
almost regular, 260
almost regular measure, 387
almost uniform measure, 306
almost-2-fully normal, 73, 108
Arhangel'skiĭ, 19
Arhangel'skiĭ's Metrization Theorem, 19
ascending collection, 63
Ascoli Theorem, 368
Axiom of Choice, 77

Baire measure, 233
Baire set, 231
Baire, R., 347
ball, 45
Banach, S., 340
Banach space, 340
Banach space isomorphism, 347
Banach-Steinhaus Theorem, 352
base, 47, 208
basis, 44
basis of a vector space, 332
basis, isomorphic, 266
Bessel's Inequality, 334
β-uniformity, 70
Birkhoff, G., 83
bonding map, 126
Borel compact, 35, 37
Borel field, 304, 305
Borel function, 245
Borel measure, 233
Borel set, 231
bounded, 21
bounded collection, 96
bounded continuous function uniformity, 53
bounded function, 3
bounded linear transformation, 341
bounded open set, 310
bounded, essentially, 319
Bourbaki, 43, 83, 119
Burdick, B., 121, 122, 139, 191, 192

$C_K(X)$, 266
$C_\infty(X)$, 325
$C(X)$, 266
$C^*(X)$, 3
C^*-imbedded, 184
C-complete, 97
C-imbedded, 184
canonical injection, 117
Carathéodory, C., 238
cardinal,
  inaccessible, 226
  strongly inaccessible, 226
categorical definition, 149
categorical property, 147
category, 146
category isomorphism, 147
category,
  abstract, 147
  concrete, 147
Cauchy, 29, 90
  almost, 121

partially, 136
Cauchy filter, 102
Čech, E., 157, 159, 165
chain, increasing, 295
characteristic function, 222
class, 147
closed class of cardinals, 224
closed function, 82
Closed Graph Theorem, 347
closed sphere, 6
closed vector subspace, 328
cluster, 29, 63, 85
cluster class, 80
cluster point, 29, 120
coarser, 44, 92
cofinal, 29, 63, 84
cofinal completeness, 28, 92
cofinal completion, 103
cofinal function, 86
cofinally Cauchy, 29, 91
cofinally Δ-Cauchy, 73, 108
cofinally Δ-complete, 108
cofinally complete, 30, 93
collectionwise normal, 21, 25
compact, 31
compact-open topology, 364
compactification, 156
compactification, Stone-Cech, 157
comparable filters, 92
compatible function, 97
complete, 30, 93
complete accumulation point, 74
complete measure, 231
complete, topologically, 175
completely additive 305
completely normal, 20, 25
completeness, 92, 28
completion, 38, 97
completion of a measure, 232
completion, topological, 103
complex measure, 370
composition of entourages, 46
concentrated, 372
concrete category, 147
conjugate exponent, 318
contained in a net, 97
content, Jordan, 305
content, Loomis, 288
continuous from above, 237
continuous from below, 237
continuous function, 3
continuous function uniformity, 53
continuous in metric topology, 4
contravariant functor, 148
converge, 29, 63, 85
converge uniformly, 27
converge weakly, 352
convergence class, 89
convex function, 317
convex vector space, 328
coordinate space, 114
coreflection, 153
coreflection functor, 153
Corson, H., 102
coset, left, 276
coset, right, 276
countably bounded, 104
countably compact, 23
countably directed, 104
countably metacompact, 63
countably paracompact, 63
countably subadditive, 238
counting measure, 334
covariant functor, 147
covering uniformity, 43
covering,
residue, 108
strongly stable, 169
CZ-maximal, 203

decomposition,
Jordan, 373, 380
Lebesgue, 380, 381
polar, 380, 382
Δ-refinement, 8
dense with respect to <, 288
derivative uniformity, 134
derived uniformity, 104
descending collection, 63
diameter, 21

Dieudonne, J., 7, 8, 163
differentiable measure, 389
directed collection, 109
directed ordering, 84
directed, countably, 104
discrete, 20
distance, 5
distance between sets, 5
distance from a point to a set, 5
distance function, 1
distance, Hausdorff, 125
distinguish points, 15
distinguish points from closed sets, 15
dominate, 371
dot product, 326
Dowker, C. H., 63
dual operator, 352
dual space, 343, 352, 376
dual space, second, 346
duality 149
duality principle, 150

e-complete, 203
e-uniformity, 67
entirely normal, 108
entourage, 46
entourage uniformity, 46
entourage, symmetric, 46
epimorphism, 148
equi-continuity, 267
equi-homogeneous uniform space, 267
equicontinuous, 366
equivalent pseudo-metrics, 5
essential supremum, 319
essentially bounded, 319
eventually, 29, 63, 85
exponent, conjugate, 318
extension of a measure, 239
extension of a set, 160
extension, proper, 161, 162

family of uniformities, 54
fee measure, 224
filter, 83, 91
  assoc. with a net, 91
  Cauchy, 102
  intersection, 91
  neighborhood, 91
finally compact, 74
fine, 55
fine space, 55
finer, 92, 44
finite measure, 234
first category, 348
Fourier coefficients, 353
Fourier series, 353
frequently, 29, 63, 85
Fubini's Theorem, 392
full subcategory, 151
fully normal, 8
function,
  Borel, 245
  Haar, 292
  Harr covering, 292
  characteristic, 222
  convex, 317
  measurable, 243
  nonexpansive, 267
  perfect, 182
  quasi-perfect, 197
  simple, 246
  simple measurable, 247
  linear, 257
functionally separated, 180
functor, 146
  contravariant, 148
  coreflection, 153
  covariant, 147
  reflection, 152
fundamental basis, 97
fundamental net, 97

generate, 45, 208
generated, 92
Gillman, L., 203
Ginsburg, S., 110, 118, 133, 134, 154
Glicksberg, I., 176
group, topological, 271

Hölder's Inequality, 318, 319

Haar, A., 264
Haar covering function, 292
Haar function, 292
Haar integral, 266
Haar measure, 266
Hahn Decomposition Theorem, 380, 385
Hahn-Banach Theorem, 344, 347
Halmos, P., 260
Hausdorff, F., 119, 347
Hausdorff distance, 125
Hausdorff uniformity, 43
heavily bounded, 96
heavy collection, 96
hedgehog space, 97
hereditarily open prering, 288
hereditarily realcompact, 216
hereditary collection, 238
Hewitt, E., 204
Hewitt realcompactification, 157, 203
Hilbert Cube, 15, 16, 20
Hilbert space, 4, 25, 34, 326, 327, 338
Hilbert space isomorphism, 337
Hilbert, D., 326
Hohti, A., 96, 181, 182
homogeneous topological space, 266
Howes, N., 66, 70, 72, 75, 76, 93, 102, 104, 188, 189, 191, 192, 219, 278, 390
hyperfunction, 125
hyperspace, 119

imbedding morphism, 151
imbedding, uniform, 150
inaccessible cardinal, 226
increasing chain, 295
indefinite integral, 391
induced uniformity, 99
inductive limit topology, 113
inductive limit uniformity, 113
infimum topology, 111
infimum uniformity, 111
infinite measure, 234
injective uniform space, 139
inner product, 326
inner product space, 326
inner regular, 260
instance, 147
integral,
  Haar, 266
  Lebesgue, 249
intersection filter, 91
interval of indicies, 299
invariance, translation, 263
invariant Loomis content, 289
invariant on compact spheres, 300
inverse image of a covering, 48
inverse limit sequence, 126, 128
inverse limit space, 126, 128
Itzkowitz, G., 264, 265, 269
Isbell, J. R., 43, 61, 110, 111, 118, 124, 133, 134, 137, 145, 154, 188, 221
Ishikawa, F., 63, 64
isogeneous,
  uniform space, 266
  perfectly, 281
  strictly, 281
isometric function, 27
isometrically imbedded, 38
isomorphic basis, 266
isomorphism, 26, 266
isomorphism,
  Banach space, 347
  category, 147
  Hilbert space, 337
iterated limit, 89
I(ω), 15, 20
I^{ω} 67

Jerison, M., 203
joint continuity, 362
joint continuity on compacta, 364
jointly continuous, 362
Jordan content, 305, 306
Jordan decomposition, 373, 380
jump 295

Kakutani, S., 260
κ uniformity

κ-bounded, 191
Kasahara, 32
Katetov, M., 7
kernel, 340
Kodaira, K., 265, 305

L^1-differentiable measure, 390
$L^2(\mu)$, 326
$L^p(\mu)$, 340
l_∞-space, 141
$l^2(X)$, 334
Lebesgue, H., 251, 254
Lebesgue Covering Theorem, 22
Lebesgue decomposition, 380, 381
Lebesgue integrable, 255
Lebesgue integral, 249
Lebesgue measure, 233, 262
Lebesgue number, 22
Lebesgue property 31, 93
Lebesgue ring, 234
left coset, 276
left uniformity, 276
lexicographic ordering, 79
limit
  inferior, 245, 248
  inferior of a net, 125
  iterated, 89
  of a monotone sequence, 314
  point-wise, 246
  superior, 245, 249
  superior of a net, 125
limit uniformity, 111
Lindelöf, E., 69, 70
Lindelöf, linearly, 74
linear combination, 331
linear function, 340
linear functional, 257, 340
linear functional, real, 343
linear operator, 340
linear transformation, 340
linear transformation, bounded, 341
linearly independent, 331
linearly Lindelöf, 74
locally fine, 133
locally fine coreflection, 133
locally finite, 5, 8
locally starring 19
Loomis, L., 284, 285, 297, 299, 305, 306
Loomis content, 288
lower semi-continuous, 258
Lusin's Theorem, 323

Mack, J., 65, 186
M-additive measure, 224
Mansfield, 73
measurable cardinal, 221, 222
measurable function, 243
measurable set, 243
measurable space, 243
measure, 231
measure,
  almost regular, 387
  almost uniform, 306
  Baire, 233
  Borel, 233
  complete, 231
  complex, 370
  counting, 334
  differentiable, 389
  extension of, 239
  finite, 234
  free, 224
  Haar, 266
  infinite, 234
  L^1-differentiable, 390
  Lebesgue, 233, 262
  M-additive, 224
  outer, 239
  regular, 387
  regular complex, 373
  zero-one, 221
measure algebra, 231
measure ring, 231
metric, 2
  discrete, 2
  trivial, 2
metric assoc. with a pseudo-metric, 27
metric property, 26

metric space, 2
metric uniformity, 56
metrization problem, 13
Miščenko, 74
Mibu, Y., 264, 265
Michael, E., 65
Minkowski's Inequality, 318, 319
Moore, E. H., 83
monomorphism, 148
monotone class, 314
monotone decreasing, 91
monotone increasing, 91
monotone sequence, 314
monotone set function, 235
Morita, K., 7, 43, 58, 59, 60, 69, 103, 129, 131, 132, 156, 163, 165, 194, 196, 197, 198
morphism, 147
  imbedding, 151
  quotient, 151
  reflection, 151
morphism function, 147
M-paracompact, 65
M-space, 103
M-space, 197
μ*-measurable, 239

Nachbin, L., 265
Nagata, J., 7, 16
Nagata-Smirnov Metrization Theorem, 14, 16, 17
negative part of a function, 246
negative variation, 373
negligible set, 256
neighborhood filter, 91
Neighborhood Principle, 75
net, 83, 84
net, universal, 89
Nikodym, O., 374
non-separating uniformity, 44
nonexpansive function, 267
norm, 341
  quotient, 353
  supremum, 325
norm topology, 340
normal covering, 13, 55, 56
normal family, 44
normal in, 45
normal sequence, 13, 44
normal subgroup, 273
normal with respect to, 45
normal, entirely, 108
normalized, 332
normed linear space, 340
nowhere dense, 348

object, 147
object function, 147
ω-directed, 104
open, 2
open function, 82
Open Mapping Theorem, 347, 349
operator, dual, 352
ordering,
  directed, 84
  lexicographic, 79
  product, 88
Ordering Lemma, 75
orthogonal (⊥), 328, 332
orthogonal measures, 372
orthogonal projection, 330
outer measure, 239
outer regular, 260

paracompact, 8
paracompactification, 156
paracompactification, uniform, 158
partial convergence, 136
partial net, 136
partially Cauchy, 136
partition, 370
partition of unity, 5
partition of unity subordinate to a covering, 5
Pasynkov, B., 132
Pelant, J., 74, 139, 145, 154
perfect mapping, 182
perfectly isogeneous, 281
perfectly normal, 20, 25
phalanxes, 83

point finite, 9
point finite uniformity, 144
point-wise limit, 246
points at infinity, 178, 179
polar decomposition, 380, 382
positive part of a function, 246
positive variation, 373
precise refinement, 57
precompact, 22, 94
precompact reflection, 154
preparacompact, 103
prering, 288
prering, hereditarily open, 288
preuniformity, 44
product (π) topology, 355, 369
product of objects, 149
product ordering, 88
product uniformity, 355
projective limit space, 112
projective limit uniformity, 113
proper extension of a covering, 162
proper extension of a set, 161
proper net, 121
pseudo-compact, 201
pseudo-metric, 1
pseudo-metric space, 2
pseudo-metric uniformity, 208
pseudo-paracompact, 201

Q-space, 203
quasi-perfect function, 197
quotient function, 82
quotient morphism, 151
quotient norm, 353
quotient normed linear space, 353
quotient of a topological group, 276
quotient topological group, 277
quotient uniform space, 115
quotient, uniform, 115, 118

radical, 171
Radon, J., 374
Radon-Nikodym derivative, 370, 373, 374
realcompact, 203
realcompact, hereditarily, 216
realcompactification, 210
realcompactification, Hewitt, 157, 203
refinement, 7
refinement, precise, 57
reflection, 151, 266
reflection, precompact, 154
reflection functor, 152
reflection morphism, 151
reflexive normed linear space, 347
regular, 260
  almost, 260
  inner, 260
  outer, 260
regular complex measure, 373
regular measure, 387
regularly bounded, 35, 37
regularly open, 54, 171
relatively fine spaces, 75
relativized uniformity, 99
residual, 63, 84, 136
residue covering, 108
residue point, 108
retract, uniform, 146
retraction, 146, 148
Rice, M., 95, 96
Riesz, F., 257, 260
Riesz Representation Theorem, 260, 261, 387
Riesz-Fischer Theorem, 335
right coset, 276
right uniformity, 276
ring, 3, 230
  measure, 231
  Lebesgue, 234
rotation, 266
Rudin, M., 74
Russell Paradox, 147

Samuel compactification, 154, 157
Samuel, P., 157
scalar product, 326
Schwarz Inequality, 327
second category, 348

second dual space, 346
Segal, I., 264, 265
separable reflection, 154
separated, uniformly, 288
sequentially compact, 23
set at infinity, 179
Shirota, T., 43, 67, 204, 211, 221, 227
Shirota's Theorem, 221
shrink, 9
shrinkable, 9
shrinking, uniformly strict, 52
σ-additive, 231
σ-algebra, 230
σ-compact, 168
σ-locally finite, 17
σ-ring, 230
simple function, 246
simple measurable function, 247
small, sufficiently, 289
Smirnov, Y. M., 7, 16, 17
Smith, H. L., 83
span, 331
spectrum of βX, 192
sphere, 2, 45
square integrable, 326
star, 7, 8
star finite, 57
star refinement, 8
Stone, A. H., 7, 10
Stone, M. H., 157, 159
Stone-Cech compactification, 70, 157
Stone's Theorem, 6, 17
strictly coarser, 92
strictly finer, 92
strictly isogeneous, 281
strongly inaccessible cardinal, 226
strongly stable covering, 169
sub-base, 47
sub-basis, 44
subadditive, 238, 239
subbase, 208
subcategory ,151
subcategory, full, 151
subfine, 139
subfine coreflection, 139, 142
subgroup, normal, 273
subnet, 86
subtractive set function, 235
sufficiently small, 289
sum of transfinite sequences, 79
supercomplete, 110
supercompletion, 191
support, 5
supremum norm, 176, 325
supremum topology, 111
supremum uniformity, 111
supremum, essential, 319
symmetric entourage, 46
symmetric set, 272

Tamano, H., 103, 156, 160, 166, 170, 171, 174, 175, 177, 180, 217, 218, 220
Tamano's Completeness Theorem, 160, 171
Tamano's Theorem, 160, 171, 178
Tamano-Morita paracompactification, 158, 182, 197, 200
topological completion, 103, 156
topological group, 271
topologically complete, 175
topologically preparacompact, 188
topology assoc. with a uniformity, 45
topology of compact convergence, 364
topology of uniform convergence, 360
topology, point-wise convergence, 355
total variation, 371
totally bounded, 94
transfinite sequence, 62
transfinite subsequence, 63
translation, 266
translation invariance, 263, 264
triangle inequality, 1
Trigonometric System, 338
trigonometric polynomial, 339

Tukey, J. W., 7, 13, 43, 56, 59, 60, 61, 83

$\mathcal{U}$-close, 119
$\mathcal{U}$-large, 95
u-uniformity, 66
$\mathcal{U}$-small, 95
uc topology, 360
uc uniformity, 360
Uniform Boundedness Principle, 351
Uniform Boundedness Theorem, 347, 351
uniform compactification, 154
uniform convergence, 27
uniform convergence on compacta, 364
uniform covering, 44
uniform derivative, 376, 389
uniform function space, 355
uniform homeomorphism, 26, 48
uniform imbedding, 150
uniform local compactness, 95, 96, 118, 133
uniform paracompactification, 103, 158
uniform paracompactness, 83, 92
uniform product space, 114
uniform property, 26
uniform quotient, 115, 118
uniform relation, 118
uniform retract, 146
uniform space, 44
uniform space,
  equi-homogeneous, 267
  injective, 139
  isogeneous, 266
uniform structure, 43
uniform sum, 117
uniform topology, 45
uniformities, family of, 54
uniformity, 43
  uniformity β, 70
  of uniform convergence, 359
  e, 67
  u, 66
  κ, 364
  bounded continuous function, 53
  continuous function, 53
  derivative, 134
  derived, 104
  entourage, 46
  induced, 99
  inductive limit, 113
  infimum, 111
  left, 276
  metric, 56
  non-separating, 44
  point finite, 144
  point-wise convergence, 355
  product, 355
  projective limit, 113
  pseudo-metric, 208
  relativized, 99
  right, 276
  supremum, 111
  universal, 55
  weak, 52
uniformizable, 52
uniformly continuous, 26, 48, 51
uniformly differentiable, 390
uniformly discrete, 133
uniformly locally uniform, 133
uniformly paracompact, 95
uniformly separated, 288
uniformly strict shrinking, 52
universal net, 89
universal uniformity, 55
upper semi-continuous, 258
Urysohn, P., 8, 14, 16, 73
Urysohn Metrization Theorem, 14, 16
Urysohn's Imbedding Theorem, 16
Urysohn's Lemma, 16

variation, positive, 373
vanish at infinity, 325
variation, negative, 373
Vietoris, L., 119

von Neumann, J., 264, 305

weak completion, 129
weak topology, 353
weak uniformity, 52
weak* topology, 353, 369
weakly Cauchy filter, 102
weakly complete, 129
Weil, A., 264, 266
Weil, A., 43
well ordered covering, 65
within, 45

Z(f), 162
zero set, 162
zero-boundary, 289
zero-one measure, 221
0(f), 162

Universitext *(continued)*

Shapiro: Composition Operators and Classical Function Theory
Smith: Power Series From a Computational Point of View
Smoryński: Self-Reference and Modal Logic
Stillwell: Geometry of Surfaces
Stroock: An Introduction to the Theory of Large Deviations
Sunder: An Invitation to von Neumann Algebras
Tondeur: Foliations on Riemannian Manifolds

9 780387 979861